New Techniques in Marine Aquaculture

New Techniques in Marine Aquaculture

Editors

Zhenhua Ma
Jianguang Qin

Basel • Beijing • Wuhan • Barcelona • Belgrade • Novi Sad • Cluj • Manchester

Editors
Zhenhua Ma
South China Sea Fisheries
Research Institute
Chinese Academy of
Fishery Sciences
Guangzhou, China

Jianguang Qin
College of Science
and Engineering
Flinders University
Bedford Park, Australia

Editorial Office
MDPI
St. Alban-Anlage 66
4052 Basel, Switzerland

This is a reprint of articles from the Special Issue published online in the open access journal *Journal of Marine Science and Engineering* (ISSN 2077-1312) (available at: https://www.mdpi.com/journal/jmse/special_issues/03Z5739284).

For citation purposes, cite each article independently as indicated on the article page online and as indicated below:

Lastname, A.A.; Lastname, B.B. Article Title. *Journal Name* **Year**, *Volume Number*, Page Range.

ISBN 978-3-0365-9680-8 (Hbk)
ISBN 978-3-0365-9681-5 (PDF)
doi.org/10.3390/books978-3-0365-9681-5

Cover image courtesy of Zhenhua Ma

Contents

About the Editors

Zhenhua Ma

Professor Zhenhua Ma graduated from Flinders University in Australia with a Ph.D. in Aquaculture. He currently serves as the Deputy Director of the South China Sea Fisheries Research Institute of the Chinese Academy of Fishery Sciences. He is a member of the eighth Committee of the Hainan Provincial Committee of the Chinese People's Political Consultative Conference, a member of the third Subcommittee on Ornamental Fish of the National Fisheries Standardization Technical Committee, a specially appointed expert of the Marine Intelligence Plan of the China Association for Science and Technology, and a member of the third Committee of the Hainan Provincial Original Good Breed Committee. He is also an external Ph.D. supervisor at Flinders University in Australia and a master's supervisor at Shanghai Ocean University, Dalian Ocean University, Tianjin Agricultural University, and Jiangsu Ocean University. He serves as the editor-in-chief of *"Insights in Aquaculture and Biotechnology"* and is a member of the editorial board of international journals such as *"The Israeli Journal of Aquaculture"*. His main research focuses on the breeding of aquatic animals and deep-sea aquaculture technology. He has published a total of 198 research papers and 9 monographs to date.

Jianguang Qin

Professor Jianguang Qin obtained his bachelor's degree at Dalian Fisheries College, China, in 1982, his master's degree at the University of Oklahoma in 1988, and his Ph.D. at Ohio State University in 1994. After a 3-year postdoc at the University of Hawaii, he joined Flinders University in 1997. Jian is a pioneer of the aquaculture program at Flinders and has led the aquaculture research at Flinders to become world-class in the area of fishery sciences. He has supervised 20 Ph.D. students and 20+ honors students to completion and has published 2 books, 13 book chapters, and 300+ journal papers. His research areas cover the food, feeding, and nutrition of aquatic animals in an attempt to discover trophic relationships in food webs and apply new findings to applied science. His research disciplines include ecology, physiology, behavior, biotechnology, and toxicology. Research organisms of interest include bacteria, algae, zooplankton, fish, oysters, abalone, and freshwater crayfish.

Preface

The rapid development of aquaculture worldwide has been driven by the emergence of new techniques. In this book, we aim to provide an update on current research and technological advancements in various aspects of aquaculture, including biotechnology, ecotechnology, feeding technology, and environmental technology.

One of the key areas covered in this book is biotechnology. Recently, researchers have been focusing on isolating and characterizing genes that provide resistance to bacterial infection and environmental stress. By understanding the genetic makeup of aquatic organisms, we can develop strategies to enhance their resistance to common diseases and stressors. This not only improves the overall health and well-being of the organisms but also ensures sustainable aquaculture practices.

Feeding technology is another important aspect addressed in this book. In recent years, scientists have been working on developing advanced feeding techniques that improve the immunity and stress resistance of aquatic organisms. By optimizing the nutritional content of their feed and incorporating additives that enhance immune response, we can ensure the proper growth and development of aquaculture species.

Additionally, this book also explores the detection of the impact of thermal stress on marine fish immunity. With increasing global temperatures and climate change, it is crucial to understand the impact of thermal stress on aquatic organisms. By studying the immune response of marine fish to thermal stress, we can develop effective mitigation strategies to protect these species and maintain the sustainability of aquaculture.

Moreover, light color manipulation to improve algal growth is another area of interest covered in this Special Issue. Algae are an essential component of aquatic ecosystems, and their growth is crucial for maintaining a healthy environment for aquaculture species. By manipulating the color and intensity of light, we can optimize algal growth, which in turn supports the growth and development of other organisms in the ecosystem.

Lastly, this book highlights the use of integrated multi-trophic systems to improve production. By combining different species in aquaculture systems, we can create a symbiotic relationship where the waste products of one species serve as nutrients for another. This integrated approach not only improves the overall efficiency of aquaculture but also reduces the environmental impact.

The research presented in this book focuses on marine fish, shrimp, sea cucumbers, sea urchins, and algae to showcase the new technological developments in aquaculture. However, the findings and techniques discussed in these papers have the potential to be extended to other species in aquaculture. They serve as a foundation for further research and innovation in the field, leading to the continued growth and improvement of aquaculture practices.

We would like to express our sincere gratitude to all of the authors whose valuable contributions have made the publication of this book possible. Their dedication and hard work have significantly enriched and enhanced the coverage of this publication. It is our hope that this book will inspire further research and collaboration, ultimately contributing to the sustainable development of aquaculture worldwide.

Zhenhua Ma and Jianguang Qin

Editors

Journal of
*Marine Science
and Engineering*

MDPI

Editorial

New Techniques in Marine Aquaculture

Zhenhua Ma [1,2],* and Jianguang Qin [2],*

1   South China Sea Fisheries Research Institute, Chinese Academy of Fishery Sciences, Guangzhou 510300, China
2   College of Science and Engineering, Flinders University, Bedford Park, SA 5001, Australia
*   Correspondence: zhenhua.ma@scsfri.ac.cn (Z.M.); jian.qin@flinders.edu.au (J.Q.)

Citation: Ma, Z.; Qin, J. New Techniques in Marine Aquaculture. *J. Mar. Sci. Eng.* **2023**, *11*, 2239. https://doi.org/10.3390/jmse11122239

Received: 15 November 2023
Accepted: 20 November 2023
Published: 27 November 2023

In recent years, the importance of marine aquaculture has been increasing globally, and new technologies are playing a significant role in this trend [1,2]. Marine aquaculture (i.e., mariculture) is the practice of farming aquatic organisms in the ocean or other marine water bodies. The practice has become increasingly important in meeting the growing demand for seafood as wild fish populations continue to decline due to overfishing and environmental degradation. Marine aquaculture has been practiced for centuries, but new technologies have recently revolutionized the industry. These technologies have enabled the farming of a broader range of species in larger quantities and with greater efficiency. They have also made it possible to reduce the environmental impact of aquaculture, making it a more sustainable practice.

The development of aquaculture technology for tuna and similar large ocean species has become increasingly important in recent years. As wild fish stocks continue to decline, the need for sustainable and efficient aquaculture practices has grown. Tuna, in particular, is a fish highly sought-after for its nutritional value and culinary versatility, making it a prime candidate for aquaculture development [3,4]. Advancements in aquaculture technology have allowed the successful farming of tuna and similar species in controlled environments. These technologies include advanced water filtration systems, specialized feeding techniques, and optimized breeding programs. These developments have increased the availability of tuna for human consumption and reduced the pressure on wild populations. Additionally, innovative monitoring and data collection systems have allowed the better management of aquaculture operations. Other emerging technologies include real-time water quality monitoring, tracking fish growth and health, and predictive analytics for optimizing feeding schedules and environmental conditions.

Another critical technology in marine aquaculture is the use of genetically modified organisms (GMOs) [5,6]. GMOs are organisms that have been genetically altered to exhibit certain traits, such as fast growth and disease resistance. In aquaculture, GMOs can be used to produce fish and other aquatic organisms that are more efficient to farm, require less feed, and are less susceptible to disease. Other new technologies in marine aquaculture include the use of automated feeding systems, advanced monitoring and control systems, and the use of renewable energy sources such as solar and wind power. These technologies are helping to make marine aquaculture more efficient, cost-effective, and environmentally sustainable.

The importance of these new technologies is not limited to meeting the growing demand for seafood. These novel technologies have food security, environmental sustainability, and economic development implications. Oceanic fish farming serves as a pivotal solution to alleviate the strain on natural fish stocks, supplying a sustainable source of superior protein to meet human dietary needs and fostering economic revitalization in shoreline locales. Advancements in aquaculture technology are instrumental in satisfying the escalating appetite for seafood, mitigating the exploitation of wild fish resources, and championing ecologically responsible practices. As the global population continues to grow and demand for seafood increases, marine aquaculture and new technologies will play an increasingly important role in meeting these challenges.

It is the basic requirement of the modern fishery to ensure the supply of high-quality aquatic animal protein and meet the increasing consumption demand. China has the longest history, the richest experience, and the most varieties and modes of comprehensive aquaculture in the world. The advent of innovative methodologies has been a catalyst for the swift expansion of the aquaculture sector on a global scale. The focal point of this Special Issue is to provide a comprehensive overview of the latest advancements in Chinese aquaculture. It encompasses a spectrum of cutting-edge fields, such as biotechnological applications, ecological management practices, nutritional tech innovations, and sustainable environmental technologies. Through this, we aim to offer insights into the technological progression that is shaping the future of aquaculture within the region. In this Special Issue, the farmed species include yellowfin tuna (*Thunnus albacares*), mackerel tuna (*Euthynnus affinis*), greater amberjack (*Seriola dumerili*), yellowtail kingfish (*Seriola aureovittata*), humphead wrasse (*Cheilinus undulatus*), bluestripe snapper (*Lutjanus Kasmira*), golden pompano (*Trachinotus ovatus*), yellow croaker (*Larimichthys crocea*), coral trout (*Plectropomus leopardus*), shrimp (*Penaeus monodon*), sea cucumber (*Apostichopus japonicus*), sea urchin (*Strongylocentrotus intermedius*), and microalgae (*Isochrysis zhanjiangensis*). The research contents and main findings in this Special Issue are as follows:

Contribution 1 conducted a preliminary study on an innovative concept that combines a floating offshore wind turbine with a fishing cage in order to optimize the utilization of ocean resources, reduce the investment return period, and directly supply energy to the fishing cage. The outcomes of their study offer insights into the conceptual design, modeling, and simulation analysis of the integrated wind turbine-fishing cage system. However, further research is needed to perform detailed structural design optimizations, strength evaluations, and experimental tests on the integrated system in the future.

In contribution 2, Zhao et al. employed the RAD-seq technology to identify the whole-genome SNPs of *Lutjanus kasmira*, a fish species. Subsequently, these SNPs were utilized to investigate the genetic diversity and structure of this species. The findings from their study offer a solid theoretical foundation for the strategic development and preservation of the germplasm resources of *Lutjanus kasmira*.

In contribution 3, Le et al. successfully cloned the MKK4 cDNA from *Penaeus monodon*, a species of shrimp. They then examined the responses of this gene to bacterial infection and low-salinity conditions. Interestingly, they found that the PmMKK4 gene was activated in both the gill tissue and hepatopancreas tissue of *P. monodon* after experiencing low-salinity stress. However, the expression change of PmMKK4 in the gill tissue was particularly significant. These findings suggest that the PmMKK4 gene plays a crucial role in both the innate immune response after pathogen infection and the shrimp's ability to adapt in a low-salt environment.

In contribution 4, Lv et al. conducted an evaluation on the impact of various LED colors on the productivity, chlorophyll (Chl-a, Chl-b, and total Chl), protein, and carbohydrate content of *Isochrysis zhanjiangensis* in indoor culture. The findings of their research indicate that the productivity of Chl (a, b, and total), as well as the protein and carbohydrate content of *Isochrysis zhanjiangensis*, can be effectively regulated by different light wavelengths. This study holds great significance in terms of selecting the appropriate light color (wavelength) to maximize the production of desired organic compounds in indoor cultures, potentially paving the way for commercially viable practices.

Contribution 5 conducted a study aimed at investigating the potential positive effects of curcumin on the ammonia nitrogen stress tolerance ability of greater amberjack (*Seriola dumerili*). The results obtained from their research demonstrated that the inclusion of curcumin in the diet of the greater amberjack can enhance their tolerance to ammonia nitrogen stress. These findings provide strong support for the promising application prospects of curcumin in promoting the well-being and stress resistance of this fish species.

In contribution 6, Liu et al. conducted an investigation to assess the impact of acute high-temperature stress (34 °C) on the physiological reactions of juvenile yellowfin tuna (*Thunnus albacares*). By measuring changes in the biochemical indexes of serum, liver, gill, and muscle, the researchers compared the experimental group under high-temperature stress with a control group (28 °C) at different time points (0 h, 6 h, 24 h, and 48 h). The findings of the study revealed that the elevated temperature negatively affected the antioxidant enzymes and metabolic indexes of the yellowfin tuna, leading to a decline in physiological indexes. These results highlight the susceptibility of juvenile yellowfin tuna to acute high-temperature stress.

Contribution 7 conducted a comparative study on the survival, growth, and behavior performances of sea urchins and sea cucumbers in an integrated multi-trophic aquaculture (IMTA) system. In the IMTA system, the experimental group (group M) consisted of *S. intermedius* and *A. japonicus* per 10,638 cm^3 of stocking density, while the control group for sea urchins (group U) had *S. intermedius* per 10,638 cm^3 of stocking density, and the control group for sea cucumbers (group C) had *A. japonicus* per 10,638 cm^3 of stocking density. The results of the study indicate that the implementation of the IMTA system significantly enhanced the feeding behavior, body growth, and survival rate of cultured *Strongylocentrotus intermedius*. This innovative aquaculture system shows great potential in improving the production efficiency of juvenile *Apostichopus japonicus* (as the primary species) and *S. intermedius* (as the subsidiary species) in China.

In contribution 8, Li et al. conducted a study on a clone of *Penaeus monodon* NHE cDNA, known as PmNHE, and assessed its effects on high-concentration ammonia nitrogen stress. The significant expression of PmNHE was observed in the intestine of *P. monodon* under high-concentration ammonia nitrogen stress. Additionally, the survival times of *P. monodon* were found to be noticeably shorter when exposed to high levels of ammonia nitrogen stress compared to the control groups. These findings suggest that PmNHE may play a crucial role in environments with elevated levels of ammonia nitrogen.

In contribution 9, Liu and colleagues performed a comprehensive DNA barcoding study on the Scaridae, a family of parrotfish, in the Hainan region. Their investigation included the analysis of 401 DNA barcode sequences, with 51 of these sequences obtained from newly collected specimens. These sequences were derived from a specific 533 base pair segment of the CO I gene, known to be a reliable marker in genetic identification. The findings from this study suggest that DNA barcoding is a valuable molecular approach for the monitoring and conservation of fishery resources. Additionally, it serves as a critical tool for resolving complex taxonomic issues, highlighting areas in need of further scientific exploration.

In contribution 10, Zhao and colleagues meticulously identified a dozen pairs of polymorphic microsatellite markers. Leveraging these genetic indicators, they engineered a duo of multiplex PCR (polymerase chain reaction) protocols to facilitate the genetic scrutiny of 30 humphead wrasse (*Cheilinus undulatus*) specimens. The implementation of these advanced multiplex PCR systems has laid the groundwork for critical applications in the realms of genetic stock identification, the evaluation of genetic variability, and the monitoring of population dynamics following the introduction and propagation of *C. undulatus*. This innovative approach promises to enhance conservation efforts and sustainable management strategies for this species.

In contribution 11, Wang and colleagues explored the impact of varying sunlight exposure—sunny versus cloudy weather—on the pattern of gene expression in the liver of the mackerel tuna (*Euthynnus affinis*). Their findings revealed a consistent daily pattern among a broad set of genes, including those governing circadian rhythms (such as CREB1, CLOCK, PER1, PER2, PER3, REVERBA, CRY2, and BMAL1), metabolism (SIRT1 and SREBP1), and immune function (NF-kB1, MHC-I, ALT, IFNA3, ISY1, ARHGEF13, GCLM, and GCLC), regardless of the weather conditions ($p < 0.05$). The study concluded that the expression of genes related to circadian rhythm, lipid metabolism, and immunity in mackerel tuna liver is influenced by diurnal cycles and environmental light conditions, undergoing significant expression fluctuations.

In contribution 12, Yu and colleagues conducted a comparative analysis of the growth dynamics and nutrient profiles of large yellow croakers cultivated within two distinct environments: a novel offshore aquaculture vessel and a conventional nearshore cage system. Their investigation revealed that the offshore aquaculture ship offers a protective haven against environmental adversities, such as typhoons and harmful algal blooms, commonly known as red tides. Remarkably, the large yellow croakers acclimatized to the ship-based farming system in a relatively short time frame, with no significant stress incidents reported throughout the cultivation period. The data gathered from this study indicate a superior growth rate and nutritional quality in the croakers bred on the offshore ship compared to those raised in traditional cages. This study substantially contributes to our comprehension of the offshore ship-based aquaculture paradigm for cultivating large yellow croakers.

In contribution 13, Zhou and colleagues delved into the effects exerted by salinity stress on the antioxidative responses in yellowfin tuna, *Thunnus albacares*. The findings from this research suggest that juvenile yellowfin tuna demonstrate alterations in their antioxidant capabilities when subjected to low-salinity environments. Conversely, evidence points to the robust resilience of these juveniles to saline conditions of 29 parts per thousand (‰), indicating a strong adaptive capacity. Nonetheless, it is important to note that excessive stress could potentially deplete the organism's energy reserves, thereby diminishing its overall resistance to environmental stressors.

In contribution 14, Guo and colleagues delved into the pathology of *Cryptocaryon irritans*, its impact on the immune enzyme activities, and the regulation of the NEMO gene within the Golden pompano (*Trachinotus ovatus*). The team discovered a marked elevation in the mRNA expression levels of the *T. ovatus* NF-kappa-B essential modulator (ToNEMO) following *C. irritans* exposure, with significant increases noted across multiple organs including the gills, skin, liver, spleen, and head kidney. These findings imply a potential role for ToNEMO in the fish's immune defense mechanisms, providing valuable insights into the biological reactions triggered by the parasitic infection.

In contribution 15, Zhou and colleagues explored how curcumin intake influences non-specific immune functions and the antioxidative capacity of *Seriola dumerili* when subjected to ammonia-induced stress, as well as during the subsequent recovery period. The experimental design included three distinct feed formulations, with curcumin concentrations set at 0, 75, and 150 mg/kg. The findings of this investigation reveal that incorporating curcumin into the diet can bolster the non-specific immune mechanisms and antioxidative potential in juvenile *S. dumerili*, thereby augmenting their resilience against the detrimental effects of elevated ammonia levels.

Contribution 16 conducted by Qian et al. explored the impact of varying water flow rates on the development and physiological health of the coral trout (*Plectropomus leopardus*) within a recirculating aquaculture system (RAS). Their research concluded that elevated water velocities disrupt the physiological balance and impair the digestive processes in the intestinal tract of the coral trout, leading to a compromised growth rate. This suggests that coral trout exhibit a heightened sensitivity to increased water flow. Practical applications in RAS operations should, therefore, maintain water flow rates at or below 1 bl/s to ensure optimal conditions for the health and growth of this species.

In contribution 17, Wang et al. delved into the impact of artificial reefs on the survival and behavioral patterns of juvenile sea cucumbers (*Apostichopus japonicus*) post air exposure and during disease proliferation at a temperature of 25 °C. The study illuminated that artificial reefs may serve as a sanctuary, mitigating direct interaction among healthy and diseased individuals, thereby potentially curtailing the spread of illness. Building on this foundation, our current study unveils an economically viable strategy to enhance the viability of young sea cucumbers within hatchery environments, particularly under elevated thermal conditions.

In contribution 18 conducted by Li and colleagues, the presence of a gene akin to the Forkhead Box O (FOXO), referred to as EsFOXO-like, was discerned within the crustacean species *Eriocheir sinensis*. The study probed into how this gene influences the 20-hydroxyecdysone (20E) signaling cascade. The team's experiments uncovered that, upon administering a combination of AS1842856 and Rapamycin to the subjects, there was a marked depletion in 20E levels and a down-regulation of the E. sinensis molt-inhibiting hormone (EsMIH) expression when contrasted with the use of AS1842856 with DMSO. In tandem, an upsurge was noted in the expression of *Eriocheir sinensis* ecdysone receptor (Es-EcR) and Retinoid X Receptor (EsRXR). These observations collectively infer that the EsFOXO-like gene modulates the 20E pathway via the mTOR signaling route, thereby enriching our comprehension of the molting mechanics in crustacean species.

In contribution 19, Cui et al. explored the genetic makeup and diversity across five distinct groups of yellowtail kingfish (*Seriola aureovittata*). They found minimal genetic variance, as indicated by low Fst values, between the yellowtail kingfish populations of China and Japan, coupled with high levels of gene flow (Nm). These findings imply a frequent genetic interchange between these collections, which points to a shared population structure. On the other hand, the Australian yellowtail kingfish showed significant genetic divergence from both the Chinese and Japanese groups, hinting at disparate evolutionary lineages and distinct population identities. The insights gleaned from this research are poised to enhance the genetic breeding programs for the Chinese yellowtail kingfish and foster the preservation and responsible management of its genetic resources.

This Special Issue delves into the forefront of aquaculture technology, placing innovative biotechnological approaches for identifying gene sequences that bolster resistance against bacterial infections and environmental challenges in the spotlight. It encompasses advancements in nutritional strategies designed to enhance the immune response and resilience of aquatic organisms. Included within this issue are insightful investigations into the thermal stress impacts on the immune systems of marine fish, along with explorations into the strategic use of light spectrum adjustments to accelerate algal proliferation. Additionally, the issue explores the benefits of employing integrated multi-trophic aquaculture systems for boosting production efficiency. This compilation of cutting-edge research not only sheds light on the technological strides in cultivating marine fish, shrimp, sea cucumbers, sea urchins, and algae, but also sets the stage for translating these breakthroughs to a broader spectrum of species within the aquaculture industry.

Acknowledgments: As Guest Editors of the Special Issue "New Techniques in Marine Aquaculture", we wish to extend our sincere gratitude to all the authors whose valuable contributions made the publication of this issue possible. Their work has significantly enriched and enhanced the coverage of this publication.

Conflicts of Interest: The authors declare no conflict of interest.

List of Contributions

1. Zhang, C.; Xu, J.; Shan, J.; Liu, A.; Cui, M.; Liu, H.; Guan, C.; Xie, S. Preliminary Study on an Integrated System Composed of a Floating Offshore Wind Turbine and an Octagonal Fishing Cage. *J. Mar. Sci. Eng.* **2022**, *10*, 1526. https://doi.org/10.3390/jmse10101526.
2. Zhao, F.; Guo, L.; Zhang, N.; Yang, J.; Zhu, K.; Guo, H.; Liu, B.; Liu, B.; Zhang, D.; Jiang, S. Genetic Diversity Analysis of Different Populations of Lutjanus kasmira Based on SNP Markers. *J. Mar. Sci. Eng.* **2022**, *10*, 1547. https://doi.org/10.3390/jmse10101547.
3. Li, Y.; Zhou, F.; Fan, H.; Jiang, S.; Yang, Q.; Huang, J.; Yang, L.; Chen, X.; Zhang, W.; Jiang, S. Molecular Technology for Isolation and Characterization of Mitogen-Activated Protein Kinase Kinase 4 from Penaeus monodon, and the Response to Bacterial Infection and Low-Salinity Challenge. *J. Mar. Sci. Eng.* **2022**, *10*, 1642. https://doi.org/10.3390/jmse10111642.

4. Lv, B.; Liu, Z.; Chen, Y.; Lan, S.; Mao, J.; Gu, Z.; Wang, A.; Yu, F.; Zheng, X.; Vasquez, H.E. Effect of Different Colored LED Lighting on the Growth and Pigment Content of Isochrysis zhanjiangensis under Laboratory Conditions. *J. Mar. Sci. Eng.* **2022**, *10*, 1752. https://doi.org/10.3390/jmse10111752.

5. Hong, J.; Fu, Z.; Hu, J.; Zhou, S.; Yu, G.; Ma, Z. Dietary Curcumin Supplementation Enhanced Ammonia Nitrogen Stress Tolerance in Greater Amberjack (*Seriola dumerili*): Growth, Serum Biochemistry and Expression of Stress-Related Genes. *J. Mar. Sci. Eng.* **2022**, *10*, 1796. https://doi.org/10.3390/jmse10111796.

6. Liu, H.; Fu, Z.; Yu, G.; Ma, Z.; Zong, H. Effects of Acute High-Temperature Stress on Physical Responses of Yellowfin Tuna (*Thunnus albacares*). *J. Mar. Sci. Eng.* **2022**, *10*, 1857. https://doi.org/10.3390/jmse10121857.

7. Hu, F.; Wang, H.; Tian, R.; Gao, J.; Wu, G.; Yin, D.; Zhao, C. A New Approach to Integrated Multi-Trophic Aquaculture System of the Sea Cucumber Apostichopus japonicus and the Sea Urchin Strongylocentrotus intermedius. *J. Mar. Sci. Eng.* **2022**, *10*, 1875. https://doi.org/10.3390/jmse10121875.

8. Li, Y.; Jiang, S.; Fan, H.; Yang, Q.; Jiang, S.; Huang, J.; Yang, L.; Zhang, W.; Chen, X.; Zhou, F. A Na+/H+-Exchanger Gene from Penaeus monodon: Molecular Characterization and Expression Analysis under Ammonia Nitrogen Stress. *J. Mar. Sci. Eng.* **2022**, *10*, 1897. https://doi.org/10.3390/jmse10121897.

9. Liu, B.; Yan, Y.; Zhang, N.; Guo, H.; Liu, B.; Yang, J.; Zhu, K.; Zhang, D. DNA Barcoding Is a Useful Tool for the Identification of the Family Scaridae in Hainan. *J. Mar. Sci. Eng.* **2022**, *10*, 1915. https://doi.org/10.3390/jmse10121915.

10. Zhao, F.; Guo, L.; Zhang, N.; Zhu, K.; Yang, J.; Liu, B.; Guo, H.; Zhang, D. Establishment and Application of Microsatellite Multiplex PCR System for Cheilinus undulatus. *J. Mar. Sci. Eng.* **2022**, *10*, 2000. https://doi.org/10.3390/jmse10122000.

11. Wang, W.; Hu, J.; Fu, Z.; Yu, G.; Ma, Z. Daily Rhythmicity of Hepatic Rhythm, Lipid Metabolism and Immune Gene Expression of Mackerel Tuna (*Euthynnus affinis*) under Different Weather. *J. Mar. Sci. Eng.* 2022, *10*, 2028. https://doi.org/10.3390/jmse101 22028.

12. Yu, Y.; Huang, W.; Yin, F.; Liu, H.; Cui, M. Aquaculture in an Offshore Ship: An On-Site Test of Large Yellow Croaker (*Larimichthys crocea*). *J. Mar. Sci. Eng.* **2023**, *11*, 101. https://doi.org/10.3390/jmse11010101.

13. Zhou, S.; Zhang, N.; Fu, Z.; Yu, G.; Ma, Z.; Zhao, L. Impact of Salinity Changes on the Antioxidation of Juvenile Yellowfin Tuna (*Thunnus albacares*). *J. Mar. Sci. Eng.* **2023**, *11*, 132. https://doi.org/10.3390/jmse11010132.

14. Guo, H.-Y.; Li, W.-F.; Zhu, K.-C.; Liu, B.-S.; Zhang, N.; Liu, B.; Yang, J.-W.; Zhang, D.-C. Pathology, Enzyme Activity and Immune Responses after *Cryptocaryon irritans* Infection of Golden Pompano *Trachinotus ovatus* (Linnaeus 1758). *J. Mar. Sci. Eng.* **2023**, *11*, 262. https://doi.org/10.3390/jmse11020262.

15. Zhou, C.; Huang, Z.; Zhou, S.; Hu, J.; Yang, R.; Wang, J.; Wang, Y.; Yu, W.; Lin, H.; Ma, Z. The Impacts of Dietary Curcumin on Innate Immune Responses and Antioxidant Status in Greater Amberjack (*Seriola dumerili*) under Ammonia Stress. *J. Mar. Sci. Eng.* **2023**, *11*, 300. https://doi.org/10.3390/jmse11020300.

16. Qian, Z.; Xu, J.; Liu, A.; Shan, J.; Zhang, C.; Liu, H. Effects of Water Velocity on Growth, Physiology and Intestinal Structure of Coral Trout (*Plectropomus leopardus*). *J. Mar. Sci. Eng.* **2023**, *11*, 862. https://doi.org/10.3390/jmse11040862.

17. Wang, H.; Wu, G.; Hu, F.; Tian, R.; Ding, J.; Chang, Y.; Su, Y.; Zhao, C. Artificial Reefs Reduce Morbidity and Mortality of Small Cultured Sea Cucumbers Apostichopus japonicus at High Temperature. *J. Mar. Sci. Eng.* **2023**, *11*, 948. https://doi.org/10.339 0/jmse11050948.

18. Li, J.; Ma, Y.; Yang, Z.; Wang, F.; Li, J.; Jiang, Y.; Yang, D.; Yi, Q.; Huang, S. FOXO-like Gene Is Involved in the Regulation of 20E Pathway through mTOR in *Eriocheir sinensis*. *J. Mar. Sci. Eng.* **2023**, *11*, 1225. https://doi.org/10.3390/jmse11061225.

19. Cui, A.; Xu, Y.; Kikuchi, K.; Jiang, Y.; Wang, B.; Koyama, T.; Liu, X. Comparative Analysis of Genetic Structure and Diversity in Five Populations of Yellowtail Kingfish (*Seriola aureovittata*). *J. Mar. Sci. Eng.* **2023**, *11*, 1583. https://doi.org/10.3390/jmse11081583.

References

1. Naylor, R.L.; Hardy, R.W.; Buschmann, A.H.; Bush, S.R.; Cao, L.; Klinger, D.H.; Little, D.C.; Lubchenco, J.; Sumway, S.E.; Troell, M. A 20-year retrospective review of global aquaculture. *Nature* **2021**, *591*, 551–563. [CrossRef]
2. Froechlich, H.E.; Gentry, R.R.; Halpern, B.S. Global change in marine aquaculture production potential under climate change. *Nat. Ecol. Evol. Vol.* **2018**, *2*, 1745–1750. [CrossRef]
3. Wang, X.; Xie, J. Evaluation of water dynamics and protein changes in bigeye tuna (*Thunnus obesus*) during cold storag. *LWT-Food Sci. Technol.* **2019**, *108*, 289–296. [CrossRef]
4. Benetti, D.D.; Partridge, G.J.; Stieglitz, J. Chapter 1—Overview on Status and Technological Advances in Tuna Aquaculture Around the World. In *Advances in Tuna Aquacultur*; Benetti, D.D., Partridge, G.J., Buentello, A., Eds.; Academic Press: San Diego, CA, USA, 2016; pp. 1–19. [CrossRef]
5. Sahoo, P.K.; Paul, A. Chapter 2—Opportunities and challenges in aquaculture biotechnology. In *Frontiers in Aquaculture Biotechnology*; Lakra, E.W., Goswami, M., Trudeau, V.L., Eds.; Academic Press: Cambridge, MA, USA, 2023; pp. 15–23.
6. Beardmore, J.A.; Porter, J.S. *Genetically Modified Organisms and Aquaculture*; FAO Fisheries Circular No. 989. FIRI/C989(E); FAO: Rome, Italy, 2003; ISSN 0429-9329.

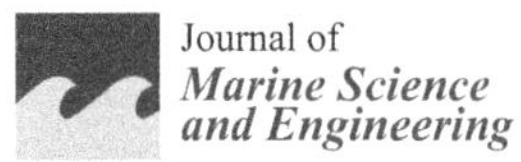
Journal of
Marine Science and Engineering

Article

Preliminary Study on an Integrated System Composed of a Floating Offshore Wind Turbine and an Octagonal Fishing Cage

Chenglin Zhang [1,2], Jincheng Xu [1], Jianjun Shan [1], Andong Liu [1], Mingchao Cui [1], Huang Liu [1], Chongwu Guan [1] and Shuangyi Xie [3,*]

[1] Fishery Machinery and Instrument Research Institute, Chinese Academy of Fishery Sciences, Shanghai 200092, China
[2] College of Engineering Science and Technology, Shanghai Ocean University, Shanghai 201306, China
[3] College of Mechanical Engineering, Chongqing University of Technology, Chongqing 400054, China
* Correspondence: xsy1986@cqut.edu.cn

Abstract: To maximize the utilization of ocean resources, shorten the return period of investment and directly supply energy to the fishing cage, this paper performs a preliminary study for a state-of-the-art concept integrating a floating offshore wind turbine with a fishing cage. An octagonal semisubmersible rigid fishing cage with a slack catenary mooring system is designed to match the NREL 5 MW offshore baseline wind turbine. Combined with the blade pitch controller, fully coupled aero-hydro-elastic-servo-mooring simulations are performed through FAST and AQWA to explore the dynamic performance of the integrated system. Free decay conditions, uniform wind with irregular and regular waves, and turbulent wind with irregular waves are tested. The results showed that the integrated system works normally at the operating conditions and exhibits different dynamic characteristics for various scenarios. Additionally, the study on the influence of mooring line length indicates that the increasing line length can significantly affect the cage surge motion and the maximum and mean values of the upwind line tension at fairlead. Specifically, the maximum surge motion with a 924-m-long line is 404.8% larger than that with an 880-m-long line. When the line length increases by 5%, the maximum and mean line tensions decrease by 45.7% and 47.7%, respectively, while when the line length increases by 10%, the maximum and mean line tension decrease by 52.9% and 54.2%, respectively. It should be noted that the main purpose of this work is to conduct a preliminary study on this integrated system, aiming to provide an idea for the conceptual design, modeling and simulation analysis of this integrated system.

Keywords: fishing cage; floating offshore wind turbine; dynamics modeling; mooring system; net system

Citation: Zhang, C.; Xu, J.; Shan, J.; Liu, A.; Cui, M.; Liu, H.; Guan, C.; Xie, S. Preliminary Study on an Integrated System Composed of a Floating Offshore Wind Turbine and an Octagonal Fishing Cage. *J. Mar. Sci. Eng.* **2022**, *10*, 1526. https://doi.org/10.3390/jmse10101526

Academic Editors: Mohamed Benbouzid and Spyros A. Mavrakos

Received: 23 September 2022
Accepted: 15 October 2022
Published: 18 October 2022

1. Introduction

With the continuous growth of the global population and worsening environment, land resources have been unable to meet the needs of human society. Additionally, the human demand for high-quality seafood is increasing strongly. The vast ocean with 360 million square kilometers is not only an important food source for human beings, but also a "blue granary" for getting high-end food and high-quality protein. Cage aquaculture is a method of cultivating aquatic products by placing a cage composed of a net, frame, buoyancy device, and fixing device in a specific sea area. This aquaculture method has been rapidly developed during the past decades [1] because of a series of advantages, such as high yield. However, nearshore fish aquaculture is facing more and more environmental problems, such as nearshore water pollution and occupation of nearshore space. In view of this, fish farm operators all over the world are considering relocating their farms to offshore locations, so as to make better use of continuous water flow and deep waters to disperse pollutants in a wider ocean space [2,3].

In recent years, some fish farming companies have put forward some designs of offshore fishing cages, and built them in selected offshore locations for testing [4]. Many ongoing projects use semisubmersible steel cages for offshore fish culture. Because the semisubmersible platform has been widely used in the oil and gas industry, it is natural to consider transforming it for marine fish farming. Ocean Farm 1, designed by Global Maritime, is a newly installed semi-submersible rigid cage (Figure 1). It is suitable for waters with water depths of 100 m to 300 m. It mainly consists of floating pontoons, slender frames and mooring systems [5]. However, it is not easy to carry out routine feeding and maintenance operations at offshore sites. Offshore fish farming, therefore, has to rely on remote technologies, such as unmanned surveillance and automated electrical equipment. These support devices require a constant power supply. Therefore, offshore fishing farms must have their own electricity supply, which can be derived from environmental power, such as solar, wind or wave energy [4].

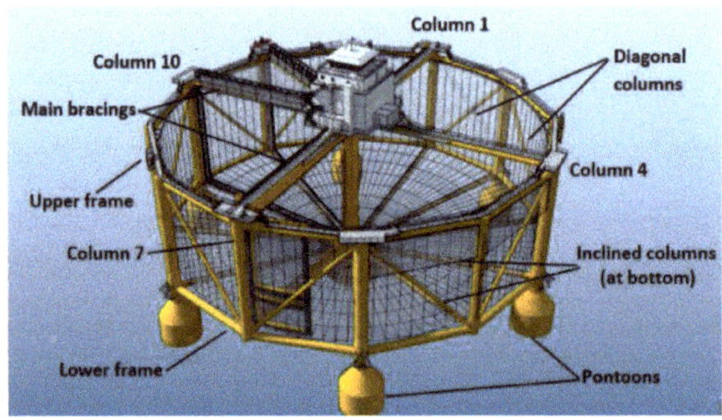

Figure 1. Illustration of Ocean Farm 1 [1]. Reprinted/adapted with permission from Ref. [1]. 2021, Jin, J.; Su, B.; Dou, R.; Luan, C.; Li, L.; Nygaard, I.; Fonseca, N.; Gao, Z.

In order to realize the sustainable development of human society, renewable energy is playing an increasingly important role in the total amount of social energy, with its advantages of large reserves, wide distribution and no pollution [6]. Among renewable energy sources, wind energy, especially offshore wind energy, is one of the most important renewable energy sources. It is considered to be a potential renewable energy resource to supplement traditional fossil fuels [7] and continues to grow rapidly around the world [8,9]. Wind turbines have become widely distributed due to advanced extraction technology. Compared with the onshore wind turbines, floating offshore wind turbines have a higher power generation efficiency due to the more abundant wind resources. Therefore, the development of the floating offshore wind turbines is considered a solution to deal with the energy crisis. Recently, research on simulations and experiments of floating offshore wind turbines have been widely carried out by many scholars [10–12]. For example, Russo et al. [11] presented new large-scale laboratory data on a physical model of a spar-type wind turbine with angular motion of control surfaces implemented. The experiments showed that the inclusion of pitch-controlled, variable-speed blades in physical tests on such types of structures is crucial. In the design of floating offshore wind turbines, in addition to considering the stability of wind turbine operation, the economy and investment return period also need to be considered emphatically.

In view of the above analyses, only a few studies have been performed integrating an aquaculture cage with a floating wind turbine, in recent years [2,13,14]. In Ref. [2], a 1 MW floating spar wind turbine and a fish cage is combined, named COSPAR. The COSPAR fish cage has four catenary mooring lines attached to the spar. Results showed that the COSPAR fish cage enhanced hydrodynamic responses compared with the floating fish cage with only four catenary lines connected to the side vertical columns of the cage. However, the influence of aerodynamic loads on the COSPAR fish cage is not considered. Ref. [13] also proposed a state-of-the-art concept integrating a floating offshore wind turbine with a steel fishing cage, named FOWT-SFFC. The aero-hydro-servo-elastic modeling and time-domain simulations were performed using FAST to study the dynamic response of FOWT-SFFC. However, some simplifications were assumed, i.e., the drag force on the fish nets was neglected. Additionally, the mooring lines were modeled using the quasi-static method. Lei et al. [14] investigated the influence of nets on the dynamic response of a floating offshore wind turbine integrated with a steel fish cage. The results showed that nets play an important role in responses when wave periods are far away from natural periods of motion.

This novel concept of integrating a floating offshore wind turbine with a fishing cage can maximize the utilization of ocean resources, and it can be regarded as a reference for constructing a new pattern of offshore wind power integration development with harmonious coexistence between humans and nature. Therefore, this integrated system is worthy of further study. Based on this, a fully coupled aero-hydro-servo-elastic-mooring model of the integrated wind turbine–fishing cage system is established in this work. A series of simulations are carried out to explore the dynamic characteristics and feasibility of the integrated system. The structure of this paper is as follows.

In Section 2, the structural model, including the fishing cage, net, mooring lines, and wind turbine, is described. In Section 3, the dynamics model of the integrated system is built. To achieve coupled simulations, a control system is designed for the integrated system in Section 4. In Section 5, free decay tests, uniform wind with irregular and regular wave tests, turbulent wind and irregular wave tests are performed. In addition, the influence of mooring line length are also investigated. Finally, the conclusions are provided in Section 6.

2. Structural Model

The integrated system comprises two parts: (1) a semisubmersible fishing cage with a slack catenary mooring system, and (2) a wind turbine on the top of the cage to provide electrical power, as shown in Figure 2. The concept of the integrated system basically aims at supporting offshore fish culture and harvesting wind energy at the same time to provide a power supply for the offshore fish culture.

2.1. Frame Structure of Fishing Cage

In this paper, in addition to the fish culture function, the fishing cage also serves as a floating foundation to support the upper wind turbine. The frame structure of the fishing cage consists of vertical columns (inner and outside, nine in total), heave plates (attached to the bottom of vertical columns, five in total), horizontal girders (top and bottom, 16 in total), X-shaped girders (sides, eight in total), radial girders (top sides and bottom sides, four and eight in total, respectively). The inner space of the fishing cage can be subdivided into four sectors to cultivate different kinds of fish. The overall shape of the fishing cage is similar to Ocean Farm 1 [1], which is the reference fishing cage design. The upper wind turbine can be installed on the top of the central column. Figure 3 shows views of the fishing cage with key dimensions.

The total draft of the fishing cage is 41 m, while the deepest culture depth in the fishing cage is 35 m, considering that a deeper water site may affect the fish growth due to the lack of light and oxygen saturation, or the significant change in water temperature. Thus, the total height of the fishing cage is 51 m (10 m above water and 41 m underwater). The inner water holding space of the fishing cage is approximately 120,000 m^3.

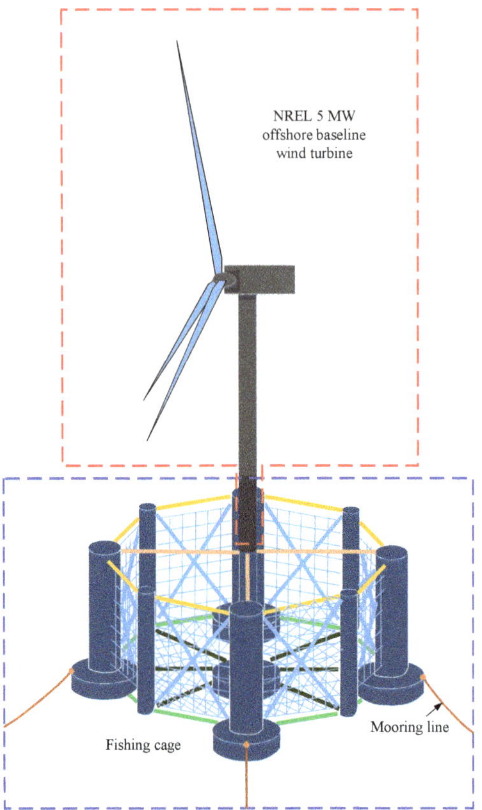

Figure 2. Schematic diagram of the integrated system of a floating offshore wind turbine and a fishing cage.

In order to match the properties of the upper wind turbine tower, the central column has a diameter of 6.5 m, and the elevation of the fishing cage above the still water level is set to 10 m. There are altogether four vertical columns with a diameter of 12 m and a thickness of 0.06 m. There are also four vertical columns with a diameter of 6 m and a thickness of 0.03 m. Seen from the top view, the eight vertical columns are evenly arranged and connected into a regular octagon by horizontal girders. The outside columns are connected to the central column by four radial girders at the top and eight radial girders at the bottom so that the whole fishing cage works as a rigid body. The radial girders have the same diameter (1.6 m) and thickness (0.0175 m). Each column has a heave plate to suppress the heave motion of the fishing cage. The height, diameter and thickness of the heave plates are 6, 24 and 0.06 m, respectively. The above dimensions of the fishing cage are obtained taking into account the hydrostatic balance of the cage in the still water condition and the sufficient amount of water for fish culture, in addition to referring to the dimensions of Ocean Farm 1 [1] and OC4-DeepCwind platform [15]. Table 1 summarizes the key dimensions of the structural members of the fishing cage.

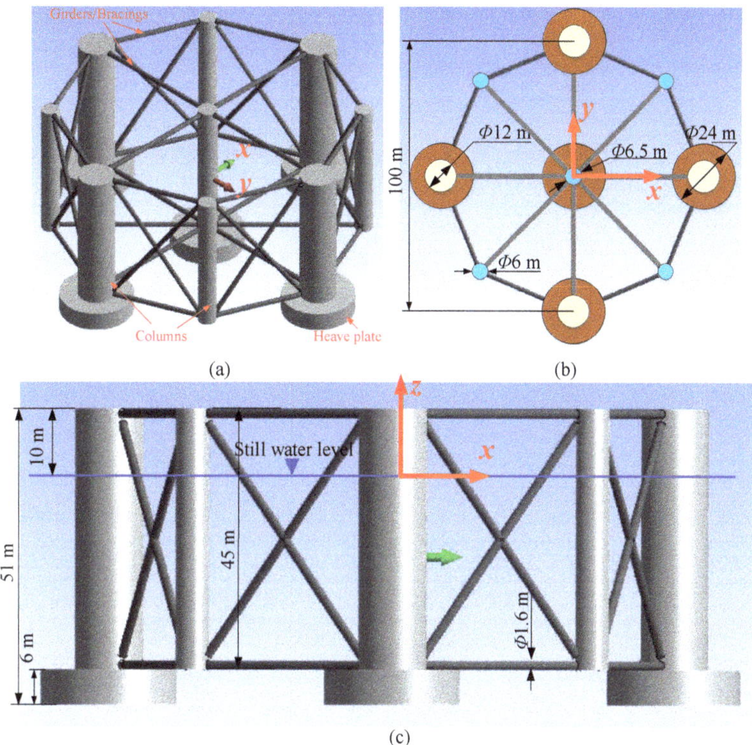

Figure 3. View of the fishing cage with key dimensions (without nets). (**a**) ISO view; (**b**) Top view; (**c**) Side view.

Table 1. Geometric properties of the fishing cage.

Property	Value
Total draft	41 m
Elevation of central column above the still water level	10 m
Elevation of outside columns above the still water level	10 m
Distance from central column centerline to outside column centerline	50 m
Height of central and outside columns	45 m
Height of heave plates	6 m
Diameter of each outside column with a heave plate	12 m
Diameter of each outside column without a heave plate	6 m
Diameter of heave plates	24 m
Diameter of central column	6.5 m
Diameter of bracings	1.6 m
Thickness of each outside column with a heave plate	0.06 m
Thickness of each outside column without a heave plate	0.03 m
Thickness of heave plates	0.06 m
Thickness of central column	0.03
Thickness of bracings	0.0175 m

The total mass of the fishing cage, including ballast, is calculated such that the combined weight of the rotor-nacelle assembly, tower and fishing cage, plus the weight of the mooring system in water should balance with the buoyancy of the undisplaced cage in still water. The concrete, with a high density of 2,400 kg/m^3, is used in heave plates for ballast, which is added from the bottom of heave plates upwards until the mass requirement is met.

The material of the fishing cage is selected as steel with an effective density of $8{,}500\ \text{kg/m}^3$. The masses of the frame structure, girders and ballast are 8,264,815, 1,304,543 and 27,449,312 kg, respectively. The center of mass of the whole fishing cage is located at about 34 m below the still water level. Table 2 summarizes the structural properties of the undisplaced cage.

Table 2. Structural properties of the fishing cage.

Property	Value
Frame structure, bracing, ballast mass	$8.265 \times 10^6, 1.305 \times 10^6, 2.745 \times 10^8\ \text{kg}$
Center of mass location below the still water level, including ballast and bracings	$-34.07\ \text{m}$
Pitch inertia about the center of mass, including ballast and without bracings	$4.1867 \times 10^{11}\ \text{kg m}^2$
Roll inertia about the center of mass, including ballast and without bracings	$4.1867 \times 10^{10}\ \text{kg m}^2$
Yaw inertia about the center of mass, including ballast and without bracings	$7.5657 \times 10^{10}\ \text{kg m}^2$

2.2. Net System

The net is stretched over the sides and bottom of the fishing cage. The type of net used in Ocean Farm 1 is EcoNet, which is a non-fiber material with a hard surface that can resist ocean pollution [5]. It is made from very strong but light PET (Polyethylene Terephthalate), and has been certified according to the Norwegian fish farming standard NS 9415 [16], and its service lifetime in the water can reach up to 14 years.

The solidity ratio is the ratio of the projected area of screen threads to the total area of the net panel. For a square net, as illustrated in Figure 4, the solidity ratio S_n is calculated by [5,17]:

$$S_n = \frac{2d_w}{l_w} - \left(\frac{d_w}{l_w}\right)^2 \tag{1}$$

where d_w is the diameter of twine, l_w is the length of twine.

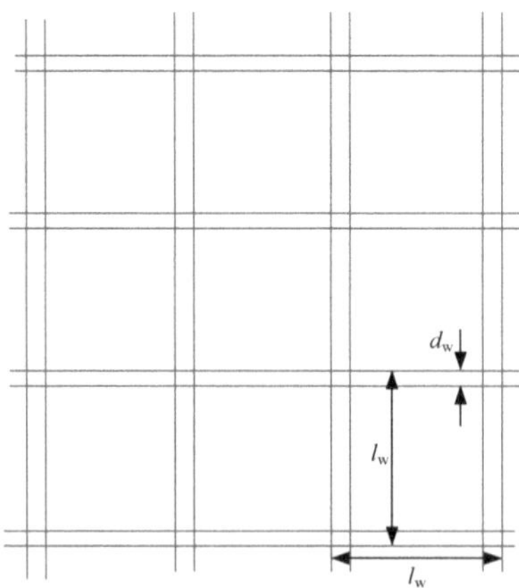

Figure 4. Illustration of net wine.

In the analysis, it is necessary to consider the wave forces applied to the net attached to the fishing cage and the total force for all net elements. Nevertheless, the number of all cells in a fish net is too large to be included in a model. In order to improve simulation efficiency, it is necessary to put forward a simplified approach, so that the total net force can be transferred to the rigid cage. Therefore, an equivalent net method, which has been studied by Duo [5] and Li and Ong [18], is used in this paper. The solidity ratio of the net on Ocean Farm 1, 0.16 [19], is applied in this work. An equivalent diameter d_w = 0.4 m is introduced, with an assumed twine length l_w = 4.8 m, to obtain the same solidity ratio based on Equation (1).

2.3. Mooring Line System

A uniformly distributed four-line spread mooring system is applied for the integrated system. The fairleads are located at the outside heave plates, at a depth of 35 m below the sea water level and at a radius of 62 m from the cage center. The anchors are located at a water depth of 200 m and at a radius of 891.6 m from the cage center. The mooring lines are made from a studless grade 4 chain, with a chain nominal diameter of 0.153 m [19]. The unstretched length of each mooring line is 880 m, a submerged unit mass of 447 kg/m, an axial stiffness of 2.1×10^6 kN and a catalogue breaking strength of 2×10^4 kN [19]. Note that the length of 880 m is determined after some study on the mooring system of the DeepCWind semisubmersible platform at the same water depth [20]. All hydrodynamic coefficients are determined according to the chain's nominal diameter. The drag and the added mass coefficients are assumed based on DNVOS-E301 [21]. Tables 3 and 4 list the configurations and properties of the mooring line system, respectively. Figure 5 shows the mooring line system configuration.

Table 3. Configurations of mooring line system.

Configuration	Value
Number of mooring lines	4
Angle between adjacent lines	90°
Water depth	200 m
Depth from fairlead to still water level	35 m
Radius from anchors to cage centerline	891.6 m
Radius from fairleads to cage centerline	62 m
Unstretched mooring line length	880 m

Table 4. Properties of mooring line system.

Property	Value
Chain type	Studless grade 4
Nominal diameter	0.153 m
Unit mass in water	447 kg/m
Axial stiffness	2.1×10^6 kN
Catalogue breaking strength	2×10^4 kN
Transversal drag coefficient	2.4
Longitudinal drag coefficient	1.15
Added mass coefficient	1

2.4. Wind Turbine System

In this work, the upper wind turbine of the integrated FOWTs is selected as the NREL offshore 5 MW baseline wind turbine [22]. For the convenience of understanding, the properties of the wind turbine will be described briefly in this section. More details can be found in Ref. [22].

To support the preliminary study aimed at assessing the integrated system composed of a floating offshore wind turbine and a fishing cage, the NREL offshore 5 MW baseline wind turbine, a representative utility-scale multi-megawatt turbine, is applied in this work. This wind turbine is a conventional three-bladed upwind variable-speed variable-pitch turbine. Table 5 shows the main properties of the NREL offshore 5 MW baseline wind turbine. More details can be referred to Refs. [22,23].

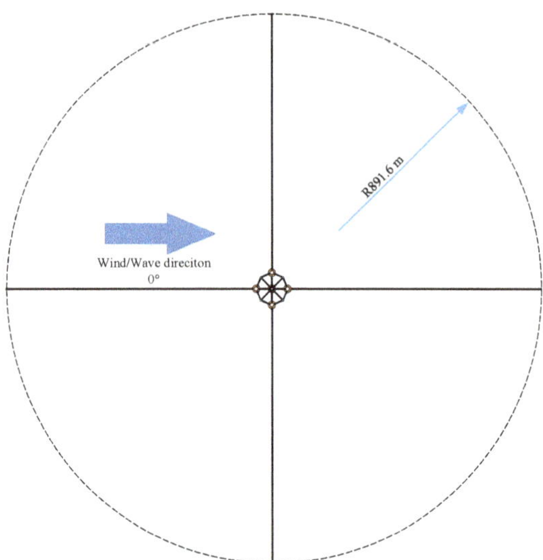

Figure 5. Top view of the mooring line system configuration.

Table 5. Main properties of the NREL offshore 5 MW wind turbine.

Property	Parameter
Rating power	5 MW
Rotor orientation, Configuration	Upwind, Three Blades
Control	Variable Speed, Collective Pitch
Drivetrain	High Speed, Multiple-Stage Gearbox
Rotor, Hub diameter	126 m, 3 m
Hub height	90 m
Cut-In, Rated, Cut-Out wind speed	3, 11.4, 25 m/s
Cut-in, Rated rotor speed	6.9 rpm, 12.1 rpm
Rated tip speed	80 m/s
Overhang, Shaft tilt, Precone	5 m, 5°, 2.5°
Rotor, Nacelle mass	110,000 kg, 240,000 kg

The tower base is coincident with the top of the central column of the fishing cage and is located at an elevation of 10 m above the still water level. The tower top is coincident with the yaw bearing and is located at an elevation of 87.6 m above the still water level. This tower-top elevation (90 m above the still water level) is consistent with the land-based version of the NREL 5 MW baseline wind turbine, as described in Ref. [23]. These properties are all relative to the undisplaced position of the fishing cage.

The diameter at the tower base for the NREL 5-MW offshore baseline wind turbine is 6.5 m, which matches the diameter of the central column of the fishing cage. The tower–base thickness, top diameter and thickness are 0.027, 3.87 and 0.019 m, respectively. The effective mechanical steel properties of the tower are determined according to the DOWEC study [24]. Young's modulus and shear modulus are set to 210 GPa and 80.8 GPa, respectively. The effective density of the steel is set as 8500 kg/m^3, which is meant to be an increase above steel's typical value of 7850 kg/m^3, to consider paint, bolts, welds and flanges that are not included in the tower thickness data [15]. The tower radius and thickness are assumed to be linearly tapered from the tower base to the tower top.

The overall tower mass is 249,718 kg and is centered at 43.4 m along the tower centerline above the still water level. This is obtained from the overall tower length of 77.6 m. A structural damping ratio of 1% critical is specified for all modes of the isolated tower, which corresponds to the values used in the DOWEC study [24]. Table 6 gives the undistributed tower properties.

Table 6. Undistributed tower properties.

Property	Parameter
Elevation to tower base above still water level	10 m
Elevation to tower top above still water level	87.6
Overall tower mass	249,718 kg
Center of mass location above still water level	43.4 m
Structural damping ration	1%

3. Dynamics Modeling

In the present work, the fully coupled aero-hydro-elastic-servo-mooring model of the integrated FOWT is established through a coupling framework based on FAST [25] and ANSYS/AQWA [26]. The upper wind turbine is modeled on FAST. The turbine dynamic responses are represented by external forces and added mass within each invocation, then are passed to the AQWA solver to be combined with the hydrodynamic loads and mooring line forces to calculate the dynamic responses of the aquaculture cage. For the convenience of understanding, the subsequent sections will briefly describe the basic theories applied in FAST and AQWA for structural modeling. Ref. [27] gives more implementation details.

Figure 6 shows the modeling overview of the integrated system through FAST and AQWA in this work.

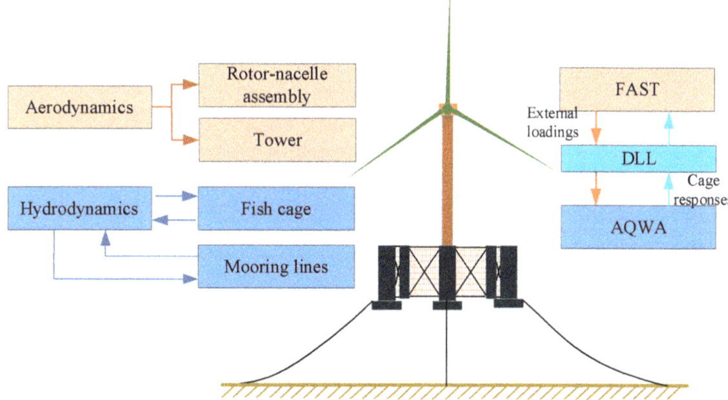

Figure 6. Modeling overview through FAST and AQWA.

3.1. Aerodynamics and Structural Dynamics

FAST is an aero-hydro-elastic-servo wind turbine simulation tool developed by the National Renewable Energy Laboratory (NREL) [25]. The aerodynamic loads on the blades are calculated by the blade element momentum (BEM) theory within the AeroDyn module [28,29]. The rotor thrust (T) and torque (M) are calculated as:

$$\begin{cases} \mathrm{d}T = \frac{1}{2}\rho_{\mathrm{air}}W^2 c(C_l \cos\varphi + C_d \sin\varphi)\mathrm{d}r \\ \mathrm{d}M = \frac{1}{2}\rho_{\mathrm{air}}W^2 c(C_l \sin\varphi - C_d \cos\varphi)r\mathrm{d}r \end{cases} \tag{2}$$

where ρ_{air} is the air density, C_l and C_d are, respectively, the lift and drag coefficients of the airfoil, W is the absolute wind speed, c is the chord length of a blade element, φ is the inflow angle and r is the local radius of a blade element.

More detailed information on the aerodynamic calculation can be found in Refs. [28,29]. When the aerodynamic loads are calculated, they are then imported into the electrodynamic module of FAST to resolve the motion equations of the wind turbine. Kane's kinetic approach is used, which is defined as

$$\mathbf{F}_r^* + \mathbf{F}_r = 0 \tag{3}$$

where $\mathbf{F}_r^*$ and $\mathbf{F}_r$ are the generalized inertia force vector and the generalized active force vector, respectively.

The generalized inertia forces are composed of the inertia forces of the nacelle, tower, hub and blades. The generalized active forces consist of aerodynamic loads $\mathbf{F}_{aero}$, elastic restoring forces $\mathbf{F}_{elas}$, gravity $\mathbf{F}_{grav}$ and damping forces:

$$\mathbf{F}_r = \mathbf{F}_{aero} + \mathbf{F}_{elas} + \mathbf{F}_{grav} + \mathbf{F}_{damp} \tag{4}$$

The wind turbine is modelled as a multi-body system including rigid bodies and flexible bodies. The hub and nacelle are modelled as rigid bodies. The tower and blades are treated as flexible bodies. The generalized inertia force of a rigid body is represented by the same formula. Further information on the motion equations can be referred to in Refs. [30,31].

3.2. Mooring Line Dynamics

The finite element approach within AQWA is applied to consider the dynamic effects of mooring lines. Each line is discretized into several finite elements, and the mass of each element is concentrated into a corresponding node, as shown in Figure 7. In the figure, S_j is the length of an unstretched line between the anchor and the jth node, and D_e represents the local segment diameter of one line. Each line is treated as a chain of Morison elements subjected to various external forces. According to Ref. [32], the motion equation of each line element is defined as:

$$\begin{cases} \frac{\partial \mathbf{T}}{\partial S_e} + \frac{\partial \mathbf{V}}{\partial S_e} + \mathbf{w} + \mathbf{F}_h = m_e \frac{\partial^2 \mathbf{R}}{\partial t^2} \\ \frac{\partial \mathbf{M}}{\partial S_e} + \frac{\partial \mathbf{R}}{\partial S_e} \times \mathbf{V} = -\mathbf{q} \end{cases} \tag{5}$$

where $\mathbf{T}$ and $\mathbf{V}$ are the tension force and shear force vectors of the first element node, respectively; $\mathbf{R}$ is the position vector of the first element node; S_e is the length of an unstretched element; $\mathbf{w}$ and $\mathbf{F}_h$ represent the weight and hydrodynamic load vectors per unit element length, respectively; m_e is the mass per unit length; $\mathbf{M}$ is the bending moment vector of the first element node; and $\mathbf{q}$ is the distributed moment load per unit length.

The bending moment and tension force vectors are calculated by:

$$\begin{cases} \mathbf{M} = EI \cdot \frac{\partial \mathbf{R}}{\partial S_e} \times \frac{\partial^2 \mathbf{R}}{\partial S_e^2} \\ \mathbf{T} = EA \cdot \varepsilon \end{cases} \tag{6}$$

where EI and EA are, respectively, the bending stiffness and axial stiffness of the line, and ε is the strain of the line.

To ensure a unique solution to Equation (5), pinned connection boundary constraints (Equation (7)) are imposed on the top and bottom ends:

$$\begin{cases} \mathbf{R}(0) = \mathbf{P}_{bot}, \mathbf{R}(L) = \mathbf{P}_{top} \\ \frac{\partial^2 \mathbf{R}(0)}{\partial S_e^2} = 0, \frac{\partial^2 \mathbf{R}(L)}{\partial S_e^2} = 0 \end{cases} \tag{7}$$

where $\mathbf{P}_{bot}$ and $\mathbf{P}_{top}$ are the position vectors of attachment points, and L is the length of the unstretched line.

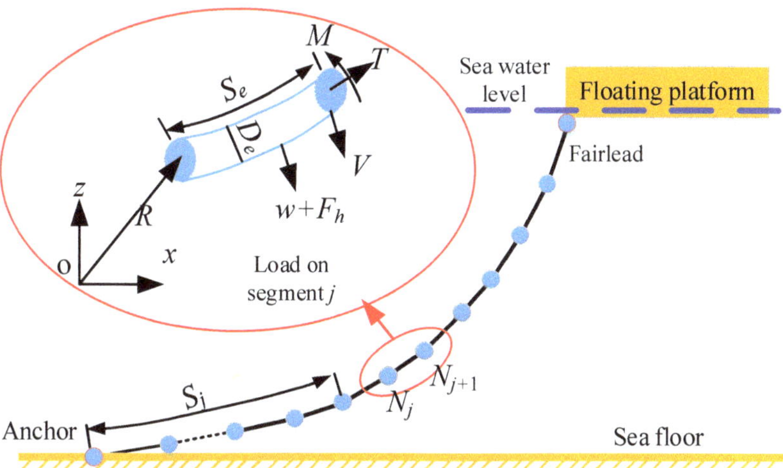

Figure 7. Schematic diagram of the dynamic model of mooring lines.

3.3. Hydrodynamics of Aquaculture Cage

The hydrodynamic loads on the aquaculture cage are obtained based on the frequency-dependent hydrodynamic coefficients, including the added mass, radiation damping and mooring line restoring forces. These coefficients can be obtained through a frequency domain analysis in AQWA.

3.3.1. Hydrodynamics of Cage Support Structures

The hydrostatic and hydrodynamic analyses are performed by the panel method to solve the radiation and diffraction problems for the interaction of surface waves with main structures in the frequency domain. Based on boundary conditions, the velocity potentials can be solved on the mean body position, and the pressure can be obtained from the linear Bernoulli's equation [2]. The governing equation for the motion of a floating body in six degrees of freedom (DOFs) is expressed by [2]:

$$\left(-\omega^2(M + A(\omega)) + i\omega B(\omega) + C\right)\zeta(\omega) = F(\omega) \tag{8}$$

where M is the mass matrix, ω is the angular frequency, $A(\omega)$ is the added mass matrix, $B(\omega)$ is the damping matrix, C is the hydrostatic stiffness matrix, $\zeta(\omega)$ is the dynamic response vector and $F(\omega)$ is the dynamic load vector.

When using the potential flow theory in AQWA, the diffraction panel elements model does not include the effect of the viscous drag force. Instead, the Morison elements are usually applied for slender structures. Therefore, the cross-bracings that act as a support role are modeled through a series of Morison elements. According to the setting in Ref. [15], the drag coefficient for the cross-bracings in this work is set to 0.63. Figure 8 shows the panel model of the aquaculture cage modeled in AQWA.

3.3.2. Net Hydrodynamics

In addition to the cage support structures, the nets of the fishing cage also need to be properly modeled. The nets are flexible and undergo various degrees of deformation depending on the wave and current conditions. The hydrodynamic force on nets and the coupling effect between nets and cage support structures have strong nonlinearity. Thus, a nonlinear model is usually needed to model the nets, and examples of flexible nets modeling can be found in Refs. [33–35].

Compared with the traditional flexible collar fishing cage, this semisubmersible rigid fishing cage has rigidity and larger supporting structures, so it is expected that the nets have less influence on the fishing cage. Therefore, in this preliminary study, the equivalent nets are modeled by Morison elements, which are considered to be stiff and rigidly connected to the cage to work as a whole rigid body [14,18]. Thus, the relative motions between the nets and the cage are neglected. It should be noted that this simplification may overestimate the hydrodynamic forces and viscous effects of the nets because deformation is not considered [18]. Nevertheless, it is still a good start to study the global dynamic responses of the integrated system with the inclusion of the nets modeled by the simplified rigid model.

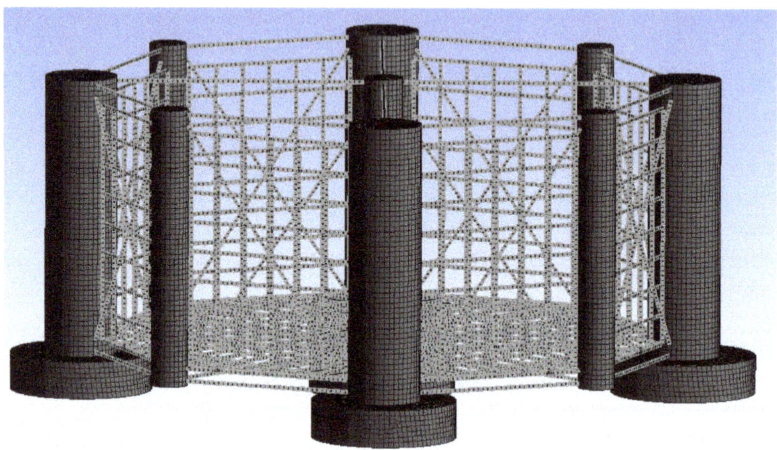

Figure 8. Panel model of the aquaculture cage in AQWA.

Hydrodynamic forces on the nets may be divided into three components: inertia force, drag force and lift force. The inertia force and drag force for the nets can be calculated by using the Morison equation. The lift force can be calculated by the same drag force term in the Morison equation, but the lift coefficient, instead of the drag coefficient, must be introduced [36]. Considering that the mass of the nets accounts for a small proportion of the whole cage mass, the inertia force of the nets may be ignored [5,37,38]. Moreover, when the nets are attached to the cage sides and bottom, the wave and flow directions are almost $0°$ or $90°$ to the normal direction of the net plane where the lift force is almost zero [15,39]. Therefore, in this paper, only the drag force on the nets is considered.

The hydrodynamic drag force of the nets is closely related to the net solidity ratio. In Refs. [36,40], a drag coefficient of nets, C_D, was estimated using the following equation:

$$C_D = 0.04 + \left(-0.04 + 0.33S_n + 6.54S_n^2 - 4.88S_n^3\right)\cos\theta \tag{9}$$

where S_n is the net solidity ratio, θ is the angle between the inflow direction and the net normal.

In this work, it is assumed that the angle of inflow to the net normal is $0°$ for both wave inflow to the side nets and vertical motion inflow to the bottom nets. The drag coefficient $C_D = 0.2$ is determined according to the selected solidity ratio, which is applied for the equivalent nets at both the side and bottom. Note that the possible influence of biofouling on the drag coefficient is neglected in this work, which requires more accurate biofouling hydrodynamic characteristics. This issue needs to be solved by performing experiments in the future.

4. Control System

As studied by Jonkman [22], the operation regions of a wind turbine are divided into five parts: 1, 1.5, 2, 2.5, and 3, as illustrated in Figure 9. In Region 1, the wind speed is lower than the cut-in wind speed, thus, no electrical power is output. In this region, the rotor is accelerating for a start-up. Region 1.5 is a linear transition between Region 1 and Region 2. In Region 2, the controller adjusts the generator torque according to the generator speed while keeping the blade pitch angles at the optimal value. In Region 3, the blade pitch angles are tuned by a collective variable-pitch strategy to maintain the rated generator speed. The demanded blade pitch angles are provided through a gain-scheduled proportional-integral (PI) controller, depending on the speed error between the filtered and the rated generator speeds. Moreover, in order to resist negative damping in the rotor speed response, the generator-torque control law in Region 3 is set to a constant generator-torque control. The constant generator torque is set to the rated value, 43,093.55 Nm [23]. Region 2.5 is a smooth transition region between Regions 2 and 3. This region is also applied to limit tip speed and noise emissions.

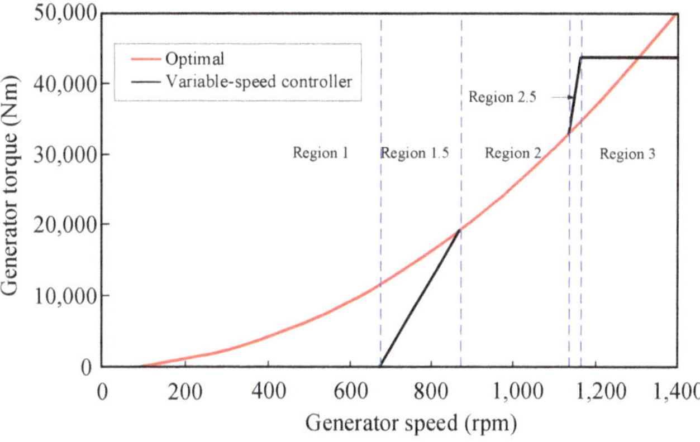

Figure 9. Relationship between generator torque and generator speed for the baseline control system.

The blade pitch control system can be represented by the following equation of motion [41]:

$$\underbrace{\left[I_{\text{drivetrain}} + \frac{1}{\Omega_0}\left(-\frac{\partial P}{\partial \theta}\right)N_{\text{gear}}K_{\text{D}}\right]}_{M_\phi}\ddot{\varphi} + \underbrace{\left[\frac{1}{\Omega_0}\left(-\frac{\partial P}{\partial \theta}\right)N_{\text{gear}}K_{\text{P}} - \frac{P_0}{\Omega_0^2}\right]}_{C_\phi}\dot{\varphi} + \underbrace{\left[\frac{1}{\Omega_0}\left(-\frac{\partial P}{\partial \theta}\right)N_{\text{gear}}K_{\text{I}}\right]}_{K_\phi}\varphi = 0 \quad (10)$$

where $I_{\text{drivetrain}}$ is the drivetrain inertia cast to the low-speed shaft; N_{gear} is the gearbox ratio; Ω_0 is the rated rotor rotational speed; P_0 is the rated mechanical power; $\partial P/\partial \theta$ is the sensitivity of aerodynamic power to the rotor collective blade pitch angle; K_{P}, K_{I} and K_{D} are the blade pitch controller proportional, integral and derivative gains, respectively; $\dot{\varphi} = \Delta\Omega$ is the rotor speed error.

The rotor speed error responds as a 1-DOF dynamic system with natural frequency $\omega_{\varphi\text{n}}$ and damping ratio ζ_φ:

$$\begin{cases} \omega_{\varphi\text{n}} = \sqrt{\dfrac{K_\varphi}{M_\varphi}} \\ \zeta_\varphi = \dfrac{C_\varphi}{2M_\varphi\omega_{\varphi\text{n}}} \end{cases} \quad (11)$$

when designing a blade pitch controller, the PI gains can be calculated by ignoring the derivative gain and negative damping term [42]:

$$\begin{cases} K_{\text{P}} = \dfrac{2I_{\text{drivetrain}}\Omega_0\zeta_\varphi\omega_{\varphi\text{n}}}{N_{\text{gear}}\left(-\frac{\partial P}{\partial \theta}\right)} \\ K_{\text{I}} = \dfrac{I_{\text{drivetrain}}\Omega_0\omega_{\varphi\text{n}}^2}{N_{\text{gear}}\left(-\frac{\partial P}{\partial \theta}\right)} \end{cases} \quad (12)$$

According to the study by Larsen [43], the smallest natural frequency of the blade pitch controller must be less than the smallest critical natural frequency of the support structure to ensure that the support structure motions of a floating offshore wind turbine with active pitch-to-feather control remain positively damped. Therefore, the blade pitch controller's natural frequency of 0.032 Hz (which is below the cage-pitch natural frequency of about 0.06 Hz) and a damping ratio of 0.7 [23] is used in this paper. The resulting proportional gain and integral gain are 0.006275604 s and 0.0008965149, respectively.

5. Results and Discussions

5.1. Free Decay Test

Free decay tests in six degrees of freedom (DOFs) are performed to determine the natural frequencies of the integrated system. The test results are shown in Figure 10 and the natural frequencies are listed in Table 7.

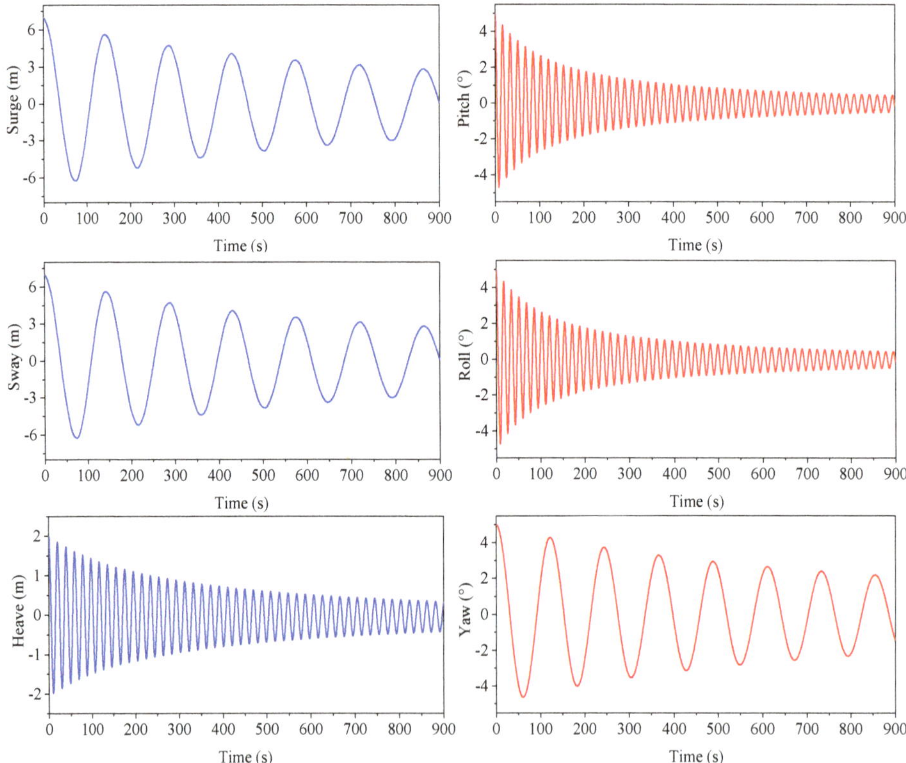

Figure 10. Free decay motions of the integrated system in six DOFs.

Table 7. Natural frequencies of the integrated system in six DOFs.

DOF	Natural Frequency (Hz)
Surge	0.00667
Sway	0.00667
Heave	0.05111
Pitch	0.05778
Roll	0.05778
Yaw	0.00778

It should be noted that because of the coupling of different DOFs, the integrated system will always generate some motions in other DOFs when performing free decay tests, especially the surge-pitch DOFs and sway-roll DOFs. For the free decay in the surge DOF (the initial surge displacement is set to 7 m), the largest pitch amplitude is almost 0.2°, while for the free decay in the pitch DOF (the initial pitch displacement is set to 5°), the largest surge amplitude is almost 3 m. Similar results are observed for the free decay tests in the sway and roll DOFs. These values are not so large when compared to those in the DOFs that are being tested. Thus, the results obtained from the free decay tests are considered reasonable.

5.2. Uniform Wind with Regular and Irregular Waves Test

The uniform wind with regular waves is firstly tested. For the uniform wind, a constant wind speed of 18 m/s, without considering wind shear, is applied. The regular waves are generated using the Airy wave theory. The significant wave height and peak period are set as 4.1 m and 10.5 s, respectively [19]. The simulation time is 700 s, and the results corresponding to the first 100 s are omitted. Figure 11 shows the time series of cage motions in six DOFs under the uniform wind with regular waves condition. It is found that the surge and have motions are more significant than the sway motion. For the rotational motions, the roll and yaw motions are close to zero. With the main action of aerodynamic loads, a mean pitch rotation of 0.48° is induced.

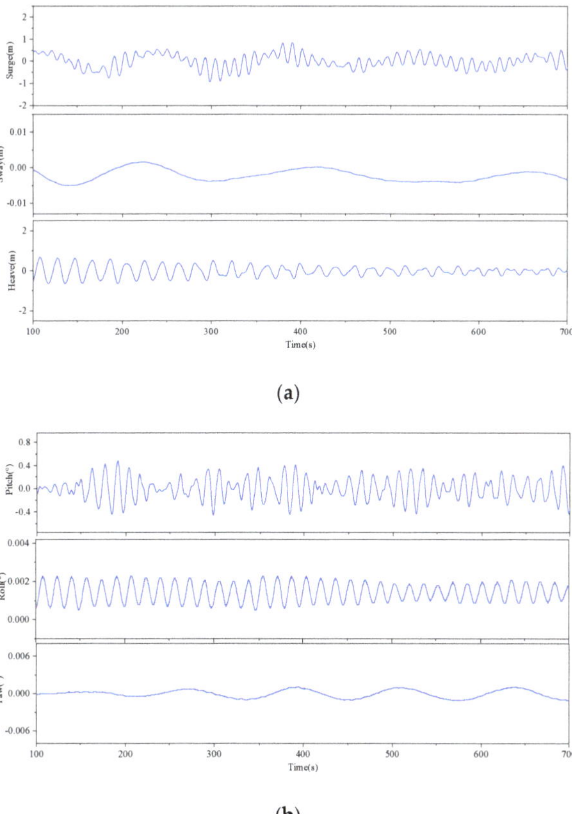

(a)

(b)

Figure 11. Time histories of cage motions under the uniform wind with regular waves. (**a**) Translational motion. (**b**) Rotational motion.

For the irregular waves, the sea state is generated by the JONSWAP spectrum. Figure 12 shows the time series of cage motions in six DOFs under the uniform wind with irregular waves condition. For the translational motions, the surge motion is more significant than the heave and sway motions. A mean surge motion of approximately 2.7 m is mainly caused by the aerodynamic loads. For the rotational motions, the roll and yaw motions are close to zero. With the main action of aerodynamic loads, a mean pitch rotation of 0.18° is induced. Additionally, it is expected that no cage-pitch resonance is found because the blade pitch controller's natural frequency is below the cage-pitch natural frequency. Overall, it is observed that, compared with the regular wave condition, the irregular condition causes more significant cage motions.

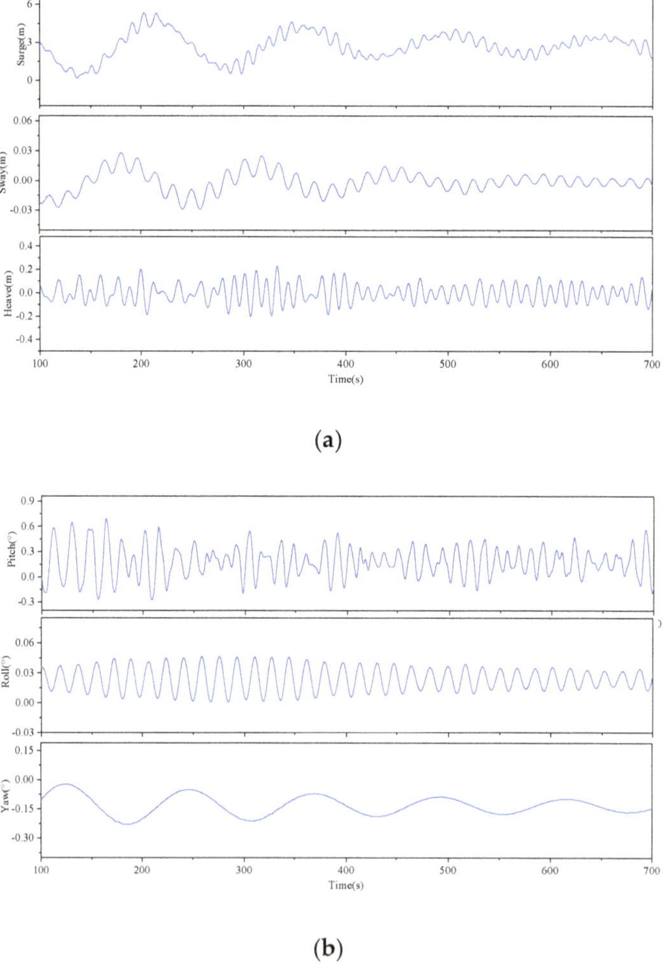

(a)

(b)

Figure 12. Time histories of cage motions under the uniform wind with irregular waves. (**a**) Translational motion. (**b**) Rotational motion.

Figure 13 illustrates the time series of blade pitch angles and wind turbine power output under the uniform wind with irregular waves condition. Due to the blade pitch angle variation tuned by the blade pitch controller, a mean generator power of 5,000 kW is output, with a small stand deviation of approximately 57 kW. The mean value of the blade pitch angle is approximately 15° and there is only slight oscillation. This is attributed to the stability of the cage pitch motion, as shown in Figure 12.

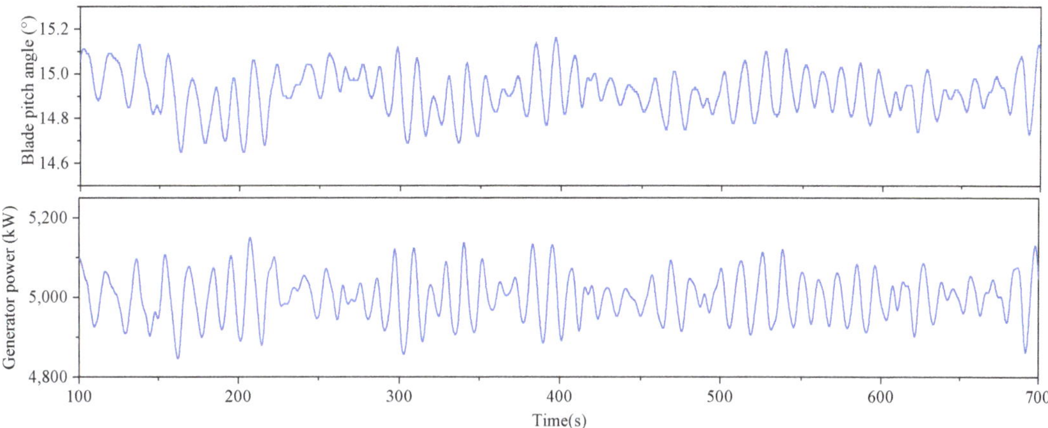

Figure 13. Time histories of pitch angles and power output of the wind turbine under the uniform wind with irregular wave.

5.3. Turbulent Wind and Irregular Wave Test

In this paper, the Norway 5 site is chosen as a representative site for checking the dynamic performance of the integrated system under the turbulent wind and irregular wave conditions, as shown in Figure 14 [19]. A generic water depth of 200 m is applied for this site.

Figure 14. Location of the Norway 5 site [19]. Reprinted/adapted with permission from Ref. [19]. 2014, 20.19. Wang, Q.

The wind and wave data at the site have been fitted with analytical joint distributions by Li [44]. Therefore, it is possible to adopt the following procedure to determine operating loads cases [19]: (1) a mean wind speed at hub height is firstly selected, and then the mean wind speed is transformed to the reference height of 10 m via a power law. (2) The conditional distribution of significant wave height for the given mean wind speed is used, and the most probable value of significant wave height is chosen. (3) The conditional distribution of the wave peak period for a given significant wave height and mean wind speed is used, and the most probable value of the wave peak period is determined.

Four load cases, including three operating cases and one parked case, are selected according to the above procedure [19], as listed in Table 8. Cases 1–3 correspond to the below-rated wind speed, rated wind speed and above-rated wind speed, respectively. Case 4 represents an extreme condition, which is carried out to examine the extreme cage motions. The cage horizontal offset should have a limitation to avoid vertical forces on the anchor, thus, an allowable offset of around 10% of the water depth [45] (that is, 20 m) is adopted in this paper.

Table 8. Environmental conditions for simulations.

Case	Mean Wind Speed (m/s)	Significant Wave Height (m)	Wave Peak Period (s)	Turbulence Intensity (%)	State
1	8	2	10.3	17	Operating
2	11.4	2.5	10.2	15	Operatng
3	18	4.1	10.5	13	Operatng
4	40	15.6	14.5	11	Parked

A normal turbulence model is applied for generating wind files for operating cases, while an extreme wind model is for the parked condition. The operating cases are the general power production cases with rotating blades and an active controller. However, all the blades are pitched to feather, and the wind turbine is shut down to avoid damage in extreme conditions; therefore, the wind turbine is parked. The Kaimal turbulence model is used to generate the wind condition. The extreme environment condition (parked) is obtained by the contour surface method with a return period of 50 years [44]. The total simulation time is 4000 s, and the results corresponding to the first 400 s are omitted.

5.3.1. Time Domain Analysis

For brevity of this paper, only the time histories of responses under cases 3 and 4 are provided as representatives, as shown in Figures 15 and 16. TwrBsMy and RootMy denote the tower-base fore-aft bending moment and the blade-root out-of-plane bending moment, respectively. Additionally, the time histories of wave elevation and horizontal wind speed are also plotted in the figures.

From the figures, it is found that the cage motions obviously oscillate in time due to the action of turbulent wind and irregular waves, especially the surge motion with a non-zero mean value. In particular, under the extreme load case, with the wind turbine parked and the blade pitched to the feather, the cage motion oscillation is much different from the operating case, because there is no damping from the controller and the oscillation is wave-dominated. Additionally, in the extreme condition, the maximum surge offset reaches up approximately 13 m, which is almost 65% of the allowable offset (20 m). It means that the stiffness of the mooring lines is just enough for the fishing cage.

The high oscillation In the curves of the blade pitch angle under condition 3 Is observed to maintain the rated generator power output of 5 MW. This indicates that the blade pitch controller is operating properly, in this case. However, there are still some places on the power curve below the rated value 5 MW, indicating that the wind speed on the rotor is dropped below the rated wind speed (11.4 m/s). Under the action of nonlinear aerodynamics, the tower-base fore-aft bending moment and blade-root out-of-plane bending moment oscillate significantly.

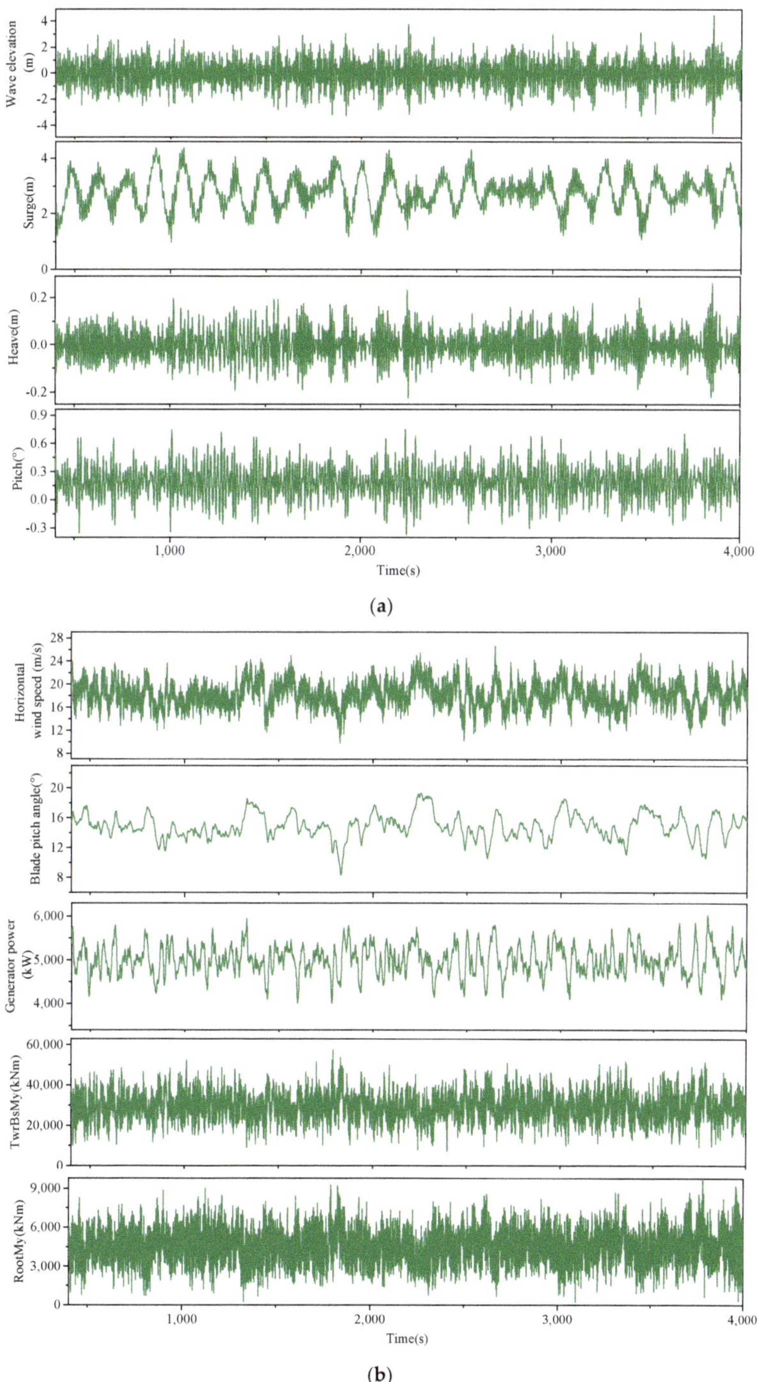

Figure 15. Time histories of responses of the cage and wind turbine under condition 3. (**a**) Cage responses. (**b**) Wind turbine responses.

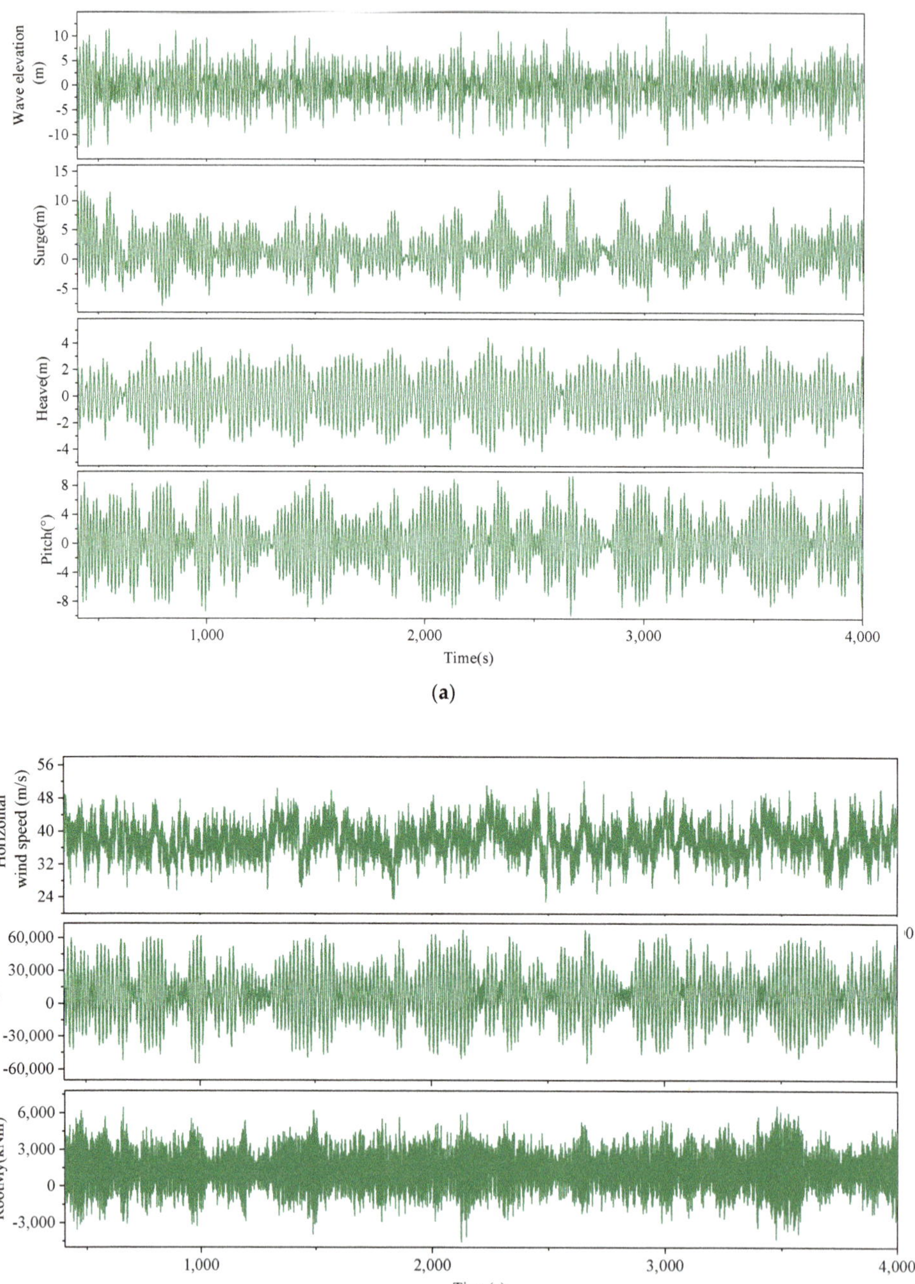

Figure 16. Time histories of responses of the cage and wind turbine under condition 4. (**a**) Cage responses. (**b**) Wind turbine responses.

Several statistical properties, including mean, absolute maximum (Abs Max) and standard deviation of the fishing cage responses (surge, heave and pitch motions) and wind turbine responses (blade pitch angle, generator power, tower-base fore-aft bending moment and blade–root out-of-plane bending moment) are calculated as evaluation indices, as shown in Figure 17. Standard deviations of responses are plotted as error bars on top of mean values.

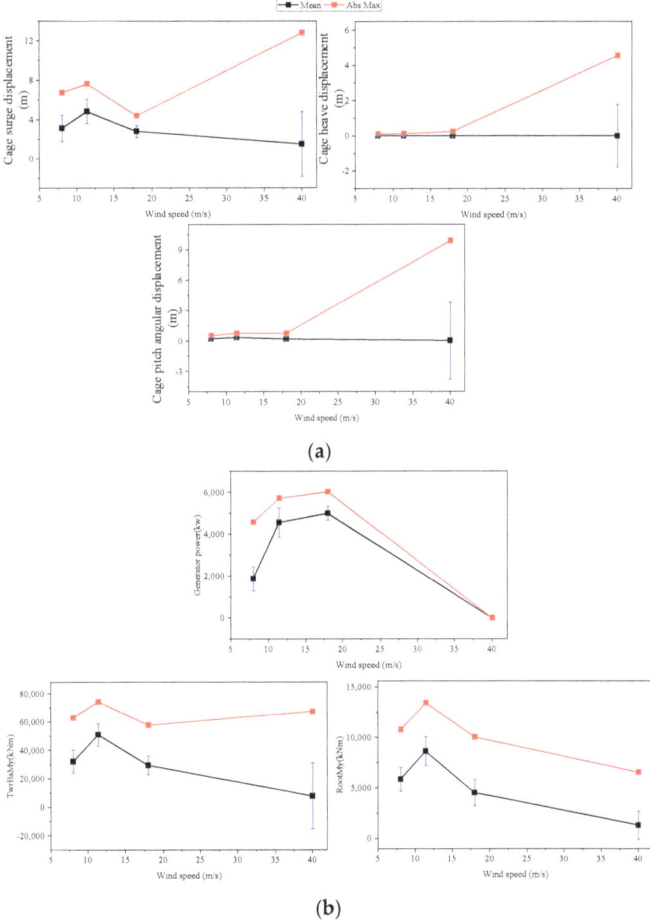

Figure 17. Statistical properties of the cage and wind turbine responses. (**a**) Cage responses. (**b**) Wind turbine responses.

For the statistical results of cage responses, it is seen that the heave and pitch motions exhibit similar patterns with almost zero mean values, while the surge motion dominates the cage response. The peak value of mean surge motions occurs at the rated wind speed condition. In addition, it is observed that the maximums and standard deviations of the surge, pitch and heave motions all occur in the extreme wind condition. For the statistical results of wind turbine responses, with the increase in wind speed, the generator power output increases, and no power is output at the parked condition. The standard deviation of generator power reaches the largest at the rated wind speed condition. This is expected because the wind speed fluctuates frequently around the rated value. The mean and maximum values of the blade-root out-of-plane bending moment show a similar pattern.

Specifically, the peaks of the mean and maximum blade-root bending moments occur at the rated wind speed condition, while the lowest values appear at the extreme condition. Moreover, the mean and maximum tower-base fore-aft bending moments have peak values at the rated wind speed condition, while the largest standard deviation occurs in the extreme condition.

5.3.2. Frequency Domain Analysis

In order to further investigate the effect of turbulent wind, a spectral analysis is performed for the cage and wind turbine responses. The smoothed spectra are shown in Figure 18. The spectra of wave elevation and horizontal wind speed are also plotted so that they can be used to compare each response spectrum and help to identify the source of the most energetic response.

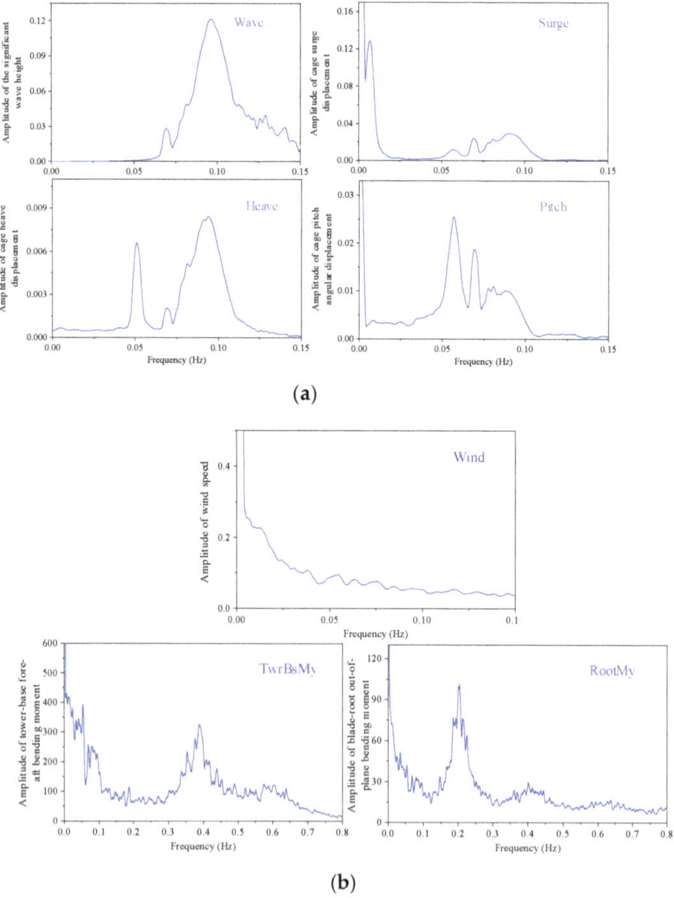

Figure 18. Smoothed spectra of the cage and wind turbine responses under condition 3. (**a**) Cage responses. (**b**) Wind turbine responses.

From Figure 18a, it is observed that the surge motion is dominated by the low-frequency responses due to the turbulent wind and surge resonant responses. A peak occurs at the surge resonant frequency, while the pitch motion and wave excitation also influence the surge motion. For the heave spectrum, two peaks occur at approximately 0.05 Hz and 0.1 Hz, which correspond to the heave natural frequency and wave natural frequency component, respectively. The surge and pitch motions have no influence on the heave motion. For the pitch response, in addition to the low-frequency component of the turbulent wind, there are also obvious pitch resonance frequency and wave frequency components. Overall, the turbulent wind has some influence on the surge and pitch motions, but with no influence on the heave motion. The possible reason is that wind inclination is not considered in simulations, which means that the wind direction is always horizontal. Moreover, the wave frequency response generates some influence in all three DOFs.

From Figure 18b, it is observed that the tower–base fore-aft bending moment (TwrBsMy) is mostly affected by turbulent wind, in addition to the cage pitch motion, irregular waves and 2P-effect. Moreover, there is a small peak at approximately 0.6 Hz, which is possible due to the 3P-effect of the blades. For the blade–root out-of-plane bending moment (RootMy), it is obvious that the turbulent wind and 1P-effect have important effect on the blade response. Moreover, the 2P-effect and 3P-effect also contribute to the blade–root out-of-plane response.

5.4. Influence of Mooring Line Length

In order to study the influence of line length on the responses of the integrated system, two additional mooring line lengths (924 m and 968 m, which are 5% and 10% longer than the original length 880 m, respectively) are chosen, with an assumption of unchanged locations of fairleads and anchors. The cross-sectional area of each line is inversely proportional to its length so that the total mass of each line is unchanged. The time histories of responses for different line lengths under case 3 are illustrated in Figure 19. It is observed that for the cage responses, the increase in mooring line length mainly influences the surge motion, while the heave and pitch motions are slightly affected. The statistics of surge motion are illustrated in Figure 20. Note that the percentages in the figure indicate the reduction of the corresponding index relative to the result with respect to the 880-m-long line. The positive and negative values indicate increase and decrease, respectively. For example, 404.8% indicates that the maximum surge motion with a 924-mlong line is 404.8% larger than that with an 880-m-long line. It is seen that with the increase in line length, the surge motion becomes remarkable.

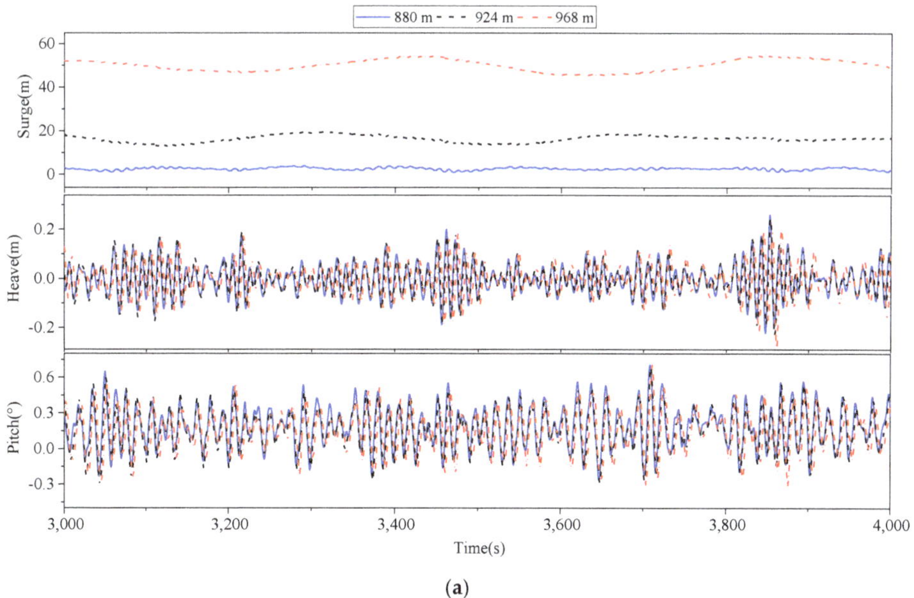

(a)

Figure 19. *Cont.*

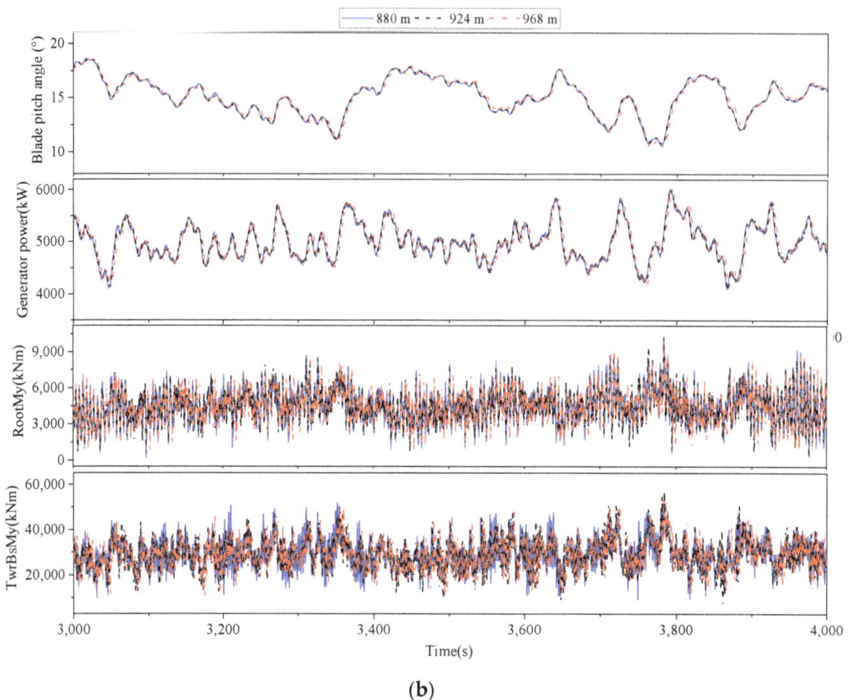

(b)

Figure 19. Comparison of time histories of the fishing cage and wind turbine responses for three mooring line lengths under case 3. (**a**) Cage responses. (**b**) Wind turbine responses.

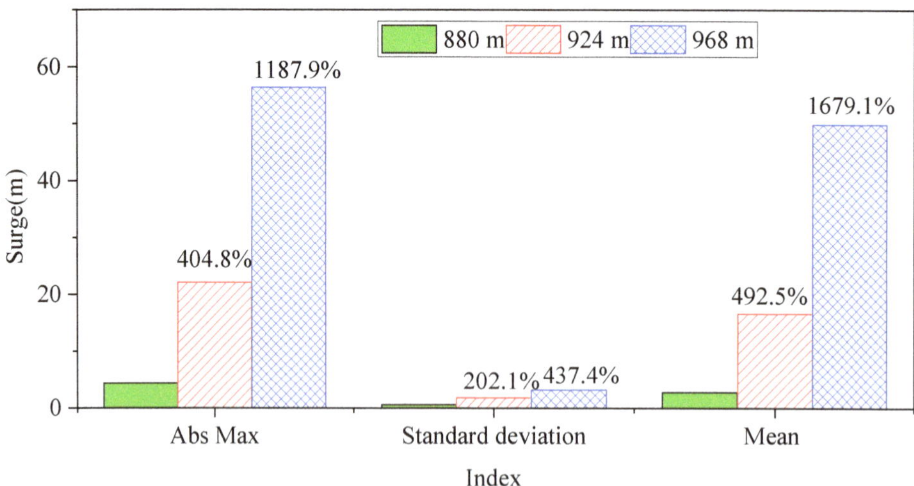

Figure 20. Statistics of surge responses for three line lengths.

As for the response of the wind turbine, although the surge movement is markedly affected, the increasing line length has a slight impact on them, which is attributed to the function of the wind turbine controller. However, the response curves have a certain offset backward. This is due to the large surge motion caused by the significant increase in line length, which delays the time for the wind to reach the rotor, thus, delaying the wind turbine responses.

Figure 21 shows the comparison of the upwind mooring line tension at the fairlead for three line lengths. It is found that with the increase in mooring line length, the maximum and mean line tensions reduce. This change in trend is opposite to that of the surge motion. The increase in line length has less effect on the standard deviation of line tension, which is related to the fact that the increasing line length has little influence on cage pitch and heave motions, as shown in Figure 19a. In addition, it is seen that a further increase in line length does not further significantly reduce the line tension. To be specific, when the line length increases by 5%, the maximum line tension decreases by 45.7%, while when the line length increases by 10%, the maximum line tension only decreases by 52.9%. The same change trend occurs in the mean line tension.

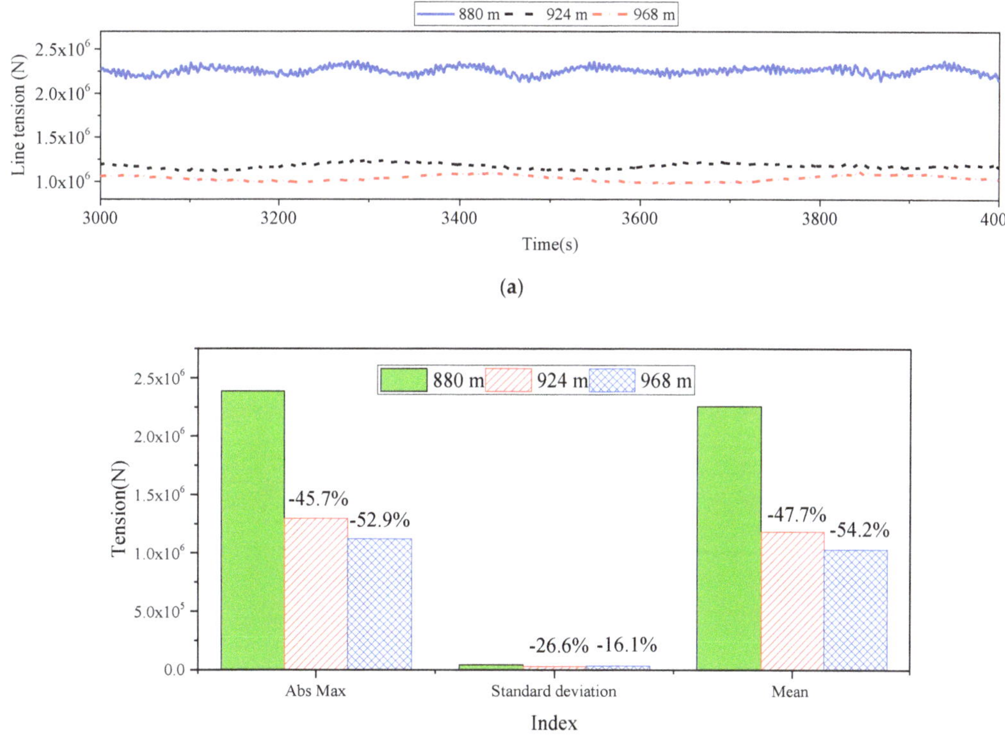

Figure 21. Comparison of upwind line tension at fairlead for three line lengths. (**a**) Time histories. (**b**) Statistics.

6. Conclusions

In this paper, a state-of-the-art concept of a floating offshore wind turbine integrated with a steel fishing cage is investigated. First, the structural configuration and dimensions of this integrated system are presented. Second, the dynamics model of the integrated system is established through FAST and AQWA. Specifically, the upper wind turbine system is modeled in FAST, while the lower fishing cage is modeled in AQWA. Information exchanges between the two codes. Third, for the designed integrated system, a blade-pitch generator-torque controller is applied. A coupled aero-hydro-elastic-servo model is then constructed. Finally, a series of simulations under selected load cases are performed to explore the dynamic response behaviors of the integrated system. The influence of mooring line length is additionally studied. Key conclusions of this work are listed as follows.

(1) For the operating conditions, the rated wind speed condition is the most important condition for the integrated system because large pitch motion occurs, which, in turn, influences the generator power production. On the other hand, in the extreme wind conditions, surge motion becomes higher and exhibits significant oscillation.

(2) Overall, the turbulent wind has some influence on the surge and pitch motions, but with no influence on the heave motion. The wave frequency component can be seen in the surge, pitch and heave responses of the fishing cage, especially in the heave response. Additionally, the surge and pitch motions have no influence on the heave motion.

(3) The tower–base fore-aft bending moment is mostly affected by turbulent wind, in addition to the cage pitch motion, irregular waves, and blade rotation effect. For the blade–root out-of-plane bending moment, the turbulent wind and 1P-effect have an important effect on the blade response.

(4) At the above-rated condition, compared with heave and pitch motions, the cage surge motion is more affected by the increase in mooring line length. The wind turbine responses are slightly influenced by the increasing line length, but exhibit delay caused by the large surge motion. In addition, with the increase in mooring line length, the maximum and mean line tensions reduce, but the standard deviation of line tension is less affected. A further increase in line length does not further remarkably reduce the line tension.

It must be pointed out that the focus of this paper is to provide an idea of conceptual design, modeling and simulation analysis for the integrated wind turbine-fishing cage system. More detailed structural design optimizations, strength checks and experimental tests for the integrated system need to be further carried out in future.

Author Contributions: C.Z.: Simulation, Data curation, Writing-original draft, Writing-review & editing; J.X.: Investigation, Data curation; J.S.: Information retrieval, Software; A.L.: Simulation; M.C.: Funding acquisition; H.L.: Supervision, Methodology; C.G.: Data curation; S.X.: Supervision, Methodology, Writing-original draft. All authors have read and agreed to the published version of the manuscript.

Funding: This paper was funded by Key R & D program of Shandong Province demonstration project of deep sea aquaculture technology (No.: 2021SFGC0701); National Natural Science Foundation of China (No.: 41976194), Project of Science and Technology Research Program of Chongqing Education Commission of China (No.: KJQN202101133), and Scientific Research Foundation of Chongqing University of Technology (No.: 2020ZDZ023).

Data Availability Statement: The data is in the article.

Conflicts of Interest: The authors declare no conflict of interest.

References

1. Jin, J.; Su, B.; Dou, R.; Luan, C.; Li, L.; Nygaard, I.; Fonseca, N.; Gao, Z. Numerical modelling of hydrodynamic responses of Ocean Farm 1 in waves and current and validation against model test measurements. *Mar. Struct.* **2021**, *78*, 103017. [CrossRef]
2. Chu, Y.I.; Wang, C.M. Hydrodynamic Response Analysis of Combined Spar Wind Turbine and Fish Cage for Offshore Fish Farms. *Int. J. Struct. Stab. Dyn.* **2020**, *20*, 2050104. [CrossRef]
3. Holm, P.; Buck, B.H.; Langan, R. Introduction: New Approaches to Sustainable Offshore Food Production and the Development of Offshore Platforms. In *Aquaculture Perspective of Multi-Use Sites in the Open Ocean*; Springer: Cham, Switzerland, 2017; pp. 1–20. [CrossRef]
4. Chu, Y.I.; Wang, C.M. Combined Spar and Partially Porous Wall Fish Cage for Offshore Site. In *EASEC16*; Springer: Singapore, 2020; pp. 569–581. [CrossRef]
5. Dou, R. Numerical Modeling and Analysis of a Semi-Submersible Fish-Cage. Master's Thesis, NTNU, Trondheim, Norway, 2018.
6. Huang, J.; Chang, C.O.; Chang, C.C. Analysis of Structural Vibrations of Vertical Axis Wind Turbine Blades via Hamilton's Principle—Part 1: General Formulation. *Int. J. Struct. Stab. Dyn.* **2020**, *20*, 2050098. [CrossRef]
7. Alam, M.; Chao, L.-M.; Rehman, S.; Ji, C.; Wang, H. Energy harvesting from passive oscillation of inverted foil. *Phys. Fluids* **2021**, *33*, 075111. [CrossRef]
8. Jahangiri, V.; Ettefagh, M.M. Multibody Dynamics of a Floating Wind Turbine Considering the Flexibility Between Nacelle and Tower. *Int. J. Struct. Stab. Dyn.* **2018**, *18*, 1850085. [CrossRef]

9. Rehman, S. Wind Power Resources Assessment at Ten Different Locations using Wind Measurements at Five Heights. *Environ. Prog. Sustain. Energy* **2022**, *41*, e13853. [CrossRef]

10. Otter, A.; Murphy, J.; Pakrashi, V.; Robertson, A.; Desmond, C. A review of modelling techniques for floating offshore wind turbines. *Wind Energy* **2021**, *25*, 831–857. [CrossRef]

11. Russo, S.; Contestabile, P.; Bardazzi, A.; Leone, E.; Iglesias, G.; Tomasicchio, G.R.; Vicinanza, D. Dynamic loads and response of a spar buoy wind turbine with pitch-controlled rotating blades: An experimental study. *Energies* **2021**, *14*, 3598. [CrossRef]

12. Bhattacharya, S.; Lombardi, D.; Amani, S.; Aleem, M.; Prakhya, G.; Adhikari, S.; Aliyu, A.; Alexander, N.; Wang, Y.; Cui, L.; et al. Physical Modelling of Offshore Wind Turbine Foundations for TRL (Technology Readiness Level) Studies. *J. Mar. Sci. Eng.* **2021**, *9*, 589. [CrossRef]

13. Zheng, X.Y.; Lei, Y. Stochastic Response Analysis for a Floating Offshore Wind Turbine Integrated with a Steel Fish Farming Cage. *Appl. Sci.* **2018**, *8*, 1229. [CrossRef]

14. Lei, Y.; Zhao, S.X.; Zheng, X.Y.; Li, W. Effects of Fish Nets on the Nonlinear Dynamic Performance of a Floating Offshore Wind Turbine Integrated with a Steel Fish Farming Cage. *Int. J. Struct. Stab. Dyn.* **2020**, *20*, 2050042. [CrossRef]

15. Robertson, A.; Jonkman, J.; Masciola, M.; Song, H.; Goupee, A.; Coulling, A.; Luan, C. *Definition of the Semisubmersible Floating System for Phase II of OC4*; NREL Technical Report; NREL/TP-5000-60601; National Renewable Energy Lab. (NREL): Golden, CO, USA, 2014.

16. *NS 9415, 2009*; Marine Fish Farms Requirement for Site Survey, Risk Analyses, Design, Dimensioning, Production, Installation and Operation. NS 9415, Stardards Norway: Norway.

17. Tsukrov, I.; Drach, A.; DeCew, J.; Swift, M.R.; Celikkol, B. Characterization of geometry and normal drag coefficients of copper nets. *Ocean Eng.* **2011**, *38*, 1979–1988. [CrossRef]

18. Li, L.; Ong, M.C. A Preliminary Study of a Rigid Semi-Submersible Fish Farm for Open Seas. In Proceedings of the International Conference on Offshore Mechanics and Arctic Engineering. American Society of Mechanical Engineers, Trondheim, Norway, 25–30 June 2017. [CrossRef]

19. Wang, Q. Design of a Steel Pontoon-Type Semi-Submersible Floater Supporting the DTU 10MW Reference Turbine. Master's Thesis, European Wind Energy Master—EWEM, Trondheim, Norway, 2014.

20. Coulling, A.J.; Goupee, A.J.; Robertson, A.N.; Jonkman, J.M. Importance of Second-Order Difference-Frequency Wave-Diffraction Forces in the Validation of a FAST Semi-Submersible Floating Wind Turbine Model. In Proceedings of the ASME 2013 32nd International Conference on Ocean, Offshore and Arctic Engineering, pages V008T09A019-V008T09A019, Nantes, France, 9 June 2013; American Society of Mechanical Engineers: New York, NY, USA, 2013.

21. *DNV-OS-E301*; Position Mooring. Det Norsk Veritas: Oslo, Norway, 2013.

22. Jonkman, J.M. Dynamics Modeling and Loads Analysis of an Offshore Floating Wind Turbine. Technical Reports (No. NREL/TP-500-41958); National Renewable Energy Laboratory (NREL): Golden, CO, USA. [CrossRef]

23. Jonkman, J.; Butterfield, S.; Musial, W.; Scott, G. *Definition of a 5-MW Reference Wind Turbine for Offshore System Development (No. NREL/TP-500-38060)*; National Renewable Energy Lab.(NREL): Golden, CO, USA, 2009.

24. Kooijman, H.J.T.; Lindenburg, C.; Winkelaar, D.; van der Hooft, E.L. *DOWEC 6 MW Pre-Design: Aero-Elastic Modeling of the DOWEC 6 MW pre-Design in PHATAS, DOWEC Dutch Offshore Wind Energy Converter 1997–2003 Public Reports, DOWEC 10046_009, ECN-CX–01-135, Petten*; Energy Research Center of the Netherlands: Petten, The Netherlands, 2003.

25. Jonkman, J.M.; Buhl, M.L., Jr. *Fast User's Guide-Updated August 2005 (No. NREL/TP-500-38230)*; National Renewable Energy Lab.(NREL): Golden, CO, USA, 2005.

26. *Ansys, AQWA Reference Manual Release 2019 R1, USA*; Ansys Inc.: Canonsburg, PA, USA,, 2019.

27. Yang, Y.; Bashir, M.; Michailides, C.; Li, C.; Wang, J. Development and application of an aero-hydro-servo-elastic coupling framework for analysis of floating offshore wind turbines. *Renew. Energy* **2020**, *161*, 606–625. [CrossRef]

28. Moriarty, P.J.; Hansen, A.C. *AeroDyn Theory Manual (No. NREL/TP-500-36881)*; National Renewable Energy Lab.: Golden, CO, USA, 2005.

29. Jonkman, J.M.; Hayman, G.J.; Jonkman, B.J.; Damiani, R.R.; Murray, R.E. *AeroDyn V15 User's Guide and Theory Manual*; NREL: Golden, CO, USA, 2015.

30. Jonkman, J.M. *Modeling of the UAE Wind Turbine for Refinement Of FAST_AD (No. NREL/TP-500-34755)*; National Renewable Energy Lab.: Golden, CO, USA, 2003. [CrossRef]

31. Jonkman, J.; Buhl, M. New Developments for the NWTC's FAST Aeroelastic HAWT Simulator. In Proceedings of the 42nd AIAA Aerospace Sciences Meeting and Exhibit, Reno, NV, USA, 5–8 January 2004. [CrossRef]

32. Newman, J.N. *Marine Hydrodynamics*; MIT Press: Cambridge, MA, USA, 2018.

33. Lader, P.F.; Fredheim, A. Dynamic properties of a flexible net sheet in waves and current—A numerical approach. *Aquac. Eng.* **2006**, *35*, 228–238. [CrossRef]

34. Lee, C.W.; Lee, J.; Park, S. Dynamic behavior and deformation analysis of the fish cage system using mass-spring model. *China Ocean Eng.* **2015**, *29*, 311–324. [CrossRef]

35. Moe, H.; Fredheim, A.; Hopperstad, O. Structural analysis of aquaculture net cages in current. *J. Fluids Struct.* **2010**, *26*, 503–516. [CrossRef]

36. Lader, P.; Enerhaug, B. Experimental Investigation of Forces and Geometry of a Net Cage in Uniform Flow. *IEEE J. Ocean. Eng.* **2005**, *30*, 79–84. [CrossRef]

37. Cifuentes, C.; Kim, M. Hydrodynamic response of a cage system under waves and currents using a Morison-force model. *Ocean Eng.* **2017**, *141*, 283–294. [CrossRef]
38. Cifuentes, C.; Kim, M.H. Numerical simulation of fish nets in currents using a Morison force model. *Ocean Syst. Eng.* **2017**, *7*, 143–155.
39. Cha, B.J.; Kim, H.Y.; Bae, J.H.; Yang, Y.S.; Kim, D.H. Analysis of the hydrodynamic characteristics of chain-link woven copper alloy nets for fish cages. *Aquac. Eng.* **2013**, *56*, 79–85. [CrossRef]
40. Løland, G. Current forces on, and water flow through and around, floating fish farms. *Aquac. Int.* **1993**, *1*, 72–89. [CrossRef]
41. Jonkman, J.M. Dynamics of offshore floating wind turbines-model development and verification. *Wind Energy* **2009**, *12*, 459–492. [CrossRef]
42. Hansen, M.H.; Hansen, A.; Larsen, T.J.; Øye, S.; Sørensen, P.; Fuglsang, P. *Control Design for a Pitch-Regulated, Variable Speed Wind Turbine*; Riso National Laboratory: Roskilde, Denmark, 2005; ISSN 0106-2840. ISBN 87-550-3409-8.
43. Larsen, T.; Hanson, T.D. A method to avoid negative damped low frequent tower vibrations for a floating, pitch controlled wind turbine. *J. Phys. Conf. Ser.* **2007**, *75*, 012073. [CrossRef]
44. Li, L.; Gao, Z.; Moan, T. Joint Environmental Data at Five European Offshore Sites for Design of Combined Wind and Wave Energy Devices. In Proceedings of the ASME 2013 32nd International Conference on Ocean, Offshore and Arctic Engineering, American Society of Mechanical Engineers Digital Collection, Nantes, France, 9–14 June 2013. [CrossRef]
45. Xue, W. Design, Numerical Modelling and Analysis of a Spar Floater Supporting the DTU 10MW Wind Turbine. Master's Thesis, NTNU, Trondheim, Norway, 2016.

Journal of
*Marine Science
and Engineering*

Article

Genetic Diversity Analysis of Different Populations of *Lutjanus kasmira* Based on SNP Markers

Fangcao Zhao [1,2,3], Liang Guo [1,3], Nan Zhang [1,3], Jingwen Yang [1,3], Kecheng Zhu [1,3] Huayang Guo [1,3], Baosuo Liu [1,3], Bo Liu [1,3], Dianchang Zhang [1,3,4,5,*] and Shigui Jiang [1,3,4,5]

[1] Key Laboratory of South China Sea Fishery Resources Exploitation and Utilization, South China Sea Fisheries Research Institute, Chinese Academy of Fishery Sciences, Ministry of Agriculture and Rural Affairs, Guangzhou 510300, China
[2] School of Fisheries and Life Science, Dalian Ocean University, Dalian 116023, China
[3] Southern Marine Science and Engineering Guangdong Laboratory (Guangzhou), Guangzhou 511458, China
[4] Guangdong Provincial Engineer Technology Research Center of Marine Biological Seed Industry, Guangzhou 510300, China
[5] Tropical Aquaculture Research and Development Center, South China Sea Fisheries Research Institute, Chinese Academy of Fishery Sciences, Sanya 572018, China
* Correspondence: zhangdch@scsfri.ac.cn

Abstract: *Lutjanus kasmira* belongs to the family Lutjanidae. Over the past 20 years, the *L. kasmira* population in the South China Sea has been shrinking due to climate change, pressure from human activities, and inadequate food supplies. In this study, single nucleotide polymorphism (SNP) data obtained from restriction site-associated DNA sequencing (RAD-seq) were used to assess the genetic diversity of *L. kasmira* in Zhubi Dao (ZB) and Meiji Dao (MJ). The genome-wide nucleotide diversity (π) of the ZB population and MJ population was 0.02478 and 0.02154, respectively. The inbreeding coefficient (Fis) of the ZB population and MJ population was -0.18729 and 0.03256, respectively. The genetic differentiation (Fst) between the ZB and MJ subpopulations was 0.00255102. The expected heterozygosity (He) of individuals from ZB and MJ was 0.33585 and 0.22098, respectively. The observed heterozygosity (Ho) of individuals from the ZB population and MJ population was 0.46834 and 0.23103, respectively. Although the ZB and MJ populations did not have significant genetic differences, the genetic differentiation between them was confirmed using population structure, phylogenetic, and principal component analyses. These results indicated that the genetic diversity of the ZB and MJ populations was relatively low at the genome level, and that their genetic differences were small.

Keywords: *Lutjanus kasmira*; SNP; genetic diversity; genetic differentiation

Citation: Zhao, F.; Guo, L.; Zhang, N.; Yang, J.; Zhu, K.; Guo, H.; Liu, B.; Liu, B.; Zhang, D.; Jiang, S. Genetic Diversity Analysis of Different Populations of *Lutjanus kasmira* Based on SNP Markers. *J. Mar. Sci. Eng.* **2022**, *10*, 1547. https://doi.org/10.3390/jmse10101547

Academic Editor: Nguyen Hong Nguyen

Received: 26 September 2022
Accepted: 14 October 2022
Published: 20 October 2022

1. Introduction

Single nucleotide polymorphisms (SNPs) are DNA sequence polymorphisms caused by the transformation or transposition of a single nucleotide at the genomic level [1]. SNP markers are a tool for studying the genetic structure of species [2]. Since any base of genomic DNA can be mutated, SNP molecular markers are ubiquitous in animals and plants [3]. SNPs have the advantages of a large number, a wide distribution, strong representativeness [4], good genetic stability, and convenience for high-throughput and highly automated detection and analysis [5]. The distribution of SNPs in the genome can comprehensively reflect a population's genetic variation [4].

L. kasmira is a typical reef-dwelling fish found throughout the Indo-Pacific region, from Australia in the east to Japan in the north [6]. Individuals of this species are mainly distributed around the South China Sea islands, Taiwan waters, and in the southern portion of the East China Sea [7]. *L. kasmira* has a bright yellow surface and a reddish underside; the side of the body bears four blue longitudinal bands, with an indistinct black spot between the third and fourth blue bands [8]. In recent years, the habitat of this species, i.e., coral reefs and mangroves, has been reduced and degraded due to anthropogenic destruction and environmental pollution [9]. Moreover, *L. kasmira* suffers from overfishing due to its delicious meat, high economic value, and spawning clusters [7]. At present, studies on *L. kasmira* mainly focus on its physiological ecology and molecular systematics [7–9]; there are no reports on the development of SNP markers and the genetic diversity level of *L. kasmira*. In this study, the whole-genome SNPs of *L. kasmira* were first discovered by restriction site-associated DNA sequencing (RAD-seq) technology [10] and then used to analyze the genetic diversity and structure of this fish species, providing a theoretical basis for the rational development and conservation of its germplasm resources.

2. Materials and Methods

2.1. Sample Collection and DNA Extraction

L. kasmira samples were collected from Meiji Dao (MJ) and Zhubi Dao (ZB), which are separated by approximately 200 km (Figure 1). A total of 30 *L. kasmira* samples were collected, comprising 14 samples from MJ and 16 samples from ZB (Table 1). Tissue was collected by clipping the fin strip of the experimental samples. The tissue samples were fixed in 75% anhydrous ethanol and stored at room temperature. A Tissue DNA Extraction CZ Kit (Mobio, Guangzhou, China) was used to extract genomic DNA from the pterygiophore of each fish according to the manufacturer's protocol.

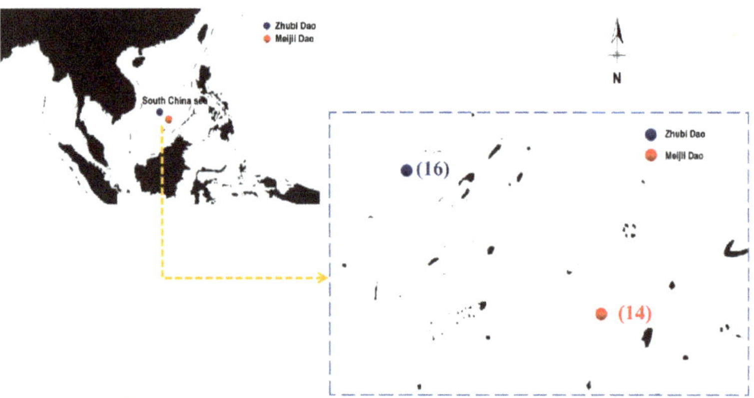

Figure 1. Locations of the 30 collected *Lutjanus kasmira*.

Table 1. Sample information for the 30 *L. kasmira* collected in Zhubi Dao and Meiji Dao.

Sample ID	Sample Location	Length (mm)	Sampling Date	Sample ID	Sample Location	Length (mm)	Sampling Date
ZB_SDDD_1	Zhubi Dao	136	2019	MJ_SDDD_1	Meiji Dao	153	2019
ZB_SDDD_2	Zhubi Dao	146	2019	MJ_SDDD_2	Meiji Dao	179	2019
ZB_SDDD_3	Zhubi Dao	150	2019	MJ_SDDD_3	Meiji Dao	153	2019
ZB_SDDD_4	Zhubi Dao	128	2019	MJ_SDDD_4	Meiji Dao	164	2019
ZB_SDDD_5	Zhubi Dao	133	2019	MJ_SDDD_5	Meiji Dao	123	2019
ZB_SDDD_6	Zhubi Dao	137	2019	MJ_SDDD_6	Meiji Dao	154	2019
ZB_SDDD_7	Zhubi Dao	146	2019	MJ_SDDD_7	Meiji Dao	126	2019
ZB_SDDD_8	Zhubi Dao	148	2019	MJ_SDDD_8	Meiji Dao	137	2019
ZB_SDDD_9	Zhubi Dao	152	2019	MJ_SDDD_9	Meiji Dao	146	2019
ZB_SDDD_10	Zhubi Dao	139	2019	MJ_SDDD_10	Meiji Dao	125	2019
ZB_SDDD_11	Zhubi Dao	120	2019	MJ_SDDD_11	Meiji Dao	143	2019
ZB_SDDD_12	Zhubi Dao	149	2019	MJ_SDDD_12	Meiji Dao	135	2019
ZB_SDDD_13	Zhubi Dao	132	2019	MJ_SDDD_13	Meiji Dao	154	2019
ZB_SDDD_14	Zhubi Dao	125	2019	MJ_SDDD_14	Meiji Dao	144	2019
ZB_SDDD_15	Zhubi Dao	165	2019				
ZB_SDDD_16	Zhubi Dao	143	2019				

2.2. RAD-Seq Library Construction and Sequencing

RAD-seq libraries were prepared as described by Yangkun Wang and Yan Hu [11]. First, genomic DNA samples were digested by the restriction endonuclease EcoRI. Then, P1 adapter was added to both ends of the digested genome fragment, which contains a complementary sequence that binds to the primer for PCR amplification and a complementary sequence that binds to the primer for Illumina sequencing. The barcode used for sample tracking and the corresponding restriction enzyme cutting site were also part of the P1 adapter. Next, the sequence of the added P1 adapter was interrupted. Through agarose gel detection, the target band was selected in the range of 400~500 bp. The DNA fragment was then attached to the P2 adapter. The multiplexed libraries were sequenced on the Illumina 1.9 platform with an average sequencing depth of $1.5\times$ for each individual.

2.3. Sequence Alignment and SNP Genotyping

Stacks (version 2.55, Julian Catchen and Nicolas Rochette, Urbana, IL, USA) software is widely used in RAD-seq data analysis [12]. First, raw reads were processed using the process_radtags program [11]. Through raw data processing, each clean read was assigned to a sample to ensure the quality of the sequencing data for use in subsequent analysis. Ustacks begins by clustering reads from a single sample obtained from the process_radtags program to produce stacks [13]. Ustacks in the Stacks package was applied to cluster the reads for each sample, with the same stack representing one enzyme cutting site (locus). The clustering parameter -m was set to 3 [14], and the loci and locus sequencing depth of each sample were statistically analyzed. Next, we used Cstacks to merge the loci of all samples and allowed up to 2 mismatches between different sample loci to obtain the catalogue consensus sequence of each locus [14]. Sstacks was used next. The minor allele frequency (MAF) was set at 0.05 [15]. Loci in each sample deviating from the catalogue consensus sequence were identified, and then populations were filtered to obtain the SNPs. Only the genotype of the first SNP for each tag and loci shared by more than 75% of individuals were used for further analysis. The loci not in Hardy–Weinberg equilibrium (HWE) were removed [16], and the selected loci were used to analyze genetic diversity and population genetic differentiation.

2.4. Genetic Diversity and Population Structure

The genetic diversity parameters of MJ and ZB were calculated by Stacks-2.55 [15], including observed heterozygosity (Ho), expected heterozygosity (He), genetic differentiation (Fst), the inbreeding coefficient (Fis), and nucleotide diversity (π).

Genetic variation was determined by population structure analysis and genetic differentiation assessment. The R package poppr [17] was used to calculate a distance matrix and draw a heatmap [18]. The R packages igraph and poppr were used to cluster multi-locus genotypes (MLGs) to generate minimum spanning networks (MSNs) [19]. A phylogenetic tree was constructed by PHYLIP (version 3.695, Phylip Team, Washington, DC, USA) software [20] to analyze the relationships between samples. Compared with Structure, Admixture software can be used to estimate the ancestral components of individuals faster [21]. Therefore, ADMIXTURE (version 1.3.0, David H. Alexander et al., Los Angeles, CA, USA) software [22] was used to consider various numbers of putative clusters (ranging from 1 to 10) and compute the ten-fold cross-validation error to select the most suitable number of clusters [23]. The PCA module of GCTA (version 1.93.2, Jian Yang et al., Hangzhou, Zhejiang, China) software [24] was used for principal component analysis (PCA).

3. Results

3.1. Genomic Data Statistics and SNP Discovery

Raw reads were obtained by RAD-seq of 30 individuals of *L. kasmira*. After trimming barcodes and filtering, a total of 286,264,230 high-quality clean reads were obtained from 30 individuals. The number of clean reads per individual ranged from 6,672,063 to 21,514,652, with an average of 9,542,141. In all the libraries, the GC content was stable between 39% and 40%, and the Phred score was 20 ≥ 93.64% (Table 2). A total of 9822 high-quality SNPs were screened out for further analysis after the selection of SNPs with an MAF < 0.05 and in HWE.

Table 2. Statistics for the genomic data obtained from 30 *L. kasmira* individuals.

Sample ID	Number of Clean Reads	Bases (Q20%)	GC%	Sample ID	Number of Clean Reads	Bases (Q20%)	GC%
ZB_SDDD_1	7,974,070	99.32	39	MJ_SDDD_1	9,265,891	95.32	39
ZB_SDDD_2	6,867,444	96.12	40	MJ_SDDD_2	13,500,657	96.75	39
ZB_SDDD_3	11,342,742	93.64	39	MJ_SDDD_3	9,120,598	99.30	39
ZB_SDDD_4	6,831,672	95.40	40	MJ_SDDD_4	12,363,573	99.64	39
ZB_SDDD_5	10,947,352	96.32	40	MJ_SDDD_5	8,022,139	98.20	39
ZB_SDDD_6	8,673,002	98.89	39	MJ_SDDD_6	10,972,365	97.67	39
ZB_SDDD_7	8,040,810	96.63	39	MJ_SDDD_7	14,935,273	94.03	39
ZB_SDDD_8	7,251,357	97.60	39	MJ_SDDD_8	10,483,102	96.43	39
ZB_SDDD_9	8,440,580	96.30	40	MJ_SDDD_9	9,245,798	94.60	39
ZB_SDDD_10	7,441,155	94.56	39	MJ_SDDD_10	6,732,763	98.75	39
ZB_SDDD_11	8,497,499	93.21	39	MJ_SDDD_11	9,673,994	95.30	39
ZB_SDDD_12	6,672,063	95.63	39	MJ_SDDD_12	9,046,087	98.00	39
ZB_SDDD_13	9,912,874	99.11	39	MJ_SDDD_13	6,814,092	95.19	39
ZB_SDDD_14	7,340,574	98.65	39	MJ_SDDD_14	7,280,377	96.32	39
ZB_SDDD_15	21,514,652	99.30	39				
ZB_SDDD_16	11,059,675	97.21	39				

3.2. Genetic Diversity and Population Structure

The genetic diversity parameters were calculated according to the original SNPs, and the statistical results were obtained. The genome-wide nucleotide diversity (π) of the ZB population and MJ population was 0.02154 and 0.02478, respectively. The inbreeding coefficient (Fis) of the ZB population and MJ population was −0.18729 and 0.03256, respectively. The genetic differentiation (Fst) between the ZB and MJ subpopulations was 0.00255102. The expected heterozygosity (He) of individuals from ZB and MJ was 0.33585 and 0.22098, respectively. The observed heterozygosity (Ho) of individuals from the ZB population and MJ population was 0.46834 and 0.23103, respectively (Table 3).

Table 3. Comparison of population genetic parameters in the ZB population and MJ population.

Population ID	π	Ho	He	Average_Fis	Fst
ZB population	0.02478	0.46834	0.33585	−0.18729	0.00255102
MJ population	0.02154	0.23103	0.22098	0.03256	

The heatmap showed positive correlations between the individuals of *L. kasmira*. In general, the correlations between individuals within the ZB population were the strongest, followed by those between the ZB population and MJ population, and the correlations between individuals within the MJ population were the weakest (Figure 2). The POP revealed the genetic distances between the 30 individuals, clustering MLGs based on genetic distance. The distances between individuals from the ZB population and MJ population were within the range of 0.002~0.004, indicating relatively short distances, consistent with other analysis results. Each MLG is a node, and the distance between the nodes represents the genetic distance between individuals (Figure 3). The phylogenetic tree revealed no significant clustering of the 30 *L. kasmira* samples (Figure 4). The population structure of *L. kasmira* was analyzed by Admixture software. Assuming that the number of groups (K value) is 1–10, the cross-validation error is maximal when K = 7 (Figure 5a,b). Therefore, the 30 samples of *L. kasmira* collected in this study were not from the same population. The genetic backgrounds of the ZB and MJ populations were complex, and there was admixture between the populations, which might be explained by the coexistence of artificial breeding and wild resources. The cross-validation results of the clustering showed that when K = 1

(Figure 5a), the error rate of cross-validation was minimal, indicating that the optimal clustering number was 1. It is speculated that *L. kasmira* individuals in the ZB and MJ populations came from the same primitive ancestor (Figure 5b). The results of PCA showed that the 30 *L. kasmira* individuals were tightly clustered (Figure 6a,b) and the genetic heterogeneity between individuals was small, which was consistent with the phylogenetic tree results.

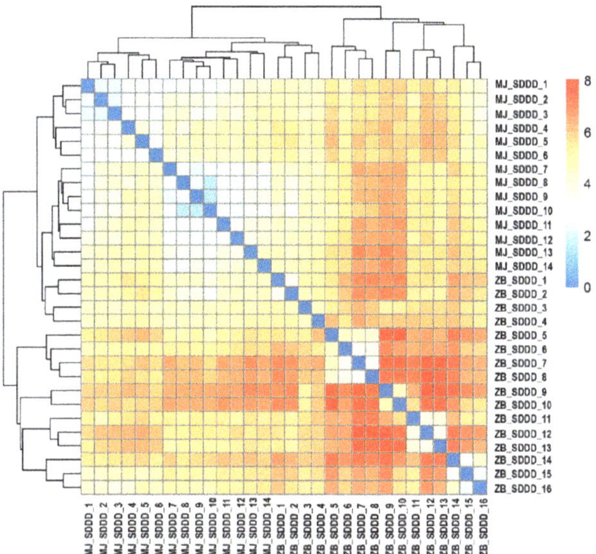

Figure 2. The distance matrix heatmap of 30 individuals of *L. kasmira* created using the R package poppr.

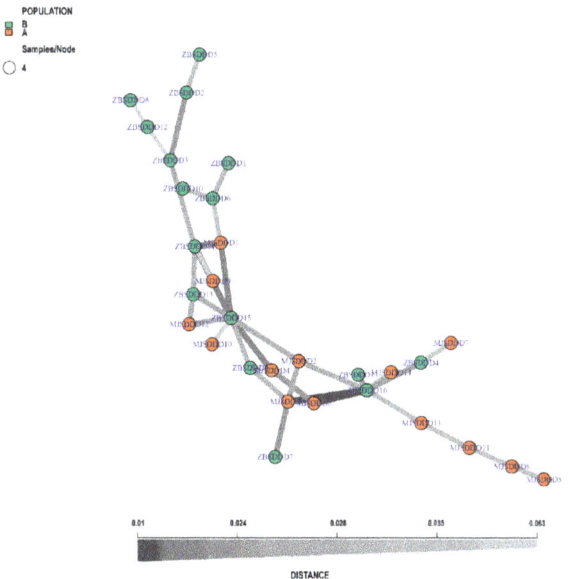

Figure 3. The minimum spanning network of 30 *L. kasmira* individuals.

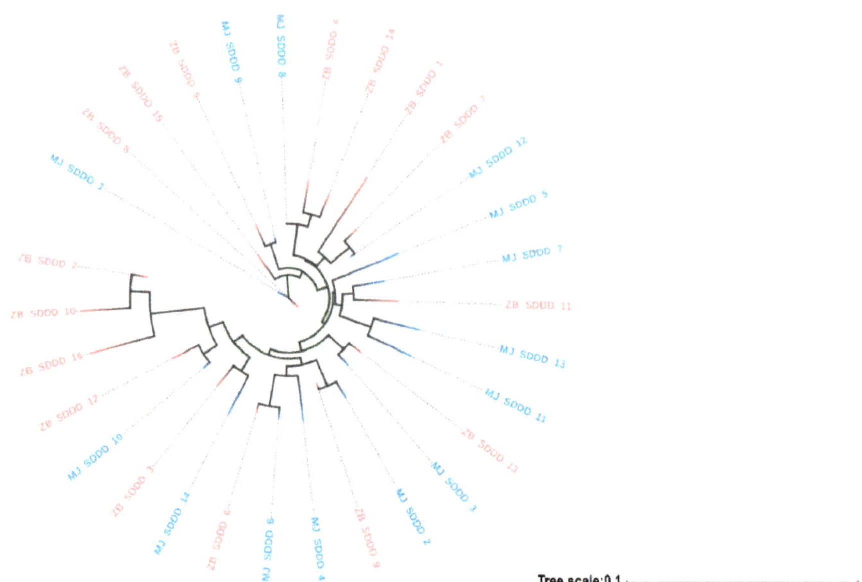

Figure 4. Phylogenetic tree of 30 *L. kasmira* individuals based on SNP loci created using the neighbor-joining method.

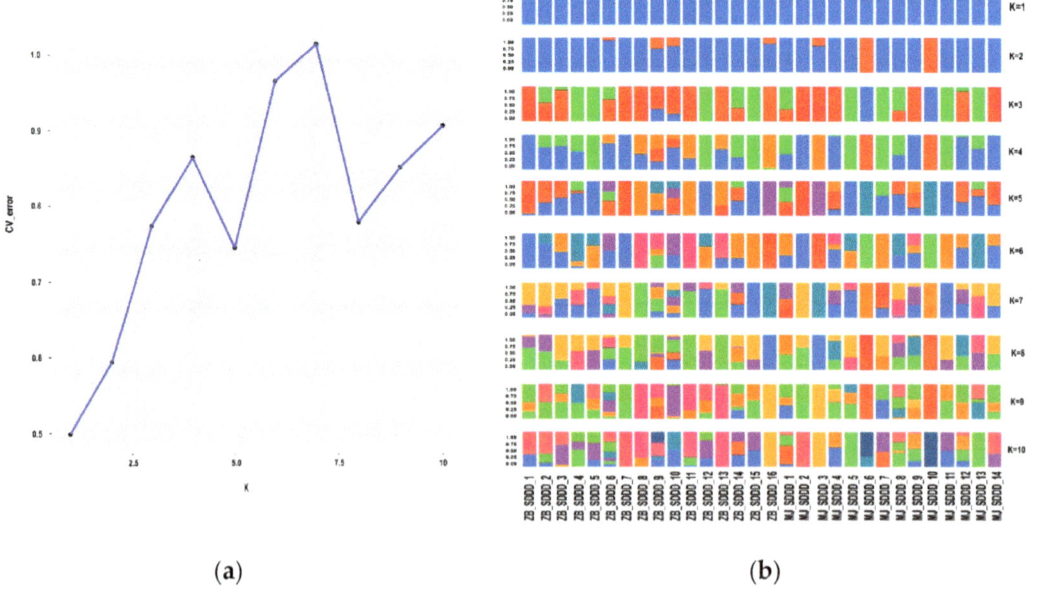

(**a**) (**b**)

Figure 5. The group structure diagrams. (**a**) The cross-validation error for *L. kasmira* according to the admixture value K; (**b**) results of Bayesian cluster analysis of *L. kasmira* based on SNP loci using ADMIXTURE software for K = 1–10 clusters.

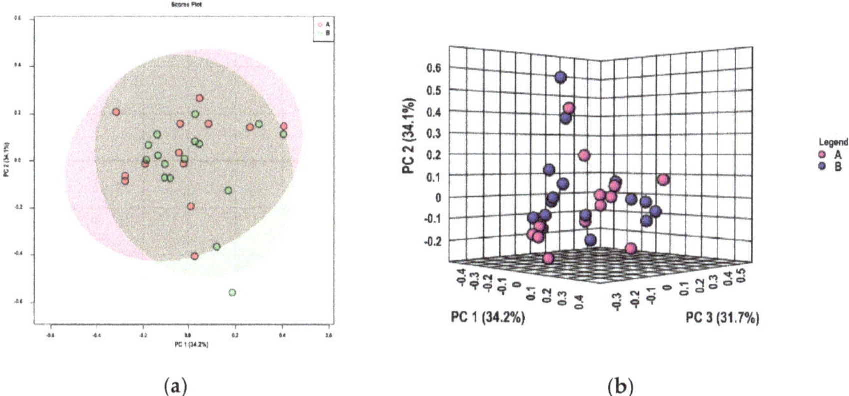

Figure 6. Population relationships of the collected *L. kasmira* individuals. (**a**) Principal component analysis (PCA) plot of 30 *L. kasmira* individuals based on all SNP loci between the ZB and MJ populations. A and B represent the MJ population and ZB population, respectively; (**b**) three−dimensional PCA clustering map of 30 *L. kasmira* individuals based on all SNP loci between the ZB and MJ populations. A and B represent the MJ population and ZB population, respectively.

4. Discussion

RAD-seq is a simplified genome sequencing technology based on whole-genome restriction sites developed on the basis of second-generation sequencing [11]. The number of RAD markers developed by this method is 10 times higher than that of traditional molecular marker development technology [11], with high accuracy and high data utilization [25]. RAD-seq shortens the marker development cycle compared to that of traditional markers and reduces experimental costs [26]. The technology can also screen for SNPs on a large scale in species without a reference genome [27]. SNPs are a new type of DNA molecular marker with broad application prospects [28]. SNPs occur at a high frequency and have a high marker density in most genomes. Compared with simple sequence repeat (SSR) markers, SNPs have higher genetic stability and are easier to automate [29]. RAD-seq technology can also be used to construct linkage maps [30]. For example, Palaiokostas et al. used RAD-seq technology to construct the first linkage map of Dicentrarchus labrax based on high-density SNPs [31].

In previous studies, genetic diversity was studied mainly by SSR analysis and D-loop sequence analysis [32], but these techniques have inherent difficulties in obtaining high-quality DNA from wild individuals [33]. By referring to the study of 29 Rhinopithecus roxellana by Zhang et al. [34], we used RAD-seq to detect SNP markers in 30 *L. kasmira* individuals in this study and further analyzed the genetic diversity levels of different populations of this species. Based on a study of *L. kasmira* in the South China Sea by microsatellite analysis [32], further exploration was performed. A comparison of the results of this study with those of other studies shows that the degree of variation of *L. kasmira* in the ZB and MJ areas is low. This reflects its p ability to adapt to environmental change and respond to natural selection. Therefore, it is necessary to investigate the wild population as soon as possible and determine how to improve its genetic diversity. A larger sample size would produce more statistically robust results. More samples should be collected in different regions for more comprehensive genetic diversity analysis in the future.

In the phylogenetic tree based on RAD-seq data, the 30 samples of *L. kasmira* did not form obvious groups, and the population structure diagram also showed that they came from the same ancestor. Similar results were obtained with PCA, which were mutually confirmed with the results of the phylogenetic tree. The genome-wide nucleotide diversity (π) of the ZB population and MJ population was 0.02478 and 0.02154, respectively, indicating high nucleotide diversity. The expected heterozygosity (He) of individuals from ZB and MJ was 0.33585 and 0.22098, respectively. However, the observed heterozygosity (Ho) of individuals from the ZB population and MJ population was 0.46834 and 0.23103, respectively. The He of the two populations was lower than the Ho, indicating a low degree of variation for *L. kasmira* in the two areas and a certain degree of heterozygote deficiency. The genetic differentiation (Fst) between the MJ and ZB subpopulations was 0.00255102. The Fst value was consistent with the results of the heatmap and MSN. The distance between individuals was short, and differentiation was not obvious. Although π was high, He and Ho were low. The overall analysis showed that the population structure of *L. kasmira* was simple and had a low degree of variation, which might lead to poor adaptability to the environment. Furthermore, the genetic diversity of the MJ population was lower than that of the ZB population.

These results may be due to overfishing and global warming, which have reduced the genetic diversity of wild populations. Therefore, there is an urgent need to protect the population genetic resources of *L. kasmira*, and we can improve its genetic diversity through fishing restriction protection measures and artificial breeding of superior varieties. It is suggested that effective population size analysis and population history tracing should be performed for *L. kasmira*.

5. Conclusions

In this study, SNP data obtained by RAD-seq technology were used to analyze the genetic diversity in two populations of *L. kasmira*, providing a new approach for genetic diversity evaluation. The results showed that the genetic diversity of the two populations was relatively low at the genome level. To ensure adequate survival of the species, it is necessary to protect existing diversity and take measures to improve genetic diversity. To promote the healthy and sustainable development of germplasm resources of *L. kasmira*, much attention should be given to improving and maintaining its population genetic diversity, to implementing methods to increase gene flow, and to breeding excellent varieties by applying molecular breeding techniques.

Finally, our results indicate that the RAD-seq technique can detect SNP markers and apply them to the research of genetic diversity, with the benefits of a huge number of markers, low costs, and simple automation. This study is useful for the conservation of aquatic germplasm resources as well as for the development and utilization of high-quality resources.

Author Contributions: Conceptualization, D.Z. and S.J.; methodology, F.Z. and L.G.; software, L.G.; validation, F.Z. and J.Y.; formal analysis, F.Z.; investigation, K.Z.; resources, N.Z.; data curation, F.Z.; writing—original draft preparation, F.Z.; writing—review and editing, D.Z.; visualization, F.Z.; supervision, H.G. and B.L. (Bo Liu); project administration, B.L. (Baosuo Liu). All authors have read and agreed to the published version of the manuscript.

Funding: This research was funded by the Financial Fund of the Ministry of Agriculture and Rural Affairs, P. R. of China (NHYYSWZZZYKZX2020) and National Marine Genetic Resource Center, and China-ASEAN Maritime Cooperation Fund.

Institutional Review Board Statement: All applicable international, national, and institutional guidelines for the care were followed by the authors.

Informed Consent Statement: Informed consent was obtained from all subjects involved in the study. Written informed consent has been obtained from the patients to publish this paper.

Data Availability Statement: All data generated or analyzed during this study are included in this published article.

Conflicts of Interest: The authors declare no conflict of interest.

References

1. Niu, W.; Huang, Y.; Zhang, C.; Pu, T.; Lu, Y.; Jia, T.; You, Y.; Du, Y.; Mao, Y.; Ding, C. Development and Characterization of 31 SNP Markers for the Crested Ibis (*Nipponia nippon*). *Conserv. Genet. Resour.* **2021**, *13*, 5–7. [CrossRef]
2. Li, Y.; Lou, F.; Song, P.; Liu, S.; Siyal, F.K.; Lin, L. New Perspective on the Genetic Structure and Habitat Adaptation of Pampus Minor off the Coast of China Based on RAD-Seq. Comp. *Biochem. Physiol. Part D Genom. Proteom.* **2021**, *39*, 100865. [CrossRef] [PubMed]
3. Catanese, G.; Trotta, J.R.; Iriondo, M.; Grau, A.M.; Estonba, A. Discovery of SNP Markers of Red Shrimp Aristeus Antennatus for Population Structure in Western Mediterranean Sea. *Conserv. Genet. Resour.* **2021**, *13*, 21–25. [CrossRef]
4. Chaves, C.L.; Blanc-Jolivet, C.; Sebbenn, A.M.; Mader, M.; Meyer-Sand, B.R.V.; Paredes-Villanueva, K.; Honorio Coronado, E.N.; Garcia-Davila, C.; Tysklind, N.; Troispoux, V.; et al. Nuclear and Chloroplastic SNP Markers for Genetic Studies of Timber Origin for Hymenaea Trees. *Conserv. Genet. Resour.* **2019**, *11*, 329–331. [CrossRef]
5. Yue, L.; Zhang, S.; Zhang, L.; Liu, Y.; Cheng, F.; Li, G.; Zhang, S.; Zhang, H.; Sun, R.; Li, F. Heterotic Prediction of Hybrid Performance Based on Genome-Wide SNP Markers and the Phenotype of Parental Inbred Lines in Heading Chinese Cabbage (*Brassica rapa* L. Ssp. Pekinensis). *Sci. Hortic.* **2022**, *296*, 110907. [CrossRef]
6. Vignon, M.; Morat, F.; Galzin, R.; Sasal, P. Evidence for Spatial Limitation of the Bluestripe Snapper Lutjanus Kasmira in French Polynesia from Parasite and Otolith Shape Analysis. *J. Fish Biol.* **2008**, *73*, 2305–2320. [CrossRef]
7. Zhang, J.; Chen, Z. Variation in the population characteristics of blue-striped snapper *Lutjanus kasmira* in the South China Sea in recent 20years. *Oceanol. Limnol. Sin.* **2020**, *51*, 114–124.
8. Tan, W.; Wang, Z. Structure and Evolution of Complete Mitochondrial Genome of *Lutjanus kasmira*. In Proceedings of the 12th Annual Conference of China Association for Science and Technology; The Transformation of Economic Development Pattern and Independent Innovation, Fuzhou, China, 1–3 November 2010; Volume 3, pp. 1143–1150.
9. Tang, C.; Xiao, L.; Zhang, Q.; Zhou, Q.; Xu, S.; Wang, Y. DNA barcoding of some *Lutjanus* species in China and its adjacent sea areas based on COI gene. *Mar. Fish.* **2019**, *41*, 129–137. [CrossRef]
10. Davey, J.W.; Blaxter, M.L. RADSeq: Next-Generation Population Genetics. *Brief. Funct. Genom.* **2010**, *9*, 416–423. [CrossRef]
11. Wang, Y.; Hu, Y. Current status and perspective of RAD-seq in genomic research. *Hereditas* **2014**, *36*, 41–49. [CrossRef]
12. Catchen, J.M.; Amores, A.; Hohenlohe, P.; Cresko, W.; Postlethwait, J.H. Stacks: Building and Genotyping Loci De Novo from Short-Read Sequences. *G3 Genes Genomes Genet.* **2011**, *1*, 171–182. [CrossRef]
13. Paris, J.R.; Stevens, J.R.; Catchen, J.M. Lost in Parameter Space: A Road Map for stacks. *Methods Ecol. Evol.* **2017**, *8*, 1360–1373. [CrossRef]
14. Mastretta-Yanes, A.; Arrigo, N.; Alvarez, N.; Jorgensen, T.H.; Piñero, D.; Emerson, B.C. Restriction Site-Associated DNA Sequencing, Genotyping Error Estimation and de Novo Assembly Optimization for Population Genetic Inference. *Mol. Ecol. Resour.* **2015**, *15*, 28–41. [CrossRef]
15. Rochette, N.C.; Catchen, J.M. Deriving Genotypes from RAD-Seq Short-Read Data Using Stacks. *Nat. Protoc.* **2017**, *12*, 2640–2659. [CrossRef]
16. Sun, L.; Gan, J.; Jiang, L.; Wu, R. Recursive Test of Hardy-Weinberg Equilibrium in Tetraploids. *Trends Genet.* **2021**, *37*, 504–513. [CrossRef]
17. Kamvar, Z.N.; Tabima, J.F.; Grunwald, N.J. Poppr: An R Package for Genetic Analysis of Populations with Clonal, Partially Clonal, and/or Sexual Reproduction. *PeerJ* **2014**, *2*, e281. [CrossRef]
18. Schaub, F.X.; Dhankani, V.; Berger, A.C.; Trivedi, M.; Richardson, A.B.; Shaw, R.; Zhao, W.; Zhang, X.; Ventura, A.; Liu, Y.; et al. Pan-Cancer Alterations of the MYC Oncogene and Its Proximal Network across the Cancer Genome Atlas. *Cell Syst.* **2018**, *6*, 282–300.e2. [CrossRef]
19. Kamvar, Z.N.; Larsen, M.M.; Kanaskie, A.M.; Hansen, E.M.; Grünwald, N.J. Spatial and Temporal Analysis of Populations of the Sudden Oak Death Pathogen in Oregon Forests. *Phytopathology* **2015**, *105*, 982–989. [CrossRef]
20. Felsenstein, J. *PHYLIP (Phylogeny Inference Package)*, version 3.6; Distributed by the author; Department of Genome Sciences, University of Washington: Seattle, DC, USA, 2005.
21. Parry, R.M.; Wang, M.D. A Fast Least-Squares Algorithm for Population Inference. *BMC Bioinform.* **2013**, *14*, 28. [CrossRef]
22. Wang, M.; Du, W.; Tang, R.; Liu, Y.; Zou, X.; Yuan, D.; Wang, Z.; Liu, J.; Guo, J.; Yang, X.; et al. Genomic History and Forensic Characteristics of Sherpa Highlanders on the Tibetan Plateau Inferred from High-Resolution InDel Panel and Genome-Wide SNPs. *Forensic Sci. Int. Genet.* **2022**, *56*, 102633. [CrossRef]
23. Wang, J.; Zhang, Z. Analysis of the Genetic Structure and Diversity of Upland Cotton Groups in Different Planting Areas Based on SNP Markers. *Gene* **2022**, *809*, 146042. [CrossRef]
24. Yang, J.; Lee, S.H.; Goddard, M.E.; Visscher, P.M. GCTA: A Tool for Genome-Wide Complex Trait Analysis. *Am. J. Hum. Genet.* **2011**, *88*, 76–82. [CrossRef]
25. Lecaudey, L.A.; Schliewen, U.K.; Osinov, A.G.; Taylor, E.B.; Bernatchez, L.; Weiss, S.J. Inferring Phylogenetic Structure, Hybridization and Divergence Times within Salmoninae (Teleostei: Salmonidae) Using RAD-Sequencing. *Mol. Phylogenet. Evol.* **2018**, *124*, 82–99. [CrossRef]
26. Zhou, W.; Ji, X.; Obata, S.; Pais, A.; Dong, Y.; Peet, R.; Xiang, Q.-Y.J. Resolving Relationships and Phylogeographic History of the Nyssa Sylvatica Complex Using Data from RAD-Seq and Species Distribution Modeling. *Mol. Phylogenet. Evol.* **2018**, *126*, 1–16. [CrossRef]

27. Díaz-Arce, N.; Arrizabalaga, H.; Murua, H.; Irigoien, X.; Rodríguez-Ezpeleta, N. RAD-Seq Derived Genome-Wide Nuclear Markers Resolve the Phylogeny of Tunas. *Mol. Phylogenet. Evol.* **2016**, *102*, 202–207. [CrossRef]
28. Singha, H.; Vorimore, F.; Saini, S.; Deshayes, T.; Saqib, M.; Tripathi, B.N.; Laroucau, K. Molecular Epidemiology of Burkholderia Mallei Isolates from India (2015–2016): New SNP Markers for Strain Tracing. *Infect. Genet. Evol.* **2021**, *95*, 105059. [CrossRef]
29. Bao, W.; Ao, D.; Wuyun, T.; Wang, L.; Bai, S.; Bai, Y. Development and Characterization of 72 SNP Markers in Armeniaca Sibirica Based on Transcriptomics. *Conserv. Genet. Resour.* **2020**, *12*, 373–378. [CrossRef]
30. Fukuda, S.; Nagano, Y.; Matsuguma, K.; Ishimoto, K.; Hiehata, N.; Nagano, A.J.; Tezuka, A.; Yamamoto, T. Construction of a High-Density Linkage Map for Bronze Loquat Using RAD-Seq. *Sci. Hortic.* **2019**, *251*, 59–64. [CrossRef]
31. Palaiokostas, C.; Bekaert, M.; Khan, M.G.Q.; Taggart, J.B.; Gharbi, K.; McAndrew, B.J.; Penman, D.J. Mapping and Validation of the Major Sex-Determining Region in Nile Tilapia (*Oreochromis niloticus* L.) Using RAD Sequencing. *PLoS ONE* **2013**, *8*, e68389. [CrossRef]
32. Li, P.; Wang, Z.; Guo, Y.; Liu, L.; Liu, C. Microsatellite analysis of *Lutjanus* bengalensis and L. Kasmira. *J. Oceanogr. Taiwan Strait* **2011**, *30*, 517–521.
33. Park, S.-H.; Scheffler, J.A.; Ray, J.D.; Scheffler, B.E. Identification of Simple Sequence Repeat (SSR) and Single Nucleotide Polymorphism (SNP) That Are Associated with the Nectariless Trait of *Gossypium hirsutum* L. *Euphytica* **2021**, *217*, 78. [CrossRef]
34. Zhang, Y.; Zhou, Y.; Liu, X.; Yu, H.; Li, D.; Zhang, Y. Genetic Diversity of the Sichuan Snub-Nosed Monkey (*Rhinopithecus roxellana*) in Shennongjia National Park, China Using RAD-Seq Analyses. *Genetica* **2019**, *147*, 327–335. [CrossRef]

Journal of
Marine Science and Engineering

Article

Molecular Technology for Isolation and Characterization of *Mitogen-Activated Protein Kinase Kinase 4* from *Penaeus monodon*, and the Response to Bacterial Infection and Low-Salinity Challenge

Yundong Li [1,2,3,4], Falin Zhou [1,2], Hongdi Fan [1], Song Jiang [1], Qibin Yang [1], Jianhua Huang [1], Lishi Yang [1], Xu Chen [2], Wenwen Zhang [1] and Shigui Jiang [1,2,4,*]

[1]   Key Laboratory of South China Sea Fishery Resources Exploitation and Utilization, Ministry of Agriculture and Rural Affairs, South China Sea Fisheries Research Institute, Chinese Academy of Fishery Sciences, Guangzhou 510300, China

[2]   Key Laboratory of Efficient Utilization and Processing of Marine Fishery Resources of Hainan Province, Sanya Tropical Fisheries Research Institute, Sanya 572018, China

[3]   Hainan Yazhou Bay Seed Laboratory, Sanya 572025, China

[4]   Key Laboratory of Tropical Hydrobiology and Biotechnology of Hainan Province, Hainan Aquaculture Breeding Engineering Research Center, College of Marine Sciences, Hainan University, Haikou 570228, China

*   Correspondence: jiangsg@21cn.com

Abstract: Mitogen-activated protein kinase kinase 4 (MKK4) is a component of the JNK signaling pathway and plays an important role in immunity and stress resistance. In this study, *MKK4* cDNA was cloned, and its bacterial infection and low-salinity challenge responses were researched. The full-length *PmMKK4* cDNA was 1582 bp long, with an 858-bp open reading frame (ORF) encoding a 285-amino acid (aa) protein. Results showed that *PmMKK-4* was expressed in all examined tissues of *P. monodon*. The *PmMKK4* expression level was found to be lowest in eyestalk ganglion and highest in muscle (approximately 41.25 times than in eyestalk ganglion). Following the infection of *Staphylococcus aureus*, *PmMKK*4 was up-regulated in both hepatopancreatic and gill tissues. However, after infection with *Vibrio harveyi*, *PmMKK4* was down-regulated for a period of time in gill tissue, with fluctuating up- and down-regulation in hepatopancreas tissue. Furthermore, after infection with *Vibrio anguillarum*, gill tissue and hepatopancreas tissue showed a continuous downward trend. The *PmMKK4* gene in the gill tissue and hepatopancreas tissue of *P. monodon* was activated after low-salinity stress. The expression change of *PmMKK4* in gill tissue was more significant. The research showed that the *PmMKK4* gene plays an important role in both innate immunities after pathogen infection and adaptation in a low-salt environment.

Keywords: *Penaeus monodon*; *MKK4*; innate immunity; salinity stress; RNAi

Citation: Li, Y.; Zhou, F.; Fan, H.; Jiang, S.; Yang, Q.; Huang, J.; Yang, L.; Chen, X.; Zhang, W.; Jiang, S. Molecular Technology for Isolation and Characterization of *Mitogen-Activated Protein Kinase Kinase 4* from *Penaeus monodon*, and the Response to Bacterial Infection and Low-Salinity Challenge. *J. Mar. Sci. Eng.* **2022**, *10*, 1642. https://doi.org/10.3390/jmse10111642

Academic Editor: Ka Hou Chu

Received: 29 September 2022
Accepted: 1 November 2022
Published: 3 November 2022

1. Introduction

Mitogen-activated protein kinase kinase 4 (*MKK4*) is an important regulatory kinase of the JNK signaling pathway. Activation of the JNK signaling pathway is achieved by dual phosphorylation of Thr and Tyr residues of the conserved Thr–Pro–Tyr (T–Y–P) motif [1,2]. *MKK4* can alternately activate the two pathways of JNK and p38 [3]. The first successful cloning of the *MKK4* gene was in *Xenopus laevis*, and since then, its role and molecular mechanism functions in both physiological and biochemical areas have been studied more and more deeply. However, the leading research focuses on mammals, such as humans and mice [4,5]. In studies on crustaceans, it was only found in *Daphnia pulex*, *Fenneropenaeus chinensis*, *Litopenaeus vannamei*, and *P. monodon* [6–8]. In a prior study of *P. monodon*, the expression of *MKK4* was only studied in spermatozoa and ovary developmental stages. The *MKK4* gene of *L. vannamei* shows significant changes under the stimulation of various bacteria and is activated by the phosphorylation of upstream genes and then activated JNK, which plays an important role in the antibacterial response [3,9]. The *MKK4* gene of *Ctenopharyngodon idella* is involved in the immune stress response to muramyl dipeptide in the intestinal tissue of *C. idella* [10].

In order to further improve the molecular function study of the *MKK4* gene in *P. monodon*, this study cloned the entire length of the *MKK4* gene in *P. monodon*, analyzed the expression of *MKK4* in various tissues of *P. monodon*, and explored the effect of different degrees of pathogen stimulation and salinity stress. It is necessary to explore the molecular function and role of *MKK4* in low-salinity stress in *P. monodon* to provide more basic data to determine the molecular mechanism of the response to salinity stress in *P. monodon*.

2. Materials and Methods

2.1. Experimental Animals and PmMKK4 cDNA Cloning

P. monodon specimens, each with a body length of 7–10 cm and a body weight of 8–12 g, were selected and cultured for 1 week in a 25–28 °C seawater environment with full aeration at the same time. Based on the unpublished transcriptome of *P. monodon* tissues constructed in our laboratory, specific primers, namely, *PmMKK4*-F1, *PmPKK4*-R1, *PmMKK4*-F2, and *PmPKK4*-R2 (Table 1), were designed by Primer Premier 5.0 (RuiBiotech, Guangzhou, China). The 5′ and 3′ ends of *PmMKK4* were acquired by using the rapid amplification of the cDNA end (RACE) method (Clonetech, Tokyo, Japan). Through the 5′ or 3′ RACE-PCR, PCR was performed initially with *PmMKK4*-5G1 (*PmMKK4*-3G1 for 3′) and universal primers UPM Long and UPM Short, followed by semi-nested PCR with *PmMKK4*-5G2 (*PmMKK4*-3G2 for 3′) and a nested universal primer NUP (Table 1). The PCR conditions were designed as follows: one cycle of 94 °C for 3 min, 35 cycles of 94 °C for 30 s, 67 °C for 30 s, and 72 °C for 45 s, followed by a final cycle of 72 °C for 10 min. The PCR products were then gel-purified and sequenced, and the sequences were determined after the analysis.

Table 1. Sequences of primers used in this study.

Primer	Sequence (5′-3′)	Function
Pm*MKK4*-F1	AATAACGACCGTCCACAGAAC	
Pm*MKK4*-R1	AATAACGACCGTCCACAGAAC	
Pm*MKK4*-F2	TGCGCTCTCCACAAGTGTCTC	ORF validation
Pm*MKK4*-R2	AAACCTCCTCTTCCAATCTCCC	
Pm*MKK4*-5G1	AAGTCAAGGTTGTTCTGTGGACGGTCG	
Pm*MKK4*-5G2	CATGCTGAGAGTCCCTGGCCTGG	5′ clone
Pm*MKK4*-3G1	TGGTGCCATATTCAAGGAGGGTGAC	
Pm*MKK4*-3G2	GCTCTGTGATTTCGGCATTTCTGGC	3′ clone
UPM Long	CTAATACGACTCACTATAGGGCAAGCAGTG GTATCAACGCAGAGT	
UPM Short	CTAATACGACTCACTATAGGGC	Universal primer
NUP	AAGCAGTGGTATCAACGCAGAGT	
q*MKK4*-F	GCCACACTAACAGCACTA	
q*MKK4*-R	GTCTACATCCAGCATCTCT	qRT-PCR
qEF-1α-F	AAGCCAGGTATGGTTGTCAACTTT	
qEF-1α-R	CGTGGTGCATCTCCACAGACT	Reference gene
MKK4-f:	TCGCAAGAGCAACACAATC	
MKK4-r	GAACATCATATCCTCGGGC	RNAi
ds*MKK4*-f	TAATACGACTCACTATAGGGTCGCAAGAGCAACACAATC	
ds*MKK4*-r	TAATACGACTCACTATAGGGGAACATCATATCCTCGGGC	RNAi
dsGFP-F	TGGAGTGGTCCCAGTTCTTGTTGA	
dsGFP-R	GCCATTCTTTGGTTTGTCTCCCAT	RNAi
qMKK7-F	TAACCTAGAAGAGACCAAGC	
qMKK7-R	CTGTGACGAAGCATCCTA	qRT-PCR
qJNK-F	CCGTCCCCGCTACCCTGGCTATTC	
qJNK-R	AGGTCTCGTGCTTGACTCGCTTTG	qRT-PCR

2.2. Bioinformatic Analysis

The ORF Finder (http://www.ncbi.nlm.nih.gov/orffinder/, accessed on 10 May 2021) was used to predict the open reading frame. The transeq in EMBOSS (http://www.bioinformatics.nl/emboss-explorer/, accessed on 10 May 2021) was utilized to translate its amino acid sequence. The predicted amino acid sequence was compared with the protein database by using the BLAST (http://blast.ncbi.nlm.nih.gov/Blast.cgi, accessed on 20 May 2021) tool in NCBI. Multiple sequence alignments were performed using Clustal X V1.83 software. ExPASy ProtParam (https://web.expasy.org/protparam/, accessed on 20 May 2021) was used to predict the isoelectric point and the theoretical molecular mass. Protein domain analysis was performed using SMART 4.0 (http://smart.embl-heidelberg.de/smart/set_mode.cgi?GENOMIC=1, accessed on 20 May 2021). Glycosylation sites were predicted using the NetNGlyc 1.0 Server (http://www.cbs.dtu.dk/services/NetNGlyc/, accessed on 20 May 2021). Phosphorylation sites were predicted using the NetPhos 3.1 Server (http://www.cbs.dtu.dk/services/NetPhos/, accessed on 20 May 2021). The phylogenetic tree was constructed based on the maximum likelihood method using Clustal X and MEGA 6.0 software.

2.3. Sample Collection and cDNA Synthesis

After 1 week of seawater culturing, healthy male and female *P. monodon* were randomly selected. Hepatopancreas, gills, intestines, stomach, lymph, heart, muscle, epidermis, eye stalk nerve, cranial nerve, thoracic nerve, ventral nerve, and ovary (testis) tissue were dissected for collection. The same tissue samples from 3 shrimps were mixed into one tube. Furthermore, during summer, the mating season of *P. monodon*, samples of shrimp larvae at different stages of development were collected. Those samples including zygote, nauplius, zoea, mysis, and postlarval stages. Zygote stage samples were collected immediately after egg laying. When 80% of the population had reached the objective stage according to their morphologies, as observed using an optical microscope, samples of the larvae of different stages were collected. The samples mentioned above were organized into three parallel groups, stored in RNAlater® RNA stabilization solution (Invitrogen, Carlsbad, CA, USA) at 4 °C overnight, and then stored at −80 °C.

Following the manufacturer's instructions of Trizol reagent (Invitrogen, USA), the total RNA of all collected samples was extracted. The ratios of ultraviolet absorbance at 260/280 nm were measured using a NanoDrop2000 device (NanoDrop Technologies, Waltham, MA, USA); 1.5% agarose gel electrophoresis was used to ensure the integrity. A template of cDNA was synthesized from the RNA using the PrimeScript II 1st strand cDNA synthesis kit (Takara, Tokyo, Japan). For the sake of real-time quantitative PCR (qRT-PCR), cDNA was synthesized in accordance with the manufacturer's instructions for the PrimeScript TM RT reagent kit with a gDNA eraser (Perfect Real-time, Takara, Japan) and diluted to 50 ng/μL for use as the template.

2.4. Bacterial Infection Challenge

In total, 200 healthy *P. monodon* with an average weight of 15–18 g were selected for immune challenge experiments. Three pathogenic bacteria provided by the Key Laboratory of South China Sea Fishery Resources Exploitation and Utilization were used. According to a previous study [11], the culture schedule and injection concentration for each strain were determined. Four experimental groups were included: PBS (shrimp specimen injected with sterile phosphate-buffered saline as control), *Staphylococcus aureus*, *Vibrio harveyi*, and *Vibrio anguillarum* groups. Each group of 50 shrimp specimens was injected into the second abdominal segment with 100 μL of sterile phosphate-buffered saline (PBS, pH 7.4) or 100 μL (1.0×10^8 cfu/mL) of *S. aureus*, *Vibrio harveyi*, and *V. anguillarum*, respectively. Healthy and intact shrimp were randomly selected at 0 h, 3 h, 6 h, 12 h, 24 h, 48 h, and 72 h after injection for the dissection of hepatopancreas and gill tissues. Hepatopancreas and gill tissues were stored overnight in RNAlater solution at 4 °C and then kept at −80 °C.

2.5. Low-Salinity Stress

The experiment site of the low-salt stress test was in the South China Sea Fisheries Research Institute (Shenzhen City, Guangdong Province, China). A total of 360 shrimps (7–10 cm) was selected for these experiments. The salinity concentration was adjusted to the target salinity by mixing the cultured seawater with freshwater using a salinity meter (AZ8371, Hengxin, Taiwan). According to the pre-test results, the 96 h half-lethal salinity was 3 psu [12]. Therefore, the stress salinity was set to 3 psu. Another experimental group with salinity of 17 psu was set up, and conventional aquaculture seawater (about 25 psu) was used as the control group [13]. Three parallel groups (n = 40/group) were created. The incubation temperature and pH were maintained at 25–28 °C and 7.0 ± 0.5, respectively, and 0 h, 3 h, 6 h, 12 h, 24 h, 48 h, 72 h, and 96 h after exposure to different salinity stresses, shrimp specimens with optimal activity during the intermolt were selected for dissection to collect gills and hepatopancreas tissues, and they were preserved in RNAlater solution. The tissues were then maintained at −80 °C after overnight storage at 4 °C.

2.6. qRT-PCR Analysis of PmMKK4 mRNA Expression

In this study, qRT-PCR was used to detect *PmMKK4* mRNA expression in different tissues at different developmental stages following bacterial challenge and low-salinity stress exposure. Since the reaction component and cycle condition for EF1a are consistent with *PmMKK*, the reference gene was chosen to be 1α (EF1a) (Table 1). The solution in each hole (12.5 μL) was a mixture of 6.25 μL of 2 × TB GreenTM Premix ExTaq (Takara, Beijing, China), 0.5 μL each of *PmMKK4*-qF and *PmMKK4*-qR (50 μmol/L), 1 μL qRT-PCR diluted cDNA, and 4.25 μL double-distilled water. Green fluorescence measurement qRT-PCR was carried out in the quantitative real-time PCR system, Roche Light Cycler® 480II. The following four steps were conducted: degeneration for 30 s at 95 °C, a quantitative analysis stage with 40 cycles of 94 °C for 5 s and 60 °C for 30 s, dissolution curve analysis for 5 s at 95 °C and 60 °C and up to 95 °C for 1 min, and an according stage of 30 s at 50 °C. The relative CT method ($2^{-\Delta\Delta CT}$) was used to obtain the PCR data. One-way ANOVA was used to work on statistical analysis. SPSS statistics version 23.0 software (IBM, Armonk, New York, USA) was used to carry out the Tukey's multiple range test. The differences were considered to be significant at $p < 0.05$. Tested data were presented as mean ± SD (standard deviation).

2.7. Low-Salinity Stress Testing in P. monodon following RNA Interference (RNAi)-Mediated Knockdown of PmMKK4

Primer Premier 5.0 was used to design the following primers: dsMKK-f, dsMKK-r, dsMKK-T7-f, and dsMKK-T7-r (Table 1). A DNA fragment containing the T7 promoter was amplified by Ex Taq using normal cDNA as a template. PCR conditions were as follows: 3 min at 94 °C, 35 cycles of 30 s at 94 °C, 30 s at 58 °C, 1 min at 72 °C, and a final cycle of 10 min at 72 °C. Excess bands were clipped after agarose gel electrophoresis, leaving clear and bright bands. DNA fragments were recovered according to the gel recovery kit instructions. dsRNA synthesis was performed according to the T7 RiboMAXTMExpress RNAi System kit instructions. Reaction system: RiboMAXTM Express T7 2X Buffer 10.0 μL, linear DNA template (total 1 μg) 1.0–8.0 μL, Nuclease-Free Water 0–7.0 μL (8 μL-DNA volume), Enzyme Mix-T7 Express 2.0 μL. A final volume of 20.0 μL was incubated at 37 °C for 30 min to obtain single-stranded RNA (ssRNA). Then, equal volumes of complementary ssRNA were mixed and incubated at 70 °C for 10 min, 65 °C for 10 min, and 25 °C for 10 min. Then, 2.0 μL of freshly diluted RNase solution (1 μL RNase solution: 199 μL Nuclease-Free Water) and 2.0 μL RQ1 RNase-Free DNase was added and incubated at 37 °C for 30 min. The obtained dsRNA was purified for later use. dsGFP-F/R primers and pD-GFP recombinant vectors were used to synthesize green fluorescent protein (GFP) double-stranded RNA in the same way.

The weight of *P. monodon* was 5.0 ± 1.0 g, and the dsRNA injection volume was 3–5 μg/g shrimp. Each shrimp was injected intramuscularly in the second abdominal segment. The injection experiments were divided into three groups: PBS group, dsGFP group, and dsMKK group. Before injection, samples of healthy shrimp in the intermolt phase were randomly collected, and their gill tissues were placed in RNAlater as a 0 h sample to check the RNAi efficiency. Twenty-four hours after injection, the three groups of shrimp specimens were transferred to plastic buckets adjusted to salinity of 3 psu. The dead shrimps of each group were collected and recorded every 3 h.

After 3 h, 6 h, 9 h, 12 h, 24 h, 48 h, 72 h, and 96 h exposure to salt stress, healthy shrimp specimens were collected from each treatment group, and their gill tissues were dissected and placed in RNAlater. Furthermore, 10 shrimp injected with dsGFP and dsMKK, respectively, were placed in two separate plastic buckets filled with seawater (about 25 psu). After 24 and 48 h, gill tissues were randomly collected and placed in RNAlater as samples to measure dsRNA interference efficiency. All RNA samples were mixed in RNAlater and stored at 4 °C overnight, then kept at −80 °C.

3. Results

3.1. PmMKK4 Sequence Analysis

The mitogen-activated protein kinase kinase 4 (Pm*MKK4*) cDNA of *P. monodon* was obtained by cloning. The full-length was 1582 bp, the GenBank accession number: MN909956, and the 5′ and 3′ non-coding regions (UTRs) were 567 bp and 157 bp, respectively; the open reading frame (ORF) was 858 bp long and encoded 285 amino acids (Figure 1). ExPASy ProtParam predicted a molecular weight of 32.87 kD and a theoretical isoelectric point of 6.46. It was predicted that *PmMKK4* would have a total of 22 phosphorylation sites, including 15 serine sites, 4 threonine sites, and 3 tyrosine sites. In addition, the predicted results showed that *PmMKK4* had no glycosylation site, no signal peptide, and no transmembrane structure. *PmMKK4* contained a conserved serine/threonine protein kinase (S–I–A–K–T) region, which is a characteristic kinase domain for MKKK phosphorylation.

3.2. Sequence Alignment and Analysis

Using NCBI BLASTP to compare the identity of the *MKK4* gene between *P. monodon* and other species, the results showed that the *PmMKK4* gene had the highest similarity with *P. chinensis* and *P. vannamei* (Figure 2), both of which were 99.65%, which were similar to *Armadillidium vulgare, Ooceraea biroi, Harpegnathos saltator, Solenopsis invicta, Nylanderia fulva, Camponotus floridanus,* and *Formica exsecta,* with a similarity of 82.39%, 82.08%, 81.49%, 81.36%, 81.49%, 82.08%, and 81.36%, respectively. Using MEGA6.06 software, the phylogenetic tree of *PmMKK4* and other species was constructed based on the NJ (neighbor-joining) method, and the Bootstrap method was repeated 1000 times (Figure 3). It can be seen from Figure 3 that the *MKK4* gene of *P. monodon* had the closest genetic relationship with *P. vannamei* and *P. chinensis* and was firstly clustered into one branch.

```
   1 CCTCCTGAGTGAGGATCGCTATTGACAACTAAGGAAAAATGTTTATTTACTGTTTATCAT 60
  61 GAAGTTTGCGCTCTCCACAAGTGTCTCAGTGGTCCATGATCCAACACAGAGTTGAAATGG 120
 121 GAACCTTAGAGCTGGTGAATGAGTAAGAGAAAGCTGAATAAGCCTTGTAGTGGGATAATA 180
 181 TCGGCGTGTTTGGCAGGCGGAGGGGATGGCCGAGCAGCAACAGCAGGGAGGAGGGCAGCC 240
 241 CAGGCCAGGGACTCTCAGCATGTCAGGGGCTGTTCCAGGTCGGCGAAGCAACCTCCGCTT 300
 301 GGCGTTTCCAGGTCAGGGAAGATCCAATAACGACCGTCCACAGAACAACCTTGACTTCCC 360
 361 TGCCTCCGGATACCCTGCGACACAGATTGCTAGGACCATGACGACTGTTCCACCCCAGCC 420
 421 AAGGGACAGGATTACTCGGGGGATTTACCAGAGCATCCAGTCATCCGGCAAACTCAAAAT 480
 481 CTCACCTGAATTGCAAGTGAGTTCACAGCGGACGACCTTCGTGATTTAGGGGAGATTGGA 540
 541 AGAGGAGGTTTTGGCACTGTCAATAAAatggtgcatcgcaagagcaacacaatcatggct 600
   1                             M  V  H  R  K  S  N  T  I  M  A    11
 601 gttaagcgaattcgttcaacagttgatgagaaggaacagaagcaactcttgatggattta 660
  12 V  K  R  I  R  S  T  V  D  E  K  E  Q  K  Q  L  L  M  D  L   31
 661 gaggtagtcatgcgaagcaacgactgcccttgtattgttcaattttatggtgccatattc 720
  32 E  V  V  M  R  S  N  D  C  P  C  I  V  Q  F  Y  G  A  I  F   51
 721 aaggagggtgactgttggatatgcatggagcttatggacacgtcacttgacaagttctat 780
  52 K  E  G  D  C  W  I  C  M  E  L  M  D  T  S  L  D  K  F  Y   71
 781 aaatttatatatgaacgtctacatgaacgtcttccagaaaacatgcttggaaaaataact 840
  72 K  F  I  Y  E  R  L  H  E  R  L  P  E  N  M  L  G  K  I  T   91
 841 gttgccacactaacagcactaaactatttaaaggagaagttgaaaataatccatcgcgat 900
  92 V  A  T  L  T  A  L  N  Y  L  K  E  L  K  I  I  H  R  D   111
 901 gtgaagccatcgaacattttattggacaagcgcggcaatatcaagctctgtgatttcggc 960
 112 V  K  P  S  N  I  L  L  D  K  R  G  N  I  K  L  C  D  F  G   131
 961 atttctggccagcttgtggattccatcgccaagacaagagatgctggatgtagaccatat 1020
 132 I  S  G  Q  L  V  D  S  I  A  K  T  R  D  A  G  C  R  P  Y   151
1021 atggcaccggaacgcatagacccagccagagcccgaggatatgatgttcgctctgacgtg 1080
 152 M  A  P  E  R  I  D  P  A  R  A  R  G  Y  D  V  R  S  D  V   171
1081 tggtccctgggcatcaccctgatggagctggccacagggagcttcccgtaccccaagtgg 1140
 172 W  S  L  G  I  T  L  M  E  L  A  T  G  S  F  P  Y  P  K  W   191
1141 aattccgtttttgagcagctaacacaagtagtgcaaggagagccaccacgtctttcaccc 1200
 192 N  S  V  F  E  Q  L  T  Q  V  V  Q  G  E  P  P  R  L  S  P   211
1201 aacgagaatggcaacactttctccgaggagtttgttagttttgttaatacatgcttaata 1260
 212 N  E  N  G  N  T  F  S  E  E  F  V  S  F  V  N  T  C  L  I   231
1261 aaagatgaaagctcacgacccaagtataagcagctcctcgatcacgagtttgtcgtgcga 1320
 232 K  D  E  S  S  R  P  K  Y  K  Q  L  L  D  H  E  F  V  V  R   251
1321 tcgagagaggaccccatggacgtgacagagtttgttagcggcattctggacaagatggca 1380
 252 S  R  E  D  P  M  D  V  T  E  F  V  S  G  I  L  D  K  M  A   271
1381 aacaacggacgtgtcatgtatacttacgattcttacaactgctgaTGAACTTACTGTCAT 1440
 272 N  N  G  R  V  M  Y  T  Y  D  S  Y  N  C  *                   285
1441 AGCACTTGTCTCATCATTTTTAACTCTGGGACTACGACTGGAACAGACTTGCTCTGGGGC 1500
1501 AAGTTAAAGCAGTGCAGCATTTATATTTTATATGTAGGCTGAATAATACCAGCTTTGCAA 1560
1561 AGAGAGAGAGTAAAAAAAAAAA 1582
```

Figure 1. A schematic diagram of full-length cDNA of *PmMKK4*. The black underscore indicates the initiation codon (ATG) and termination codon (TGA), the black box indicates the phosphorylation site, and the red underscore indicates the "S–I–A–K–T" kinase region.

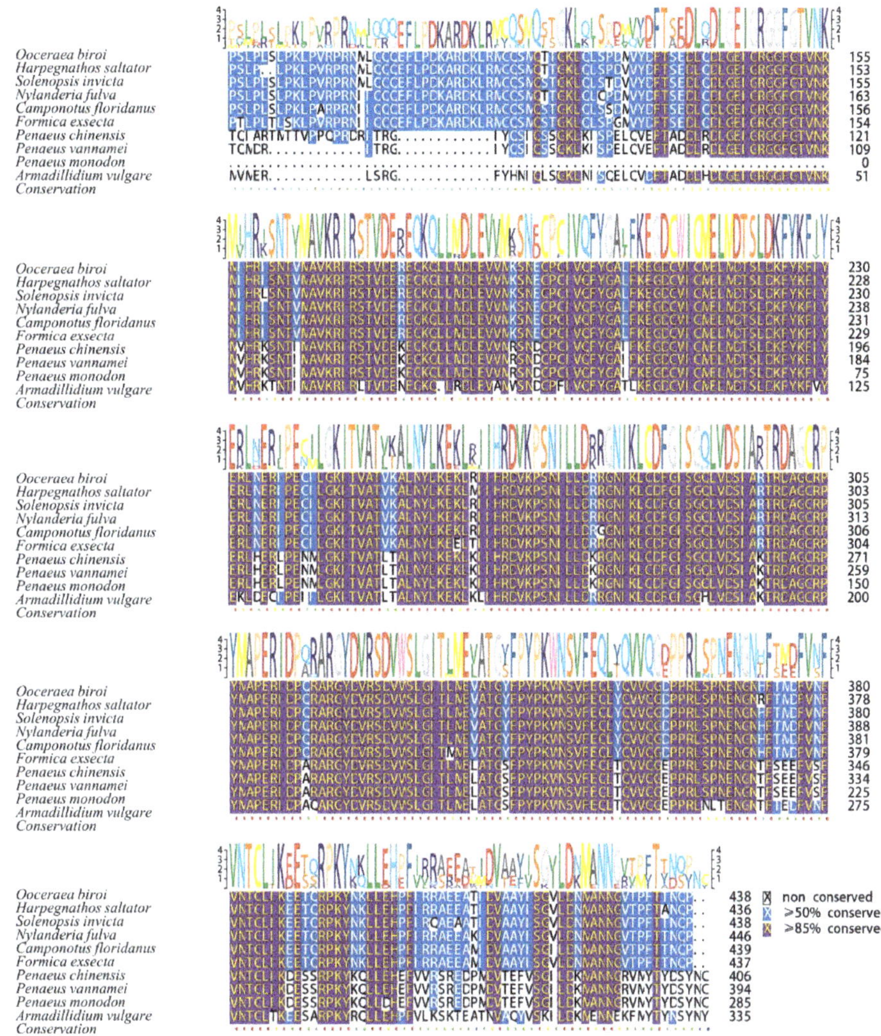

Figure 2. Multiple alignment of the amino acid sequences of *MKK4* in *P. monodon* and other species. Thymosins sequence numbers used in multiple alignment included *L. vanname* (AQY45917.1), *F. chinensis* (AIY23114.1), *A. vulgare* (RXG53605.1), *O. biroi* (XP_011345314.1), *H. saltator* (XP_011144111.1), *S. invicta* (XP_011160881.1), *N. fulva* (XP_029170034.1), *C. floridanus* (XP_011252612.1), and *F. exsecta* (XP_029664029).

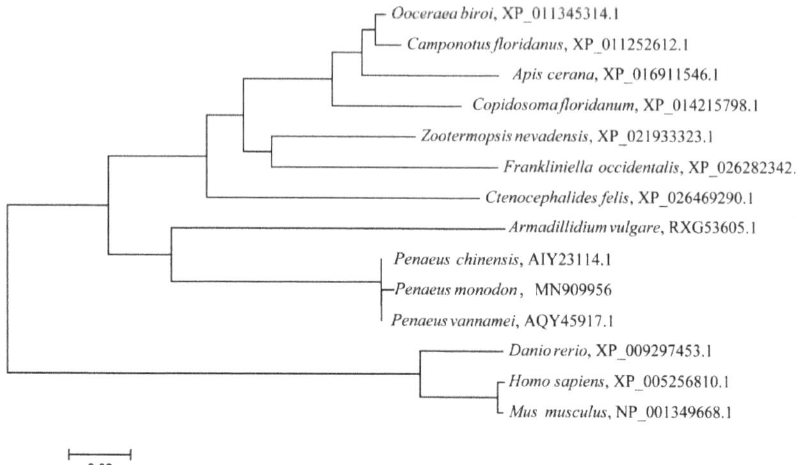

Figure 3. Phylogenetic analysis of *P. monodon MKK4* with other reported *MKK4*.

3.3. Tissue Expression Analysis of PmMKK4

Using real-time quantitative PCR technology, the expression differences of the *PmMKK4* gene in different tissues of *P. monodon* were explored. The results showed that the *PmMKK4* gene was expressed in all the tested tissues. Among them, the expression level was the highest in muscle tissue, and the lowest expression level occurred in the eye stalk nerve. The expression level in muscle was 41.25 times of that in the eye stalk nerve. Secondly, the expression was also higher in the thoracic nerve, intestine, epidermis, heart, gill, and lymphoid tissues, which was higher than that in other tested tissues (Figure 4).

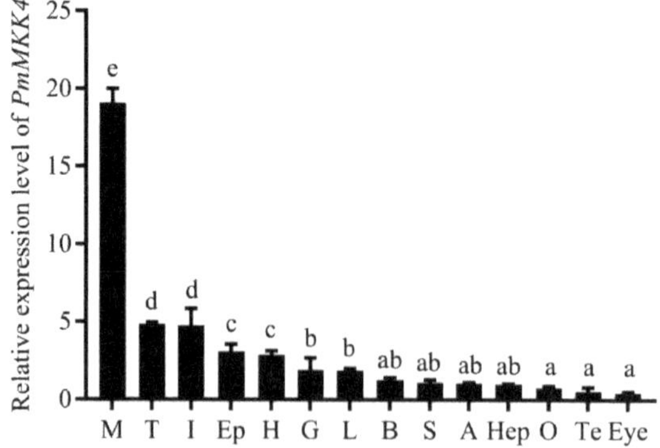

Figure 4. mRNA relative expression of *PmMKK4* in different tissues. Abbreviations: M: muscle, T: thoracic nerve, I: intestines, Ep: epidermis, H: heart, G: gill, L: lymph, B: brain nerve, S: stomach, A: abdominal nerve, Hep: hepatopancreas, O: ootheca, Te: testis, Eye: eyestalk ganglion. Different letters between different tissues mean significant difference ($p < 0.05$).

3.4. Expression Analysis of PmMKK4 under Bacterial Stimulation

In gill tissues, the expression of *PmMKK4* showed different changes after injection of PBS, *S. aureus*, *Vibrio harveyi*, and *V. anguillarum* (Figure 5A). Taking the PBS group as the control, the *S. aureus* infection group was significantly up-regulated at 6 h and 24 h and reached the highest value at 24 h, which was 2.42 times that of the PBS group at this time, and there was a significant difference ($p < 0.05$). Significant down-regulation occurred at 48h, and there was no significant difference between the rest of the time and the PBS group. The group infected with *Vibrio harveyi* showed down-regulation from 12 h and returned to normal level at 72 h. In the group infected with *V. anguillarum*, the expression level remained down-regulated from 6 h, which was significantly different from that in the control group ($p < 0.05$).

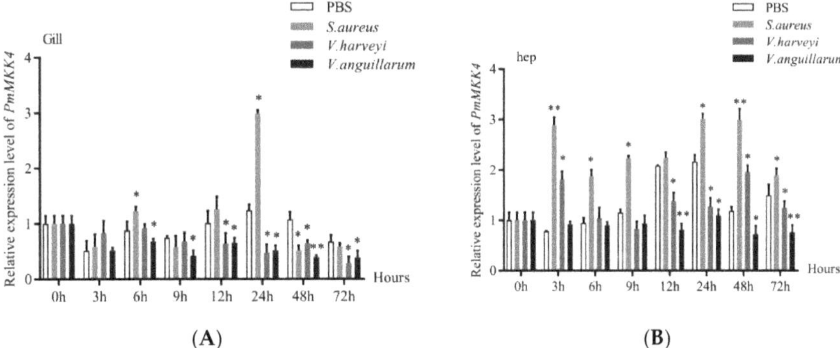

Figure 5. Expression of *PmMKK4* in *P. monodon* after bacterial stimulation. Expression of *PmMKK4* in gill (**A**) and hepatopancreas (**B**) after stimulation by *S. aureus*, *V. harvey* and *V. anguillus*. Vertical bars represent the mean ± S.E (n = 3). Significant differences are indicated by * ($p < 0.05$) and ** ($p < 0.01$).

In the hepatopancreas tissue, each experimental group showed different trends (Figure 5). Taking the PBS group as the control, after infection with *S. aureus*, except for 12 h, it maintained an upward trend, and there was a very significant difference at 3 h and 48 h. Compared with the control group, the *V. harveyi* infection group fluctuated; it was up-regulated at 3 h and 48 h and down-regulated at 12 h, 24 h, and 72 h ($p < 0.05$). There was no significant change within 9 h after infection with *V. anguillarum*, and it was down-regulated from 12 h.

3.5. Expression Analysis of PmMKK4 under Low Salt Stress

The overall expression level of *PmMKK4* after low-salt stress showed that gill tissue was more sensitive to low-salt stress than hepatopancreas tissue. In the gill tissue (Figure 6A), the overall change trend of salinity, which dropped sharply to 17 groups, was a very significant up-regulation and then turned to a significant up-regulation, which remained up-regulated except for 72 h. *PmMKK4* was up-regulated within 96 h after salinity suddenly dropped to the three groups, and the difference was extremely significant at 12 h, 24 h, and 72 h. In the hepatopancreas tissue (Figure 6B), the salinity 7 group was significantly up-regulated at 3 h, 12 h, and 72 h, and there was no significant difference between the expression levels at other time points and the control group. The salinity 3 groups were significantly up-regulated at 12 h and 96 h, and there was no significant difference at other time points.

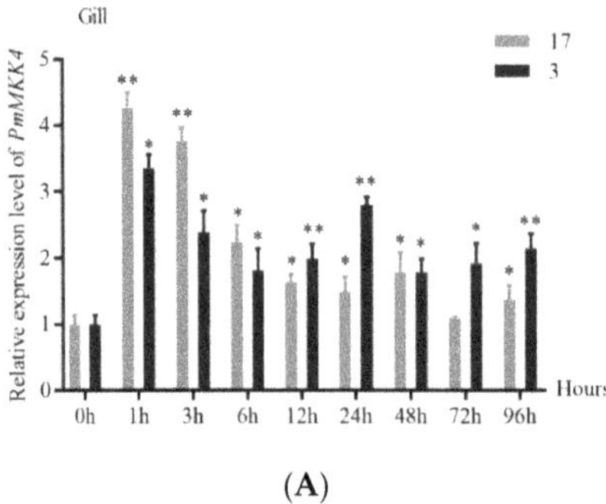
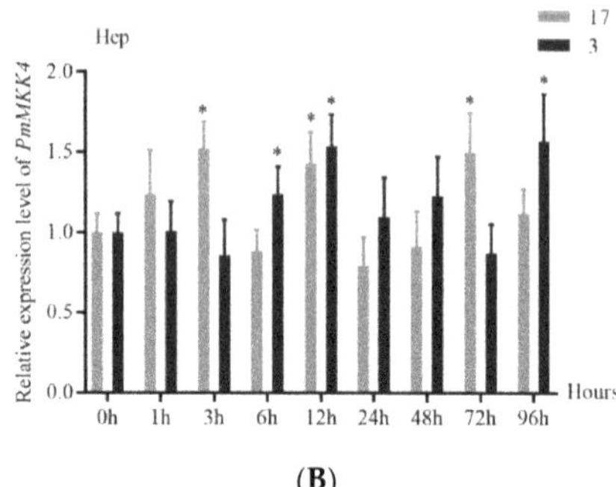

Figure 6. Relative expression levels of *PmMKK4* in hepatopancreas (**A**) and gill (**B**) under acute low-salinity stress. Quantitative RT-PCR was performed to determine the expression of *PmMKK4* in gill (**A**) and hepatopancreas (**B**) of *P. monodon* at different time intervals (n = 3 for each group) after the salinity drops sharply to 17 and 3, ranging from 0 to 96 h. Vertical bars represent the mean ± S.E. (n = 3). Significant differences are indicated by * ($p < 0.05$) and ** ($p < 0.01$).

3.6. Expression Analysis after PmMKK4 Interference

According to the quantitative results of whole tissue expression, the expression of *PmMKK4* was the highest in muscle, so after the RNA interference test, muscle tissue was selected for subsequent quantitative research. Quantitative detection of the *PmMKK4* gene was performed at 24 h, 48 h, and 72 h after RNA interference to verify the interference efficiency. Taking the non-injected group as the control, as shown in Figure 7, the expression of *PmMKK4* in the ds*MKK4* group was only 35.7% of that in the non-injected group at 24 h after injection. At 48 h, the expression level of the injected ds*MKK4* group was 54.1% of that of the non-injected group, and at 72 h, the expression level of the injected ds*MKK4* group was 16.9% of that of the non-injected group, indicating that the interference efficiency of dsPm*MKK4* was obvious, and the interference was effective during the experiment. The mortality rate of each group in the RNAi experiment at each time point after low salt stress is shown in Figure 8. The mortality rate of the dsPm*MKK4* injection group was always higher than that of the other three groups. At 18 h, the mortality rate started to exceed 20%, and the final mortality rate was 25.58%. The lowest mortality rate was in the PBS injection group, with a final mortality rate of 13.04%. The three experimental groups under low-salt stress had more death in the first 24 h, and less death after that.

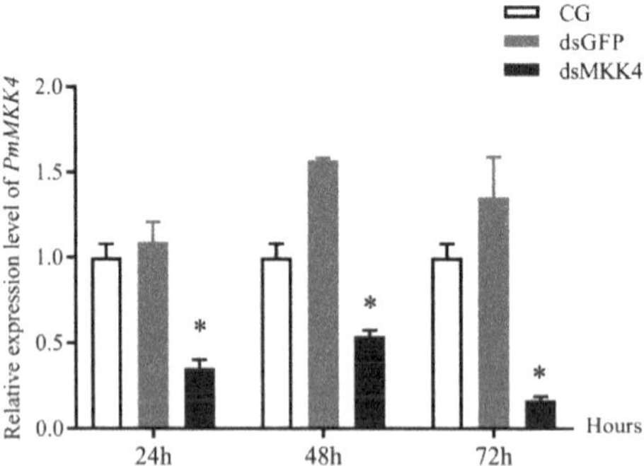

Figure 7. *PmMKK4* mRNA expression profiles after silencing by RNA interference. Quantitative RT-PCR was performed to determine the expression of *PmMKK4* in gills of *P. monodon* at different time intervals (n = 3 for each group) after dsRNA injection. Vertical bars represent the mean ± S.E. (n = 3). Significant differences are indicated by * ($p < 0.05$).

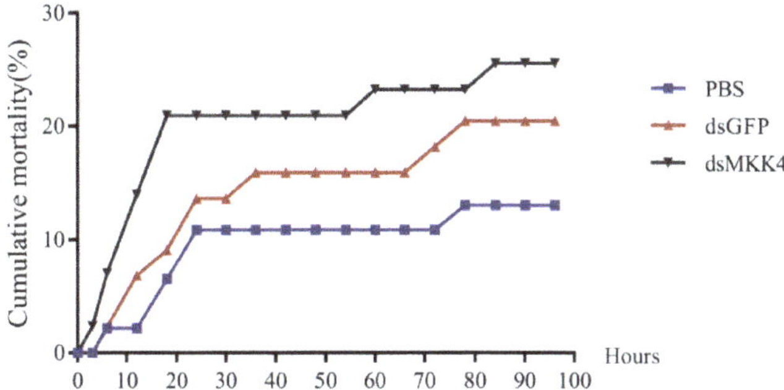

Figure 8. Mortality changes of *Penaeus monodon* in different gene silencing groups under acute low salinity stress.

Figure 9 shows the expression levels of *PmMKK7* and *PmJNK* in the muscles under acute low-salt stress after the injection of dsPm*MKK4*. Taking the PBS injection group as the control, the expression of the *MKK7* gene in *P. monodon* showed a significant upward-regulated trend as a whole, except for 3 h and 24 h, and the expression was significantly up-regulated at other time points. The expression level of *PmJNK* fluctuated throughout the experiment. After acute low-salt stress, the dsMKK4 group first increased at 3 h, then decreased at 6 h, then increased expression, down-regulated at 24 h, and then continued to increase until the end of the experiment.

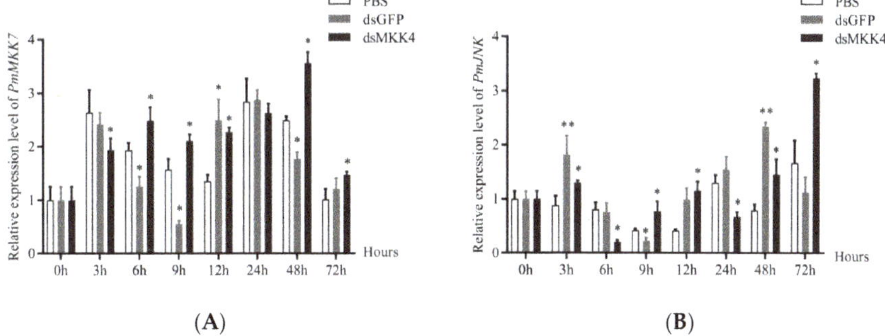

(A)   **(B)**

Figure 9. Relative expression levels of *PmMKK7* (**A**) and *PmJNK* (**B**) after dsRNA-*MKK4* under acute low salinity stress. * ($p < 0.05$), ** ($p < 0.01$). (Significant difference from the CG group).

4. Discussion

The *MKK4* gene of *P. monodon* was cloned in this study. The results of amino acid sequence analysis showed that *PmMKK4* has 22 phosphorylation sites. The phosphate groups can regulate different functions of the protein and may play a synergistic or reverse role in various reactions [14]. Structural prediction indicated that *PmMKK4* contains a conserved serine/threonine protein kinase (S–T–T–K–C) region, which is a potential double phosphorylation site. The *MKK4* gene has two downstream pathways: the JNK signaling pathway, and the p38 signaling pathway. These two branch pathways play important roles in immunity and anti-stress processes in organisms. Phylogenetic tree analysis showed that the *MKK4* gene of *P. monodon* was most closely related to *P. vannamei* and *P. chinensis*, clustered into a clade. The results of multiple sequence alignment showed that *PmMKK4* has a high similarity with *MKK4* genes of other species, among which the similarity with *P. vannamei* and *P. chinensis* was the highest (99.65%), indicating that *MKK4* genes were relatively similar among closely related species. It is speculated that its conserved protein kinase domain may play an important role in many aspects such as physiology and biochemistry.

Through quantitative tissue analysis, it was determined that *PmMKK4* was expressed in all tested tissues, which is the same as the tissue distribution results of other species. In the detected tissues, the expression level of Pm*MKK4* in muscle was the highest, which was significantly higher than other tissues. Similarly, in the study of *P. chinensis*, the expression level of *MKK4* in muscle tissue was also significantly higher than that in other tissues [7], suggesting that muscle may be an important tissue for the function of the *MKK4* gene. In addition to muscle, *PmMKK4* was also highly expressed in tissues such as intestine, heart, gill, and lymph. Intestine and lymph were important immune tissues, and gills were important tissues for crustaceans to exchange ions with the environment and regulate osmotic pressure. Therefore, it was speculated that *PmMKK4* plays an important role in the immune and salinity stress responses of *P. monodon*.

In order to explore the role of the *PmMKK4* gene in the innate immunity of *P. monodon*, we carried out pathogen infection experiments. Although the expression of *PmMKK4* was the highest in muscle, considering that it had been widely accepted that gill and hepatopancreas were important immune organs in previous studies, while muscle and nerve were rarely considered immune-related tissues, we chose gill and hepatopancreas. Tissue responses to bacterial stimulation were studied. The experimental results showed that *PmMKK4* was up-regulated in hepatopancreas and gill tissues after infection with *S. aureus*, which is similar to the *MKK4* gene in *L. vannamei* in response to *S. aureus* [3]. After infection with *Vibrio harveyi*, *PmMKK4* was down-regulated for a period of time in gill tissue, while fluctuating between up- and down-regulation in hepatopancreas tissue. After *P. monodon* was infected with *Vibrio harveyi*, the expression levels of three related genes in the JNK pathway were significantly reduced, indicating that some effector molecules of the JNK signaling pathway may be involved in the process of immune regulation, which is consistent with the research results of Shi et al. [15]. After infection with *V. anguillarum*, gill tissue and hepatopancreas tissue showed a continuous downward trend. However, the *MKK4* gene was significantly up-regulated after *L. vannamei* infection with the Gram-negative bacteria *Vibrio parahaemolyticus* [3]. The physiological role of the post-*MKK4* gene is different, and therefore the response mode is different. In addition, when *Pinctada fucata* are infested by exogenous pathogens, the expression level of *MKK4* gene is significantly changed and phosphorylated, suggesting that it is involved in the self-protection mechanism of *Pinctada fucata* to defend against the occurrence of diseases [16,17].

Changes in salinity can cause changes in the osmotic pressure and the activities of various non-specific immune enzymes in the shrimp, affecting the immune defense ability of the shrimp. The experimental results of acute low-salt stress in this study showed that the *PmMKK4* gene in the gill tissue and hepatopancreas tissue of *P. monodon* was activated after stress, resulting in different degrees of up-regulation. Among them, the expression change of *PmMKK4* in gill tissue was more significant. It can be speculated that under low salt stress, the gill tissue undertakes more physiological and biochemical reactions related to the *MKK4* gene. In the process of ammonia nitrogen stress in *P. chinensis*, the expression of the *MKK4* gene in muscle, hepatopancreas, gill, and other tissues is significantly increased, suggesting that the *MKK4* gene is involved in the stress resistance process of *P. chinensis*. *MKK4* can activate downstream JNK signaling pathway. Studies have confirmed that the JNK branch pathway plays an important role in the salinity adaptation process of aquatic animals. Under salinity stress, the expression of *MAPK 8 (JNK1)* and *MAPK 9 (JNK 2)* was significantly up-regulated [18–20]. To further explore the role of the *PmMKK4* gene in low-salt stress, we performed double-stranded RNA injection experiments. In a low-salt environment, the mortality rate after knocking down the *PmMKK4* gene increased rapidly in a short period of time, while the mortality rates of the other two groups were lower. It is speculated that the low expression of the *MKK4* gene reduces the adaptability of *P. monodon* to the low-salt environment and causes more deaths. The *PmMKK4* gene may play an important role in adaptation to the low-salt environment. After the interference, we further conducted a quantitative study on other genes in the JNK signaling pathway. The results showed that the expression of the *MKK7* gene in *P. monodon* showed a significant upward trend as a whole, while the expression of *PmJNK* fluctuated in the early stage and stably increased in the later stage. It is speculated that the *JNK* signaling pathway is further activated under low-salt stress, and the *PmMKK7* gene maintains high expression, while the *PmMKK4* gene is expressed less to ensure the stable functioning of the JNK signaling pathway.

5. Conclusions

In conclusion, in this study, we cloned the *PmMKK4* gene sequence of *P. monodon*, which was ubiquitously expressed in all tissues of *P. monodon* and play an important role in the innate immunity after pathogen infection and the adaptation process in low-salt environments. The role of the mitogen-activated protein kinase signaling pathway in the process of low-salt stress and the immune response in *P. monodon* was provided.

Author Contributions: Conceptualization, S.J. (Shigui Jiang) and F.Z.; methodology, Y.L.; software, H.F.; investigation, S.J. (Song Jiang); resources, Q.Y.; data curation, L.Y. and X.C.; writing—original draft preparation, Y.L.; writing—review and editing, W.Z.; visualization, J.H.; supervision, F.Z.; project administration, S.J. (Shigui Jiang) All authors have read and agreed to the published version of the manuscript.

Funding: This research was supported by China Agriculture Research System (CARS-48); Hainan Yazhou Bay Seed Laboratory (Project of B21Y10701 and B21HJ0701); Central Public Interest Scientific Institution Basal Research Fund, South China Sea Fisheries Research Institute, CAFS (2020ZD01, 2021SD13); Hainan Provincial Natural Science Foundation of China (320QN359, 322RC806, 320LH008); Guangdong Basic and Applied Basic Research Foundation (2020A1515110200); Guangzhou Science and Technology Planning Project (202102020208); and Hainan Provincial Association for Science and Technology of Young Science and Technology Talents Innovation Plan Project (QCQTXM202206).

Institutional Review Board Statement: The use of all the shrimps in these experiments was approved by the Animal Care and Use Committee at the Chinese Academy of Fishery Sciences (CAFS), and we also applied the national and institutional guidelines for the care and use of laboratory animals at the CAFS.

Informed Consent Statement: Not applicable.

Conflicts of Interest: The authors have declared no conflict of interest.

References

1. Zhang, Y.L.; Chen, D. Map kinases in immune responses. *Cell. Mol. Immunol.* **2005**, *1*, 22–29. Available online: https://pubmed.ncbi.nlm.nih.gov/16212907/ (accessed on 20 September 2022).
2. Krzyzowska, M.; Swiatek, W.; Fijalkowska, B.; Niemialtowski, M.; Schollenberger, A. The role of map kinases in immune response. *Adv. Cell Biol.* **2010**, *2*, 125–138. [CrossRef]
3. Wang, S.; Yin, B.; Li, H.Y.; Xiao, B.; Lǚ, K.; Feng, C.G.; He, J.G.; Li, C.Y. Mkk4 from *litopenaeus vannamei* is a regulator of p38 mapk kinase and involved in anti-bacterial response. *Dev. Comp. Immunol.* **2017**, *78*, 61–70. [CrossRef] [PubMed]
4. Derijard, B.; Raingeaud, J.; Barrett, T.; Wu, I.H.; Han, J.H.; Ulevitch, R.J. Independent human map kinase signal transduction pathways defined by mek and mkk isoforms. *Science* **1995**, *267*, 682–685. [CrossRef] [PubMed]
5. Sanchez, I.; Hughes, R.T.; Mayer, B.J.; Yee, K.; Woodgett, J.R.; Avruch, J.; Kyrlakls, J.M.; Zon, L.I. Role of sapk/erk kinase-1 in the stress-activated pathway regulating transcription factor c-jun. *Nature* **1994**, *372*, 794–798. [CrossRef]
6. Colbourne, J.K.; Pfrender, M.E.; Gilbert, D.; Thomas, W.K.; Tucker, A.; Oakley, T.H.; Tokishita, H.; Aerts, A.; Arnold, G.J.; Basu, M.K.; et al. The ecoresponsive genome of *Daphnia pulex*. *Science* **2011**, *331*, 555–561. [CrossRef] [PubMed]
7. Yao, W.L.; He, Y.Y.; Liu, P.; Li, J.; Wang, Q.Y. cDNA cloning and expression analysis of *MKK4* gene under ammonia-N stress in *Fenneropenaeus Chinensis*. *J. Fish. China* **2015**, *39*, 779–789. [CrossRef]
8. Su, W.W.; Yang, L.S.; Su, T.F.; Yang, Q.B.; Zhu, C.Y.; Zhou, F.L. Expression analysis of *MKK4* gene in *Penaeus monodon*. *Guangdong Agric. Sci.* **2012**, *39*, 155–157. [CrossRef]
9. Wang, S.; Qian, Z.; Li, H.Y.; Lǚ, K.; Xu, X.P.; Weng, S.P.; He, J.G.; Li, C.Z. Identification and characterization of mkk7 as an upstream activator of jnk in *litopenaeus vannamei*. *Fish Shellfish Immunol.* **2016**, *48*, 285–294. [CrossRef] [PubMed]

10. Qu, F.F.; Tang, J.Z.; Peng, X.Y.; Zhang, H.; Shi, L.P.; Huang, Z.Z.; Xu, W.Q.; Chen, H.Q.; Shen, Y.; Yan, J.P.; et al. Two novel mkks (mkk4 and mkk7) from *ctenopharyngodon idella* are involved in the intestinal immune response to bacterial muramyl dipeptide challenge. *Dev. Comp. Immunol.* **2019**, *93*, 103–114. [CrossRef]

11. Qin, Y.K.; Jiang, S.G.; Huang, J.H.; Zhou, F.L.; Yang, Q.B.; Jiang, S.; Yang, L.S. C-type lectin response to bacterial infection and ammonia nitrogen stress in tiger shrimp (*penaeus monodon*). *Fish Shellfish Immunol.* **2019**, *90*, 188–198. [CrossRef]

12. He, P.; Jiang, S.G.; Li, Y.D.; Yang, Q.B.; Jiang, S.; Yang, L.S.; Huang, J.H.; Zhou, F.L. Molecular cloning and expression pattern analysis of *GLUT1* in black tiger shrimp (*Penaeus monodon*). *South China Fish. Sci.* **2019**, *15*, 72–82. Available online: https://www.cabdirect.org/cabdirect/welcome/?target=%2fcabdirect%2fabstract%2f20193290256 (accessed on 20 September 2022).

13. Si, M.R.; Li, Y.D.; Jiang, S.G.; Yang, Q.B.; Jiang, S.; Yang, L.S.; Huang, J.H.; Chen, X.; Zhou, F.L. A CSDE1/Unr gene from *Penaeus monodon*: Molecular characterization, expression and association with tolerance to low salt stress. *Aquaculture* **2022**, *561*, 738660. [CrossRef]

14. Grindheim, A.K.; Saraste, J.; Vedeler, A. Protein phosphorylation and its role in the regulation of annexin a2 function. *Biochim. Biophys. Acta (BBA)-Gen. Subj.* **2017**, *1861*, 2515–2529. [CrossRef]

15. Shi, G.F.; Zhao, C.; Fu, M.J.; Qiu, L.H. The immune response of the c-jun in the black tiger shrimp (*penaeus monodon*) after bacterial infection. *Fish Shellfish Immunol.* **2017**, *61*, 181–186. [CrossRef]

16. Zhang, H.; Huang, X.D.; Shi, Y.; Liu, W.G.; He, M.X. Identification and analysis of an mkk4 homologue in response to the nucleus grafting operation and antigens in the pearl oyster, pinctada fucata. *Fish Shellfish Immunol.* **2017**, *73*, 279–287. [CrossRef]

17. Fan, H.; Li, Y.; Yang, Q.; Jiang, S.; Yang, L.; Huang, J.; Jiang, S.; Zhou, F. Isolation and characterization of a MAPKK gene from *Penaeus monodon* in response to bacterial infection and low-salinity challenge. *Aquac. Rep.* **2021**, *20*, 100671. [CrossRef]

18. Tian, Y.; Wen, H.S.; Qi, X.; Zhang, X.Y.; Li, Y. Identification of mapk gene family in *lateolabrax maculatus* and their expression profiles in response to hypoxia and salinity challenges. *Gene* **2018**, *684*, 20–29. [CrossRef] [PubMed]

19. Fan, R.; Li, Y.; Yang, Q.; Jiang, S.; Huang, J.; Yang, L.; Chen, X.; Zhou, F.; Jiang, S. Expression Analysis of a Novel Oxidoreductase Glutaredoxin 2 in Black Tiger Shrimp, *Penaeus monodon*. *Antioxidants* **2022**, *11*, 1857. [CrossRef] [PubMed]

20. Wei, W.-Y.; Huang, J.-H.; Zhou, F.-L.; Yang, Q.-B.; Li, Y.-D.; Jiang, S.; Jiang, S.-G.; Yang, L.-S. Identification and Expression Analysis of Dsx and Its Positive Transcriptional Regulation of IAG in Black Tiger Shrimp (*Penaeus monodon*). *Int. J. Mol. Sci.* **2022**, *23*, 12701. [CrossRef]

Journal of
*Marine Science
and Engineering*

Article

Effect of Different Colored LED Lighting on the Growth and Pigment Content of *Isochrysis zhanjiangensis* under Laboratory Conditions

Bu Lv [1,†], Ziling Liu [1,†], Yu Chen [1], Shuaiqin Lan [1], Jing Mao [1], Zhifeng Gu [1,2], Aimin Wang [1,2], Feng Yu [1], Xing Zheng [1,2,*] and Hebert Ely Vasquez [1,2,*]

[1] Department of Aquaculture, College of Marine Sciences, Hainan University, Haikou 570228, China
[2] State Key Laboratory of Marine Resource Utilization in South China Sea, Hainan University, 58 Renmin Avenue, Haikou 570228, China
* Correspondence: zhengxing_edu@163.com (X.Z.); hebertely@163.com (H.E.V.)
† These authors contributed equally to this work.

Abstract: Light is one of the most important environmental factors affecting the growth and reproduction of algae. In this study, the effect of various LED colors on the productivity, chlorophyll (Chl-*a*, Chl-*b*, and total Chl), protein, and carbohydrate content of *Isochrysis zhanjiangensis* in indoor culture was investigated. Microalgae monocultures were cultivated under five different colors (red, green, blue, yellow, and white) for twenty-one days. The microalgae cultured under red light exhibited a higher specific growth rate ($0.4431 \pm 0.0055\ \mu$ day^{-1}), and under white light a higher productivity (0.0728 ± 0.0013 g L^{-1} day^{-1}). The poorest performance was observed under yellow and green lights. Interestingly, green light exhibited the highest levels of chlorophylls (Chl-*a*, 1.473 ± 0.037 mg L^{-1}; Chl-*b*, 1.504 ± 0.001 mg L^{-1}; total Chl, 2.827 ± 0.083 mg L^{-1}). The highest protein content was observed under the white light (524.1935 ± 6.5846 mg L^{-1}), whereas the carbohydrate content was remarkably high under the blue light (24.4697 ± 0.0206 mg L^{-1}). This study is important in terms of the selection of light at the appropriate color (wavelength) to increase the content of organic compounds desired to be obtained indoors with the potential for commercially produced cultures.

Keywords: *Isochrysis zhanjiangensis*; light-emitting diodes; growth; chlorophyll; organic compounds

Citation: Lv, B.; Liu, Z.; Chen, Y.; Lan, S.; Mao, J.; Gu, Z.; Wang, A.; Yu, F.; Zheng, X.; Vasquez, H.E. Effect of Different Colored LED Lighting on the Growth and Pigment Content of *Isochrysis zhanjiangensis* under Laboratory Conditions. *J. Mar. Sci. Eng.* **2022**, *10*, 1752. https://doi.org/10.3390/jmse10111752

Academic Editors: Sang Heon Lee and Azizur Rahman

Received: 10 October 2022
Accepted: 10 November 2022
Published: 15 November 2022

1. Introduction

The microalgae *Isochrysis zhanjiangensis*, a marine single-cellular golden-brown flagellated species isolated from Nansan Island of Zhanjiang of Guangdong Province, China [1], is an important species in the aquaculture economy and commonly used in the fodder industry and various mariculture systems [2]. Due to the small size, fast-growing, cellulose-free cell walls, and nutrient-richness, especially in polyunsaturated fatty acid (Omega-3), chlorophylls, and carotenoids, *Isochrysis zhanjiangensis* has been mass-produced for feeding fish, shrimp, shellfish seedlings, as well as larvae of a variety of aquaculture animals [3,4]. In particular, *I. zhanjiangensis* is used as a food supply worldwide during broodstock hatchery conditioning. Additionally, it is especially used to culture suspension-feeding larvae and early juvenile bivalve mollusks, because of its nutritional properties that support shell growth and the high survival rates when used as a mono-species diet [5,6].

The growth and propagation of microalgae are affected by several environmental factors, such as temperature, salinity, light, and pH. Light, the main source of energy for algae growth, is one important key factor in regulating its growth and development [7]. Photosynthetic microorganisms do not utilize the whole solar spectrum but only a fraction of it, in particular from 400 to 700 nm. The absorption wavelengths of visible light by algae and plants are mainly concentrated in the blue-violet light region of 400–510 nm and the red-orange light region of 610–720 nm. However, the wavelengths absorbed by microalgae differ according to species [8]. Several investigations have been focused on either the single or combined influence of light quality (meaning the different wavelengths which are absorbed by water to various extents) [9–12], light quantity (different light intensities) [13–19], and light periodicity (different photoperiods) [20–23], thus indicating that illumination is a complex external factor for microalgae cultivation. Moreover, researchers have found that light quality plays an important role in regulating the growth and development, morphology, photosynthesis, and metabolism of algae [24–27]. For instance, red light is an efficient light quality for the growth of *Arthrospira (Spirulina) platensis*, while it has a significant inhibitory effect on the chlorophyll content [28]. The cultivation of chlorophytes under a mix of green and blue LEDs may prove optimal for growth, biomass productivity, pigments, proteins, and lipids [29–32]. Green light enhanced growth rates, protein, and lipid contents in *Brachiomonas submarina*, and pigment content in *Kirchneriella aperta*. High- and low-intensity green LEDs enhanced lutein biosynthesis compared to red or blue LEDs in *B. submarina* and *Scenedesmus obliquus* [33,34]. High-intensity blue LEDs increased the carotenoid zeaxanthin, and white light was optimal for phycobiliprotein in *Rhodella* sp. and fucoxanthin content for *Stauroneis* sp. and *Phaeothamnion* sp. [33]. Although, the use of any white light sources (fluorescent lamps, RGB LEDs, and white LEDs) for the cultivation of green algae seems to not affect growth. A species-specific response of algae to light intensity has been described in *Desmodesmus quadricauda*, *Parachlorella kessleri*, and *Chlamydomonas reinhardtii* [35]. In *P. kessleri* cells, the concentration of pigments decreased with increasing light intensity, a response found not only in the genus *Chlorella* [13,36–38], but also in other green algae [39].

Light quality, intensity, and photoperiod also affect the growth, biochemical composition, and physiology of *Isochrysis* sp. [40–42]. Microalgal pigments change with algal variety. Therefore, the influences of different light qualities on the physiological properties of algae, such as growth, photosynthesis, and cellular metabolism, are diverse [43]. The ability of the microalgae to utilize different light qualities is determined by this composition of pigments in their cells, and different pigments absorb different light qualities. The growth and development of microalgae and the generation of metabolites are related to light quality, and the light quality that is most suitable for the growth of one microalgae species may not be suitable for another [33]. Therefore, it is of great significance to explore the optimum light quality for the growth of *I. zhangjiangensis*. Thus, this study aims to examine the effects of different LED light qualities on the productivity, chlorophyll, protein, and carbohydrate content of *I. zhanjiangensis* in indoor culture. Our results will aid the optimization of the light conditions for the growth of *I. zhanjiangensis*. Additionally, the results will provide the basis for the optimization of microalgae propagation in indoor conditions and other systems that require artificial illumination in general.

2. Materials and Methods

2.1. Microalgae Culture Condition

The stock of the microalgae species *I. zhanjiangensis* was obtained from the Microalgae Laboratory of the College of Oceanography at Hainan University (Haikou, China). For the enrichment of the culture media, the nutrient medium Ningbo 3$^{\#}$ was dissolved in filtered and sterilized seawater (29 PSU) with the composition per 1 L of: 100 mg NaNO$_3$, 10 mg NaH$_2$PO$_4$, 2.5 mg FeSO$_4$, 10 mg EDTA-2Na, 0.25 mg MnSO$_4$, 0.5×10^{-3} µg vitamin B$_{12}$, and 6 µg vitamin B$_1$.

The microalgae were cultured in 5 L flat-bottom glass flasks at 25 ± 1 °C, with the illumination provided by light-emitting diodes (LED, 191.8 µmoles/m^2/s) with a 14:10 h light/dark photoperiod. To enhance growth and prevent the algae from settling, continuous aeration was applied using air stones at 20 L/min. The initial culture media inoculation of the microalgae was 100,000 cells/mL, and the illumination was immediately provided using green (495–530 nm), blue (450–480 nm), red (615–650 nm), white (450–465 nm), and yellow (580–595 nm) light-emitting diodes. Each illumination treatment was set separately to avoid any light interference from the neighboring treatments and set in triplicate.

2.2. Measurement of I. Zhanjiangensis Growth

The growth of *I. zhanjiangensis* cultured in the experiment was measured using cell density and biomass (dry weight) to precisely determine the growth pattern of the microalgae. Microalgae cell density was determined daily by counting the cell number using a hemacytometer under a white-light field microscope and was presented as the number of cells/mL. Cell dry weight was estimated by filtering 20 mL of cultured microalgae using a pre-weighed Whatman GF/C filter (47 mm ⊘), washed three times with filtered seawater, and dried to constant weight. Samples were always collected at the same time of the day.

Productivity (P_x) and specific growth rate (μ) of *I. zhangjiangensis* were calculated according to the following Formulae [44]:

$$P_x = (X_m - X_i)\,(T_c)^{-1} \tag{1}$$

$$\mu = (lnX_m - lnX_i)\,(T_c)^{-1} \tag{2}$$

where, X_i = initial biomass concentration (g/L), X_m = maximum biomass concentration (g/L), and T_c = cultivation time related to the maximum biomass concentration (days).

2.3. Measurement of Photosynthetic Pigments, Protein, and Carbohydrate Content

Chlorophyll concentration (Chl) content was determined using a modified method from Jeffrey and Humphrey [45]. Briefly, 20 mL samples were withdrawn from the culture flasks and transferred to opaque plastic bottles, kept away from light, and warmed to room temperature, then filtered using Whatman GF/C filters (25 mm ⊘). After filtration, the filters containing the microalgae biomass were folded and stored separately at $-2\ ^\circ$C. Chl extraction was carried out by grinding the filters in 90% acetone (2–4 mL) in a glass homogenizer on an ice bath under low-light conditions for up to 1 min. After grinding, the Chl extracts were transferred to a graduated and stoppered centrifuge tube and rinsed with 10 mL of acetone 90% (10 mL + dead volume of filter). The extract was then centrifuged for 10 min at 500 g. After completion of centrifugation, the absorbance of the supernatant (OD) was measured at 750, 664, 647, and 630 nm against a 90% acetone blank. The concentrations of Chl-*a*, Chl-*b*, and the total Chl were calculated according to the following Equations [45,46]:

$$C_{chla} = (11.85\,E_{664} - 1.54\,E_{647} - 0.08\,E_{630}) \times 10/V \tag{3}$$

$$C_{chlb} = (21.03\,E_{647} - 5.43\,E_{664} - 2.66\,E_{630}) \times 10/V \tag{4}$$

$$C_{tch1} = (20.21\,E_{647} + 8.02\,E_{664}) \times 10/V \tag{5}$$

where, C_{chla}, C_{chlb}, and C_{tch1} are the concentrations of Chl-*a*, Chl-*b*, and the total Chl (mg/L), and V is the filtered volume (mL).

The protein content of each sample was determined by the Lowry method [47]. Carbohydrates were measured by the phenol sulfuric acid method [48].

2.4. Statistical Analysis

Significant differences ($p < 0.05$) among variables were first identified using the *t*-test. Before analysis, data were tested for normality using Kolmogorov–Smirnov's test and for homogeneity of variance using Cochran's C test. All statistical analyses were performed using DPS14.5 software (Hangzhou Rui Feng Information Technology Co., Ltd., Hangzhou, China).

3. Results

3.1. Specific Growth Rate and Biomass Productivity

The dry weight, specific growth rate, and productivity of *I. zhangjiangensis* under different light conditions are shown in Table 1. The highest and lowest accumulated biomass (dry weight) were observed in the microalgae cultured under the white and red lights, with 1.1761 ± 0.0212 and 0.5683 ± 0.0284 g/L, respectively. Significant differences

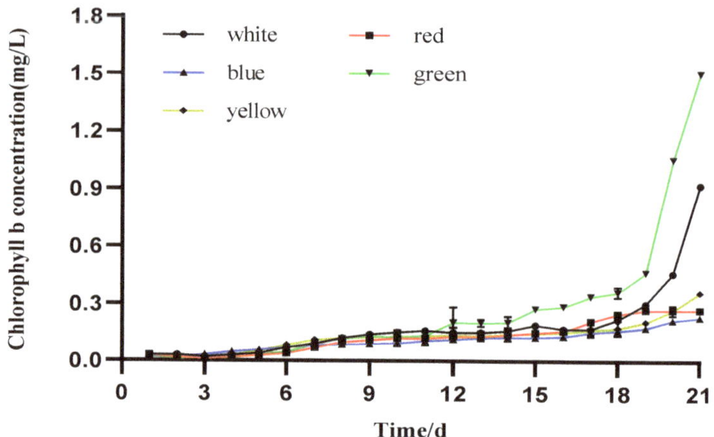

Figure 3. Daily mean values of chlorophyll-*b* in *I. zhangjiangensis* cultured under different colored LEDs (n = 3, *p* < 0.05 *t*-test).

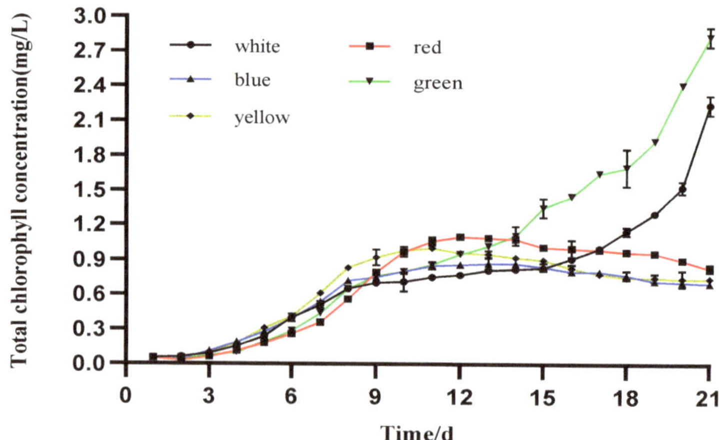

Figure 4. Daily mean values of total chlorophyll in *I. zhangjiangensis* cultured under different colored LEDs (n = 3, *p* < 0.05 *t*-test).

3.4. Protein and Soluble Carbohydrates' Production

Significant differences were observed in the total protein content of *I. zhangjiangensis* grown under different light colors (Table 2). A particularly high content (524.1935 ± 6.5846 mg/L) was observed in the microalgae cultured under the white light. No significance was observed between the blue and yellow lights, in which both lights exhibited lower content than in the white light. The microalgae cultivated under green and red lights observed the lowest values.

Table 2. Protein and carbohydrate content in *I. zhangjiangensis* cultured under different colored LEDs.

Light Color	Maximum Total Protein Content (mg/L)	Maximum Carbohydrate Content (mg/L)
White	524.1935 ± 6.5846 [a]	8.5859 ± 0.0206 [d]
Red	403.2258 ± 6.5846 [c]	9.1667 ± 0.0206 [c]
Blue	440.0325 ± 3.3723 [b]	24.4697 ± 0.0206 [a]
Green	406.6346 ± 3.3548 [c]	9.3687 ± 0.0206 [b]
Yellow	454.9462 ± 3.4375 [b]	7.8455 ± 0.1247 [e]

Note: Different lowercase letters represent significant differences between different colored LEDs ($p < 0.05$).

The mean values of the carbohydrate content were also significant among all light color qualities (Table 2). The highest mean value was 24.4697 ± 0.0206 mg/L, observed in the microalgae cultured under blue light. The green, red, white, and yellow lights exhibited significant mean values within them, varying from 7.8455 ± 0.1247 to 9.3687 ± 0.0206 mg/L.

4. Discussion

4.1. Specific Growth Rate and Biomass Productivity of I. zhangjiangensis

The present study examined the effect of different colored LED illumination on the productivity, Chl, protein, and carbohydrate content in the indoor culture of microalgae *I. zhangjiangensis*. The highest specific growth rate was observed under the red light, 1.5-fold higher when compared to the white light. Whereas the highest biomass observed under the white light was 2.0-fold higher than that under the red light (Table 1). This observation seems contradictory; however, it is possible that the size of the cells in the microalgae cultured under white light was larger than under the other light colors. Unfortunately, cell size data were not collected at any stage during the experimental period to fully support this idea. Nevertheless, it has been reported that light quality regulates the cell size of the microalgae [49]. For instance, the cell sizes of the green microalgae *C. reinhardtii* grown under blue light were 1.3 and 1.6 times larger than under white and red light sources, respectively, due to a delay in cellular division processes [50]. Additionally, the light intensity has resulted in cell enlargement by a factor of 2.5 in these species, compared to 1.9 in *D. quadricauda*, and only 1.3 in *P. kessleri*. The smaller increase in cell size in *P. kessleri* was compensated for by a 13.6-fold daily increase in cell number under optimal conditions, as compared with a 9.7-fold increase in *C. reinhardtii* and *D. quadricauda* [35]. Another study using different strains of *Chlorella* sp. revealed that the light spectrum had a significant influence on microalgae cell size. The largest and smallest cells were observed under blue and red lights, respectively [51]. Such an increase in cell size is a specific response of organisms that divide by multiple fission and thus can respond to better growth conditions beyond a simple increase in the cell number. The larger cell size is a mechanism that supports better growth in the next cell cycle, possibly leading to better productivity [35]. Nevertheless, it should be emphasized that although the value of the specific growth rate is important, the most relevant parameter related to potential large-scale microalgae production systems is biomass productivity, which was highest when *I. zhangjiangensis* was cultivated under white light. It has been proven that multi-chromatic white light, whether provided by a fluorescent lamp or by LED, is more advantageous than monochromatic light for promoting the growth of *Isochrysis* sp. [42,52,53], which can also be supported by our results. A similar observation has already been reported in the productivity of *Nannochloropsis* sp. cultured under different colored lights, where pink and white lights exhibited higher biomass productivity [54]. These results are probably related to the differences in energy provided by light and captured by the photosynthetic apparatus of the photosynthetic microorganisms [55]. Particularly, between 380 and 750 nm, the energy content is sufficient to produce chemical changes in the absorbing molecules, as happens throughout the photosynthetic pathways prevailing in the microalgae [56].

4.2. The Effect of Different LED Colors on the Pigment Content of I. zhangjiangensis

Microalgae, similar to plants, capture light energy (light-harvesting antennas) and produce electrons in the reaction center of the photosystems. For efficient photosynthesis, preserving an excitation balance between the two photosystems (PSI and PSII) is of prime importance. To serve this purpose, microalgae possess specific light-harvesting antennas to expand the available light wavelength. Certain groups of algae contain accessory pigments that help in efficiently harvesting light for photosynthesis [54]. Green algae, in particular, possess a chlorophyll–protein complex which is comprised of Chl-*a* and *b* and carotenoids for carrying out the photosynthesis [57]. Chl molecules absorb light energy and transfer this energy to the photochemical reaction centers presented in algae, cyanobacteria, and higher plants by PSI and PSII, where charge separation occurs. Upon illumination, two electrons are extracted from water, mostly (O_2 is evolved), and transferred through a chain of electron carriers to produce one molecule of $NADPH_2$ (nicotinamide adenine dinucleotide hydrogen). Simultaneously, protons are transported from an external space (stroma) into the intra-thylakoid space (lumen), forming a pH gradient. According to Mitchel's chemiosmotic hypothesis, the gradient drives ATP synthesis, which is catalyzed by the protein complex called ATPase or ATP synthase—a reaction called photophosphorylation [58]. In the present investigation, for *I. zhangjiangensis*, the highest Chl content was attained under green light, while blue light resulted in the lowest Chl content, indicating that different light qualities can evoke different levels of photosynthetic pigments. Green light has promoted the production of Chl in *Chlorella vulgaris* [46], and also allowed improvement of *S. platensis* growth [10,59]. A different result has been reported for *I. galbana*. The highest contents of Chl-*a*, Chl-*c*, and Car were obtained under white light, while blue light resulted in the lowest pigment contents [53]. The same report described the highest light absorption in cells cultivated under blue light, but the photochemical reaction was lowered. Cells cultivated under blue and red lights were, respectively, restricted by downregulated photosynthetic efficiency and sufficient light absorption. Meanwhile, green light showed an increase in photosynthetic efficiency, associated with a light absorption close to that for cells exposed to white light, suggesting that green light promotes the photosynthesis of *I. galbana* by balancing the light absorption and utilization [53]. Light with a shorter wavelength, for example, blue light, has a higher probability to cause growth photo-inhibition by striking the light-harvesting complex of cells at its peak electrical energy due to its high energy [60].

Studies have shown that infrared light can cause cell damage [61], while in some multicellular algae, blue light can significantly increase the content of algal photosynthetic pigments, increasing photosynthetic efficiency and ultimately, the growth rate [62,63]. The content of photosynthetic pigments in *I. zhangjiangensis* is the highest under green light, which is conducive to the accumulation of its biomass, while the content of photosynthetic pigments is the lowest under blue light. A previous study reported a contrasting result: low-intensity blue light reduced the pigment content of *Chaetoceros gracilis* but increased it in *I. galbana* [64]. The effect of light qualities on the high-value pigments has also been reported in five microalgae strains from three distinct lineages [33]. In the Rhodophyte *Rhodella* sp., the Chl-*a* levels obtained under red and white LEDs were higher than those reached under green and blue illumination for medium and high intensity. Similarly, the diatom *Stauroneis* sp. also returned a higher Chl-*a* content under medium white light intensities. Contrariwise, in the chlorophyte *K. aperta* and *B. submarina*, the Chl content was significantly higher under blue and green lights at high and medium intensities, returning two-fold higher Chl-*a* compared to red and white LEDs. The cultivation of *Phaeothamnion* sp. under high-intensity blue LEDs also induced a significant increase in Chl-*a*. In general, responses of each strain to different colored LEDs were generally species-specific. These results indicate that the growth performance of different microalgae under different light qualities is not consistent, which may be due to the different compositions of the pigment system of different microalgae, resulting in different requirements for light quality in photosynthesis. It is also known that the content of the photosynthetic pigments increases as light intensity decreases [65]. In the case of *P. kessleri*, the concentrations of Chl-*a* and *b* and carotenoids decreased with increasing light intensity. Cultures of *C. reinhardtii* and *D. quadricauda* maintained similar levels of photosynthetic pigments at low light intensities, but their concentration increased with the time of cultivation at the highest light intensity [35]. This increase in the photosynthetic pigments is attributed to the need of the microorganisms to improve their photosynthetic efficiency and capture as much energy as possible from light [28]. Considering this, a conclusion can be drawn that when Chl content is high under a specific light color, it does not necessarily imply that this illumination provides adequate amounts of energy for biomass synthesis, and this may be the case of *I. zhangjiangensis* cultured under green light. In addition, light-harvesting ability and energy-transfer processes also differed between green algae species. This has been demonstrated by observing the delayed fluorescence spectra in *C. reinhardtii* and *C. variabilis* cells grown under different light qualities. Both types of green algae primarily modified the associations between light-harvesting chlorophyll protein complexes (LHCs) and photosystems (PSII and PSI) [57].

4.3. The Effect of Different LED Colors on the Protein and Carbohydrate Content of I. Zhangjiangensis

The effect of light intensity on microalgal cultures has been extensively studied, and it has been demonstrated that light intensity controls not only the growth rate, but also the lipid storage [17,18,20,21,25,41], structural distribution [66], cellular composition (such as proteins and essential fatty acids) [8], and pigment synthesis [67]. While most research has been focused on the effects of light intensity, it has been shown that light quality also plays a key role in algal metabolism and the effects of light wavelengths on growth are species-specific because of the differences in metabolic pathways, pigmentation, and photoreceptors between species [8]. Moreover, the biochemical composition is a pivotal factor in determining the nutritional value of microalgae. In this investigation, the protein and carbohydrate contents of *I. zhangjiangensis* were influenced by the different LED colors used in the indoor cultures (Table 2). The maximum protein content was measured in the microalgae cultured under white light, followed by the yellow and blue lights, and the lowest level in the red light. Similar results have been reported in microalgal conglomerates of *Chlorella variabilis* and *Scenedesmus obliquus* when the protein content of the microalgal consortia was highest under a cool-white light [68]. Contrasting results have been observed when blue light fluorescent tubes were closely related to protein enhancement in *Isocrhysis* sp. cultured in a bioreactor [42]. However, cell concentration and productivity did not change substantially upon changing the light spectrum during steady-state growth. In addition, experiments conducted with *Tisochrysis lutea* (previously named *I.* aff. *galbana*) under white, blue, green, and red fluorescent lamps in batch cultures revealed that the growth rate and cell density were highest with white light, followed by blue light. Meanwhile, cells under green light had a greater dry weight during exponential growth in comparison with the other light colors, and this monochromatic light also increased the eicosapentaenoic acid and protein contents [69]. In axenic cultures of *Dunaliella tertiolecta* and *Thalassiosira rotula*, blue light also allows higher photosynthetic carbon incorporation into protein than white light [70]. It can be inferred that the optimal light color for the cultivation of *Isochrysis* varies depending on the algal strains and light sources used.

In this study, carbohydrate content was higher in the blue light and the lowest content was in the yellow light. These results concur with the one reported for *A. platensis*, where the highest carbohydrate content was also measured under blue light [28]. However, there are conflicting results on the influence of blue light on microalgae carbohydrate content: In *T. lutea*, it did not change [69], but it decreased in *Isochrysis* sp. [42]. Blue light also enhanced dark respiration, as previously reported for *Scenedesmus obliquus* [71], *Rhodomonas salina* [72], or *D. tertiolecta* and *T. rotula* [70], confirming a higher rate of carbohydrate degradation under blue light. The evidence presented in this study confirms that light quality can affect the biochemical composition of microalgae cells of *I. zhangjiangensis* when they are cultured under different light colors.

5. Conclusions

The productivity, Chl (*a*, *b*, and total), protein, and carbohydrate content of *I. zhangjiangensis* can be regulated by different light wavelengths. White light increased the productivity and protein content, and red light increased the specific growth rate. Pigment content was higher under the green light but possibly does not provide adequate amounts of energy for biomass synthesis. The blue light highly promoted carbohydrate content, suggesting that the light influence is a complex phenomenon that is still far from being completely understood. This study is important in terms of the selection of light at the appropriate wavelength to increase the number of metabolites desired in indoor production levels, with the possibility of adaptation to other culture systems that require artificial illumination.

6. Ethics Statement

The experiment complied with the regulations and guidelines established by the Animal Care and Use Committee of Hainan University.

Author Contributions: B.L., investigation, data curation, visualization, formal analysis, writing—original draft; Z.L., investigation, conceptualization, formal analysis; Y.C., investigation, methodology; S.L., investigation, methodology; J.M., investigation, methodology; Z.G., funding acquisition; A.W., project administration; F.Y., formal analysis; X.Z., conceptualization, methodology, formal analysis, writing—original draft, writing—review and editing; H.E.V., methodology, formal analysis, writing—original draft, writing—review and editing. All authors have read and agreed to the published version of the manuscript.

Funding: This work was funded by the Key Research and Development Project of Hainan Province (Grant No. ZDYF2021XDNY277), the Natural Science Foundation for Young Scholars of Hainan Province (Grant No. 320QN207), the Starting Research Fund from Hainan University (Grant No. KYQD-ZR 20061), and the Academician Innovation Center of Hainan Province, China.

Institutional Review Board Statement: Not applicable.

Informed Consent Statement: Not applicable.

Conflicts of Interest: The authors declare no conflict of interest.

References

1. Hu, H.; Lu, S.; Liu, R. A New Species of *Isochrysis* (Isochrysidales)-*I. Zhanjiangensis* and Its Observation on the Fine Structure. *Acta Oceanol. Sin.* **2007**, *29*, 111–119.
2. Phatarpekar, P.V.; Sreepada, R.A.; Pednekar, C.; Achuthankutty, C.T. A Comparative Study on Growth Performance and Biochemical Composition of Mixed Culture of *Isochrysis galbana* and *Chaetoceros calcitrans* with Monocultures. *Aquaculture* **2000**, *181*, 141–155. [CrossRef]
3. Cao, J.-Y.; Kong, Z.-Y.; Ye, M.-W.; Ling, T.; Chen, K.; Xu, J.-L.; Zhou, C.-X.; Liao, K.; Zhang, L.; Yan, X.-J. Comprehensive Comparable Study of Metabolomic and Transcriptomic Profiling of *Isochrysis galbana* Exposed to High Temperature, an Important Diet Microalgal Species. *Aquaculture* **2020**, *521*, 735034. [CrossRef]
4. Zhu, C.; Han, D.; Li, Y.; Zhai, X.; Chi, Z.; Zhao, Y.; Cai, H. Cultivation of Aquaculture Feed *Isochrysis zhangjiangensis* in Low-Cost Wave Driven Floating Photobioreactor without Aeration Device. *Bioresour. Technol.* **2019**, *293*, 122018. [CrossRef] [PubMed]
5. Huang, L.; Xu, J.; Zong, C.; Zhu, S.; Ye, M.; Zhou, C.; Chen, H.; Yan, X. Effect of High Temperature on the Lipid Composition of *Isochrysis galbana* Parke in Logarithmic Phase. *Aquac. Int.* **2017**, *25*, 327–339. [CrossRef]
6. Yang, S.; Li, X.; Zang, Z.; Li, J.; Wang, A.; Shi, Y.; Zhang, X.; Gu, Z.; Zheng, X.; Vasquez, H.E. Effect of Fresh and Spray-Dried Microalgal Diets on the Growth, Digestive Enzymatic Activity, and Gut Microbiota of Juvenile Winged Pearl Oyster *Pteria penguin*. *Aquac. Rep.* **2022**, *25*, 101251. [CrossRef]
7. Takada, J.; Murase, N.; Abe, M.; Noda, M.; Suda, Y. Growth and Photosynthesis of *Ulva prolifera* under Different Light Quality from Light Emitting Diodes. *Aquac. Sci.* **2011**, *59*, 101–107.
8. Kwan, P.P.; Banerjee, S.; Shariff, M.; Yusoff, F.M. Influence of Light on Biomass and Lipid Production in Microalgae Cultivation. *Aquac. Res.* **2021**, *52*, 1337–1347. [CrossRef]
9. Li, Y.; Li, R.; Yi, X. Effects of Light Quality on Growth Rates and Pigments of *Chaetoceros gracilis* (Bacillariophyceae). *J. Ocean. Limnol.* **2020**, *38*, 795–801. [CrossRef]
10. Ravelonandro, P.H.; Ratianarivo, D.H.; Joannis-Cassan, C.; Isambert, A.; Raherimandimby, M. Influence of Light Quality and Intensity in the Cultivation of *Spirulina platensis* from Toliara (Madagascar) in a Closed System. *J. Chem. Technol. Biotechnol.* **2008**, *83*, 842–848. [CrossRef]

11. Shu, C.-H.; Tsai, C.-C.; Liao, W.-H.; Chen, K.-Y.; Huang, H.-C. Effects of Light Quality on the Accumulation of Oil in a Mixed Culture of *Chlorella* sp. and *Saccharomyces cerevisiae*. *J. Chem. Technol. Biotechnol.* **2012**, *87*, 601–607. [CrossRef]

12. You, T.; Barnett, S.M. Effect of Light Quality on Production of Extracellular Polysaccharides and Growth Rate of *Porphyridium cruentum*. *Biochem. Eng. J.* **2004**, *19*, 251–258. [CrossRef]

13. Metsoviti, M.N.; Papapolymerou, G.; Karapanagiotidis, I.T.; Katsoulas, N. Effect of Light Intensity and Quality on Growth Rate and Composition of *Chlorella vulgaris*. *Plants* **2019**, *9*, 31. [CrossRef]

14. Pang, N.; Fu, X.; Fernandez, J.S.M.; Chen, S. Multilevel Heuristic LED Regime for Stimulating Lipid and Bioproducts Biosynthesis in *Haematococcus pluvialis* under Mixotrophic Conditions. *Bioresour. Technol.* **2019**, *288*, 121525. [CrossRef] [PubMed]

15. Hallenbeck, P.C.; Grogger, M.; Mraz, M.; Veverka, D. The Use of Design of Experiments and Response Surface Methodology to Optimize Biomass and Lipid Production by the Oleaginous Marine Green Alga, *Nannochloropsis gaditana* in Response to Light Intensity, Inoculum Size and CO_2. *Bioresour. Technol.* **2015**, *184*, 161–168. [CrossRef] [PubMed]

16. Solovchenko, A.E.; Khozin-Goldberg, I.; Didi-Cohen, S.; Cohen, Z.; Merzlyak, M.N. Effects of Light Intensity and Nitrogen Starvation on Growth, Total Fatty Acids and Arachidonic Acid in the Green Microalga *Parietochloris incisa*. *J. Appl. Phycol.* **2008**, *20*, 245–251. [CrossRef]

17. Takeshita, T.; Ota, S.; Yamazaki, T.; Hirata, A.; Zachleder, V.; Kawano, S. Starch and Lipid Accumulation in Eight Strains of Six *Chlorella* Species under Comparatively High Light Intensity and Aeration Culture Conditions. *Bioresour. Technol.* **2014**, *158*, 127–134. [CrossRef]

18. Yeesang, C.; Cheirsilp, B. Effect of Nitrogen, Salt, and Iron Content in the Growth Medium and Light Intensity on Lipid Production by Microalgae Isolated from Freshwater Sources in Thailand. *Bioresour. Technol.* **2011**, *102*, 3034–3040. [CrossRef]

19. Mandotra, S.K.; Kumar, P.; Suseela, M.R.; Nayaka, S.; Ramteke, P.W. Evaluation of Fatty Acid Profile and Biodiesel Properties of Microalga *Scenedesmus abundans* under the Influence of Phosphorus, PH and Light Intensities. *Bioresour. Technol.* **2016**, *201*, 222–229. [CrossRef]

20. Sirisuk, P.; Ra, C.-H.; Jeong, G.-T.; Kim, S.-K. Effects of Wavelength Mixing Ratio and Photoperiod on Microalgal Biomass and Lipid Production in a Two-Phase Culture System Using LED Illumination. *Bioresour. Technol.* **2018**, *253*, 175–181. [CrossRef]

21. Babuskin, S.; Radhakrishnan, K.; Babu, P.A.S.; Sivarajan, M.; Sukumar, M. Effect of Photoperiod, Light Intensity and Carbon Sources on Biomass and Lipid Productivities of *Isochrysis galbana*. *Biotechnol. Lett.* **2014**, *36*, 1653–1660. [CrossRef] [PubMed]

22. Krzemińska, I.; Pawlik-Skowrońska, B.; Trzcińska, M.; Tys, J. Influence of Photoperiods on the Growth Rate and Biomass Productivity of Green Microalgae. *Bioprocess Biosyst. Eng.* **2014**, *37*, 735–741. [CrossRef] [PubMed]

23. Huang, Y.; Ding, W.; Zhou, X.; Jin, W.; Han, W.; Chi, K.; Chen, Y.; Zhao, Z.; He, Z.; Jiang, G. Sub-Pilot Scale Cultivation of *Tetradesmus dimorphus* in Wastewater for Biomass Production and Nutrients Removal: Effects of Photoperiod, CO2 Concentration and Aeration Intensity. *J. Water Process Eng.* **2022**, *49*, 103003. [CrossRef]

24. De Mooij, T.; de Vries, G.; Latsos, C.; Wijffels, R.H.; Janssen, M. Impact of Light Color on Photobioreactor Productivity. *Algal Res.* **2016**, *15*, 32–42. [CrossRef]

25. Rugnini, L.; Rossi, C.; Antonaroli, S.; Rakaj, A.; Bruno, L. The Influence of Light and Nutrient Starvation on Morphology, Biomass and Lipid Content in Seven Strains of Green Microalgae as a Source of Biodiesel. *Microorganisms* **2020**, *8*, 1254. [CrossRef]

26. Jeon, Y.-C.; Cho, C.-W.; Yun, Y.-S. Measurement of Microalgal Photosynthetic Activity Depending on Light Intensity and Quality. *Biochem. Eng. J.* **2005**, *27*, 127–131. [CrossRef]

27. Elisabeth, B.; Rayen, F.; Behnam, T. Microalgae Culture Quality Indicators: A Review. *Crit. Rev. Biotechnol.* **2021**, *41*, 457–473. [CrossRef]

28. Markou, G. Effect of Various Colors of Light-Emitting Diodes (LEDs) on the Biomass Composition of *Arthrospira platensis* Cultivated in Semi-Continuous Mode. *Appl. Biochem. Biotechnol.* **2014**, *172*, 2758–2768. [CrossRef]

29. Atta, M.; Idris, A.; Bukhari, A.; Wahidin, S. Intensity of Blue LED Light: A Potential Stimulus for Biomass and Lipid Content in Fresh Water Microalgae *Chlorella vulgaris*. *Bioresour. Technol.* **2013**, *148*, 373–378. [CrossRef]

30. Hultberg, M.; Jönsson, H.L.; Bergstrand, K.-J.; Carlsson, A.S. Impact of Light Quality on Biomass Production and Fatty Acid Content in the Microalga *Chlorella vulgaris*. *Bioresour. Technol.* **2014**, *159*, 465–467. [CrossRef]

31. Teo, C.L.; Atta, M.; Bukhari, A.; Taisir, M.; Yusuf, A.M.; Idris, A. Enhancing Growth and Lipid Production of Marine Microalgae for Biodiesel Production via the Use of Different LED Wavelengths. *Bioresour. Technol.* **2014**, *162*, 38–44. [CrossRef] [PubMed]

32. Ra, C.-H.; Kang, C.-H.; Jung, J.-H.; Jeong, G.-T.; Kim, S.-K. Effects of Light-Emitting Diodes (LEDs) on the Accumulation of Lipid Content Using a Two-Phase Culture Process with Three Microalgae. *Bioresour. Technol.* **2016**, *212*, 254–261. [CrossRef] [PubMed]

33. McGee, D.; Archer, L.; Fleming, G.T.A.; Gillespie, E.; Touzet, N. Influence of Spectral Intensity and Quality of Led Lighting on Photoacclimation, Carbon Allocation and High-Value Pigments in Microalgae. *Photosynth. Res.* **2020**, *143*, 67–80. [CrossRef] [PubMed]

34. Ho, S.-H.; Chan, M.-C.; Liu, C.-C.; Chen, C.-Y.; Lee, W.-L.; Lee, D.-J.; Chang, J.-S. Enhancing Lutein Productivity of an Indigenous Microalga *Scenedesmus obliquus* Fsp-3 Using Light-Related Strategies. *Bioresour. Technol.* **2014**, *152*, 275–282. [CrossRef] [PubMed]

35. Bialevich, V.; Zachleder, V.; Bišová, K. The Effect of Variable Light Source and Light Intensity on the Growth of Three Algal Species. *Cells* **2022**, *11*, 1293. [CrossRef] [PubMed]

36. Li, S.F.; Fanesi, A.; Martin, T.; Lopes, F. Biomass Production and Physiology of *Chlorella vulgaris* during the Early Stages of Immobilized State Are Affected by Light Intensity and Inoculum Cell Density. *Algal Res.* **2021**, *59*, 102453. [CrossRef]

37. He, Q.; Yang, H.; Wu, L.; Hu, C. Effect of Light Intensity on Physiological Changes, Carbon Allocation and Neutral Lipid Accumulation in Oleaginous Microalgae. *Bioresour. Technol.* **2015**, *191*, 219–228. [CrossRef]
38. Beale, S.I.; Appleman, D. Chlorophyll Synthesis in *Chlorella*: Regulation by Degree of Light Limitation of Growth. *Plant Physiol.* **1971**, *47*, 230–235. [CrossRef]
39. Da Silva Ferreira, V.; Sant'Anna, C. Impact of Culture Conditions on the Chlorophyll Content of Microalgae for Biotechnological Applications. *World J. Microbiol. Biotechnol.* **2017**, *33*, 20. [CrossRef]
40. Yoshioka, M.; Yago, T.; Yoshie-Stark, Y.; Arakawa, H.; Morinaga, T. Effect of High Frequency of Intermittent Light on the Growth and Fatty Acid Profile of *Isochrysis galbana*. *Aquaculture* **2012**, *338–341*, 111–117. [CrossRef]
41. Che, C.A.; Kim, S.H.; Hong, H.J.; Kityo, M.K.; Sunwoo, I.Y.; Jeong, G.-T.; Kim, S.-K. Optimization of Light Intensity and Photoperiod for *Isochrysis galbana* Culture to Improve the Biomass and Lipid Production Using 14-L Photobioreactors with Mixed Light Emitting Diodes (LEDs) Wavelength under Two-Phase Culture System. *Bioresour. Technol.* **2019**, *285*, 121323. [CrossRef] [PubMed]
42. Marchetti, J.; Bougaran, G.; Jauffrais, T.; Lefebvre, S.; Rouxel, C.; Saint-Jean, B.; Lukomska, E.; Robert, R.; Cadoret, J.P. Effects of Blue Light on the Biochemical Composition and Photosynthetic Activity of *Isochrysis sp.* (T-Iso). *J. Appl. Phycol.* **2013**, *25*, 109–119. [CrossRef]
43. Fernandes, B.D.; Dragone, G.M.; Teixeira, J.A.; Vicente, A.A. Light Regime Characterization in an Airlift Photobioreactor for Production of Microalgae with High Starch Content. *Appl. Biochem. Biotechnol.* **2010**, *161*, 218–226. [CrossRef]
44. Sukumaran, P.; Nulit, R.; Halimoon, N.; Simoh, S.; Omar, H.; Ismail, A. Formulation of Cost-Effective Medium Using Urea as a Nitrogen Source for *Arthrospira platensis* Cultivation under Real Environment. *ARRB* **2018**, *22*, 1–12. [CrossRef]
45. Jeffrey, S.W.; Humphrey, G.F. New Spectrophotometric Equations for Determining Chlorophylls a, b, C1 and C2 in Higher Plants, Algae and Natural Phytoplankton. *Biochem. Physiol. Pflanz.* **1975**, *167*, 191–194. [CrossRef]
46. Mohsenpour, S.F.; Willoughby, N. Luminescent Photobioreactor Design for Improved Algal Growth and Photosynthetic Pigment Production through Spectral Conversion of Light. *Bioresour. Technol.* **2013**, *142*, 147–153. [CrossRef]
47. Lowry, O.H. Protein Measurement with the Folin Phenol Reagent. *J. biol. Chem.* **1951**, *193*, 265–275. [CrossRef]
48. Somani, B.L.; Khanade, J.; Sinha, R. A Modified Anthrone-Sulfuric Acid Method for the Determination of Fructose in the Presence of Certain Proteins. *Anal. Biochem.* **1987**, *167*, 327–330. [CrossRef]
49. Koc, C.; Anderson, G.A.; Kommareddy, A. Use of Red and Blue Light-Emitting Diodes (LED) and Fluorescent Lamps to Grow Microalgae in a Photobioreactor. *Isr. J. Aquac.-Bamidgeh* **2013**, *65*, 20661. [CrossRef]
50. Oldenhof, H.; Zachleder, V.; Ende, H. Blue Light Delays Commitment to Cell Division in *Chlamydomonas Reinhardtii*. *Plant Biology.* **2004**, *6*, 689–695. [CrossRef]
51. Izadpanah, M.; Gheshlaghi, R.; Mahdavi, M.A.; Elkamel, A. Effect of Light Spectrum on Isolation of Microalgae from Urban Wastewater and Growth Characteristics of Subsequent Cultivation of the Isolated Species. *Algal Res.* **2018**, *29*, 154–158. [CrossRef]
52. Gómez-Loredo, A.; Benavides, J.; Rito-Palomares, M. Growth Kinetics and Fucoxanthin Production of *Phaeodactylum tricornutum* and *Isochrysis galbana* Cultures at Different Light and Agitation Conditions. *J. Appl. Phycol.* **2016**, *28*, 849–860. [CrossRef]
53. Li, Y.; Liu, J. Analysis of Light Absorption and Photosynthetic Activity by *Isochrysis galbana* under Different Light Qualities. *Aquac. Res.* **2020**, *51*, 2893–2902. [CrossRef]
54. Vadiveloo, A.; Moheimani, N.R.; Cosgrove, J.J.; Bahri, P.A.; Parlevliet, D. Effect of Different Light Spectra on the Growth and Productivity of Acclimated *Nannochloropsis* sp. (Eustigmatophyceae). *Algal Res.* **2015**, *8*, 121–127. [CrossRef]
55. Carvalho, A.P.; Silva, S.O.; Baptista, J.M.; Malcata, F.X. Light Requirements in Microalgal Photobioreactors: An Overview of Biophotonic Aspects. *Appl. Microbiol. Biotechnol.* **2011**, *89*, 1275–1288. [CrossRef]
56. Kommareddy, A.; Anderson, G. Study of Light as a Parameter in the Growth of Algae in a Photo-Bio Reactor (PBR). In Proceedings of the 2003 ASAE Annual Meeting, Las Vegas, NV, USA, 27–30 July 2003. [CrossRef]
57. Ueno, Y.; Aikawa, S.; Kondo, A.; Akimoto, S. Adaptation of Light-Harvesting Functions of Unicellular Green Algae to Different Light Qualities. *Photosynth. Res.* **2019**, *139*, 145–154. [CrossRef]
58. Masojídek, J.; Torzillo, G.; Koblížek, M. *Handbook of Microalgal Culture: Applied Phycology and Biotechnolog*, 2nd ed.; John Wiley & Sons, Ltd.: Hoboken, NJ, USA, 2004; ISBN 9781118567166.
59. Wang, C.-Y.; Fu, C.-C.; Liu, Y.-C. Effects of Using Light-Emitting Diodes on the Cultivation of *Spirulina platensis*. *Biochem. Eng. J.* **2007**, *37*, 21–25. [CrossRef]
60. Das, P.; Lei, W.; Aziz, S.S.; Obbard, J.P. Enhanced Algae Growth in Both Phototrophic and Mixotrophic Culture under Blue Light. *Bioresour. Technol.* **2011**, *102*, 3883–3887. [CrossRef] [PubMed]
61. Holzinger, A.; Lütz, C. Algae and UV Irradiation: Effects on Ultrastructure and Related Metabolic Functions. *Micron* **2006**, *37*, 190–207. [CrossRef]
62. Kuwano, K.; Abe, N.; Nishi, Y.; Seno, H.; Nishihara, G.N.; Iima, M.; Zachleder, V. Growth and Cell Cycle of *Ulva Compressa* (Ulvophyceae) under LED Illumination. *J. Phycol.* **2014**, *50*, 744–752. [CrossRef]
63. Tsekos, I.; Niell, F.X.; Aguilera, J.; López-Figueroa, F.; Delivopoulos, S.G. Ultrastructure of the Vegetative Gametophytic Cells of Porphyra Leucosticta (Rhodophyta) Grown in Red, Blue and Green Light. *Phycol. Res.* **2002**, *50*, 251–264. [CrossRef]
64. Gorai, T.; Katayama, T.; Obata, M.; Murata, A.; Taguchi, S. Low Blue Light Enhances Growth Rate, Light Absorption, and Photosynthetic Characteristics of Four Marine Phytoplankton Species. *J. Exp. Mar. Biol. Ecol.* **2014**, *459*, 87–95. [CrossRef]

65. Danesi, E.D.G.; Rangel-Yagui, C.O.; Carvalho, J.C.M.; Sato, S. Effect of Reducing the Light Intensity on the Growth and Production of Chlorophyll by *Spirulina platensis*. *Biomass Bioenergy* **2004**, *26*, 329–335. [CrossRef]
66. George, B.; Pancha, I.; Desai, C.; Chokshi, K.; Paliwal, C.; Ghosh, T.; Mishra, S. Effects of Different Media Composition, Light Intensity and Photoperiod on Morphology and Physiology of Freshwater Microalgae *Ankistrodesmus falcatus*-A Potential Strain for Bio-Fuel Production. *Bioresour. Technol.* **2014**, *171*, 367–374. [CrossRef] [PubMed]
67. Fu, W.; Guðmundsson, Ó.; Paglia, G.; Herjólfsson, G.; Andrésson, Ó.S.; Palsson, B.Ø.; Brynjólfsson, S. Enhancement of Carotenoid Biosynthesis in the Green Microalga *Dunaliella salina* with Light-Emitting Diodes and Adaptive Laboratory Evolution. *Appl. Microbiol. Biotechnol.* **2013**, *97*, 2395–2403. [CrossRef]
68. Gatamaneni Loganathan, B.; Orsat, V.; Lefsrud, M.; Wu, B.S. A Comprehensive Study on the Effect of Light Quality Imparted by Light-Emitting Diodes (LEDs) on the Physiological and Biochemical Properties of the Microalgal Consortia of *Chlorella variabilis* and *Scenedesmus obliquus* Cultivated in Dairy Wastewater. *Bioprocess Biosyst. Eng.* **2020**, *43*, 1445–1455. [CrossRef]
69. Del Pilar Sánchez-Saavedra, M.; Maeda-Martínez, A.N.; Acosta-Galindo, S. Effect of Different Light Spectra on the Growth and Biochemical Composition of *Tisochrysis lutea*. *J. Appl. Phycol.* **2016**, *28*, 839–847. [CrossRef]
70. Rivkin, R. Influence of Irradiance and Spectral Quality on the Carbon Metabolism of Phytoplankton I. Photosynthesis, Chemical Composition and Growth. *Mar. Ecol. Prog. Ser.* **1989**, *55*, 291–304. [CrossRef]
71. Brinkmann, G.; Senger, H. The Development of Structure and Function in Chloroplasts of Greening Mutants of *Scenedesmus* IV. Blue Light-Dependent Carbohydrate and Protein Metabolism. *Plant Cell Physiol.* **1978**, *19*, 1427–1437. [CrossRef]
72. Hammer, A.; Schumann, R.; Schubert, H. Light and Temperature Acclimation of *Rhodomonas salina* (Cryptophyceae): Photosynthetic Performance. *Aquat. Microb. Ecol.* **2002**, *29*, 287–296. [CrossRef]

Journal of
Marine Science and Engineering

Article

Dietary Curcumin Supplementation Enhanced Ammonia Nitrogen Stress Tolerance in Greater Amberjack (*Seriola dumerili*): Growth, Serum Biochemistry and Expression of Stress-Related Genes

Jiawei Hong [1,2,3], Zhengyi Fu [1,2], Jing Hu [1,2], Shengjie Zhou [1,2], Gang Yu [1,2] and Zhenhua Ma [1,2,*]

[1] Key Laboratory of Efficient Utilization and Processing of Marine Fishery Resources of Hainan Province, Sanya Tropical Fisheries Research Institute, Sanya 572018, China
[2] Tropical Aquaculture Research and Development Center, South China Sea Fisheries Research Institute, Chinese Academy of Fishery Science, Sanya 572018, China
[3] State Key Laboratory of Marine Environmental Science, College of Ocean and Earth Sciences, Xiamen University, Xiamen 361005, China
* Correspondence: zhenhua.ma@hotmail.com

Abstract: This study was conducted to determine whether curcumin has a positive effect in greater amberjack (*Seriola dumerili*), especially the ammonia nitrogen stress tolerance ability. The results showed that the stress recovery process of digestive enzymes amylase and trypsin, as well as absorptive enzymes Na^+/K^+-ATPase, γ-GT and CK, was accelerated. Lysozyme activity increased in the fish fortified with both curcumin diets. Aspartate aminotransferase activity restriction was activated at a low curcumin level. However, alanine aminotransferase activity restriction happened only at 0.02% dietary curcumin. Facilitation of lipid metabolism by curcumin was very clear, as triglyceride and total cholesterol content was basically maintained at the original level or even showed a slight decrease after recovery. HSP70 and HSP90 genes were not evidently stimulated to express in liver, kidney and spleen tissues. In addition, curcumin showed its inhibition capacity on IL1β and IFN-γ and a promoting effect on TGF-β1. The expression of NF-κB1 decreased in a higher degree in fish fed with 0.02% dietary curcumin, while 0.01% dietary curcumin accelerated the recovery pace of C3 and lgT after stress. This study showed that dietary curcumin supplementation can enhance ammonia nitrogen stress tolerance in greater amberjack, and its application prospect can be confirmed.

Keywords: *Seriola dumerili*; curcumin; ammonia nitrogen tolerance

Citation: Hong, J.; Fu, Z.; Hu, J.; Zhou, S.; Yu, G.; Ma, Z. Dietary Curcumin Supplementation Enhanced Ammonia Nitrogen Stress Tolerance in Greater Amberjack (*Seriola dumerili*): Growth, Serum Biochemistry and Expression of Stress-Related Genes. *J. Mar. Sci. Eng.* **2022**, *10*, 1796. https://doi.org/10.3390/jmse10111796

Academic Editor: Azizur Rahman

Received: 27 September 2022
Accepted: 16 November 2022
Published: 21 November 2022

1. Introduction

Aquaculture has developed so far that intensification is thought to be a great solution to increase production and revenue. However, it comes at a cost, such as water pollution and health issues [1–5]. Furthermore, the appearance of antibiotic resistance caused by drug abuse has brought aquaculture into an adverse cycle [6]. Ultimately, once all these negative factors gradually accumulate to a certain extent, they can restrict the development of aquaculture production [7]. Greater amberjack (*Seriola dumerili*) is greatly favored in many countries and regions, and it is considered a significant species for promoting aquaculture diversification around the world [8,9]. Greater amberjack has a high commercial value due to its high quality of meat and large body size [10–13]. In addition, it also has excellent production characteristics, such as high feed utilization, a rapid growth rate and low mortality in any kind of culture method [8,14–16]. As a consequence, its culture has been growing at a remarkable speed in recent years. However, as an important part of cultured fish, greater amberjack is also facing the same dilemma of the threat of water quality deterioration caused by high-density aquaculture as other species as a consequence of efforts to increase yield.

In addition to the efforts in the breeding of disease- and stress-resistant varieties as well as culture equipment and methods, the exploration of immune boosters in diet is also a key breakthrough in strengthening fish health; it is also another way to increase aquaculture production. In recent years, much research involving the great functions of vegetative dietary supplementation has emerged, especially into Chinese herbs [17]. Many essential effective ingredients of Chinese herbs have been screened and identified and have proved to be sufficient to become excellent promoters of immune functions in many species. Meanwhile, the sources of these herb extracts are natural and easily available [18–20]. These advantages further increase their research potential and value. Therefore, our study focused on vegetative immune boosters based on their previous performance reported in much research. For example, a dietary addition of a mixture of Astragalus and Lonicera extracts to *Nile tilapia* exhibited enhanced phagocytosis, as well as a rise in respiratory burst and lysozyme activity [21]. Similarly, black rockfish treated with green tea ethanol extract in their diet also witnessed an increase in lysozyme activity; moreover, stress recovery capacity was strengthened at the same time [22]. Ref. [23] found that an *Azadirachta indica* extract showed an effect equal to that of antibiotics on treating *Citrobacter freundii* infection in *Oreochromis mossambicus*. As to our study, the supplementation we tested is curcumin.

Curcumin is a polyphenol derived from the roots of *Curcuma longa* L. or other *Curcuma* spp. plants [24,25], which has a long history of use by humans [26]. It is well-known not only for the functions of colouring and flavouring [27], but also for its extensive health care effects, such as antioxidant actions, anti-inflammation, inhibition of carcinogenesis, oxygen radical scavenging, liver protection, etc. [28–32]. In view of its numerous positive effects, curcumin was soon studied and applied in mammals, and it has also attracted great attention in aquaculture in recent years. Curcumin and its active molecules have been proved to have effects on fish health similar to those on human and other mammals [33–35]. For example, Ji et al., found that 0.04% curcumin supplementation can reduce hepatic lipid deposition, improve antioxidant activity and increase PUFA of large yellow croaker; and finally, to eliminate the side effects caused by a high-fat diet [36]. Curcumin supplements in carp diet can increase the activity of antioxidant enzymes by promoting the release of Nrf2 and ultimately protect carp from liver damage caused by CCl4 [37]. Hemat et al., also found that dietary curcumin showed great performance in improving growth, feed utilization, oxidative status, immune responses, and disease resistance in *Nile tilapia* [38].

Based on all the known information, the application prospect of dietary curcumin supplementation in greater amberjack is highly anticipated. Thus, whether curcumin can show great performance in improving the health characteristic of greater amberjack is well worth investigating.

2. Materials and Methods

2.1. Experimental Fish and Feeding Experiment

The greater amberjack used in the study were bred by the Tropical Fisheries Research and Development Center, South China Sea Fisheries Research and Development Center, South China Sea Fisheries Research Institute, Chinese Academy of Fishery Science (Lingshui, Hainan, China). After being cultivated temporarily for one week, 135 individual fish appearing to be healthy and energetic and showing no injuries were selected and randomly divided into three experimental groups. Every treatment was in triplicates, and each triplicate had 15 fish. Initial mean weights of fish in each group were 149.02 ± 12.59 g, 153.51 ± 4.09 g and 151.78 ± 3.84 g, respectively. The experiment was conducted in a circulating mariculture tank (5000 L). Water temperature, salinity, and the pH of the aquaculture seawater during the period were 27–31.5 °C, 35‰ and 7.5–8, respectively. Ammonia nitrogen content was maintained below 0.01 mg/L, nitrite content below 0.02 mg/L, and dissolved oxygen was always above 7.0 mg/L. The fish were fed to apparent satiety twice a day (at 8:00 am and 16:00 pm). Feces were siphoned out to avoid water quality deterioration at 1 h after feeding. The feeding experiment lasted eight weeks.

2.2. Experimental Diet

The addition amounts of curcumin in the three experimental diets were 0, 100 and 200 mg/kg, respectively. All diets were designed and produced by the Tropical Fisheries Research and Development Center, South China Sea Fisheries Research and Development Center, South China Sea Fisheries Research Institute, Chinese Academy of Fishery Science. The added curcumin (purity > 95%) was provided by Xi'an Feida biotechnology Co., Ltd. (Xi'an, China). The formulation of experimental diet is shown in Table 1.

Table 1. Formulation and proximate composition of the experimental diet.

Ingredients	Proportion (g/kg)		
	Control	0.01% CUR [1]	0.02% CUR [1]
Fish meal	590	590	590
Corn gluten meal	70	70	70
Soybean meal	80	80	80
Corn starch	80	80	80
Microcrystalline Cellulose	50	50	50
Fish oil	70	70	70
Phospholipids	10	10	10
[2] Vitamin premix	5	5	5
[3] Mineral premix	5	5	5
Choline chloride	5	5	5
Betaine	5	5	5
Carboxymethyl cellulose	30	30	30
Curcumin	0	0.1	0.2
Analyzed nutrients			
Dry Matter	878.98	878.98	878.98
Crude protein	497	497	497
Crude lipids	126.98	126.98	126.98
Crude ash	106.85	106.85	106.85

[1] 0.01% CUR and 0.02% CUR indicate the treatment groups in which the proportion of curcumin supplementation in the diet are 0.01% and 0.02%. [2] Vitamin premix (mg/kg diet): vitamin A—9,000,000 (IU/kg diet), vitamin K3—600 (IU/kg diet), vitamin D—2,500,000 (IU/kg diet), vitamin E—500 (IU/kg diet), vitamin B—(B1 3200 (IU/kg diet), B2—10,900 (IU/kg diet), B5—20,000 (IU/kg diet), B6—5000 (IU/kg diet), B12—1160 (IU/kg diet), vitamin C—50,000 (IU/kg diet), phaseomannite—1500 mg/kg diet, calcium pantothenate—200 mg/kg diet, niacin—400 mg/kg diet, folic acid—50 mg/kg diet, biotin—2 mg/kg diet. [3] Mineral premix (mg/kg diet): KCl—70; KI—1.5; $MgSO·7H_2O$—300; $MnSO_4·4H_2O$—3; $CuCl_2$—5; $ZnSO_4·7H_2O$—14; $CoCl_2·6H_2O$—0.5; $FeSO_4·7H_2O$—15; $CaCl_2$—2.8 (g/kg diet); $KH_2 PO_4·H_2O$—4.5 (g/kg diet). The dietary energy was calculated as carbohydrate: 17.15 MJ/kg, protein: 23.64 MJ/kg, lipid: 39.54 MJ/kg.

2.3. Ammonia Nitrogen Challenge Assay

After being fed with experimental diets for eight weeks, a challenge assay was conducted on the fish. Ammonium chloride (NH_4Cl) was added into the aquaculture water at the concentration of 1 g/L. At six minutes, all the fish showed either manic agitation or lack of apparent consciousness; based on our pre-experiment, two fish from each tank were selected. One was sampled as the challenge assay sample, and the other was placed in water with no ammonia poison to recover. It was sampled after recovery from lack of apparent consciousness. Recovery time was about 5 min.

2.4. Sampling and Analysis

The body length and weight of the fish were measured at the beginning and the end of the feeding experiment. Then the specific growth rate, weight gain percentage, feed coefficient and condition factor were calculated. The calculation formulas are as follows:

Weight gain percentage (%) = (final average weight − initial average weight)/initial average weight × 100

Specific growth rate (%/d) = (Ln final average weight − Ln initial average weight)/breeding test days × 100

Condition factor = body weight/body length^3 × 100

Feed coefficient = feed weight/fish weight gain

In the formula, the unit of body length is cm, the unit of final average weight and initial average weight is g, and the unit of breeding test days is d.

At the two sampling times, fish blood was taken from the tail vein with a 1-mL syringe and was left to clot at room temperature for 2 h, then separated by centrifugation (3000× g, 10 min). Serum was collected and kept at −80 °C until analysis. Serum was used to detect enzyme activity. Total cholesterol (TC) was measured by the CHOD-PAP enzymatic-colorimetric method. Triglycerides (TG) were determined enzymatically by the Triglyceride method [39]. Albumin (ALB) was analyzed according to the method described by Doumas et al., (1971) [40]. Aspartate aminotransferase (AST/GOT) and alanine aminotransferase (ALT/GPT) were evaluated using a colorimetric method according to Reitman and Frankel (1957) [41]. Lysozyme (LZM) activity was determined using a turbidimetric assay as in Ref. [42].

In addition, the liver, head kidney, intestine, gill and spleen of fish were sampled and put into centrifuge tubes separately, followed by immediate storage at −80 °C until analyzed. A nine-fold volume of 0.86% saline solution equal to weight was added in tissue samples. The tissue was then homogenized using the handheld homogenizer (Gloucester, Prima PB100, England) on ice. The homogenate was centrifuged for 20 min at 2500× g, and the supernatant was used for the further analysis. Amylase, Lipase and Trypsin were analyzed according to the method described by Reimer (1982) [43]. Na$^+$K$^+$ adenosinetriphosphatase (Na$^+$/K$^+$–ATPase) activity determination was performed according to Ay et al., (1999) [44]. Creatine kinase (CK) activity was determined using the method of Weng et al., (2002) [45]. γ-glutamyl transpeptidase (γ-GT) activities were measured using the method of Rosalki et al., (1970) [46].

All assays were determined using commercial assay kits (Nanjing Jiancheng Bioengineering Research Institute, Nanjing, China) according to the manufacturer's instruction.

2.5. Gene Expression

All fish tissue samples were taken to extract total RNA, according to the manufacturer's protocol for Trizol reagent (Invitrogen, Thermo Fisher Scientific Co., Ltd., Shanghai, China). A micro ultraviolet spectrophotometer (ND5000, Bioteke Corporation, Beijing, China) and agarose gel electrophoresis were used to test the concentration and purity of RNA. RNA was considered valid when OD260/280 (Optical density) was in the range of 1.9~2.1. 1 μg total RNA, drawn off for the reverse transcription process following the manufacturer's instruction for One-Step gDNA Removal and cDNA Synthesis SuperMix (EasyScript, Beijing TransGen Biotech Co., Ltd., Beijing, China).

Quantitative real-time PCR analysis was run by Real-Time PCR System (Q1000, Hangzhou LongGene Scientific Instruments Co., Ltd., Hangzhou, China). The thermal cycling program was performed as follows: initial denaturation at 95 °C for 15 min, 40 cycles of denaturation at 95 °C for 10 s, annealing at 60 °C for 20 s, extension at 72 °C for 30 s. All PCR amplifications were performed in triplicate on 20 μL of reaction mixture containing 10 μL of 2× RealUniversal PreMix, 0.6 μL (10 μmol/L) of each primer, 2 μL of template cDNA, and 6.8 μL of RNase-free ddH2O. EF1 α and β-actin were used as the reference genes. A dissociation analysis was performed to determine the absence of nonspecific products at the end of each PCR reaction. A single peak was seen on the melt curve analysis, which indicated a single PCR product. A standard curve was established for 10-fold serial dilutions of cDNA for each primer pair. PCR reaction efficiency reached the range of 90–110%. The information on primers used for the analysis is provided in Table 2. The relative expression of HSP70 and HSP90 were determined in liver, head kidney, gill and spleen tissues. The relative expression of NF-κB1, TNF-α, IL-1β, IL-8, TGF-β1, C3, C4, IgT, Hepc, IFN-γ, and Mx were determined in intestine tissue.

Table 2. Primer sequence table.

Gene Abbreviation	Primer Sequence (5′-3′)	Amplicon Size (bp)	Accession Number
NF-κB1	F:CACAGACAGTTCGCCATCG	185	XM_022761336.1
	R:AGCGTCTTCTGCCTCTTCC		
[1] TNF-α	F:GAAAACGCTTCATGCCTCTC	212	XM_022746377.1
	R:GTTGGTTTCCGTCCACAGTT		
[1] IL-1β	F:TGATGGAGAACATGGTGGAA	205	XM_022753745.1
	R:GTCGACATGGTCAGATGCAC		
[1] IL-8	F:GAAGCCTGGGAGTAGAGCTG	164	XM_022758559.1
	R:GGGGTCTAGGCAGACCTCTT		
TGF-β1	F:CGGAGCTGCGGATGTTAA	111	XM_022738547.1
	R:TGGTGATGAAGCGGGAAG		
C3	F:CATCGTTCCGCATCATAGC	81	XM_022755728
	R:AGTCCTTGACATCCACCCA		XM_022755434
C4	F:ACATCGCAATGGAGGAGAAC	170	XM_022768450.01
	R:CAGTCCCGTGATAGGCTTTA		
[2] IgT	F:TGGACCAGTCGCCATCTGAG	196	XM_022756471.1
	R:GGGAAACGGCTTTGAAAGGA		
[2] Hepc	F:GATGATGCCGAATCCCGTCAGG	99	XM_022764299.1
	R:CAGAAACCGCAGCCCTTGTTGGC		
IFN-γ	F:TCTGTCTGACCCTCTGGTTTTC	136	LC146385.1
	R:AAGATGGGCTTCCCGCTA		
Mx	F:GACTTGGCTCTACCTGCTATCG	177	XM_022744797.1
	R:GCTTATCTTTCCGTACCACTCC		
HSP70	F:CACGTATTCTTGCGTTGGG	146	XM_022741879.1
	R:TCATGGCGACCTGGTTCT		
HSP90	F:GGCTACATGGCCGCTAAA	187	XM_022764360.01
	R:TGCGATTGGAGTGGGTCTG		
[3] EF1 α	F:ATCGTTGCCGCTGGTGTT	134	XM_022744048.1
	R:TCGGTGGAGTCCATCTTGTT		
[1,3] β-actin	F:TCTGGTGGGGCAATGATCTTGATCTT	212	XM_022757055.1
	R:CCTTCCTTCCTCGGTATGGAGTCC		

[1] Primers of TNF-α, IL-1β, IL-8 and β-Actin were based on the study of [47]. [2] Primers of IgT and Hep were based on the study of [48]. [3] EF1-α and β-actin were used as reference genes.

2.6. Statistical Analyses

The relative quantity of gene expression was calculated according to the $2^{-\Delta\Delta Ct}$ method [49]. The results were analyzed using the SPSS 19.0 statistical software packages. All data are expressed as the standard deviation (mean $\pm$ SD). Comparisons between different groups were conducted by one-way ANOVA and Tukey's test. Significant difference was considered at $p < 0.05$.

3. Results

Growth performance indicators (specific growth rate, weight gain percentage, feed coefficient and condition factor) after the feeding experiment are shown in Table 3. The condition factor of *Seriola dumerili* in the 0.01% treatment group was significantly higher than in the other two groups ($p < 0.05$), while there was no significant difference between the control group and the 0.02% treatment group ($p > 0.05$). The feed coefficient of *Seriola dumerili* in the control group was significantly higher than that in all treatment groups ($p < 0.05$), and there was no significant difference between treatment groups ($p > 0.05$). Moreover, different levels of dietary curcumin supplementation had no significant effects on weight gain percentage or specific growth rate of *Seriola dumerili* ($p > 0.05$).

Table 3. Growth performance of *Seriola dumerili* at different levels of curcumin in feed.

Growth Performance	Group		
	Control Group	0.01% Treatment Group	0.02% Treatment Group
Initial weight (g)	149.02 ± 12.59	153.51 ± 4.09	151.78 ± 3.85
Final weight (g)	263.59 ± 83.29	327.38 ± 48.38	306.55 ± 23.95
Weight gain percentage (%)	73.89 ± 53.73	116.44 ± 34.07	102.30 ± 12.83
Specific growth rate (%/d)	1.26 ± 0.09	1.34 ± 0.25	1.25 ± 0.12
Feed coefficient	3.45 ± 0.33 [a]	2.21 ± 0.21 [b]	2.30 ± 0.31 [b]
Condition factor	2.36 ± 0.05 [b]	2.48 ± 0.06 [a]	2.27 ± 0.05 [b]

Data are presented as mean ± SD. In the same row, the same lowercase letters on the right side of the data indicated no significant difference ($p > 0.05$). Different lowercase letters indicated a significant difference ($p < 0.05$).

Digestive enzyme activities in the intestine of greater amberjack are shown in Figure 1. Only the 0.02% CUR group exhibited a significant difference with the control as to AMS activity ($p < 0.05$). Regarding LPS activity, two treatment groups had a similar difference with the control. There was no remarkable difference among all the groups regarding TPS activity ($p > 0.05$).

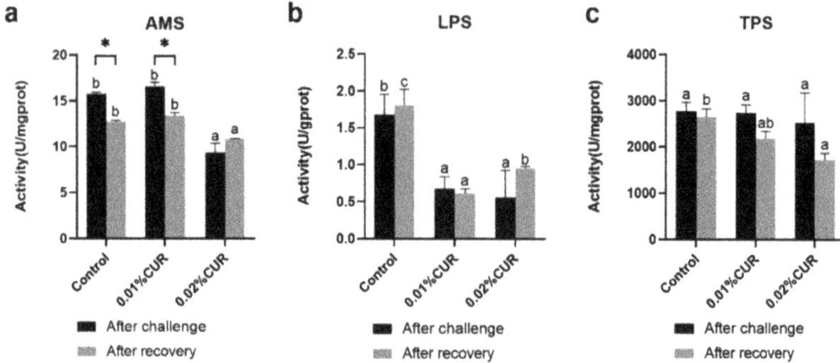

Figure 1. (**a**) Amylase (AMS), (**b**) Lipase (LPS), and (**c**) Trypsin (TPS) activities of greater amberjack fed with dietary curcumin supplementation. (Different letters above bars and asterisk sign (*) indicate significant differences at the 0.05 level).

The activity of Na^+/K^+–ATPase, γ-GT, and CK in intestinal tissue was also determined. The results are described in Figure 2. Na^+/K^+–ATPase activity in 0.02% CUR group is apparently lower than in the other two groups after the ammonia nitrogen challenge. An evident lower level was found in γ-GT activity in the 0.01% CUR group. The CK activity of both treatment groups was significantly higher than in the control.

LZM and ALB (Figure 3), AST and ALT (Figure 4) activity, as well as TG and TC (Figure 5) content of serum were determined after the ammonia nitrogen challenge. According to Figure 3, LZM activity of both two treatments groups was higher than that of the control, and there was no significant difference between them. In addition, ALB content of serum showed no difference among groups. The variations in AST and ALT activity were quite different. In particular, no sharp distinction was seen in AST activity among all groups. With respect to ALT activity, the 0.02% CUR group compared to other groups was lower after the ammonia nitrogen challenge. In addition, different change characteristics were also revealed in the TG and TC content of serum. The TG content of all treatment groups was obviously higher than that in the control. By contrast, the TC contents of both treatment groups were lower.

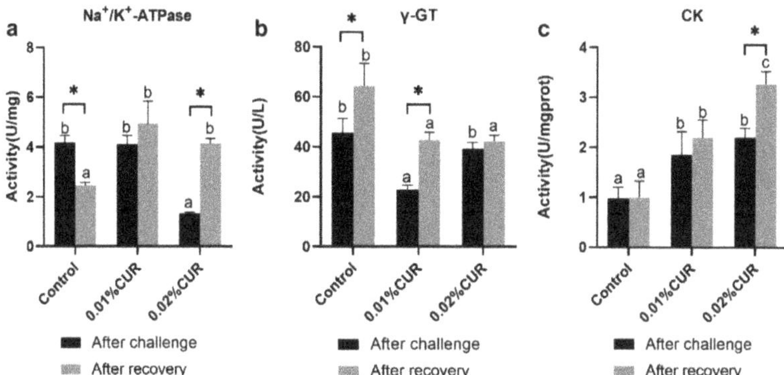

Figure 2. (**a**) Na$^+$K$^+$ adenosinetriphosphatase (Na$^+$K$^+$–ATPase), (**b**) γ-Glutamyl transferase (γ-GT), (**c**) creatine kinase (CK) activities of greater amberjack fed with dietary curcumin supplementation. (Different letters above bars and asterisk sign (*) indicate significant differences at the 0.05 level).

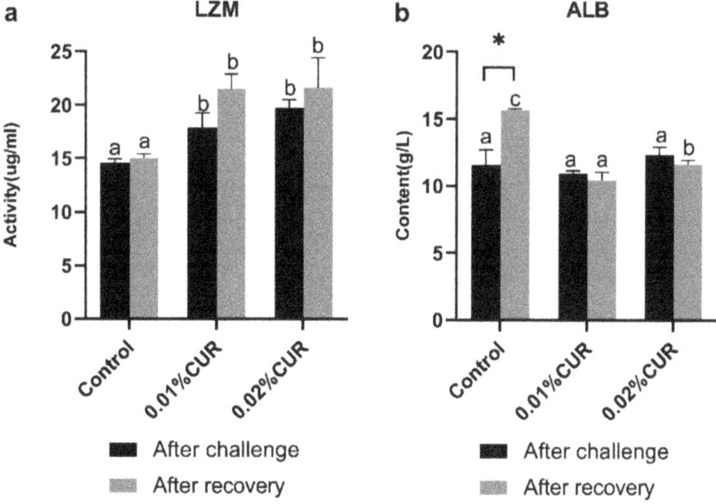

Figure 3. (**a**) Lysozyme (LZM) activity and (**b**) albumin (ALB) content of greater amberjack fed with dietary curcumin supplementation. (Different letters above bars and asterisk sign (*) indicate significant differences at the 0.05 level).

After recovery from apparent lack of consciousness, fish were sampled and the same in vivo indexes were determined as in the ammonia nitrogen challenge experiment. The activity of three digestive enzymes all showed a significant decline phenomenon compared with the control. The declines in AMS and TPS activity were seen in the 0.02% CUR group and the decline in LPS activity was found in both two curcumin-added groups. The levels of Na$^+$/K$^+$–ATPase and CK activity of the intestine in the two treatment groups after recovery were significantly higher than in the control. γ-GT activity in the two treatment groups was obviously lower than that of the control. Moreover, in serum, LZM activity in the two treatment groups was still maintained at a higher level than in the control. The ALB content of the control reached the highest level after recovery. Figure 4 provides the results of the AST and ALT activity. The lowest AST activity level appeared in the 0.01% CUR group, while in contrast the ALT activity in the 0.01% CUR group was the highest. In addition, the TG in 0.01% CUR and 0.02% CUR groups were both significantly lower than in the control according to Figure 5. Regarding the TC content, the same significant difference was seen only in 0.01% CUR.

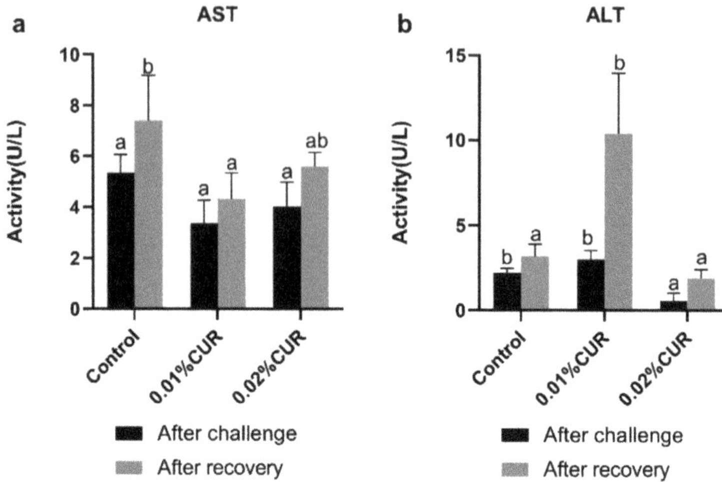

Figure 4. (a) Aspartate aminotransferase (AST) and (b) alanine aminotransferase (ALT) activities of greater amberjack fed with dietary curcumin supplementation. (Different letters above bars indicate significant differences at the 0.05 level).

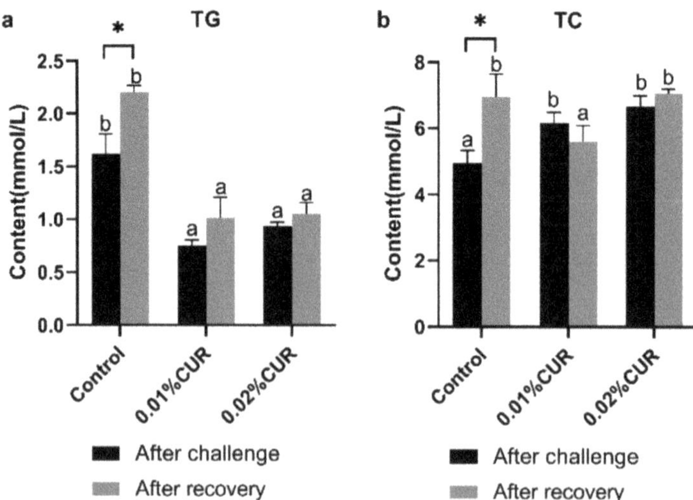

Figure 5. (a) Triglyceride (TG) and (b) total cholesterol (TC) content of greater amberjack fed with dietary curcumin supplementation after ammonia nitrogen challenge assay. (Different letters above bars and asterisk sign (*) indicate significant differences at the 0.05 level).

Heat shock protein genes, containing HSP70 and HSP90, were quantified in liver (a), kidney (b), spleen (c) and gill (d) tissues of greater amberjack after the challenge assay (Figures 6 and 7). Regarding the HSP70 gene, the relative expression of the HSP70 gene of experimental groups in the liver was significantly lower than that of the control after the treatment challenge assay. In contrast, the relative expression of the HSP70 gene in the kidney and spleen of two curcumin-added groups were both higher than that of the control.

In the gill, only the relative expression of the HSP70 gene of the 0.02% CUR group was at the highest level.

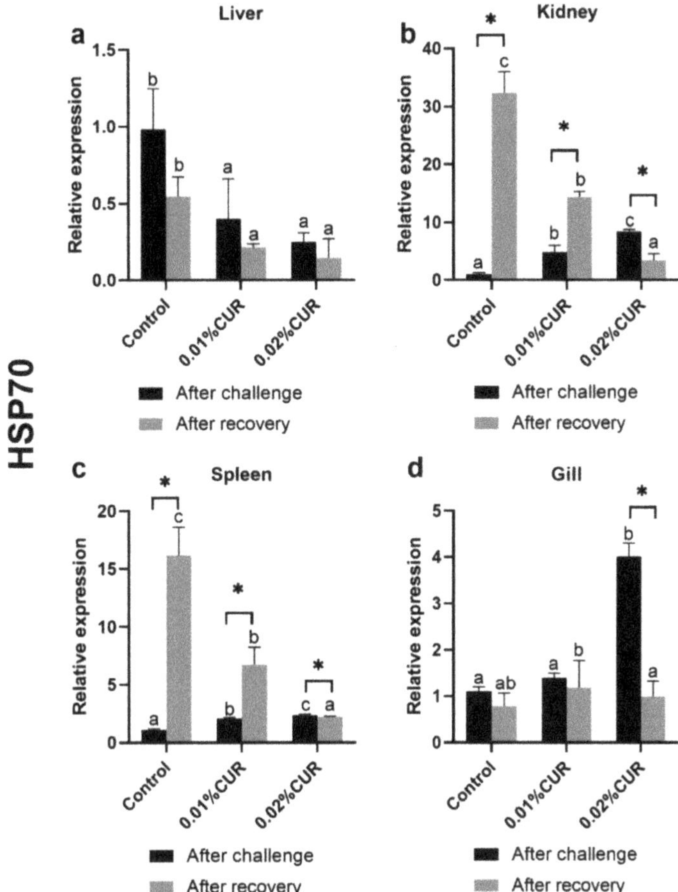

Figure 6. The relative expression of heat shock protein (HSP70) genes in (**a**) liver, (**b**) kidney, (**c**) spleen and (**d**) gill tissues of greater amberjack fed with dietary curcumin supplementation. (Different letters above bars and asterisk sign (*) indicate significant differences at the 0.05 level).

The rules of the relative expression of HSP90 gene in each tissue were similar with those of HSP70. In kidney and spleen tissues, the relative expression of the HSP90 gene in both the two curcumin-added groups was significantly higher than that of the control, but the HSP90 expression level of the 0.02% CUR group was significantly lower than that of the 0.01% CUR group. The relative expressions of the HSP90 gene in liver were higher with the increase in the amount of added curcumin. Moreover, the change situation of the relative expressions in gills showed a huge difference between the two heat shock protein genes. The relative expression of the HSP90 gene in gills of the 0.02% CUR group was the highest; however, in HSP70, the relative expression of the HSP90 gene of the control was the highest.

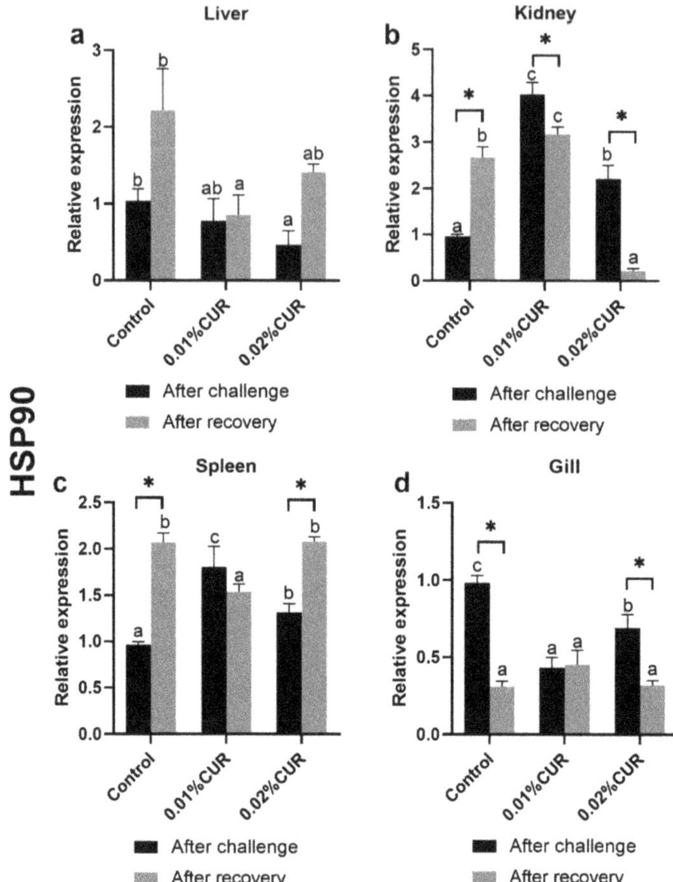

Figure 7. The relative expression of heat shock protein (HSP90) genes in (**a**) liver, (**b**) kidney, (**c**) spleen and (**d**) gill tissue of greater amberjack fed with dietary curcumin supplementation. (Different letters above bars and asterisk sign (*) indicate significant differences at the 0.05 level).

There have been some changes in the relative expression of the HSP70 and HSP90 genes in four tissues after recovery compared to the state after the ammonia nitrogen challenge (Figures 6 and 7). As for the HSP70 gene, its relative expression in liver, kidney and spleen was obviously less than that of the control group, but in gill, its relative expression had no difference among all the groups. It was noted that the situation of the relative expression of HSP90 in gill was also quite similar with that of HSP70; experimental groups also showed no difference with the control group. Furthermore, the relative expression of HSP90 in liver was still kept at a lower level than the control group after recovery. However, the relative expression of HSP90 in kidney and spleen fell into a different situation. In spleen in particular, only the relative expression of HSP90 in the 0.01% CUR group different from other groups. In kidney, however, a higher value was found in the 0.01% CUR group and a lower value in the 0.02% CUR group compared with the control.

The relative expression of cytokine IL8, IL1β, TNF-α, IFN-γ and TGF-β1 genes in intestines of greater amberjack after the challenge assay are given in Figure 8. A lower level occurred both in the relative expression of the IL8 and TGF-β1 genes in two curcumin-added groups compared with the control. Regarding the TNF-α gene, only the high curcumin-added group showed the lower level. However, IL1β gene and IFN-γ gene relative expressions of the curcumin-added group were all significantly higher than that of the control. The difference is that the highest level of the IL1β gene occurred in the 0.02% CUR group, and that of the IFN-γ gene occurred in the 0.01% CUR group.

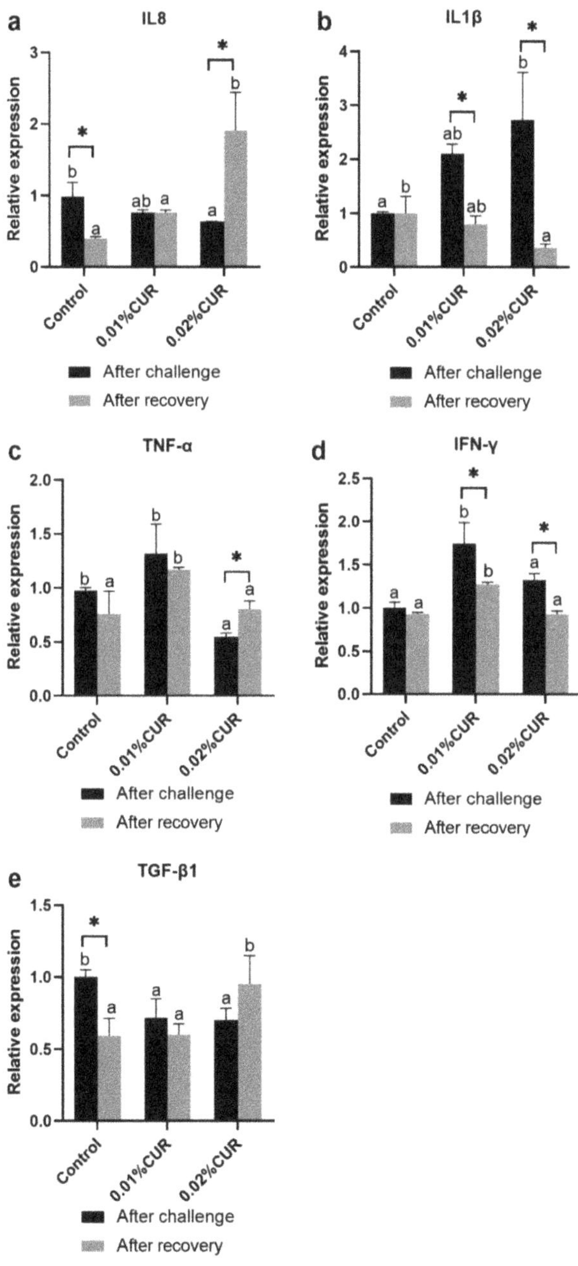

Figure 8. The relative expression of (**a**) cytokine IL8, (**b**) IL1β, (**c**) TNF-α, (**d**) IFN-γ and (**e**) TGF-β1 genes in intestine of greater amberjack fed with dietary curcumin supplementation. (Different letters above bars and asterisk sign (*) indicate significant differences at the 0.05 level).

The relative expression of other genes (C3, C4, NF-κB1, Mx, Hepc and lgT) after challenge are analyzed in Figure 9. C3 and lgT genes were both found to have a higher relative expression in the 0.01% CUR group and a lower relative expression in the 0.02% CUR group. The relative expression of C4 and Hepc genes showed the same phenomenon; in both there appeared a distinct lower value in the 0.02% CUR group compared with other groups. In addition, NF-κB1 and Mx genes were on the opposite sides. Treatment groups in particular were significantly higher than the control regarding the NF-κB1gene and lower regarding the Mx gene after the ammonia nitrogen challenge compared with the control.

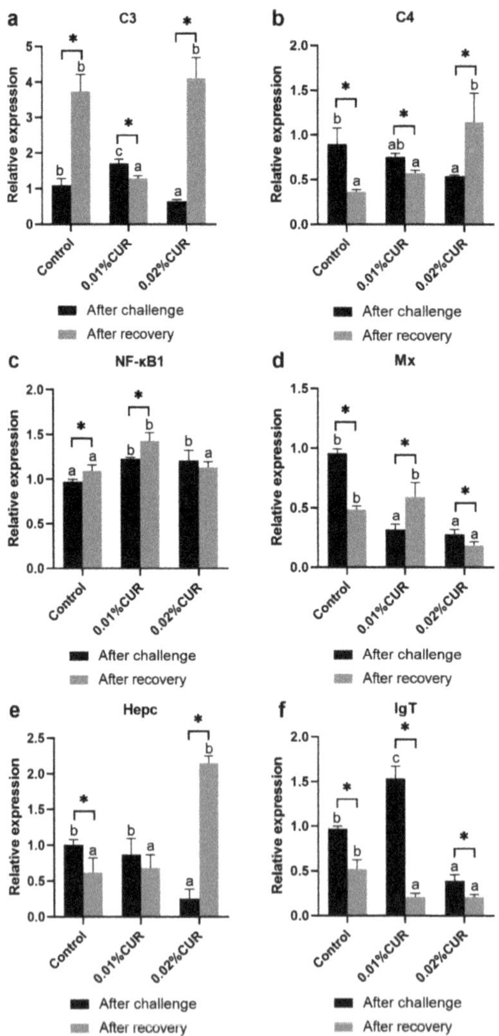

Figure 9. The relative expression of complements (**a**) (C3), (**b**) C4, (**c**) NF-κB1, (**d**) Mx, (**e**) Hepc and (**f**) lgT genes in intestine of greater amberjack fed with dietary curcumin supplementation. (Different letters above bars and asterisk sign (*) indicate significant differences at the 0.05 level).

The results of the relative expression of some cytokine genes and other related genes after recovery are also provided in Figures 8 and 9. As for cytokine genes, the highest relative expression of IL8 and TGF-β1 genes was witnessed in the 0.02% CUR group and the highest relative expression of TNF-α and IFN-γ genes in the 0.01% CUR group instead. In addition, the IL1β gene was expressed the least in the 0.03% CUR group while the relative expression of IL1β in the 0.01% CUR group showed no significant difference to others.

The relative expression of the C3 gene in the 0.01% CUR group was remarkably lower than in other groups. On the contrary, the NF-κB1 gene's relative expression in the 0.01% CUR group was the highest among the groups. The relative expression of the C4 and Hepc genes had the same tendency: their level of expression in the 0.02% CUR group rose to a higher value compared with the control after recovery. Moreover, only the relative expression of the Mx gene in the 0.02% CUR group was significantly lower than in other groups, but both two curcumin-added groups witnessed a markedly lower value of the relative expression of the lgT gene than in the control.

4. Discussion

In order to determine whether the addition of curcumin in the diet has an effect on the resistance to external environmental stress of greater amberjack, or whether it effectively protects the in vivo multiple functions of fish and enhances the capacity of recovery from stress, we analyzed and confirmed the results from multiple perspectives, such as nutrition, organ protection and the immune system, etc.

The role of dietary curcumin supplementation in animal nutrition is our first concern, on account of the digestion and absorption ability of fish having direct and important links with final growth performance [50–52]. Digestive enzymes including amylase, lipase, and trypsin were determined after two phases of the experiment. Comparisons of digestive enzyme activity after acute ammonia nitrogen stress and recovery provided the evidence of the promotion of recovery of digestive functions owing to the dietary curcumin supplementation. In particular, the activity of amylase and lipase of greater amberjack fed with 0.02% curcumin additive was not only maintained at the previous level but also increased. In contrast, the counterparts of the control and the 0.01% additive amount group decreased together and showed no change in their relative relationship, indicating that the high content of the curcumin additive maintained and even strengthened the activity of amylase and trypsin during the recovery following stress. Similar findings occurred in other fish. For instance, in crucian carp (*Carassius auratus*), significant increases were found in trypsin and lipase activities in the intestine after supplemental doses of dietary curcumin at 5 g/kg [35]. In addition, Ref. [53] demonstrated the ascent of these two digestion enzymes in another fish, *Ctenopharyngodon idella*, along with a rise in amylase activity. However, in the present study, the same positive assistance was not seen in trypsin activity in the fish fed at both dosage rates of curcumin additive, which is inconsistent with previous studies. This contradiction may be due to an insufficient dose, or there may be another possibility that should be further investigation by experiment. As for absorption capacity, in our experiment the enhanced effect of curcumin supplementation on Na^+/K^+–ATPase, γ-GT and CK activity was more obvious. The recovery mechanism was started earlier, compared with the control, regardless of the additive amount; in other words, the stress recovery process was accelerated. The studies mentioned above on crucian carp and grass carp also witnessed a positive increase in these three enzymes related to nutrition absorption with additional curcumin supplementation in the fish diet. Furthermore, a great deal of research has provided more direct enhanced-growth data more intuitively (increases in WG, SGR, DWG, FCR and FE, etc.) [54–58]. Collectively, it can be concluded that dietary curcumin supplementation can not only increase growth in daily culture by improving digestion and absorption, but can also raise the performance in a stress environment.

The role of curcumin as an immunostimulant that stimulates the activity of the non-specific immune system in fish is also worthy of attention. The activity of lysozyme is an important marker for representing in vivo non-specific immunity of fish [59,60]. In this study, when there was no significant change in the control group, lysozyme activity increased in the fish fortified with both curcumin diets. The improving effect of curcumin on lysozyme activity was also reported by Leya et al. [61], who fed *Labeo rohita* fingerlings a diet with 0.5%, 1%, 1.5% and 2% curcumin for 60 days; a significant rise was observed in lysozyme activity from 15 days to the end. Furthermore, another study also showed an increase in the lysozyme activity of *Oreochromis niloticus* fed the 50 mg and 100 mg/kg curcumin diet, but not in all the groups [38]. That indicated that the benefits of the additive amount of curcumin did not improve with higher amounts; however, the proportion of 0.01% and 0.02% in our present study remained positive for lysozyme activity. Albumin (ALB) should be another important component of a non-specific immune system [62,63]. Ref. [64] studied the effect of a curcumin-supplemented diet in *Cirrhinus mrigala* and found that fishes fed a curcumin-enriched diet (1–1.5%) showed significantly higher serum albumin. Similarly, research into the effects of two kinds of Supplementary Materials on *Nile tilapia* immunity also witnessed a remarkable increase in the curcumin-addition group [65]. However, the ALB of the two treatment groups showed no obvious change, while the ALB of the control pulled away from the state of no significant difference among the three groups. The results of the ALB were inconsistent with previous studies, and whether it was related to the amount of curcumin addition should be further explored.

Aspartate aminotransferase (AST) and alanine aminotransferase (ALT) are produced and released by hepatocytes, and they are considered biomarkers of liver damage injury [66–69]. As for the protective effect of curcumin on the liver, the involvement of AST and ALT has been reported in some research. A significantly lower level of ALT and AST activities were observed in *Megalobrama amblycephala* from the 60 mg/kg curcumin diet group [70]. In addition, a study on grass carp showed that ALT and AST activities decreased significantly on account of curcumin addition [71]. The protective effect of curcumin on the liver was also found in this study, and a greater reduction in serum AST or a quicker reduction in serum AST were observed. However, no similar trend occurred in ALT activity. This phenomenon illustrated that, at least at the curcumin level of this experiment, a protective effect of restricting ALT activity was not activated.

Curcumin has been proved to have lipid-lowering and anti-obesity effects on mammals [72–76]. Therefore, its function of improving lipid metabolism in fish was also what we wanted to understand, and triglyceride (TG) and total cholesterol (TC) were tested in this study for verification. The present study showed that TG and TC content were basically maintained at the original level or even showed a slight decrease when the control group increased significantly after recovery. Based on that, facilitation of lipid metabolism on greater amberjack by curcumin has been very clear. These results are consistent with the results of a study [36] in which TC, TG and LDL-c were observed in the curcumin treatment group.

Heat-shock proteins (HSPs) are a highly conserved protein family; commonly analyzed HSP70 and HSP90 are included. These can be induced by a variety of stressors and perform their protective roles. Many heat shock proteins have molecular chaperone activity to preserve the tertiary structure of other proteins. Therefore, the proteins are an important part of the cellular stress response and considered excellent markers of stress response [77–81]. The relative expressions of HSP70 and HSP90 in four tissues (liver, kidney, spleen and gill) were determined in this study. Regarding HSP70, through the comparison of stress state and recovery state, curcumin obtained an outstanding performance in the relative expressions of HSP70 in fish, but its effect on liver cannot be clarified from the data. Regarding HSP90, its situation was very different from that of HSP70. HSP90 in all tissues except gill remained in the state of stimulated stress and did not fully recover. Nonetheless, the positive effect of curcumin also can be found in liver, kidney and spleen tissues. It should be noted, however, that different additive amounts of curcumin resulted in great differences.

Cytokine is a class of small molecular proteins with a wide range of biological activities. The class mainly includes interleukins (ILs), tumor necrosis factors (TNFs), transforming growth factor-β family (TGF-β family) and interferon (IFN), etc. [82]. Cytokine also can be roughly divided into two types: pro-inflammatory cytokines and anti-inflammatory cytokines. Pro-inflammatory cytokines determined in this study were IL8, IL1β, TNF-α and IFN-γ, and an anti-inflammatory cytokine was TGF-β1. The function related to the anti-inflammation role of curcumin has been widely reported [83,84]. The study in rats reported that the expression of TNF-α and IL-6 in hypercholesterolemic rats decreased after a 200 mg/kg curcumin treatment. In fish, Ref. [85] investigated the protective effects of curcumin on liver-damaged *Cyprinus carpio*; the results indicated that 0.5% and 1.0% curcumin reduced the CCl4-induced damage in liver by inhibiting NF-kB, IL-1β, TNF-α, and IL-12 expression. Regarding pro-inflammatory cytokines in the present study, curcumin showed only its inhibition capacity on IL1β and IFN-γ for their sharper declines during recovery. In addition, the sole anti-inflammatory cytokine TGF-β1 we determined to be obviously improved in the high curcumin-dosage group. Transcription factor nuclear factor kappa B (NF-κB) has a central role in promoting the process of inflammation and cancer, and cytokine expressions can be activated via this signaling pathway [86,87]. Curcumin inhibits the expression of cytokine via blocking the NF-κB pathway. Ref. [88] provided the evidence with the study on *Oncorhynchus mykiss*; NF-κB expression declined dramatically after curcumin was added to the diet. Another test on *Cyprinus carpio* showed an attenuation effect on the expression of NF-κB caused by Pb exposure [89]. The result of this study was similar with theirs. The expression of NF-κB1 decreased to a higher degree in fish fed with 0.02% dietary curcumin, thus avoiding the continuation of inflammation response more quickly.

Some other immune-related genes were determined in this study; both effector molecules related to the innate immune system and the adaptive immune system were included (C3, C4, Hepc, Mx and lgT). The complement system comprises more than 35 proteins and plays an important role in the innate immune system of fish. It can exert positive effects on phagocytosis, antigen-antibody aggregate clearance, inflammatory reactions, and antibody production [90–93]. Antimicrobial peptides (AMPs) are also a vital component of the innate immune system which cannot be ignored. Hepcidin (Hepc), as a member of AMPs which are rich in cysteine, also performs well in antimicrobial activity. Coupled with its specific iron regulation function, it is of great importance to organism [94–96]. In addition, proteins encoded by the Mx gene are high-molecular-weight GTPases induced by interferon; they have strong intrinsic GTP hydrolysis, broad-spectrum antiviral activity and can inhibit a variety of negative-strand RNA viruses [97,98]. Immunoglobulins (Igs) play a key role in the adaptive immune system to help the organism against pathogens. Immunoglobulin T (lgT) was the last found among all immunoglobulins in teleost fish. It has been proved to have a more specific focus on gut mucosal immunity compared to other immunoglobulins (IgM, IgD, IgG, IgE, and IgA) [99–101]. The expression of these genes was sufficient to reflect the status of many important parts of the fish immune system, which enabled us to glimpse the truth of immune changes. In the present study, no striking enhancement was found in the restorative ability of C4, Hepc and Mx under both proportions of dietary curcumin treatments. However, 0.01% dietary accelerated the recovery pace of C3 and lgT after stress. Although positive results were not acquired for all the immune-related genes, the possibility that curcumin improves the immune functions of greater amberjack is assured, and more exciting results can be obtained on more indexes after more investigations into additive schemes.

Recent research into dietary curcumin supplementation on more fish species than our present study covers further broadens its application. For example, dietary curcumin proved to promote gilthead seabream larvae digestive capacity and modulate the oxidative status [102], and it was also found to work at the post-larval stage [103]. Dietary curcumin was also found to relieve the lipopolysaccharide (LPS)-induced and deltamethrin (DEL)-induced stress response (oxidative stress, inflammation and cell apoptosis) in *Channa argus* via Nrf2 and NF-κB signaling pathways [104,105]. With more related studies combined with the results of our research, it can be predicted that dietary curcumin supplementation will play an important role in the healthcare and maintenance of robustness of fish.

5. Conclusions

In brief, the application prospects for dietary curcumin supplementation in greater amberjack can be confirmed based on the results of this study. Curcumin has shown enhanced capability of protective effect and stress recovery in nutritional function, anti-inflammation, immune system, disease resistance, tissue protection, etc. These results are consistent with previous studies. In addition, our research shows that the relationship between the effect and the additive amount was not linear, and that a greater amount of additive did not necessarily lead to a better effect; therefore, a wider range of additive-amount experiments should be pursued. Eventually, the curcumin effect curve in greater amberjack will be obtained through these works, in order to provide a better reference for future practical applications.

Supplementary Materials: The following supporting information can be downloaded at: https://www.mdpi.com/article/10.3390/jmse10111796/s1.

Author Contributions: Z.M. and J.H. (Jiawei Hong) conceptualization of the experiment; J.H. (Jiawei Hong) and Z.F. methodology and performance of the experiment; J.H. (Jing Hu) performed the statistical analysis; J.H. (Jiawei Hong) and Z.F. writing—original draft preparation; S.Z. writing, review and editing; G.Y. and Z.M. supervised the final version of the paper. All authors have read and agreed to the published version of the manuscript.

Funding: This study was supported by Central Public-interest Scientific Institution Basal Research Fund, CAFS (NO. 2020TD55), Guangxi Innovation Driven Development Special Fund Project [Guike AA18242031].

Institutional Review Board Statement: This study was carried out in strict accordance with the recommendation in the Animal Welfare Committee of Chinese Academy of Fishery Sciences. No protected species were used during the experiment.

Informed Consent Statement: Not applicable (Present study did not involve humans). Written informed consent has been obtained from all the patients to publish this paper.

Data Availability Statement: The data presented in this study are available on request from the corresponding author.

Conflicts of Interest: The authors declare no conflict of interest.

References

1. Angel, D.L.; Eden, N.; Breitstein, S.; Yurman, A.; Katz, T.; Spanier, E. In situ biofiltration: A means to limit the dispersal of effluents from marine finfish cage aquaculture. *Hydrobiologia* **2002**, *469*, 1–10. [CrossRef]
2. Luis, A.I.S.; Campos, E.V.R.; de Oliveira, J.L.; Fraceto, L.F. Trends in aquaculture sciences: From now to use of nanotechnology for disease control. *Rev. Aquac.* **2019**, *11*, 119–132. [CrossRef]
3. Terlizzi, A.; Tedesco, P.; Patarnello, P. Spread of pathogens from marine cage aquaculture-a potential threat for wild fish assemblages under protection regimes. *Health Environ. Aquac.* **2012**, *12*, 403–414.
4. Tovar, A.; Moreno, C.; Mánuel-Vez, M.P.; García-Vargas, M. Environmental impacts of intensive aquaculture in marine waters. *Water Res.* **2000**, *34*, 334–342. [CrossRef]
5. Zhang, X.; Zhang, Y.; Zhang, Q.; Liu, P.; Guo, R.; Jin, S.; Liu, J.; Chen, L.; Ma, Z.; Liu, Y. Evaluation and analysis of water quality of marine aquaculture area. *Int. J. Environ. Res. Public Health* **2020**, *17*, 1446. [CrossRef]
6. Wu, J.; Su, Y.; Deng, Y.; Guo, Z.; Cheng, C.; Ma, H.; Liu, G.; Xu, L.; Feng, J. Spatial and temporal variation of antibiotic resistance in marine fish cage-culture area of Guangdong, China. *Environ. Pollut.* **2019**, *246*, 463–471. [CrossRef] [PubMed]
7. Ahmed, N.; Thompson, S.; Glaser, M. Global aquaculture productivity, environmental sustainability, and climate change adaptability. *Environ. Manag.* **2019**, *63*, 159–172. [CrossRef] [PubMed]
8. Mazzola, A.; Favaloro, E.; Sarà, G. Cultivation of the Mediterranean amberjack, *Seriola dumerili* (Risso, 1810), in submerged cages in the Western Mediterranean Sea. *Aquaculture* **2000**, *181*, 257–268. [CrossRef]
9. Rigos, G.; Katharios, P.; Kogiannou, D.; Cascarano, C.M. Infectious diseases and treatment solutions of farmed greater amberjack *Seriola dumerili* with particular emphasis in Mediterranean region. *Rev. Aquac.* **2021**, *13*, 301–323. [CrossRef]
10. Fernández-Montero, A.; Torrecillas, S.; Tort, L.; Ginés, R.; Acosta, F.; Izquierdo, M.; Montero, D. Stress response and skin mucus production of greater amberjack (*Seriola dumerili*) under different rearing conditions. *Aquaculture* **2020**, *520*, 735005. [CrossRef]
11. Mylonas, C.; Papandroulakis, N.; Corriero, A.; Montero, D.; Hernández-Cruz, C.; Banovic, M.; Tacken, G.; Robles, R. Rearing of greater amberjack (*Seriola dumerili*). *Aquac. Eur.* **2019**, *44*, 18–25.

12. Yilmaz, E.; Şereflişan, H. Offshore Farming of the Mediterranean Amberjack (*Seriola dumerili*) in the Northeastern Mediterranean. *Isr. J. Aquac. Bamidgeh* **2011**, *63*, 7. [CrossRef]

13. Zupa, R.; Rodríguez, C.; Mylonas, C.C.; Rosenfeld, H.; Fakriadis, I.; Papadaki, M.; Pérez, J.A.; Pousis, C.; Basilone, G.; Corriero, A. Comparative study of reproductive development in wild and captive-reared greater amberjack *Seriola dumerili* (Risso, 1810). *PLoS ONE* **2017**, *12*, e0169645. [CrossRef] [PubMed]

14. Mazzora, A. Sistemi di maricoltura open-sea l'allevamento di *Seriola dumerili* (Pisces, Osteichthyes) nel Golfo di Catellammare (Sicilia Occidentale). *Biol. Mar. Med.* **1996**, *3*, 176–185.

15. Porrello, S.; Andaloro, F.; Vivona, P.; Marino, G. Rearing trial of Seriola dumerili in a floating cage. In *Production, Environment and Quality, Proceedings of the International Conference Bordeaux Aquaculture '92, Bordeaux, France, 25–27 March 1992*; EAS Special Publication: Bordeaux, France, 1993; pp. 299–307.

16. Jover, M.; Garcıa-Gómez, A.; Tomas, A.; de la Gándara, F.; Pérez, L. Growth of Mediterranean yellowtail (*Seriola dumerilii*) fed extruded diets containing different levels of protein and lipid. *Aquaculture* **1999**, *179*, 25–33. [CrossRef]

17. Wang, W.; Sun, J.; Liu, C.; Xue, Z. Application of immunostimulants in aquaculture: Current knowledge and future perspectives. *Aquac. Res.* **2017**, *48*, 1–23. [CrossRef]

18. Khanna, D.; Sethi, G.; Ahn, K.S.; Pandey, M.K.; Kunnumakkara, A.B.; Sung, B.; Aggarwal, A.; Aggarwal, B.B. Natural products as a gold mine for arthritis treatment. *Curr. Opin. Pharmacol.* **2007**, *7*, 344–351. [CrossRef]

19. Divyagnaneswari, M.; Christybapita, D.; Michael, R.D. Enhancement of nonspecific immunity and disease resistance in *Oreochromis mossambicus* by *Solanum trilobatum* leaf fractions. *Fish Shellfish Immun.* **2007**, *23*, 249–259. [CrossRef]

20. Jian, J.; Wu, Z. Influences of traditional Chinese medicine on non-specific immunity of Jian carp (*Cyprinus carpio* var. Jian). *Fish Shellfish Immun.* **2004**, *16*, 185–191. [CrossRef]

21. Ardó, L.; Yin, G.; Xu, P.; Váradi, L.; Szigeti, G.; Jeney, Z.; Jeney, G. Chinese herbs (*Astragalus membranaceus* and *Lonicera japonica*) and boron enhance the non-specific immune response of Nile tilapia (*Oreochromis niloticus*) and resistance against *Aeromonas hydrophila*. *Aquaculture* **2008**, *275*, 26–33. [CrossRef]

22. Hwang, J.-H.; Lee, S.-W.; Rha, S.-J.; Yoon, H.-S.; Park, E.-S.; Han, K.-H.; Kim, S.-J. Dietary green tea extract improves growth performance, body composition, and stress recovery in the juvenile black rockfish, *Sebastes schlegeli*. *Aquac. Int.* **2013**, *21*, 525–538. [CrossRef]

23. Thanigaivel, S.; Vijayakumar, S.; Gopinath, S.; Mukherjee, A.; Chandrasekaran, N.; Thomas, J. In vivo and in vitro antimicrobial activity of *Azadirachta indica* (Lin) against *Citrobacter freundii* isolated from naturally infected Tilapia (*Oreochromis mossambicus*). *Aquaculture* **2015**, *437*, 252–255. [CrossRef]

24. Lestari, M.L.; Indrayanto, G. Curcumin. *Profiles Drug Subst. Excip. Relat. Methodol.* **2014**, *39*, 113–204. [PubMed]

25. Singh, S.; Khar, A. Biological effects of curcumin and its role in cancer chemoprevention and therapy. *Anti-Cancer Agents Med. Chem.* **2006**, *6*, 259–270. [CrossRef]

26. Brouk, B. *Plants Consumed by Man*; Academic Press: London, UK, 1975.

27. WHO. *Evaluation of Certain Food Additives and Contaminants: Sixty-First Report of the Joint FAO/WHO Expert Committee on Food Additives*; World Health Organization: Geneva, Switzerland, 2004.

28. Kunchandy, E.; Rao, M. Oxygen radical scavenging activity of curcumin. *Int. J. Pharm.* **1990**, *58*, 237–240. [CrossRef]

29. Goel, A.; Boland, C.R.; Chauhan, D.P. Specific inhibition of cyclooxygenase-2 (COX-2) expression by dietary curcumin in HT-29 human colon cancer cells. *Cancer Lett.* **2001**, *172*, 111–118. [CrossRef]

30. Ammon, H.P.; Wahl, M.A. Pharmacology of *Curcuma longa*. *Planta Med.* **1991**, *57*, 1–7. [CrossRef]

31. Sharma, R.; Gescher, A.; Steward, W. Curcumin: The story so far. *Eur. J. Cancer* **2005**, *41*, 1955–1968. [CrossRef]

32. Li, M.; Zhang, Z.; Hill, D.L.; Wang, H.; Zhang, R. Curcumin, a dietary component, has anticancer, chemosensitization, and radiosensitization effects by down-regulating the MDM2 oncogene through the PI3K/mTOR/ETS2 pathway. *Cancer Res.* **2007**, *67*, 1988–1996. [CrossRef]

33. Abdel-Tawwab, M.; Abbass, F.E. Turmeric powder, *Curcuma longa* L., in common carp, *Cyprinus carpio* L., diets: Growth performance, innate immunity, and challenge against pathogenic *Aeromonas hydrophila* infection. *J. World Aquac. Soc.* **2017**, *48*, 303–312. [CrossRef]

34. Adeshina, I.; Adewale, Y.; Tiamiyu, L. Growth performance and innate immune response of *Clarias gariepinus* infected with *Aeromonas hydrophila* fed diets fortified with *Curcuma longa* leaf. *West Afr. J. Appl. Ecol.* **2017**, *25*, 87–99.

35. Jiang, J.; Wu, X.-Y.; Zhou, X.-Q.; Feng, L.; Liu, Y.; Jiang, W.-D.; Wu, P.; Zhao, Y. Effects of dietary curcumin supplementation on growth performance, intestinal digestive enzyme activities and antioxidant capacity of crucian carp *Carassius auratus*. *Aquaculture* **2016**, *463*, 174–180. [CrossRef]

36. Ji, R.; Xiang, X.; Li, X.; Mai, K.; Ai, Q. Effects of dietary curcumin on growth, antioxidant capacity, fatty acid composition and expression of lipid metabolism-related genes of large yellow croaker fed a high-fat diet. *Br. J. Nutr.* **2021**, *126*, 345–354. [CrossRef] [PubMed]

37. Zhang, Y.; Li, F.; Yao, F.; Ma, R.; Zhang, Y.; Mao, S.; Hu, B.; Ma, G.; Zhu, Y. Study of dietary curcumin on the restorative effect of liver injury induced by carbon tetrachloride in common carp *Cyprinus carpio*. *Aquac. Rep.* **2021**, *21*, 100825. [CrossRef]

38. Mahmoud, H.K.; Al-Sagheer, A.A.; Reda, F.M.; Mahgoub, S.A.; Ayyat, M.S. Dietary curcumin supplement influence on growth, immunity, antioxidant status, and resistance to *Aeromonas hydrophila* in *Oreochromis niloticus*. *Aquaculture* **2017**, *475*, 16–23. [CrossRef]

39. Garcia-Garrido, L.; Muñoz-Chapuli, R.; de Andres, A. Serum cholesterol and triglyceride levels in *Scyliorhinus canicula* (L.) during sexual maturation. *J. Fish Biol.* **1990**, *36*, 499–509. [CrossRef]
40. Doumas, B.T.; Watson, W.A.; Biggs, H.G. Albumin standards and the measurement of serum albumin with bromcresol green. *Clin. Chim. Acta* **1971**, *31*, 87–96. [CrossRef]
41. Reitman, S.; Frankel, S. A colorimetric method for the determination of serum glutamic oxalacetic and glutamic pyruvic transaminases. *Am. J. Clin. Pathol.* **1957**, *28*, 56–63. [CrossRef]
42. Obach, A.; Quentel, C.; Laurencin, F.B. Dicen trarch us la brax. *Dis. Aquat. Org.* **1993**, *15*, 175–185. [CrossRef]
43. Reimer, G. The influence of diet on the digestive enzymes of the Amazon fish Matrincha, *Brycon* cf. *melanopterus. J. Fish Biol.* **1982**, *21*, 637–642. [CrossRef]
44. Ay, Ö.; Kalay, M.; Tamer, L.; Canli, M. Copper and lead accumulation in tissues of a freshwater fish *Tilapia zillii* and its effects on the branchial Na, K-ATPase activity. *Bull. Environ. Contam. Toxicol.* **1999**, *62*, 160–168. [CrossRef] [PubMed]
45. Weng, C.F.; Chiang, C.C.; Gong, H.Y.; Chen, M.H.; Lin, C.J.; Huang, W.T.; Cheng, C.Y.; Hwang, P.P.; Wu, J.L. Acute changes in gill Na+-K+-ATPase and creatine kinase in response to salinity changes in the euryhaline teleost, tilapia (*Oreochromis mossambicus*). *Physiol Biochem. Zool* **2002**, *75*, 29–36. [CrossRef] [PubMed]
46. Rosalki, S.; Rau, D.; Lehmann, D.; Prentice, M. Gamma-glutamyl transpeptidase in chronic alcoholism. *Lancet* **1970**, *296*, 1139. [CrossRef]
47. Fernández-Montero, Á.; Torrecillas, S.; Izquierdo, M.; Caballero, M.J.; Milne, D.J.; Secombes, C.J.; Sweetman, J.; da Silva, P.; Acosta, F.; Montero, D. Increased parasite resistance of greater amberjack (*Seriola dumerili* Risso 1810) juveniles fed a cMOS supplemented diet is associated with upregulation of a discrete set of immune genes in mucosal tissues. *Fish Shellfish Immun.* **2019**, *86*, 35–45. [CrossRef] [PubMed]
48. Fernández-Montero, Á.; Torrecillas, S.; Acosta, F.; Kalinowski, T.; Bravo, J.; Sweetman, J.; Roo, J.; Makol, A.; Docando, J.; Carvalho, M. Improving greater amberjack (*Seriola dumerili*) defenses against monogenean parasite *Neobenedenia girellae* infection through functional dietary additives. *Aquaculture* **2021**, *534*, 736317. [CrossRef]
49. Livak, K.J.; Schmittgen, T.D. Analysis of relative gene expression data using real-time quantitative PCR and the 2$-\Delta\Delta$CT method. *Methods* **2001**, *25*, 402–408. [CrossRef] [PubMed]
50. Wen, Z.P.; Zhou, X.Q.; Feng, L.; Jiang, J.; Liu, Y. Effect of dietary pantothenic acid supplement on growth, body composition and intestinal enzyme activities of juvenile Jian carp (*Cyprinus carpio* var. Jian). *Aquac. Nutr.* **2009**, *15*, 470–476. [CrossRef]
51. Blier, P.; Pelletier, D.; Dutil, J.D. Does aerobic capacity set a limit on fish growth rate? *Rev. Fish Sci.* **1997**, *5*, 323–340. [CrossRef]
52. Ling, J.; Feng, L.; Liu, Y.; Jiang, J.; Jiang, W.D.; Hu, K.; Li, S.H.; Zhou, X.Q. Effect of dietary iron levels on growth, body composition and intestinal enzyme activities of juvenile Jian carp (*Cyprinus carpio* var. Jian). *Aquac. Nutr.* **2010**, *16*, 616–624. [CrossRef]
53. Li, G.; Zhou, X.; Jiang, W.; Wu, P.; Liu, Y.; Jiang, J.; Kuang, S.; Tang, L.; Shi, H.; Feng, L. Dietary curcumin supplementation enhanced growth performance, intestinal digestion, and absorption and amino acid transportation abilities in on-growing grass carp (*Ctenopharyngodon idella*). *Aquac. Res.* **2020**, *51*, 4863–4873. [CrossRef]
54. Sruthi, M.; Nair, A.B.; Arun, D.; Thushara, V.; Sheeja, C.; Vijayasree, A.S.; Oommen, O.V.; Divya, L. Dietary curcumin influences leptin, growth hormone and hepatic growth factors in Tilapia (*Oreochromis mossambicus*). *Aquaculture* **2018**, *496*, 105–111. [CrossRef]
55. Yonar, M.E.; Yonar, S.M.; İspir, Ü.; Ural, M.Ş. Effects of curcumin on haematological values, immunity, antioxidant status and resistance of rainbow trout (*Oncorhynchus mykiss*) against *Aeromonas salmonicida* subsp. *achromogenes. Fish Shellfish Immun.* **2019**, *89*, 83–90. [CrossRef] [PubMed]
56. Cui, H.; Liu, B.; Ge, X.; XiE, J.; Xu, P.; Miao, L.; Sun, S.; Liao, Y.; Chen, R.; Ren, M. Effects of dietary curcumin on growth performance, biochemical parameters, HSP70 gene expression and resistance to *Streptococcus iniae* of juvenile Gift Tilapia, *Oreochromis niloticus. Isr. J. Aquac.* **2013**, *66*, 986–996.
57. Manal, I. Detoxification and antioxidant effects of garlic and curcumin in *Oreochromis niloticus* injected with aflatoxin B 1 with reference to gene expression of glutathione peroxidase (GPx) by RT-PCR. *Fish Physiol. Biochem.* **2016**, *42*, 617–629.
58. Manal, I. Impact of garlic and curcumin on the hepatic histology and cytochrome P450 gene expression of aflatoxicosis *Oreochromis niloticus* using RT-PCR. *Turk. J. Fish. Aquat. Sci.* **2018**, *18*, 405–415.
59. Tort, L.; Balasch, J.; Mackenzie, S. Fish immune system. A crossroads between innate and adaptive responses. *Inmunología* **2003**, *22*, 277–286.
60. Saurabh, S.; Sahoo, P. Lysozyme: An important defence molecule of fish innate immune system. *Aquac. Res.* **2008**, *39*, 223–239. [CrossRef]
61. Leya, T.; Dar, S.A.; Kumar, G.; Ahmad, I. Curcumin supplement diet: Enhanced growth and down-regulated expression of pro-inflammatory cytokines in *Labeo rohita* fingerlings. *Aquac. Res.* **2020**, *51*, 4785–4792. [CrossRef]
62. Misra, C.; Das, B.; Mukherjee, S.; Meher, P. The immunomodulatory effects of tuftsin on the non-specific immune system of Indian Major carp, *Labeo rohita. Fish Shellfish Immun.* **2006**, *20*, 728–738. [CrossRef]
63. Wiegertjes, G.F. Immunogenetics of Disease Resistance in Fish. Doctoral Thesis, Wageningen Agricultural University, Wageningen, The Netherlands, 1995.
64. Leya, T.; Raman, R.P.; Srivastava, P.P.; Kumar, K.; Ahmad, I.; Gora, A.H.; Poojary, N.; Kumar, S.; Dar, S.A. Effects of curcumin supplemented diet on growth and non-specific immune parameters of *Cirrhinus mrigala* against *Edwardsiella tarda* infection. *Int J. Curr. Microbiol. Appl. Sci.* **2017**, *6*, 1230–1243.
65. Diab, A.M.; Saker, O.; Eldakroury, M.; Elseify, M. Effects of garlic (*Allium sativum*) and curcumin (Turmeric, *Curcuma longa* Linn) on *Nile tilapia* immunity. *Vet. Med. J.* **2014**, *60*, C1–C19.

66. Kwo, P.Y.; Cohen, S.M.; Lim, J.K. ACG clinical guideline: Evaluation of abnormal liver chemistries. *Off. J. Am. Coll. Gastroenterol. ACG* **2017**, *112*, 18–35. [CrossRef] [PubMed]
67. Bao, J.-W.; Qiang, J.; Tao, Y.-F.; Li, H.-X.; He, J.; Xu, P.; Chen, D.-J. Responses of blood biochemistry, fatty acid composition and expression of microRNAs to heat stress in genetically improved farmed tilapia (*Oreochromis niloticus*). *J. Therm. Biol.* **2018**, *73*, 91–97. [CrossRef] [PubMed]
68. Pakhira, C.; Nagesh, T.; Abraham, T.; Dash, G.; Behera, S. Stress responses in rohu, *Labeo rohita* transported at different densities. *Aquac. Rep.* **2015**, *2*, 39–45. [CrossRef]
69. Hong, J.; Zhou, S.; Yu, G.; Qin, C.; Zuo, T.; Ma, Z. Effects of transporting stress on the immune responses of Asian seabass *Lates calcarifer* fry. *Aquac. Res.* **2021**, *52*, 2182–2193. [CrossRef]
70. Xia, S.-L.; Ge, X.-P.; Liu, B.; Xie, J.; Miao, L.-H.; Ren, M.-C.; Zhou, Q.-L.; Zhang, W.-X.; Jiang, X.-J.; Chen, R.-L. Effects of supplemented dietary curcumin on growth and non-specific immune responses in juvenile wuchang bream (*Megalobrama amblycephala*). *Isr. J. Aquac. Bamid.* **2015**, *67*. [CrossRef]
71. Yu, S.; Yu, Y.; Ai, T.; Li, Q. Study of the recovery effect of curcumin on liver injury in grass carp. *Chin. J. Vet. Drug* **2013**, *47*, 29–31.
72. Shao, W.; Yu, Z.; Chiang, Y.; Yang, Y.; Chai, T.; Foltz, W.; Lu, H.; Fantus, I.G.; Jin, T. Curcumin prevents high fat diet induced insulin resistance and obesity via attenuating lipogenesis in liver and inflammatory pathway in adipocytes. *PLoS ONE* **2012**, *7*, e28784. [CrossRef]
73. Yang, Y.S.; Su, Y.F.; Yang, H.W.; Lee, Y.H.; Chou, J.I.; Ueng, K.C. Lipid-lowering effects of curcumin in patients with metabolic syndrome: A randomized, double-blind, placebo-controlled trial. *Phytother. Res.* **2014**, *28*, 1770–1777. [CrossRef]
74. Arafa, H.M. Curcumin attenuates diet-induced hypercholesterolemia in rats. *Med. Sci. Monit.* **2005**, *11*, BR228–BR234.
75. Jang, E.-M.; Choi, M.-S.; Jung, U.J.; Kim, M.-J.; Kim, H.-J.; Jeon, S.-M.; Shin, S.-K.; Seong, C.-N.; Lee, M.-K. Beneficial effects of curcumin on hyperlipidemia and insulin resistance in high-fat–fed hamsters. *Metabolism* **2008**, *57*, 1576–1583. [CrossRef] [PubMed]
76. Panahi, Y.; Kianpour, P.; Mohtashami, R.; Jafari, R.; Simental-Mendía, L.E.; Sahebkar, A. Curcumin lowers serum lipids and uric acid in subjects with nonalcoholic fatty liver disease: A randomized controlled trial. *J. Cardiovasc. Pharm.* **2016**, *68*, 223–229. [CrossRef] [PubMed]
77. Roberts, R.; Agius, C.; Saliba, C.; Bossier, P.; Sung, Y. Heat shock proteins (chaperones) in fish and shellfish and their potential role in relation to fish health: A review. *J. Fish Dis.* **2010**, *33*, 789–801. [CrossRef] [PubMed]
78. Newton, J.; de Santis, C.; Jerry, D. The gene expression response of the catadromous perciform barramundi *Lates calcarifer* to an acute heat stress. *J. Fish Biol.* **2012**, *81*, 81–93. [CrossRef] [PubMed]
79. Stitt, B.C.; Burness, G.; Burgomaster, K.A.; Currie, S.; McDermid, J.L.; Wilson, C.C. Intraspecific variation in thermal tolerance and acclimation capacity in brook trout (*Salvelinus fontinalis*): Physiological implications for climate change. *Physiol. Biochem. Zool* **2014**, *87*, 15–29. [CrossRef] [PubMed]
80. Templeman, N.M.; LeBlanc, S.; Perry, S.F.; Currie, S. Linking physiological and cellular responses to thermal stress: β-adrenergic blockade reduces the heat shock response in fish. *J. Comp. Physiol. B* **2014**, *184*, 719–728. [CrossRef]
81. Currie, S. Temperature | Heat Shock Proteins and Temperature. *Encycl. Fish Physiol.* **2011**, *3*, 1732–1737.
82. Savan, R.; Sakai, M. Genomics of fish cytokines. *Comp. Biochem. Phys. D* **2006**, *1*, 89–101. [CrossRef]
83. Kohli, K.; Ali, J.; Ansari, M.; Raheman, Z. Curcumin: A natural antiinflammatory agent. *Indian J. Pharm.* **2005**, *37*, 141. [CrossRef]
84. Zhou, H.; Beevers, C.S.; Huang, S. The targets of curcumin. *Curr. Drug Targets* **2011**, *12*, 332–347. [CrossRef]
85. Cao, L.; Ding, W.; Du, J.; Jia, R.; Liu, Y.; Zhao, C.; Shen, Y.; Yin, G. Effects of curcumin on antioxidative activities and cytokine production in Jian carp (*Cyprinus carpio* var. Jian) with CCl4-induced liver damage. *Fish Shellfish Immun.* **2015**, *43*, 150–157. [CrossRef] [PubMed]
86. Alagawany, M.; Farag, M.R.; Abdelnour, S.A.; Dawood, M.A.; Elnesr, S.S.; Dhama, K. Curcumin and its different forms: A review on fish nutrition. *Aquaculture* **2021**, *532*, 736030. [CrossRef]
87. Anthwal, A.; Thakur, B.K.; Rawat, M.; Rawat, D.; Tyagi, A.K.; Aggarwal, B.B. Synthesis, characterization and in vitro anticancer activity of C-5 curcumin analogues with potential to inhibit TNF-α-induced NF-κB activation. *BioMed. Res. Int.* **2014**, *2014*, 524161. [CrossRef] [PubMed]
88. Akdemir, F.; Orhan, C.; Tuzcu, M.; Sahin, N.; Juturu, V.; Sahin, K. The efficacy of dietary curcumin on growth performance, lipid peroxidation and hepatic transcription factors in rainbow trout *Oncorhynchus mykiss* (Walbaum) reared under different stocking densities. *Aquac. Res.* **2017**, *48*, 4012–4021. [CrossRef]
89. Giri, S.S.; Kim, M.J.; Kim, S.G.; Kim, S.W.; Kang, J.W.; Kwon, J.; Lee, S.B.; Jung, W.J.; Sukumaran, V.; Park, S.C. Role of dietary curcumin against waterborne lead toxicity in common carp *Cyprinus carpio*. *Ecotoxicol. Environ. Saf.* **2021**, *219*, 112318. [CrossRef] [PubMed]
90. Holland, M.C.H.; Lambris, J.D. The complement system in teleosts. *Fish Shellfish Immun.* **2002**, *12*, 399–420. [CrossRef] [PubMed]
91. Boshra, H.; Li, J.; Sunyer, J. Recent advances on the complement system of teleost fish. *Fish Shellfish Immun.* **2006**, *20*, 239–262. [CrossRef]
92. Carroll, M.C. The role of complement and complement receptors in induction and regulation of immunity. *Annu. Rev. Immunol.* **1998**, *16*, 545–568. [CrossRef]
93. Gasque, P. Complement: A unique innate immune sensor for danger signals. *Mol. Immunol.* **2004**, *41*, 1089–1098. [CrossRef]
94. Chen, S.-L.; Li, W.; Meng, L.; Sha, Z.-X.; Wang, Z.-J.; Ren, G.-C. Molecular cloning and expression analysis of a hepcidin antimicrobial peptide gene from turbot (*Scophthalmus maximus*). *Fish Shellfish Immun.* **2007**, *22*, 172–181. [CrossRef]

95. Cuesta, A.; Meseguer, J.; Esteban, M.A. The antimicrobial peptide hepcidin exerts an important role in the innate immunity against bacteria in the bony fish gilthead seabream. *Mol. Immunol.* **2008**, *45*, 2333–2342. [CrossRef] [PubMed]

96. Lehrer, R.I.; Ganz, T. Antimicrobial peptides in mammalian and insect host defence. *Currt. Opin. Immunol.* **1999**, *11*, 23–27. [CrossRef]

97. Haller, O.; Kochs, G. Interferon-induced mx proteins: Dynamin-like GTPases with antiviral activity. *Traffic* **2002**, *3*, 710–717. [CrossRef] [PubMed]

98. Haller, O.; Kochs, G. Human MxA protein: An interferon-induced dynamin-like GTPase with broad antiviral activity. *J. Interf. Cytok Res.* **2011**, *31*, 79–87. [CrossRef] [PubMed]

99. Zhang, Y.-A.; Salinas, I.; Sunyer, J.O. Recent findings on the structure and function of teleost IgT. *Fish Shellfish Immun.* **2011**, *31*, 627–634. [CrossRef] [PubMed]

100. Zhang, Y.-A.; Salinas, I.; Li, J.; Parra, D.; Bjork, S.; Xu, Z.; LaPatra, S.E.; Bartholomew, J.; Sunyer, J.O. IgT, a primitive immunoglobulin class specialized in mucosal immunity. *Nat. Immunol.* **2010**, *11*, 827–835. [CrossRef] [PubMed]

101. Giacomelli, S.; Buonocore, F.; Albanese, F.; Scapigliati, G.; Gerdol, M.; Oreste, U.; Coscia, M.R. New insights into evolution of IgT genes coming from *Antarctic teleosts*. *Mar. Genom.* **2015**, *24*, 55–68. [CrossRef] [PubMed]

102. Xavier, M.J.; Dardengo, G.M.; Navarro-Guillén, C.; Lopes, A.; Colen, R.; Valente, L.M.; Conceição, L.E.; Engrola, S. Dietary curcumin promotes gilthead seabream larvae digestive capacity and modulates oxidative status. *Animals* **2021**, *11*, 1667. [CrossRef]

103. Xavier, M.J.; Navarro-Guillén, C.; Lopes, A.; Colen, R.; Teodosio, R.; Mendes, R.; Oliveira, B.; Valente, L.M.; Conceição, L.E.; Engrola, S. Effects of dietary curcumin in growth performance, oxidative status and gut morphometry and function of gilthead seabream postlarvae. *Aquac. Rep.* **2022**, *24*, 101128. [CrossRef]

104. Kong, Y.; Li, M.; Guo, G.; Yu, L.; Sun, L.; Yin, Z.; Li, R.; Chen, X.; Wang, G. Effects of dietary curcumin inhibit deltamethrin-induced oxidative stress, inflammation and cell apoptosis in *Channa argus* via Nrf2 and NF-κB signaling pathways. *Aquaculture* **2021**, *540*, 736744. [CrossRef]

105. Li, M.; Kong, Y.; Wu, X.; Guo, G.; Sun, L.; Lai, Y.; Zhang, J.; Niu, X.; Wang, G. Effects of dietary curcumin on growth performance, lipopolysaccharide-induced immune responses, oxidative stress and cell apoptosis in snakehead fish (*Channa argus*). *Aquac. Rep.* **2022**, *22*, 100981. [CrossRef]

Journal of
*Marine Science
and Engineering*

Article

Effects of Acute High-Temperature Stress on Physical Responses of Yellowfin Tuna (*Thunnus albacares*)

Hongyan Liu [1,2,3,†], Zhengyi Fu [2,3,4,†], Gang Yu [2,3], Zhenhua Ma [1,2,3,*] and Humin Zong [5,*]

1　College of Fisheries and Life, Shanghai Ocean University, Shanghai 201306, China
2　Key Laboratory of Efficient Utilization and Processing of Marine Fishery Resources of Hainan Province, Sanya Tropical Fisheries Research Institute, Sanya 572018, China
3　South China Sea Fisheries Research Institute, Chinese Academy of Fishery Sciences, Guangzhou 510300, China
4　College of Science and Engineering, Flinders University, Adelaide 5001, Australia
5　National Marine Environmental Center, Dalian 116023, China
*　Correspondence: zhenhua.ma@scsfri.ac.cn (Z.M.); hmzong@nmemc.org.cn (H.Z.)
†　These authors contributed equally to this work.

Abstract: To understand the physiological reactions of juvenile yellowfin tuna (*Thunnus albacares*) under acute high-temperature stress, this study measured the changes in biochemical indexes of serum, liver, gill, and muscle of yellowfin tuna under acute high-temperature stress (HT, 34 °C) and a control group (28 °C) for 0 h and 6 h, 24 h and 48 h. The rising speed of water temperature in the HT group was 2 °C/h and the timing started when the temperature reached 34 °C. In the HT group, there was no significant difference between the four adjacent times in cortisol and lactic acid concentration. Serum triglyceride, cholesterol, and alkaline phosphatase concentration were significantly different from the four adjacent times. The superoxide dismutase (SOD) activity in the liver and gills increased at 6 h and 24 h, and the gills and liver had antioxidant reactions in a short time. The malondialdehyde (MDA) concentration in the gills changed significantly at 6 h, while that in the liver did not change significantly. The gills were more sensitive to temperature stress than the liver and muscle. Acute high-temperature stress affected yellowfin tuna's antioxidant enzymes and metabolic indexes, resulting negative trend in physiological indexes, indicating that yellowfin tuna juveniles are susceptible to elevated temperature.

Keywords: biochemical indexes; metabolism; serum ionic concentration; immune; oxidative stress parameters

Citation: Liu, H.; Fu, Z.; Yu, G.; Ma, Z.; Zong, H. Effects of Acute High-Temperature Stress on Physical Responses of Yellowfin Tuna (*Thunnus albacares*). *J. Mar. Sci. Eng.* **2022**, *10*, 1857. https://doi.org/10.3390/jmse10121857

Academic Editor: Valerio Zupo

Received: 3 November 2022
Accepted: 24 November 2022
Published: 2 December 2022

1. Introduction

Fish often live in various water environments. It is a significant physiological process for fish to adjust to environmental pressure. Temperature change is one of the most common environmental changes, which is considered the main abiotic factor affecting aquatic animals [1]. At present, there are many related studies on the impact of temperature on fish, including killifish (*Oryzias latipes*) [2], sardine (*Sardine pilchardus*) [3], grass carp (*Ctenophryngodon Idella*) [4], black head minnow (*Fathead minnow*) [5]. The change in water temperature can lead to changes in the immune system, metabolism, oxidative stress, and other physiological changes to adjust to the environment [6–9]. Temperature fluctuations cause physiological changes in fish that generally begin by causing stress and then metabolic rates change [1]. Metabolic changes cause the production of reactive oxygen species [10]. Excessive reactive oxygen species will damage DNA, protein, and lipids [11], resulting in oxidative damage. At the same time, temperature also affects fish feeding and digestive processes, thus affecting their metabolism [12]. Unsuitable water environment temperature also leads to slow growth and poor appetite of fish. Therefore, studying the impact of temperature on the physical responses in fish is crucial.

The temperature change can cause stress, composed of components characterized by sympathetic nerve activation and the secretion of adrenaline and cortisol [13]. The second stage is the increase in plasma glucose and the disorder of osmotic pressure regulation [13]. Cortisol (COR) is a steroid hormone [14], which has many biological activities, including maintaining osmotic pressure, regulating blood glucose, and inhibiting immunity [15]. Fish can also adapt to temperature changes by changing the superoxide dismutase (SOD) activity and then influencing the content of malondialdehyde (MDA) [16]. SOD is an important antioxidant enzyme that catalyzes the disproportionation of free superoxide anion radicals to hydrogen peroxide, which is then converted into water and oxygen by catalase or glutathione peroxidase [17]. Elevated MDA levels are an indicator of lipid peroxidation, which results from oxidative stress damage caused by exposure of fish to environmental changes or pollutants [18]. SOD can alleviate the oxidative damage caused by MDA produced by lipid peroxidation. Heat stress can also lead to significant changes in the metabolism-related indicators of fish, such as triglyceride, cholesterol [19] and plasma components calcium and magnesium, alanine aminotransferase (ALT), and alkaline phosphatase (ALP) [20].

Tuna is a highly demanded marine fish. Its back muscle contains 26.2% crude protein and 0.2% fat. It has rich nutrition [21]. Additionally, it is one of the most important economic fish in the world [22,23]. Tuna products are exported to more than 60 countries worldwide, of which three major markets are Japan, the European Union, and the United States [24]. In the 50 years from 1950 to 2000, the total catch of commercial tuna stocks increased from 4000 tons to 3.9 million tons [25]. Yellowfin tuna is an important species caught by fisheries in the Pacific region [26] and the global average annual capture production has increased year by year since the 1960s, but with significant volatility after 2004 [27]. The reason was mainly due to the exhaustion of wild resources. Resource survey results show that since the 1970s, wild yellowfin tuna spawning has been in a long-term decline, and the fishing mortality of adults and juveniles has continued to increase [28]. Wild resources of yellowfin tuna in the Central and Western Pacific Oceans have been fully exploited [29], and resources are declining. Therefore, it is urgent to conduct research related to the artificial culture of yellowfin tuna.

Yellowfin tuna (*Thunnus albacares*) belongs to the mackerel family and the tuna genus [30]. It is a highly migratory fish species in the ocean. It can swim at high speed and in deep water. It can quickly dive to the cold-water area below the thermocline (20 °C isotherms) to feed, and the maximum depth exceeds 1000 m [31]. Tuna can automatically adjust the active water depth when encountering temperature changes in the sea [32], but in the cage or land-based culture, the active space is limited, and high temperatures cannot be avoided. The appropriate temperature for yellowfin tuna larval is 28.0 ± 1.0 °C [33], and after our observation, we found that the summer temperature of the land-based culture pond in the tropics can be as high as 34 °C. In fish farming, the aquaculture water temperature is easy to maintain at a high level in summer, and summer is the season of increased fish diseases [34]. Therefore, it is essential to explore the expression and change of fish's physiological and biochemical indicators, immune function, and oxidative stress parameters under acute high-temperature conditions and analyze fish's response to high temperature. In this experiment, the water temperature is raised to 34 °C. By measuring the relevant indicators of serum, gills, liver, and muscle of young yellowfin tuna at 0 h and 6 h, 24 h and 48 h after the change of environmental conditions, the effects of acute temperature rise on osmotic physiology and oxidative stress parameters of young yellowfin tuna are discussed, it provides a theoretical basis for in-depth study of the stress response of the organism caused by environmental changes, provides a reference for yellowfin tuna aquaculture.

2. Materials and Methods

2.1. Experiment Design and Sample Collection

Juvenile yellowfin tuna used in the experiment were cultured in offshore sea cages near Xincun Harbour, Xincun Town, Lingshui County, Hainan Province. It was temporarily raised for seven days in the Tropical Aquaculture Research and Development Center, South China Sea Fisheries Research Institute, Chinese Academy of Fishery Sciences. A total of 60 fish were randomly collected before the experiment and transferred into two indoor cement tanks (5 m in diameter and 2.5 m deep) with a recirculating water system. The water temperature was controlled at 28.0 ± 0.5 °C, dissolved oxygen > 5.27 mg/L, ammonia nitrogen < 0.1 mg/L, pH 7.57 ± 0.12, salinity 32‰. Fish were fed daily from 08:30 to 09:00. Fresh miscellaneous fish (4 cm × 2 cm pieces; *Trachurus japonicus, Mene maculata*) were fed with 5–8% body weight daily by satiety.

Upon the experiment conducted, the mean body length and wet weight were 28.03 ± 1.78 cm and 503.23 ± 36.78 g, respectively. At the beginning of the experiment, 60 fish were randomly collected and divided into two groups. The temperature of the control group and the high-temperature group were set at 28 °C (control group) and 34 °C (HT group) in 3000 L tanks with three replicates each. The rising speed of water temperature is 2 °C/h, and the timing starts when the temperature reaches 34 °C. The water temperature of the HT group was raised and maintained by heating rods (Sensen Group Co., Ltd., Zhoushan, China)

The sampling time was at 0 h, 6 h, 24 h, and 48 h after stress. When the stress time was up, the samples were taken immediately. Three fish from each rearing tank were randomly collected and anesthetized with 0.03% MS-222, and body length and weight were measured. Liver, gill, red muscle, white muscle, and blood were taken. Blood was taken from the caudal vein, the white muscle from 2 cm below the dorsal fin, and the red muscle from near the spine perpendicular to the dorsal fin. Each tissue was stored in 2 mL RNA-free tubes at -80 °C until enzyme activity measurement. Store the extracted blood in a 2 mL centrifuge tube and use a desktop high-speed freezing centrifuge (EXPERT 18K-R) for centrifugation (temperature 4 °C, rotating speed 3000 R·min^{-1}, lasting for 10 min) after standing for 1 h, and store it at -80 °C until analysis.

2.2. Enzyme Activity Measurement

All the tissue samples were partially thawed and homogenized mechanically using a tissue homogenizer on ice. The suspensions were centrifuged according to the requirements of the kits and the protein content in the supernatant was determined by the BCA Protein Assay kit (A045-4-2). The activities of SOD (A001-3-2) and MDA (A003-1-2) contents in the gill, liver, red muscle, and white muscle were measured. The concentration of COR (H094), triglyceride, cholesterol, ALP, osmotic blood glucose, lactic acid, K$^+$, Na$^+$, Cl$^-$, C3, and C4 in the serum were measured, and the concentration of ALP in the liver was measured. All of the above assays were determined using commercial kits (Nanjing Jiancheng Bioengineering Institute, Nanjing, China) in triplicates. Ion concentration, osmotic pressure, glucose, and lactic acid in serum were measured by a blood gas analyzer (PL2000 Plus, Nanjing Perlang Medical Equipment Co., Ltd., Nanjing, China), triglyceride, cholesterol, ALP, C3, C4, LDH, and liver antioxidant index ALP, LDH was measured by the biochemical analyzer (PVZS-300X, Beijing Prolong New Technology Co., Ltd., Beijing, China). The above measurements using the blood gas analyzer and biochemical analyzer were carried out in accordance with the instructions in triplicates.

2.3. Calculations and Statistical Analysis

Excel 2021 software was used for data sorting, Origin 2021 was used for drawing, and SPSS 26.0 software was used for significant difference analysis. Comparisons between different groups at the same time were conducted by independent *t*-test, comparisons between different times in the HT group were conducted by independent one-way ANOVA and least significant difference (LSD) test, and significant difference was set at $p < 0.05$.

3. Results

3.1. Changes in Serum Indexes of Juvenile Yellowfin Tuna under an Acute Temperature Rise

3.1.1. Changes in Ion Concentration, Osmotic Pressure, Blood Glucose, and Lactic Acid in the Serum of Juvenile Yellowfin Tuna under an Acute Temperature Rise

The osmotic pressure of the HT group (Figure 1a) showed a trend of increasing first and then decreasing with time, reaching the highest value at 24 h. There was a significant difference between 6 h, 24 h, and 48 h. At 6 h and 48 h, the osmotic pressure of the HT group was lower than the concentration in the control group and higher than the control group at 0 h and 24 h (Figure 1b). At 0 h, the osmotic pressure between the HT group and 28 °C had no significant difference, but there was a significant difference at 6 h, 24 h, and 48 h. The blood glucose in the HT group did not change with time (Figure 1c), there was a gradual increase, but it was not significant. At 24 h, the blood glucose in the HT group was higher than the concentration in the control group and lower than the concentration in the control group at other time points (Figure 1d). At 0 h and 24 h, there was no significant difference in blood glucose levels between the HT group and control group, but there was a significant difference at other time points. There was no significant difference between the lactic acid groups in the HT group (Figure 1e), but it fluctuated up and down, but not significantly (Figure 1f).

The K^+ concentration in the HT group (Figure 1g) showed a trend of first decreasing and then increasing with the prolongation of time. There was no significant difference between 0 h and 6 h, 24 h, and 48 h. At 48 h, K^+ in the HT group was lower than in the control group and higher than in the control group at other time points (Figure 1h). The concentration of Na^+ and Cl^- increased first and then decreased with the extension of stress time, with the same trend, and reached the peak at 24 h. Na^+ concentration in the HT group (Figure 1i) had significant differences at 6 h, 24 h, and 48 h. At 24 h, the Na^+ concentration in the HT group was higher than that in the control group and lower than that in the control group at other times (Figure 1j). The concentration of Cl^- in the HT group (Figure 1k) was significantly different between adjacent groups. At 24 h, the Cl^- concentration in the HT group was higher than that in the control group and lower than that in the control group at other times (Figure 1l). The concentration order of three ions in each group was $Na^+ > Cl^- > K^+$.

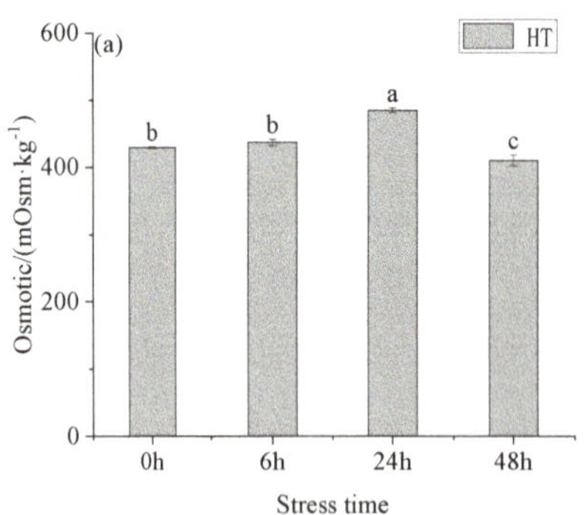

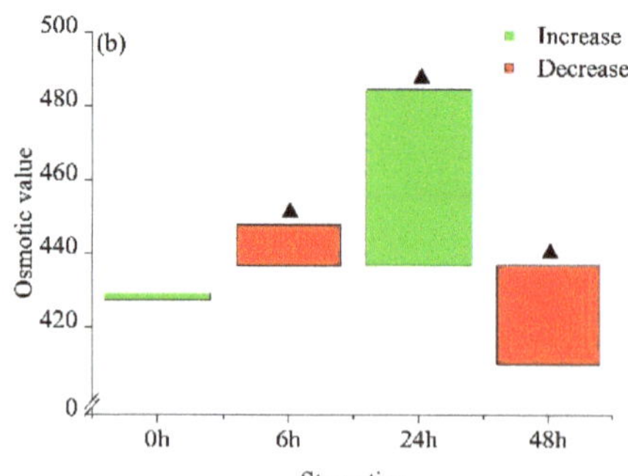

Figure 1. *Cont.*

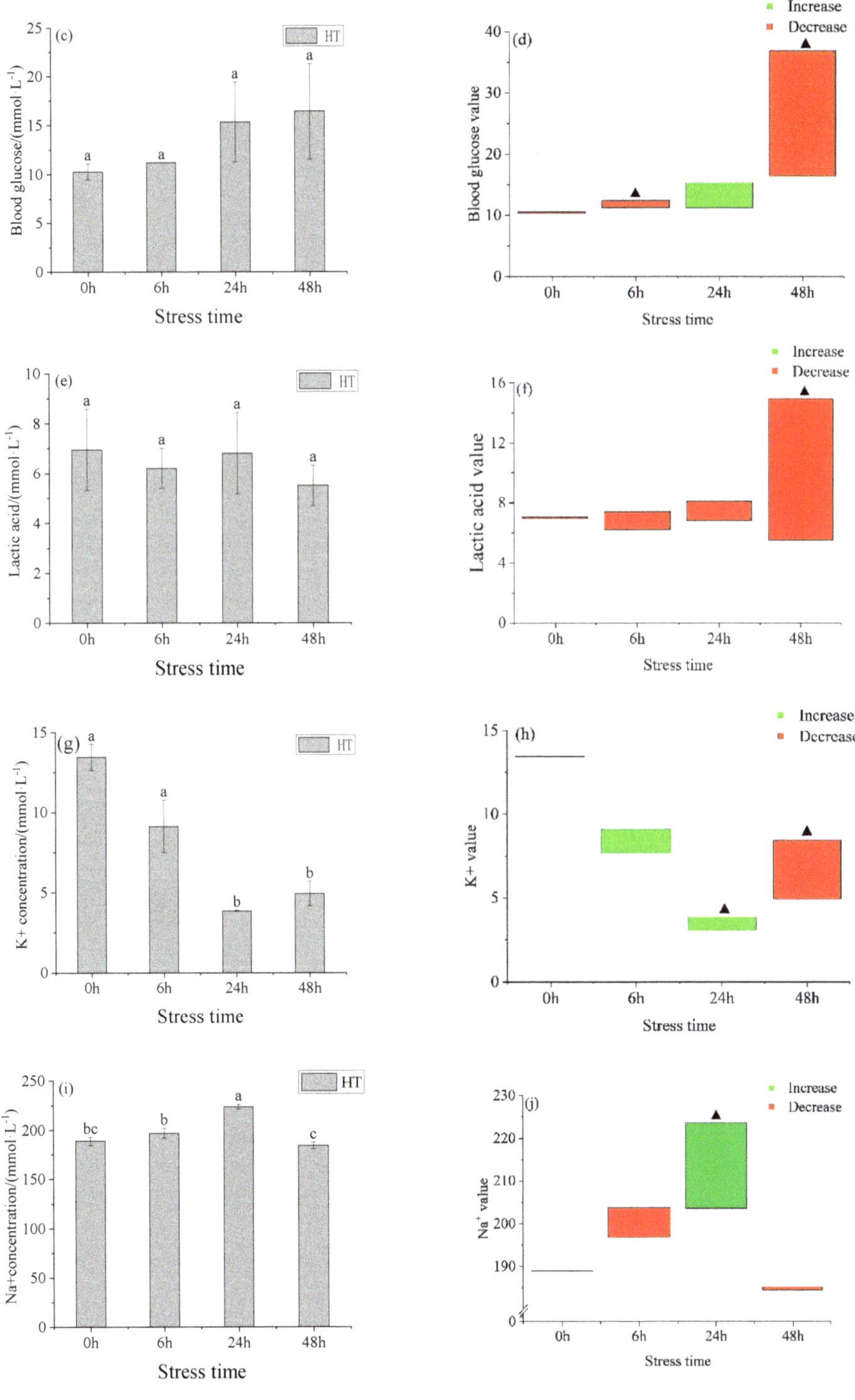

Figure 1. *Cont.*

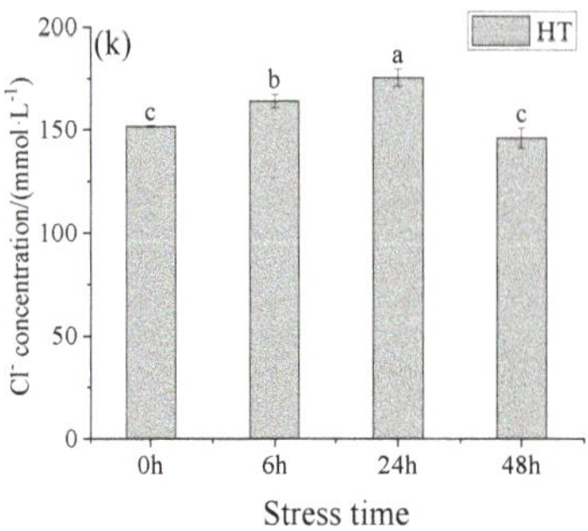
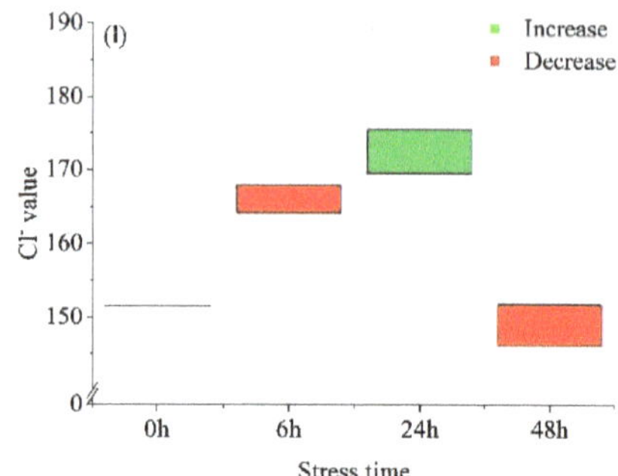

Figure 1. Changes in osmotic pressure (**a**), osmotic pressure value (**b**), blood glucose (**c**), blood glucose value (**d**), lactic acid (**e**), lactic acid value (**f**), K⁺ concentration (**g**), K⁺ value (**h**), Na⁺ concentration (**i**), Na⁺ value (**j**), Cl⁻ concentration (**k**) and Cl⁻ value (**l**) in the serum of young yellowfin tuna under acute high–temperature stress. The value is the gap of the experimental group minus the control group. Red means down and green means up. The significantly different between the control group and the HT group was represented by a ▲. Different and the same letters indicate a significant difference ($p < 0.05$) and insignificant difference ($p > 0.05$).

3.1.2. Changes in Serum Cortisol in Juvenile Yellowfin Tuna under an Acute Temperature Rise

After the acute temperature increase, the cortisol in the HT group first decreased and then increased with the prolongation of stress (Figure 2a), and there was no significant difference between adjacent groups. Among them, the difference between the HT group and the control group (Figure 2b), during 48 h, the HT group was higher than the control group. There was no significant difference in serum cortisol concentration between the control group and the HT group at 0 h, 6 h, and 24 h, but there was a significant difference between the control group and the HT group at 48 h.

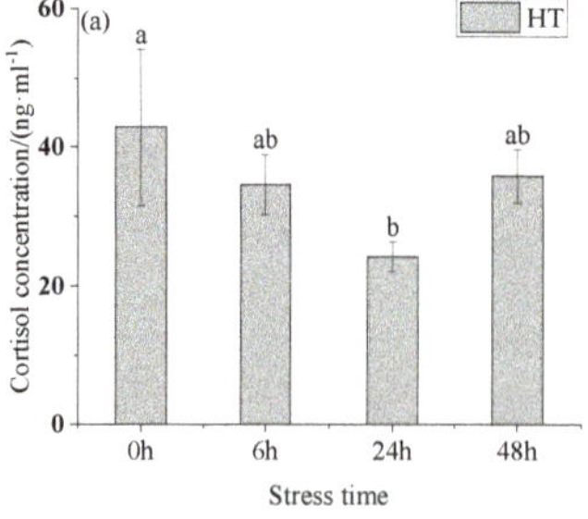
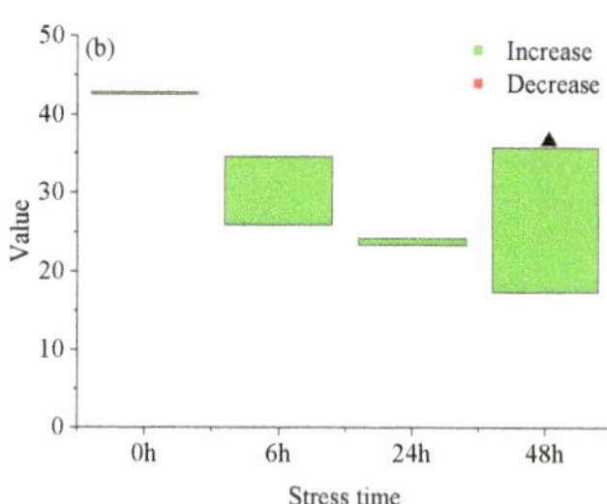

Figure 2. Changes of serum cortisol (**a**) and cortisol value (**b**) in young yellowfin tuna under acute high-temperature stress. The value is the gap of the experimental group minus the control group. Red means down and green means up. The significantly different between the control group and the HT group was represented by a ▲. Different and the same letters indicate a significant difference ($p < 0.05$) and insignificant difference ($p > 0.05$).

3.1.3. Changes of Metabolic Indexes in Serum of Juvenile Yellowfin Tuna under an Acute Temperature Rise

The serum triglyceride and cholesterol concentration in the HT group (Figure 3a,c) showed a trend of first decreasing and then increasing with time, there were significant differences between the groups. The concentration of serum triglycerides in the HT group was lower than the corresponding concentration in the control group at four times point (Figure 3b). There was no significant difference between the HT group and the control group at 0 h, but there were significant differences at the other points. The cholesterol concentration in the HT group was lower than the control group at 6 h and higher than at the other points (Figure 3d). There was no significant difference in serum cholesterol concentration between the HT group and 28 °C at 0 h, but there was a significant difference between 6 h, 24 h, and 48 h. The activity of alkaline phosphatase in the serum of the HT group (Figure 3e) showed a trend of first decreasing and then increasing. The lowest value was reached at 6 h, and it reached to gradually stable after 6 h. The activity of alkaline phosphatase of the HT group was higher than the control group all the time. There was no significant difference in serum alkaline phosphatase concentration between the HT group and the control group at 0 h, but there was a significant difference at the other times point.

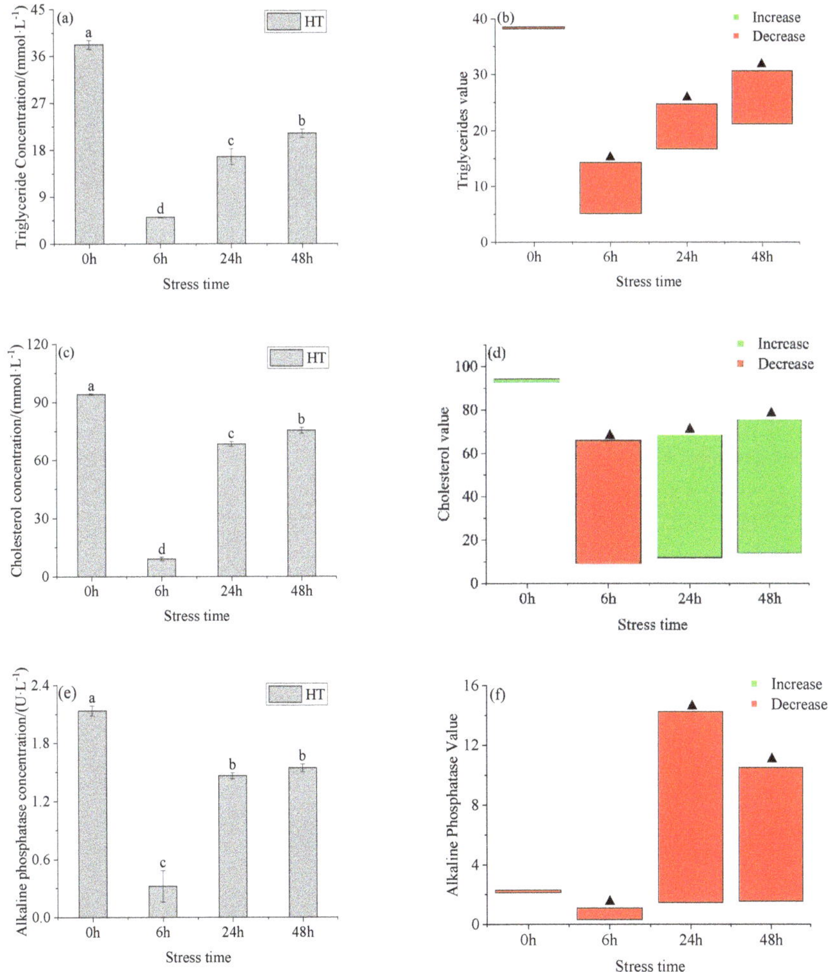

Figure 3. Changes in triglyceride (**a**), triglyceride value (**b**), cholesterol (**c**), cholesterol (**d**), alkaline phosphatase (**e**), and alkaline phosphatase value (**f**) in serum of young yellowfin tuna under acute high–temperature stress. The value is the gap of the experimental group minus the control group. Red means down and green means up. The significantly different between the control group and the HT group was represented by a ▲. Different and the same letters indicate a significant difference ($p < 0.05$) and insignificant difference ($p > 0.05$).

3.1.4. Changes of Immune Indexes in Serum of Juvenile Yellowfin Tuna under an Acute Temperature Rise

After acute temperature increase, the concentration of C3 (Figure 4a) in serum in the HT group showed a downward trend, and it remained stable with time, there was no significant difference between 6 h, 24 h, and 48 h. At 0 h, there was no significant difference between the HT group and 28 °C, and there was a significant difference at other time points. The C3 concentration in the serum of the HT group was higher than the corresponding concentration in the control group at 0 h and 6 h, and lower than the corresponding concentration in the control group at 24 h and 48 h. The concentration of complement C4 (Figure 4c) in the serum of the HT group showed a trend of first decreasing and then increasing with the prolongation of time, and there was a significant difference between each time point. At 0 h, there was no significant difference between the HT group and 28 °C, and there was a significant difference between 6 h, 24 h, and 48 h.

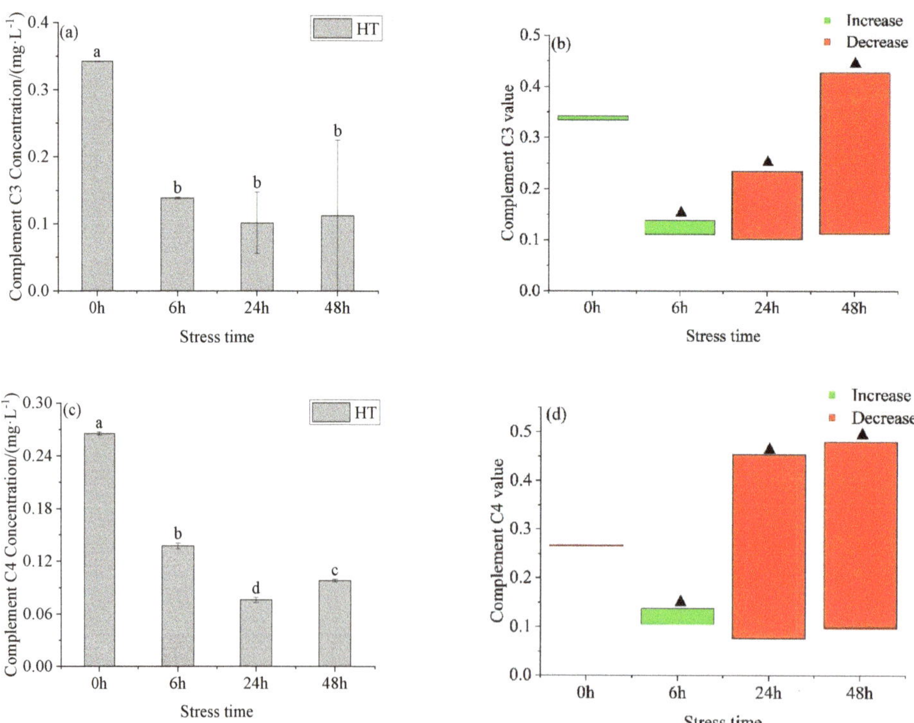

Figure 4. Changes in complement C3 concentration (**a**), C3 value (**b**), complement C4 concentration (**c**) and C3 value (**d**) in serum of young yellowfin tuna under acute high–temperature stress. The value is the gap of the experimental group minus the control group. Red means down and green means up. The significantly different between the control group and the HT group was represented by a ▲. Different and the same letters indicate a significant difference ($p < 0.05$) and insignificant difference ($p > 0.05$).

3.2. Changes in Oxidative Stress Parameters in Organs of Juvenile Yellowfin Tuna under an Acute Temperature Rise

The superoxide dismutase (SOD) in the gills of the HT group (Figure 5a) showed an increasing and gradually stable trend, and there was no significant difference at 6 h, 24 h, and 48 h. The activity of SOD in the gills in the HT group gills was higher than in the control group at all times (Figure 5b). There was no significant difference between the HT group and the 28 °C at all times. The activity of SOD in the liver (Figure 5c) in the HT group first increased and then decreased with time, and the activity reached the highest value at 24 h, and there was no significant difference between 6 h and 24 h. At 6 h and 24 h, the liver SOD activity of the HT group and control group was significantly different (Figure 5d).

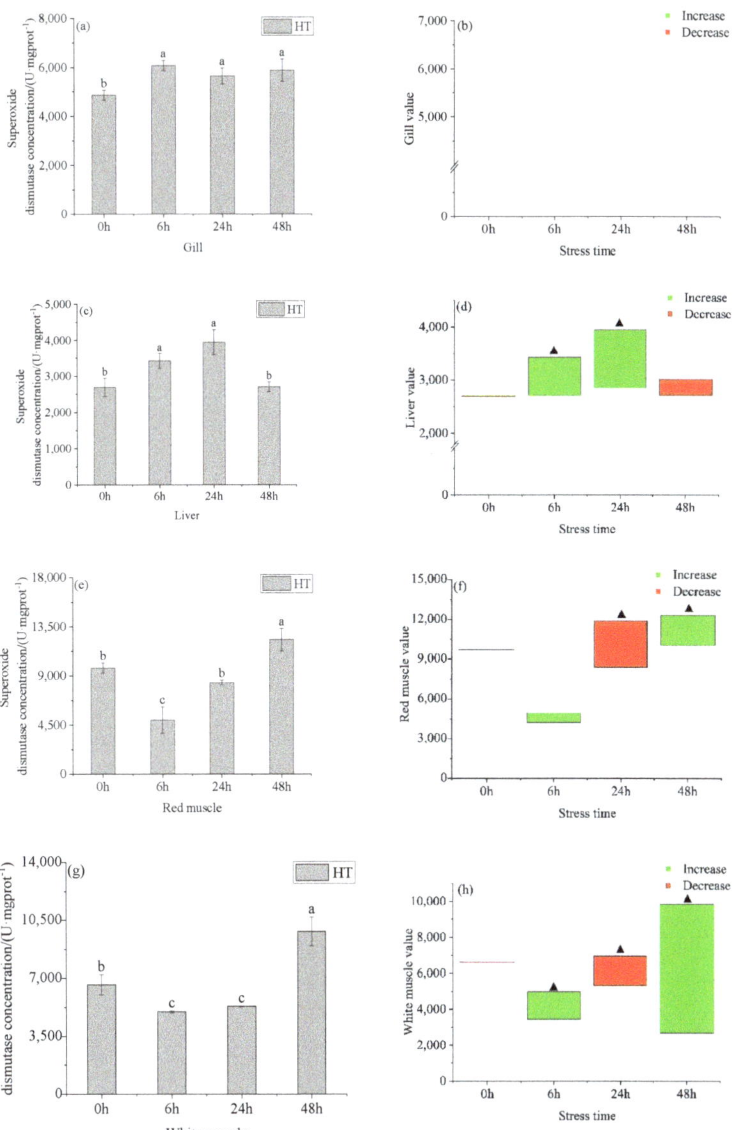

Figure 5. Changes in gill superoxide dismutase (SOD) (**a**), gill SOD value (**b**), liver SOD (**c**), liver SOD value (**d**), red muscle SOD (**e**), red muscle SOD value (**f**), white muscle SOD (**g**) and white muscle SOD value (**h**) in organs of young yellowfin tuna under acute high–temperature stress. The value is the gap of the experimental group minus the control group. Red means down and green means up. The significantly different between the control group and the HT group was represented by a ▲. Different and the same letters indicate a significant difference ($p < 0.05$) and insignificant difference ($p > 0.05$).

The activity of SOD in the red and white muscle of the HT group (Figure 5e,g) showed a decreasing and then increasing trend, with a significant difference between adjacent groups in the red muscle. The activity in the red muscle of the HT group was higher than the corresponding time concentration in the control group at 0 h, 6 h, and 48 h, and lower than the control group at 24 h (Figure 5f). There was no significant difference in the activity of SOD in red muscle between the HT group and control group at 0 h and 6 h, but there was a significant difference between 24 h and 48 h. The activity in the white muscle was no significant difference at 6 h and 24 h. The activity in the white muscle of the HT group was higher than the control group at 0 h and 24 h, and lower than the control group at 6 h and 48 h. There was no significant difference in SOD activity between the HT group and control group at the corresponding time at 0 h, but there had a significant difference at the other times.

After acute heating, the concentration of malondialdehyde in the gills of the HT group (Figure 6a) increased first and then decreased with time, and there was no significant difference between 6 h and 24 h. The concentration of malondialdehyde in the gills of the HT group was higher than that in the control group at 0 h, 6 h, and 24 h, and lower than 48 h (Figure 6b). There was no significant difference in the concentration of malondialdehyde in the gills of the HT group and control group all the time.

The MDA concentration of the liver in the HT group (Figure 6c) showed an upward trend with the prolongation of the temperature stress time. There was no significant difference between the concentrations at 0 h and 6 h, 24 h, and 48 h. The MDA concentration in the liver of the HT group was higher than that in the control group at 0 h, 24 h, and 48 h, and lower than that in the control group at 6 h (Figure 6d). At 0 h and 24 h, there was no significant difference in the concentration of malondialdehyde in the liver between the HT group and control group, but there was a significant difference at 24 h and 48 h. There was no significant difference in the concentration of MDA between 0 h and 6 h (Figure 6e) in the red muscle of the HT group, there was a significant difference between 24 h and 48 h. Concentrations in the HT group were higher than those in the control group at 0 h, 6 h, and 48 h, and lower than those in the control group at 24 h (Figure 6f). The MDA in the red muscle of the HT group was not significantly different from that in the control group at all times. The concentration of MDA in the white muscle of the HT group (Figure 6g) showed a trend of first decreasing and then increasing with the prolongation of stress time, and there was a significant difference between 0 h and 6 h, and 24 h and 48 h had no significant difference. At 6 h, 24 h, and 48 h, the concentration of MDA in the white muscle of the HT group was lower than that in the control group (Figure 6h). At 6 h, the concentration of malondialdehyde in the white muscle of the HT group had significant difference from that in the control group, but there had no significant difference at the other times.

3.3. Changes in Immune Indexes in the Liver of Juvenile Yellowfin Tuna under an Acute Temperature Rise

The alkaline phosphatase activity in the liver in the HT group (Figure 7a) decreased at first and then increased with prolonged stress time. The concentration at 24 h was significantly lower than the other three time points, and there was no significant difference between 6 h and 48 h. After temperature stress, the hepatic alkaline phosphatase activity of yellowfin tuna in the HT group was higher than the corresponding concentration in the control group (Figure 7b) at 0 h and lower than the control group at the other points. At 0 h and 48 h, there was no significant difference in the concentration of hepatic alkaline phosphatase between the HT group and control group, but there was a significant difference at 6 h and 24 h.

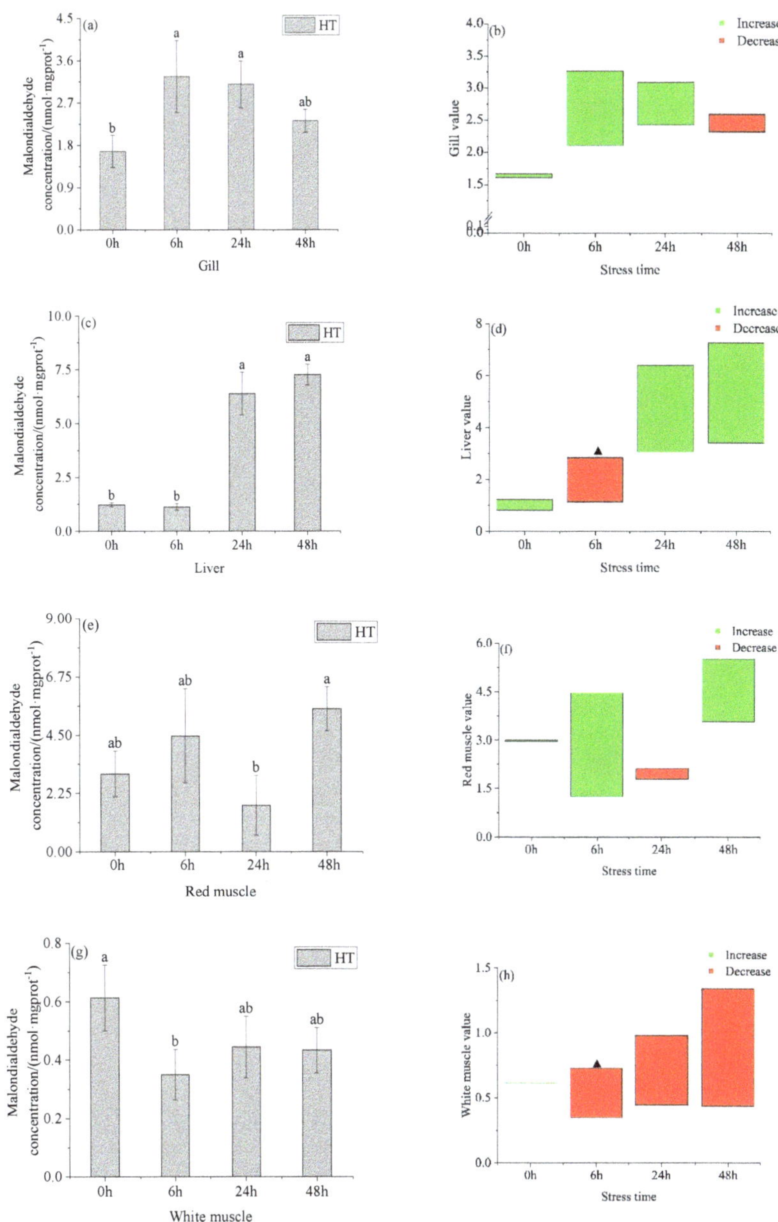

Figure 6. Changes in gill malondialdehyde (MDA) (**a**), gill MDA value (**b**), liver MDA (**c**), liver MDA value (**d**), red muscle MDA (**e**), red muscle MDA value (**f**), white muscle MDA (**g**) and white muscle MDA value (**h**) in organs of young yellowfin tuna under acute high–temperature stress. The value is the gap of the experimental group minus the control group. Red means down and green means up. The significantly different between the control group and the HT group was represented by a ▲. Different and the same letters indicate a significant difference (*p* < 0.05) and insignificant difference (*p* > 0.05).

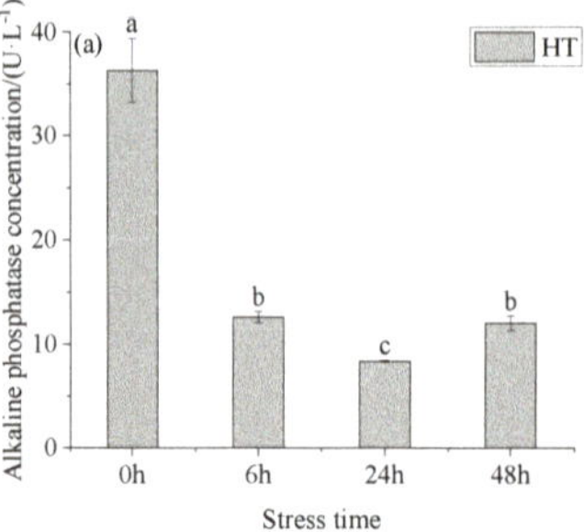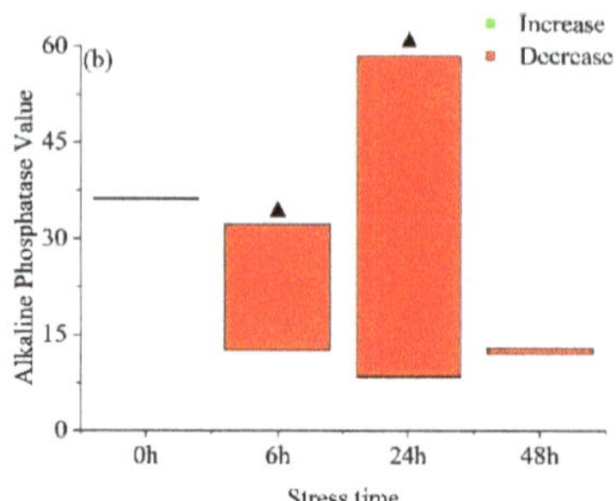

Figure 7. Changes in alkaline phosphatase (**a**) and alkaline phosphatase value (**b**) in the liver of young yellowfin tuna under acute high–temperature stress. The value is the gap of the experimental group minus the control group. Red means down and green means up. The significantly different between the control group and the HT group was represented by a ▲. Different and the same letters indicate a significant difference ($p < 0.05$) and insignificant difference ($p > 0.05$).

4. Discussion

4.1. Changes in Serum Indexes of Juvenile Yellowfin Tuna under an Acute Temperature Rise

4.1.1. Changes in Ion Concentration, Osmotic Pressure, Blood Glucose, Lactic Acid, and Cortisol in Serum of Juvenile Yellowfin Tuna under an Acute Temperature Rise

The change trend of osmotic pressure in the HT group was first increased and then decreased. This was because the fish maintained its own pressure by regulating its own osmotic pressure to regulate the concentration of ions after stress. The research showed that [35] the effect of environmental temperature stress on juvenile catfish plasma ion concentration and osmotic pressure concentration, and found that Na^+, Cl^- and osmotic pressure showed an upward trend, while K^+ remained roughly unchanged. The results of Na^+, Cl^-, and osmotic stress were similar to those of this study, but K^+ was different. It is found that the ionic permeability increases [36] when the temperature rises, so Na^+, Cl^-, and osmotic pressure show an upward trend. The change in K^+ concentration may be due to the fact that after the increase in cell membrane permeability, K^+ needs to enter the cell to play a balancing role, leading to the decrease in the concentration. The Na^+ and Cl^- concentration change at 48 h, which may be due to the balance of K^+.

The research reported that blood glucose increased with the increase in temperature in carp and Senegalese sole (*Solea senegalensis* Kaup) under high temperatures [37–39]. The changing trend in this study is the same as that in previous studies, indicating that the metabolism is more vigorous and more glycogen is needed. In this study, the lactic acid content in the HT group is lower than the control group most of the time, and lactic acid in the HT group shows a downward trend. This indicates that to adapt to the environment under high-temperature stress, lactic acid was taken to the liver and decomposed into CO_2 and H_2O, reducing the lactic acid content. The research showed that the plasma cortisol concentration of the fish adapted to high temperature was about twice that of the control fish, which was consistent with the results of carp [40,41] and black snapper (*Acantopagrus schlegelii*) [42] adapting to the high temperature previously reported. The result of this study is that it decreases first and then increases, which is different from the results of other studies. This may be because the yellowfin tuna is a kind of temperate fish, which has a tolerance to temperature for a short period of time. The HT group rises at 48 h, which may be due to the stress response gradually enhanced with the extension of time.

4.1.2. Changes of Metabolic Indexes in Serum of Juvenile Yellowfin Tuna under an Acute Temperature Rise

The research showed that [43] the growth performance and metabolism of Roche Labeo (*Labeo rohita*) rohita under heat stress and found that the triglycerides and cholesterol concentrations in serum decreased gradually. The research showed [19] the physiological and biochemical reactions of fat short cap carp (*Piaractus brachypomus*) under temperature stress. The results showed that the contents of triglycerides and cholesterol in plasma decreased after heat shock. The results of this study showed that the triglyceride concentration of the control group and the experimental group decreased gradually after 6 h, which was consistent with the previous study. However, the concentration increased after 24 h, which may be due to the prolonged stress time, the body's adaptation to environmental changes, and the gradual recovery of triglyceride metabolism, leading to the increase in triglyceride concentration. Alkaline phosphatase (ALP) is an essential non-specific immune marker enzyme and metabolic regulator enzyme. This enzyme has detoxification, defense, and digestion functions and is also an essential metabolic regulating enzyme involved in phosphate group transport and metabolism and is a sign of fish health [44]. Environmental changes affect its content, reflecting the stress state of fish [45]. The research showed that [46] rainbow trout's physiological and biochemical reactions to heat stress. The results showed that the alkaline phosphatase decreased significantly after the temperature increased. This study showed that the serum alkaline phosphatase concentration in the HT group decreased first and then increased, similar to the previous study. This is because alkaline phosphatase promotes fat degradation and reduces concentration under high-temperature stress. Under long-term pressure, the body's triglyceride, cholesterol, and alkaline phosphatase concentrations increase. The results showed that acute heat stress could promote the lipid metabolism of fish in a short time.

4.2. Changes in Oxidative Stress Parameters and Immune Indexes of Juvenile Yellowfin Tuna under an Acute Temperature Rise

Superoxide dismutase can maintain the superoxide anion free radicals produced under normal conditions and maintain balance [47]. The research showed that [48] the changes of oxidative stress and related enzyme activities in goldfish tissues caused by temperature rise, indicating that the activity of SOD in the liver has increased by about twice, showing a similar trend in the muscle, which is consistent with the results of this study. The activity of SOD in the liver decreases at 48 h, this may result in the decreased antioxidant capacity in the liver of juvenile yellowfin tuna due to the prolonged stress time. The research showed that [49] the antioxidant response of carp liver under temperature stress. The results showed that under the same conditions, the activity of SOD in the liver decreased with the increase in temperature, under the same salinity, SOD in gills gradually decreased with the temperature increase. This study may differ from the results of our study because the heat stressors previously studied by researchers hinder intestinal digestion and local immunity, thereby inhibiting systemic immunity. In this study, heat stress improved the antioxidant capacity of yellowfin tuna, but did not inhibit it. The results showed that the time after the high temperature had little effect on the SOD in gills, and the SOD in gills tended to be stable after gradually adapting. The change trend of SOD in the red and white muscle of the experimental group was the same, but the red power was more sensitive to temperature. In this study, the muscle content is the highest, which may be because different experimental times have different effects on each organ or other species.

Malondialdehyde (MDA) is a substance with strong toxicity to organisms after the decomposition of lipid peroxide. It can directly reflect the degree of lipid peroxidation and cell damage, and indirectly reflect the ability of cells to remove free radicals [50]. The research showed [51] the tissue oxidative stress response of Nile tilapia under temperature shock. The results showed that the level of malondialdehyde in gills increased significantly. In this study, the concentration of MDA in the gills of the experimental group gradually increased and then decreased, which is similar to the previous study results. Dawood, MAO [51] found that under the same salinity, MDA in gills gradually increased with the increase in temperature, which is also similar to the results of this study. With the extension of time, fish gradually adapt to the environment, SOD increases, and MDA decreases. The MDA had no significant change at the 6th hour in the liver. It may be that the gills are more sensitive to changes in environmental conditions than the liver, so the MDA content in the gills is significantly increased. In this study, the change of MDA in red muscle is more obvious than that in white muscle, and the MDA content in white muscle gradually decreases and tends to be stable because red muscle is more sensitive to temperature changes than white muscle. According to the analysis of MDA concentration in various organs, acute warming affects the content of the liver, and MDA in gills and red muscles is more sensitive to temperature.

The research showed [52] the effect of different temperature stress on the immune indexes of crucian carp. The results showed that the alkaline phosphatase activity first increased and then decreased with the temperature increase, which was different from the results of this study. This may be because acute warming causes yellowfin tuna to consume a large amount of alkaline phosphatase for non-specific immunity. Additionally, it increases over 48 h, this is because the fish adapt to the temperature change for 48 h, and the non-specific immune function is enhanced. This shows that acute temperature stress significantly changes the non-specific immunity of juvenile yellowfin tuna. The research showed [53] the effect of temperature on the immunity of juvenile carp. The results showed that the activities of C3 and C4 in the liver increased significantly with the temperature increase. This study showed that C3 and C4 decreased first and then increased after high-temperature stress, which was different from previous studies. This may be because heat stress causes cannot be synthesized in a short time, and with the extension of non-specific immune time, the addition of complement gradually increases. These results indicate that the ability of yellowfin tuna to synthesize complement is weak in a short time.

5. Conclusions

The result shows that the serum ion concentration and osmotic pressure, biochemical indicators of blood glucose, and lactate were changed after acute temperature stress. Acute temperature stress can cause excessive free radicals in the body and reduce immune indicators (e.g., ALP in the liver and the complement C3 and C4 in serum). Under acute temperature rise stress, antioxidant enzymes and metabolic indicators of the juvenile yellowfin tuna changed significantly. The triglycerides, cholesterol, and alkaline phosphatase in serum have changed significantly and gradually adapted to the environment with time. The gills and liver also improve the activities of SOD to eliminate free radicals, but still could not stop the increase in MDA, which may cause peroxidation damage to the body. In this study, the juvenile yellowfin tuna is sensitive to temperature rise, and the tendency of physiological activity disorder was aggravated over time within 48 h. Therefore, in actual production and large-scale intensive aquaculture, it is necessary to avoid sharp temperature changes as much as possible, reduce the frequency and duration of acute temperature stress, and make it a suitable breeding and growing environment.

Supplementary Materials: The following supporting information can be downloaded at: https://www.mdpi.com/article/10.3390/jmse10121857/s1, Table S1: Statistical values of HT group (including the squares of F, df, P and eta). Table S2: Statistical values between the control group and experimental group (including the squares of df, P and Sig).

Author Contributions: Conceptualization, G.Y. and H.Z.; methodology, Z.M.; software, Z.M.; validation, H.Z. and H.L.; formal analysis, G.Y.; investigation, Z.F.; resources, Z.M.; data curation, H.L.; writing—original draft preparation, H.L.; writing—review and editing, Z.M. and Z.F.; visualization, H.L.; supervision, G.Y.; project administration, Z.M.; funding acquisition, Z.M. All authors have read and agreed to the published version of the manuscript.

Funding: This work was supported by Hainan Major Science and Technology Project (ZDKJ2021011); Central Public-interest Scientific Institution Basal Research Fund, CAFS (2020TD55), and Central Public-Interest Scientific Institution Basal Research Fund South China Sea Fisheries Research Institute, CAFS (2021SD09).

Institutional Review Board Statement: The animal study protocol was approved by the Institutional Review Board (or Ethics Committee) of Animal Care and Use Committee of South China Sea fisheries Research Institute, Chinese Academy of Fishery Sciences (BIOL5346, 9 May 2022).

Data Availability Statement: The original contributions presented in the study are included in the article/Supplementary Materials. Further inquiries can be directed to the corresponding authors.

Acknowledgments: The authors would like to thank Jicai Liu for his support and help in sample determination.

Conflicts of Interest: The authors declare no conflict of interest.

References

1. Joy, S.; Alikunju, A.P. Oxidative stress and antioxidant defense responses of *Etroplus suratensis* to acute temperature fluctuations. *J. Therm. Biol.* **2017**, *70*, 20–26. [CrossRef] [PubMed]
2. Leggatt, R.A.; Brauner, C.J. Effects of acclimation and incubation temperature on the glutathione antioxidant system in killifish and RTH-149 cells. *Comp. Biochem. Physiol. Mol. Integr. Physiol.* **2007**, *146*, 317–326. [CrossRef] [PubMed]
3. Maltar-Strmecki, N.; Ljubic-Beer, B. Effect of the gamma radiation on histamine production, lipid peroxidation and antioxidant parameters during storage at two different temperatures in sardine (*Sardina pilchardus*). *Food Control* **2013**, *34*, 132–137. [CrossRef]
4. Li, Z.H.; Li, P. Regulation of glutathione-dependent antioxidant defense system of grass carp *Ctenopharyngodon idellaunder* the combined stress of mercury and temperature. *Environ. Sci. Pollut. Res.* **2020**, *28*, 1689–1696. [CrossRef]
5. Clotfelter, E.D.; Lapidus, S.J.H. The effects of temperature and dissolved oxygen on antioxidant defences and oxidative damage in the fathead minnow *Pimephales promelas*. *J. Fish Biol.* **2013**, *82*, 1086–1092. [CrossRef]
6. Wen, X.; Chu, P. Combined effects of low temperature and salinity on the immune response, antioxidant capacity and lipid metabolism in the pufferfish (*Takifugu fasciatus*). *Aquaculture* **2020**, *531*, 735866. [CrossRef]
7. Pereira, L.A.L.; Amanajas, R.D. Health of the Amazonian fish tambaqui (*Colossoma macropomum*): Effects of prolonged photoperiod and high temperature. *Aquaculture* **2021**, *541*, 736836. [CrossRef]
8. Ainsworth, A.J.; Dexiang, C. Effect of temperature on the immune system of channel catfish (*Ictalurus punctatus*)—I. Leucocyte distribution and phagocyte function in the anterior kidney at 10 degrees C. Comparative biochemistry and physiology. *Comp. Physiol.* **1991**, *100*, 907–912.
9. Zapata, A.G.; Varas, A. Seasonal variations in the immune system of lower vertebrates. *Immunol. Today* **1992**, *13*, 142–147. [CrossRef]
10. Boveris, A.; Chance, B. The mitochondrial generation of hydrogen peroxide. General properties and effect of hyperbaric oxygen. *Biochem. J.* **1973**, *134*, 707–716. [CrossRef]
11. Rubbo, H.; Radi, R. Nitric oxide regulation of superoxide and peroxynitrite-dependent lipid peroxidation. Formation of novel nitrogen-containing oxidized lipid derivatives. *J. Biol. Chem.* **1994**, *269*, 26066–26075. [CrossRef] [PubMed]
12. Volkoff, H.; Ronnestad, I. Effects of temperature on feeding and digestive processes in fish. *Temperature* **2020**, *7*, 307–320. [CrossRef] [PubMed]
13. Davis, K.B. Temperature affects physiological stress responses to acute confinement in sunshine bass (*Morone chrysops* x *Morone saxatilis*). *Comp. Biochem. Physiol. Mol. Integr. Physiol.* **2004**, *139*, 433–440. [CrossRef] [PubMed]
14. Mommsen, T.P.; Vijayan, M.M. Cortisol in teleosts: Dynamics, mechanisms of action, and metabolic regulation. *Rev. Fish Biol. Fish.* **1999**, *9*, 211–268. [CrossRef]
15. Iversen, M.; Finstad, B. Stress responses in Atlantic salmon (*Salmo salar* L.) smolts during commercial well boat transports, and effects on survival after transfer to sea. *Aquaculture* **2005**, *243*, 373–382. [CrossRef]
16. Liu, H.Y.; Fu, Z.Y. Effect of Transport Density on Greater Amberjack (*Seriola dumerili*) Stress, Metabolism, Antioxidant Capacity and Immunity. *Front. Mar. Sci.* **2020**, *9*, 931816. [CrossRef]
17. Heink, A.E.; Parrish, A.N. Oxidative stress among SOD-1 genotypes in rainbow trout (*Oncorhynchus mykiss*). *Aquat. Toxicol.* **2014**, *144*, 75–82. [CrossRef]
18. Garcia, D.; Lima, D. Decreased malondialdehyde levels in fish (*Astyanax altiparanae*) exposed to diesel: Evidence of metabolism by aldehyde dehydrogenase in the liver and excretion in water. *Ecotoxicol. Environ. Saf.* **2020**, *190*, 110107. [CrossRef]
19. Ale, A.; Bacchetta, C. Low temperature stress in a cultured fish (*Piaractus mesopotamicus*) fed with *Pyropia columbina* red seaweed-supplemented diet. *Fish Physiol. Biochem.* **2021**, *47*, 829–839. [CrossRef]
20. Kim, J.H.; Park, H.J. Changes in Hematological Parameters and Heat Shock Proteins in Juvenile Sablefish Depending on Water Temperature Stress. *J. Aquat. Anim. Health* **2019**, *31*, 147–153. [CrossRef]
21. Nakamura, Y.N.; Ando, M. Changes of proximate and fatty acid compositions of the dorsal and ventral ordinary muscles of the full-cycle cultured Pacific bluefin tuna *Thunnus orientalis* with the growth. *Food Chem.* **2007**, *103*, 234–241. [CrossRef]
22. Pang, J.H.; Cheng, Q.Q. The sequence and organization of complete mitochondrial genome of the yellowfin tuna, *Thunnus albacares* (Bonnaterre, 1788). *Mitochondrial DNA Part A* **2016**, *27*, 3111–3112. [CrossRef] [PubMed]
23. Jeyasekaran, G.; Arunkumar, G. Authentication of commercially important tuna species landed in Tuticorin coast of Tamil Nadu, India by SE-AFLP method. *Indian J. Fish.* **2018**, *64*, 254–259. [CrossRef]
24. Nguyen, K.Q.; Phan, H.T. Length-length, Length-weight, and Weight-weight Relationships of Yellowfin (*Thunnus Albacares*) and Bigeye (*Thunnus Obesus*) Tuna Collected from the Commercial Handlines Fisheries in the South China Sea. *Thalass. Int. J. Mar. Sci.* **2022**, 1–7. [CrossRef]
25. Shen, H.H. Study on Management System of Tuna Fishery Resources. Ph.D. Thesis, Shanghai Ocean University, Shanghai, China, 2019.

26. Benetti, D.D.; Partridge, G.J. Overview on status and technological advances in tuna aquaculture around the world. In *Advances in Tuna Aquaculture*; Benetti, D.D., Partridge, G.J., Buentello, A., Eds.; Academic Press: Cambridge, MA, USA, 2016; pp. 1–19.
27. FAO. *Thunnus albacares* (Bonnaterre, 1788). 2022. Available online: https://www.fao.org/fishery/en/aqspecies/2497/en (accessed on 2 November 2022).
28. WCPFC. WCPO Yellowfin Tuna (*Thunnus albacares*) Stock Status and Management Advice. Available online: https://www.wcpfc.int/doc/02/yellowfin-tuna (accessed on 17 February 2021).
29. Zhang, P.; Yang, L. Current situation and prospect of tuna and squid resources exploitation in South China Sea. *South. Fish.* **2010**, *6*, 68–74.
30. Liu, H.Y.; Fu, Z.Y. The Complete Mitochondrial Genome of *Pennella* sp. Parasitizing *Thunnus albacares*. *Front. Cell. Infect. Microbiol.* **2022**, *12*, 945152. [CrossRef] [PubMed]
31. Schaefer, K.M.; Fuller, D.W. Movements, behavior, and habitat utilization of yellowfin tuna (*Thunnus albacares*) in the northeastern Pacific Ocean, ascertained through archival tag data. *Mar. Biol.* **2007**, *152*, 503–525. [CrossRef]
32. Aoki, Y.; Aoki, A. Physiological and behavioural thermoregulation of juvenile yellowfin tuna Thunnus albacares in subtropical waters. *Mar. Biol.* **2020**, *167*, 1–14. [CrossRef]
33. Kim, Y.S.; Delgado, D.I. Effect of temperature and salinity on hatching and larval survival of yellowfin tuna *Thunnus albacares*. *Fish. Sci.* **2015**, *81*, 891–897. [CrossRef]
34. Woo, K.J.; Jeong, S.-H. Monitoring of Pathogens in Cultured Fish of Korea for the Summer Period from 2000 to 2006. *J. Fish Pathol.* **2006**, *19*, 207–214.
35. Nordlie, F.G. Influence of environmental temperature on plasma ionic and osmotic concentrations in *Mugil cephalus* Lin. *Comp. Biochem. Physiol. Comp. Physiol.* **1976**, *55*, 379–381. [CrossRef] [PubMed]
36. Rasio, E.A.; Bendayan, M.; Goresky, C.A. Effect of temperature change on the permeability of eel rete capillaries. *Circ. Res.* **1992**, *70*, 272–284. [CrossRef] [PubMed]
37. Shahjahan, M.; Khatun, M.S. Nuclear and Cellular Abnormalities of Erythrocytes in Response to Thermal Stress in Common Carp *Cyprinus carpio*. *Front. Physiol.* **2020**, *11*, 543. [CrossRef]
38. Costas, B.; Aragao, C. Different environmental temperatures affect amino acid metabolism in the eurytherm teleost Senegalese sole (*Solea senegalensis Kaup*, 1858) as indicated by changes in plasma metabolites. *Amino Acids* **2012**, *43*, 327–335. [CrossRef] [PubMed]
39. Fan, X.P.; Qin, X.M. Effects of temperature on metabolism function and muscle quality of grouper during process of keeping alive with water. *Trans. Chin. Soc. Agric. Eng.* **2018**, *24*, 241–248.
40. Arends, R.J.; van der Gaag, R. Differential expression of two pro-opiomelanocortin mRNAs during temperature stress in common carp (*Cyprinus carpio* L.). *J. Endocrinol.* **1998**, *159*, 85–91. [CrossRef] [PubMed]
41. Metz, J.R.; van den Burg, E.H. Regulation of branchial Na+/K+-ATPase in common carp *Cyptinus carpio* L. acclimated to different temperatures. *J. Exp. Biol.* **2003**, *206*, 2273–2280. [CrossRef]
42. Choi, C.Y.; Min, B.H. Molecular cloning of PEPCK and stress response of black porgy (*Acanthopagrus schlegeli*) to increased temperature in freshwater and seawater. *Gen. Comp. Endocrinol.* **2007**, *152*, 47–53. [CrossRef]
43. Roychowdhury, P.; Aftabuddin, M. Thermal stress altered growth performance and metabolism and induced anaemia and liver disorder in *Labeo rohita*. *Aquac. Res.* **2020**, *51*, 1406–1414. [CrossRef]
44. Hoseinifar, S.H.; Roosta, Z. The effects of *Lactobacillus acidophilus* as feed supplement on skin mucosal immune parameters, intestinal microbiota, stress resistance and growth performance of black swordtail (*Xiphophorus helleri*). *Fish Shellfish Immunol.* **2015**, *42*, 533–538. [CrossRef]
45. Silva, M.J.D.; da Costa, F.F.B. Biological responses of Neotropical freshwater fish *Lophiosilurus alexandri* exposed to ammonia and nitrite. *Sci. Total Environ.* **2018**, *616*, 1566–1575. [CrossRef] [PubMed]
46. Li, S.W.; Liu, Y.J. Physiological responses to heat stress in the liver of rainbow trout (*Oncorhynchus mykiss*) revealed by UPLC-QTOF-MS metabolomics and biochemical assays. *Ecotoxicol. Environ. Saf.* **2020**, *242*, 113949. [CrossRef] [PubMed]
47. Cheng, C.H.; Guo, Z.X. Effect of dietary astaxanthin on the growth performance, non-specific immunity, and antioxidant capacity of pufferfish (*Takifugu obscurus*) under high temperature stress. *Fish Physiol. Biochem.* **2018**, *44*, 209–218. [CrossRef] [PubMed]
48. Lushchak, V.I.; Bagnyukova, T.V. Temperature increase results in oxidative stress in goldfish tissues. 2. Antioxidant and associated enzymes. *Comp. Biochem. Physiol. Toxicol. Pharmacol.* **2006**, *143*, 36–41. [CrossRef]
49. Dawood, M.A.O.; Alkafafy, M. The antioxidant responses of gills, intestines and livers and blood immunity of common carp (*Cyprinus carpio*) exposed to salinity and temperature stressors. *Fish Physiol. Biochem.* **2022**, *48*, 397–408. [CrossRef] [PubMed]
50. Duran, A.; Talas, Z.S. Biochemical changes and sensory assessment on tissues of carp (*Cyprinus carpio*, Linnaeus 1758) during sale conditions. *Fish Physiol. Biochem.* **2009**, *35*, 709–714. [CrossRef]
51. Phrompanya, P.; Panase, P. Histopathology and oxidative stress responses of Nile tilapia Oreochromis niloticus exposed to temperature shocks. *Fish. Sci.* **2021**, *87*, 491–502. [CrossRef]
52. Wang, B.; Ma, G.X. Effects of Different Temperatures on the Antibacterial, Immune and Growth Performance of Crucian Carp Epidermal Mucus. *Fishes* **2022**, *6*, 66. [CrossRef]
53. Huang, J.F.; Xu, Q.Y. Effects of temperature and dietary protein on gene expression of *Hsp70* and *Wap65* and immunity of juvenile mirror carp (*Cyprinus carpio*). *Aquac. Res.* **2015**, *46*, 2776–2788. [CrossRef]

Journal of
Marine Science and Engineering

Article

A New Approach to Integrated Multi-Trophic Aquaculture System of the Sea Cucumber *Apostichopus japonicus* and the Sea Urchin *Strongylocentrotus intermedius*

Fangyuan Hu [†], Huiyan Wang [†], Ruihuan Tian, Jujie Gao, Guo Wu, Donghong Yin * and Chong Zhao *

Key Laboratory of Mariculture & Stock Enhancement in North China's Sea, Ministry of Agriculture and Rural Affairs, Dalian Ocean University, Dalian 116023, China
* Correspondence: yindh@dlou.edu.cn (D.Y.); chongzhao@dlou.edu.cn (C.Z.)
† These authors contributed equally to this work.

Abstract: The sea cucumber *Apostichopus japonicus* and the sea urchin *Strongylocentrotus intermedius* are two commercially important species and are widely cultured in China. Here, a laboratory experiment was conducted for 34 days to assess whether the survival, growth and behavior performances are better in the new commercially valuable integrated multi-trophic aquaculture (IMTA) system (group M, 90 *S. intermedius* and 37 *A. japonicus*/10,638 cm^3 of stocking density) than those in the control group for sea urchins (group U, 90 *S. intermedius*/10,638 cm^3 of stocking density) and the control group for sea cucumbers (group C, 37 *A. japonicus*/10,638 cm^3 of stocking density). We found that feeding behavior, crawling behavior, body length and body weight of sea cucumbers were significantly greater in group M than those in group C. These results suggest that the new IMTA system improves fitness-related behaviors and consequently leads to a better growth in *A. japonicus* while maintaining a high biomass. We further found that group M showed significantly larger body size and Aristotle's lantern reflex as well as significantly lower mortality and morbidity in sea urchins, compared to those in group U. This suggests that the new IMTA system greatly improves feeding behavior and body growth, and survival of cultured *S. intermedius*. This IMTA system is a promising candidate to promote the production efficiency of juvenile *A. japonicus* (as primary species) and *S. intermedius* (as subsidiary species) in China.

Keywords: echinoderm; IMTA; feeding behavior; growth; survival

Citation: Hu, F.; Wang, H.; Tian, R.; Gao, J.; Wu, G.; Yin, D.; Zhao, C. A New Approach to Integrated Multi-Trophic Aquaculture System of the Sea Cucumber *Apostichopus japonicus* and the Sea Urchin *Strongylocentrotus intermedius*. *J. Mar. Sci. Eng.* **2022**, *10*, 1875. https://doi.org/10.3390/jmse10121875

Academic Editor: Ka Hou Chu

Received: 14 October 2022
Accepted: 20 November 2022
Published: 3 December 2022

1. Introduction

Integrated Multi-Trophic Aquaculture (IMTA) system attracts wide attentions in recent years [1], because it brings higher profits and more diversification of commercial production with less environmental pollution [2–4]. The land-based IMTA of sea urchins (as primary product species) and sea cucumbers (as subsidiary product species) has been successfully developed due to their high commercial value and feeding habits [5–7]. For example, sea cucumbers *Holothuria tubulosa* ingested 54% of organic matters from the feces of sea urchins *Paracentrotus lividus* [8], which greatly reduces the total waste and increases additional value (i.e., sea cucumbers). However, the aquaculture production of sea cucumber is higher than that of sea urchins in China. The annual production, for example, is 196,564 metric tons in sea cucumbers and 7952 metric tons in sea urchins in China in 2020 [9]. This suggests that the above approach probably is not applicable to producing echinoderms in China. It is, therefore, essential to develop a new approach to IMTA with sea cucumbers, as the primary species, and sea urchins as the subsidiary species with a high stocking biomass in China.

The sea cucumber *Apostichopus japonicus* and the sea urchin *Strongylocentrotus intermedius* are two commercially important species and are widely cultured in China [10,11]. Seed productions of *A. japonicus* and *S. intermedius* are both feasible in May and June [12,13].

The body weight of *A. japonicus* reaches 1–2.5 g in November while the test diameter of *S. intermedius* reaches 6–15 mm at that time [12]. In the production of juvenile *A. japonicus*, polyethylene nets were commonly used as the substrate on the top of nursery tanks to increase culture biomass and avoid intraspecific competition [11]. In this context, many sea cucumber diets (mainly commercial powdered diets) fail to adhere to the nets and settle at the bottom of tanks, making it difficult to use the water space and deposited diets below the nursery tanks for *A. japonicus* situated on the nets. Sea urchins are omnivorous animals and their unique feeding organ, Aristotle's lantern, is adapted for omnivorous diets such as macroalgae, hard calcified surfaces, and soft sediments [14–16]. Many edible macroalgae for sea urchins, such as *Saccharina japonica* (the most commonly food used for sea urchin aquaculture), *Sargassum thunbergii*, *S. polycystum*, are made into powdered diets and used for sea cucumber aquaculture [17–19]. Thus, it is reasonable to assume that the powdered diets wasted in nursery tanks can be effectively used by *S. intermedius*. However, the sufficient food source for *S. intermedius* is not enough to support an effective IMTA with *A. japonicus*. Physical interactions between sea urchins and sea cucumbers display negative effects on the fitness of both species. For example, Sun et al. [20] found that sea cucumbers showed significantly higher escaping speed when behavioral interactions existed between sea cucumbers and sea urchins. Mass mortality occurred in *A. japonicus* with the increased stocking density of *S. intermedius* when they were not cultured separately [21], which may be due to the injuries caused to the sea cucumbers by sea urchin spines. This suggests that a specialized culture facility is further required to separate *S. intermedius* from *A. japonicus* in the IMTA system. Our previous studies found that segregation in multi-layer culture significantly improved survival [22], food utilization and body growth [23] of *S. intermedius*, compared with those without the multi-layer culture. We assumed that a plastic box divided into three layers (each layer has many compartments) full of holes would represent a promising candidate for the *S. intermedius* aquaculture, because it not only allows the powdered diets to enter the facility through the holes for *S. intermedius* consumption, but also provides an additional substrate for *A. japonicus* in the IMTA system.

Here, the main purposes of the present study are to investigate: (1) whether juvenile *S. intermedius* show better survival, feeding behaviors and growth in the IMTA system; (2) whether the survival, fitness-related behaviors and growth of juvenile *A. japonicus* are better in the IMTA system; (3) what is the application potential of the new IMTA system with *A. japonicus* as the primary species and *S. intermedius* as the subsidiary species.

2. Materials and Methods

2.1. Experimental Animals

Juvenile *S. intermedius* (7.4 ± 1.0 mm of test diameter, 0.2 ± 0.1 g of wet body weight, mean ± SD) (n = 20) and *A. japonicus* (green type, 30.1 ± 3.5 mm of relaxed body length, 1.1 ± 0.4 g of wet body weight, mean ± SD) (n = 10) were randomly chosen from an aquafarm of Lvshun, Dalian (121°13′ E, 38°88′ N) on 13 April 2021 and Dalian Zhuang Yuanhai Ecological Seedling Industry Co., Ltd. (122°69′ E, 39°27′ N) on 16 April 2021, respectively. They were subsequently maintained in fiberglass tanks (length × width × height: 1150 × 750 × 600 mm) with aeration in the Key Laboratory of Mariculture & Stock Enhancement in north China's Sea, Ministry of Agriculture and Rural Affairs at Dalian Ocean University (121°56′ E, 38°87′ N). The incandescent light intensity was ~30 lx with the photoperiod (12 light: 12 dark), according to the culture management commonly used for seed production in China. Sea urchins and sea cucumbers were fed the leaf blade of fresh kelp *S. japonica* and a commercial sea cucumber powdered diet (mainly composed of algal powder, the grain size of 0.125 mm) (Anyuan Industrial Co., Yantai, China.), respectively. Two-thirds of seawater was changed daily. Water temperature and salinity were monitored daily using a portable water quality monitoring meter (Xylem Co., OH, USA). They were 15.4 ± 0.2 °C and 30.8 ± 0.1‰, respectively.

Experimental animals were fasted for three days to standardize their nutritional status, and their initial body sizes were subsequently assessed at the beginning of the trial.

2.2. Experimental Design

There were three groups in this study: the control group for *A. japonicus* (group C, Figure 1A), the control group for *S. intermedius* (group U, Figure 1B) and the integrated multi-trophic aquaculture of *A. japonicus* and *S. intermedius* (group M, Figure 1C). The stocking density of sea urchins and sea cucumbers in group M was consistent with those in groups U and C, respectively. In group C, three pieces of polyethylene nets (length × width: 100×100 mm for each net; 1 mm diameter of mesh size) were tied as substrate under a plastic ball (to float those nets) in a cylindrical plastic bucket (diameter × height: 220×280 mm; 37 *A. japonicus*/10,638 cm^3 of stocking density). Thirty-seven *A. japonicus* were randomly chosen and placed in the plastic bucket according to the culture management commonly used for the seed production of sea cucumbers. In group U, 90 juvenile *S. intermedius* (~7 mm of test diameter) were randomly selected and maintained in the plastic cage (length × width × height: $150 \times 150 \times 130$ mm, 3 mm diameter of mesh size) in a cylindrical plastic bucket (diameter × height: 220×280 mm; 90 *S. intermedius*/10,638 cm^3 of stocking density) according to the culture practice for the seed production of sea urchins. In group M, culture areas existed for *A. japonicus* on top and *S. intermedius* at the bottom in the plastic bucket (diameter × height: 220×280 mm; 90 *S. intermedius* and 37 *A. japonicus* /10,638 cm^3 of stocking density). A dismountable plastic box (length × width × height: $150 \times 150 \times 130$ mm, 30 holes of 4 mm diameter/100 cm^2), divided into three layers and six compartments in each layer (Figure 1D), was put into the culture areas of *S. intermedius*. Five *S. intermedius* were randomly selected and put into each compartment of the plastic box (90 *S. intermedius* in total for each box).

Each group had 6 replicates under ~30 lx of incandescent light intensity with the photoperiod (12 h light: 12 h dark) in this study from 24 April 2021 to 28 May 2021. The leaf blade of wild fresh kelp *S. japonica* collected from Heishijiao, Dalian (121°58′ E, 38°87′ N) was provided ad libitum to *S. intermedius* in group U. Animals in groups C and M were fed a commercial powdered diet (Anyuan Industrial Co., Yantai, China). The feces and dead individuals were removed daily for all the groups. Water temperature was not controlled, ranging from 14.3 to 16.6 °C (ambient temperatures). Salinity was 30.7 ± 0.7 ‰, according to the weekly measurement by a portable water quality monitoring meter (Xylem Co., Yellow Springs, OH, USA). All the experimental tanks were kept in still water with aeration, because this does not cause the loss of powdered diet. Two-thirds of the seawater was renewed daily to avoid deterioration of water quality.

2.3. Mortality and Morbidity

Black-mouth disease, which is characterized by the blackened peristomial membrane (Figure 1E), is one of the most serious diseases in *S. intermedius* aquaculture [24]. The performance of sea urchins without disease is shown in Figure 1E. Sea cucumbers with skin ulceration syndrome are indicated by the white spots on the integument, and these spots quickly fill the whole integument and consequently leads to death [25] (Figure 1F). The performance of sea cucumbers without disease is shown in Figure 1F. Mortality and morbidity were evaluated at the end of the experiment.

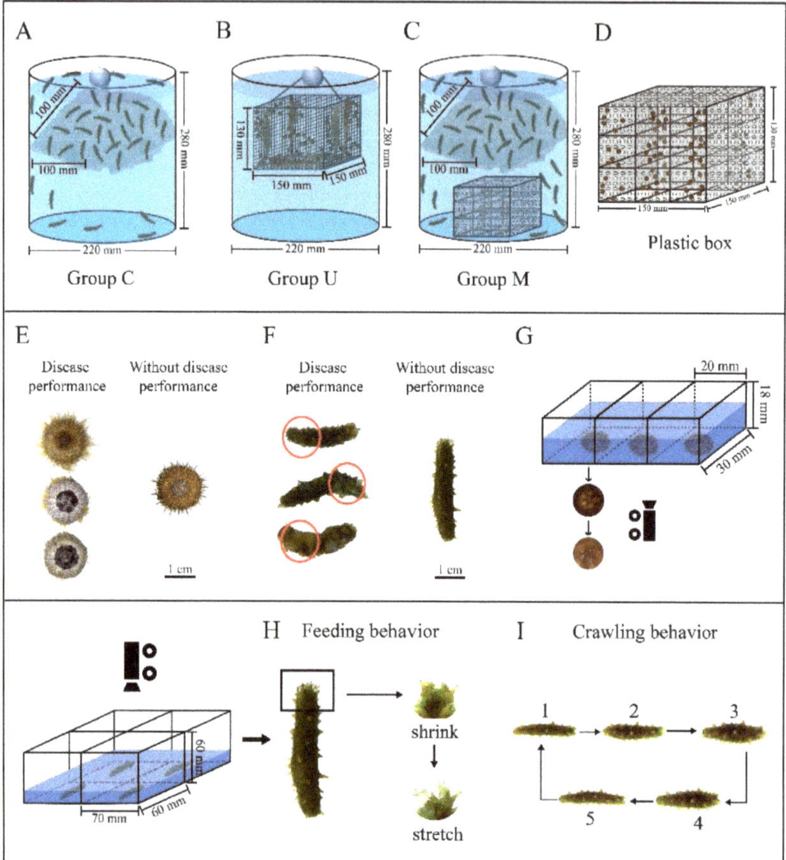

Figure 1. Experimental facilities used for group C (**A**), group U (**B**), and group M (**C**). Group C: the control group for sea cucumbers. Group U: the control group for sea urchins. Group M: the integrated multi-trophic aquaculture of sea cucumbers and sea urchins. A dismountable plastic box divided into three layers and six compartments in each layer was used to culture sea urchins in group M (**D**). Sea urchins with the performance of black-mouth disease and without the disease performance (**E**). Sea cucumbers with the performance of skin ulceration syndrome and without the disease performance (**F**). Red circles indicate the ulcerative skin in sea cucumbers (**F**). Aristotle's lantern reflex of sea urchin (**G**). Feeding behavior of sea cucumbers refers to the process, by which the tentacles reach the food sediments and deliver them to the mouth (**H**). Crawling behavior indicates the movement ability of sea cucumbers, and it was divided into five stages (**I**).

2.4. Growth

Final growth traits were assessed at the end of the experiment. Sea cucumbers (both inside and outside nets) from groups C and M were randomly selected for the subsequent measurement. Sea urchins of group M were randomly selected from ten haphazardly chosen compartments for the following measurement [22]. Test diameter and lantern length of *S. intermedius* were measured using a digital vernier caliper (Mahr Co., Ruhr, Germany). Body and lantern were weighted wet by an electric balance (G & G Co., San Diego, CA, USA). To evaluate the growth traits of *A. japonicus*, they were randomly selected and placed in a small tank (length × width × height: 227 × 157 × 61 mm) filled with fresh seawater according to Broeke et al. [26]. The software ImageJ 1.51 n was used to measure their naturally relaxed body length as a polygonal line after being photographed by a digital video (Canon Co., Shenzhen, China). Wet body weight was assessed using an electric balance (G & G Co., San Diego, CA, USA). The average of all the ten animals was considered as one value for each of the six replicates ($n = 6$).

2.5. Aristotle's Lantern Reflex of S. intermedius

Aristotle's lantern reflex is defined as one cycle of the teeth from opening to closing, indicating the ability to obtain food in *S. intermedius* [24]. A simple device with three compartments (length × width × height: 30 × 20 × 18 mm for each compartment) with agar film (2 g kelp powder with 3 g agar powder for feeding of sea urchins) at the bottom was used to assess Aristotle's lantern reflex according to Ding et al. [27] (Figure 1G). Five *S. intermedius* were randomly selected from each group and put into the experimental device at the end of the rearing experiment. The number of Aristotle's lantern reflex within 10 min were counted using a digital camera (Canon Co., Shenzhen, China). The average of all the five *S. intermedius* was considered as one value for each of the six replicates ($n = 6$).

2.6. Feeding and Crawling Behaviors of A. japonicus

Feeding behavior, which is closely related to the food intake of *A. japonicus* [28], was evaluated according to Sun et al. [29]. It refers to the process by which the tentacles reach the food sediments and deliver them to the mouth in *A. japonicus* [29] (Figure 1H). Completing one cycle was recorded as one tentacle activity frequency in this study.

Crawling behavior indicates the movement ability of *A. japonicus* [30,31]. Crawling behavior includes the processes below according to Lin [32]: (1) sea cucumbers remain at the quiescent condition; (2) they contract from the anus to the back; (3) the contraction gradually transits to the ostium; (4) the contraction subsides, and sea cucumbers remain at the quiescent condition (Figure 1I). Completing one cycle was recorded as one crawling frequency in this study.

Four *A. japonicus* of each replicate were randomly selected and individually put into experimental tanks (length × width × height: 70 × 60 × 60 mm) filled with a commercial powdered diet (Anyuan Industrial Co., Yantai, China) at the end of the experiment. Tentacle activity frequency and crawling frequency were recorded within one hour using a digital camera (Canon Co., Shenzhen, China). The average of all the four *A. japonicus* was considered as one value for each of the six replicates ($n = 6$)

2.7. Statistical Analysis

The normal distribution and homogeneity of variance were analyzed using the Kolmogorov–Smirnov test and Levene test, respectively. Body weight and Aristotle's lantern reflex of *S. intermedius*, body weight and tentacles activity frequency of *A. japonicus* were analyzed using the Mann–Whitney *U* test, because of the non-normal distribution and/or heterogeneity of variance. Differences in the rest traits between groups were compared using the independent sample *t*-test. The replicate means were calculated for all variables. All statistical analyses were performed using SPSS 19.0 statistical software. $p < 0.05$ was considered statistically significant.

3. Results

3.1. Mortality and Morbidity

The mortality ($53.3 \pm 5.0\%$) and morbidity ($54.4 \pm 5.5\%$) of the group M were significantly lower than those in the group U ($67.8 \pm 12.1\%$, $t = 7.349$, $p = 0.022$ for mortality, Figure 2A) ($68.0 \pm 12.0\%$, $t = 6.297$, $p = 0.031$ for morbidity, Figure 2B). However, there were no significant differences in mortality ($14.0 \pm 13.0\%$ for group M and $21.6 \pm 11.8\%$ for group C) and morbidity ($15.3 \pm 15.0\%$ for group M and $23.9 \pm 12.5\%$ for group C) between groups M and C ($t = 1.141$, $p = 0.310$ for mortality, Figure 2C) ($t = 1.156$, $p = 0.307$ for morbidity, Figure 2D).

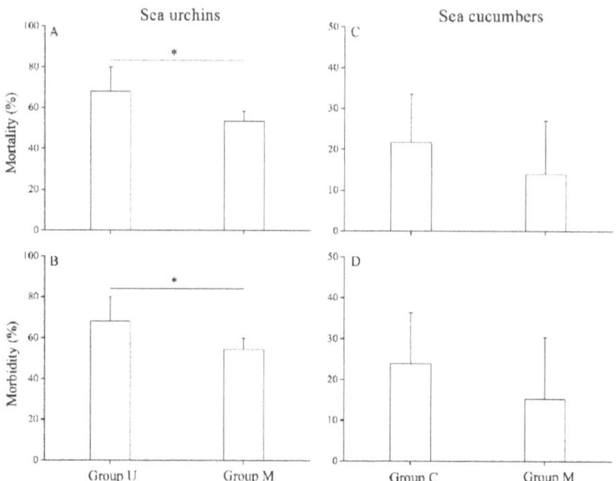

Figure 2. Mortality of *Strongylocentrotus intermedius* (**A**) and *Apostichopus japonicus* (**C**) as well as the morbidity of *S. intermedius* (**B**) and *A. japonicus* (**D**) between groups (mean ± SD, *n* = 6). Group C: the control group for sea cucumbers. Group U: the control group for sea urchins. Group M: the integrated multi-trophic aquaculture of sea cucumbers and sea urchins. The asterisk * means *p* < 0.05.

3.2. Growth

Test diameter (12.8 ± 2.2 mm) and body weight (0.8 ± 0.3 g) of *S. intermedius* in group M were significantly higher than those in group U (10.4 ± 2.1 mm, *t* = 6.197, *p* < 0.001 for test diameter, Figure 3A) (0.5 ± 0.2 g, Mann–Whitney *U* = 669, *p* < 0.001 for body weight, Figure 3B). Further, body length (38.2 ± 3.1 mm) and body weight (1.4 ± 0.8 g) of *A. japonicus* in group M were significantly larger than those in group C (33.9 ± 1.3 mm, *t* = 9.500, *p* = 0.012 for body length, Figure 3C) (1.0 ± 0.4 g, Mann–Whitney *U* = 1260, *p* = 0.005 for body weight, Figure 3D).

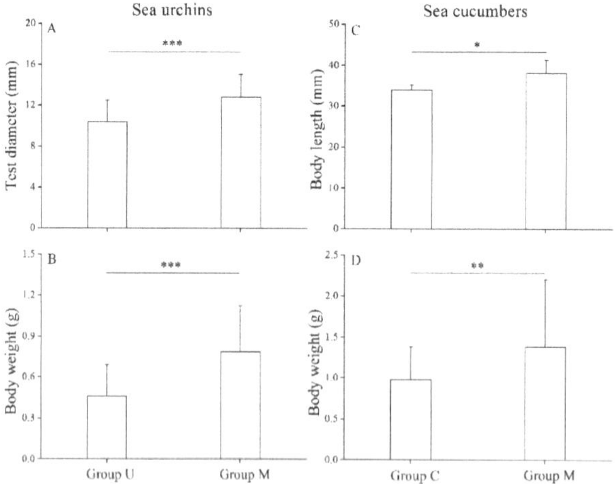

Figure 3. Test diameter (**A**) and wet body weight (**B**) of *Strongylocentrotus intermedius* between groups U and M. Body length (**C**) and wet body weight (**D**) of *Apostichopus japonicus* between groups C and M (mean ± SD, *n* = 6). Group C: the control group for sea cucumbers. Group U: the control group for sea urchins. Group M: the integrated multi-trophic aquaculture of sea cucumbers and sea urchins. The asterisks *, ** and *** mean *p* < 0.05, *p* < 0.01, *p* < 0.001, respectively.

No significant difference in lantern length/test diameter was found between groups U (0.033 ± 0.006) and M (0.037 ± 0.005) ($t = 0.002$, $p = 0.999$, Figure 4A)

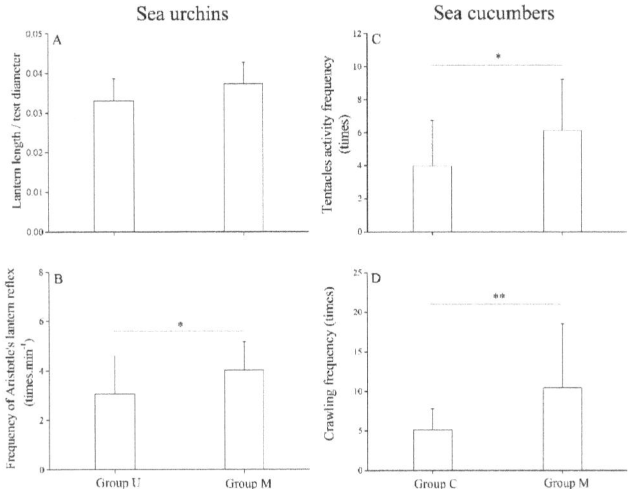

Figure 4. Lantern length/test diameter (**A**) and Aristotle's lantern reflex (**B**) of *Strongylocentrotus intermedius* between groups U and M (mean ± SD, $n = 6$). Tentacles activity frequency (**C**) and crawling frequency (**D**) of *Apostichopus japonicus* between groups C and M (mean ± SD, $n = 6$). Group C: the control group for sea cucumbers. Group U: the control group for sea urchins. Group M: the integrated multi-trophic aquaculture of sea cucumbers and sea urchins. The asterisks * and ** mean $p < 0.05$ and $p < 0.01$, respectively.

3.3. Aristotle's Lantern Reflex of S. intermedius

Significantly higher Aristotle's lantern reflex was observed in group M (4.0 ± 1.2 times·min⁻¹) than that in group U (3.1 ± 1.5 times·min⁻¹) (Mann–Whitney $U = 611.5$, $p = 0.017$, Figure 4B).

3.4. Feeding and Crawling Behaviors of A. japonicus

The tentacles activity frequency (6.1 ± 3.1 times) and crawling frequency (10.5 ± 8.1 times) of group M were both significantly higher than those of group C (4.0 ± 2.8 times, Mann–Whitney $U = 176$, $p = 0.020$ for tentacles activity frequency, Figure 4C) (5.2 ± 2.6 times, $t = 3.045$, $p = 0.005$ for crawling frequency, Figure 4D).

4. Discussion

4.1. Growth Performance of A. japonicus at a High Biomass

The sea cucumber *A. japonicus* (>1 g of body weight) are seeded into the bottom of the sea in the pond culture and stock enhancement in China [33]. This suggests that juvenile *A. japonicus* require an intermediate culture in land-based nursery tanks. However, slow growth and serious food wastage occur in the production of juvenile *A. japonicus* (<1 g of body weight), which greatly hampers the development of the aquaculture industry. Significantly better body size and body weight were found in group M (90 *S. intermedius* and 37 *A. japonicus*/10,638 cm³) than those in group C (37 *A. japonicus*/10,638 cm³). High stocking density is generally considered to display negative impacts on the growth of cultured animals [34,35]. Reducing density is therefore a general practice for the trade-off between animal welfare and economic benefits [36,37]. This study suggests an effective approach that greatly improves body growth while maintaining a high biomass. More research should be conducted to support the potential extension to large-scale aquaculture using this new IMTA system.

4.2. Fitness-Related Behaviors and Growth in A. japonicus

Feeding and crawling are essential fitness-related behaviors, displaying a strong relationship with the growth performance of *A. japonicus* [38]. Feeding behavior is defined as the process of sea cucumbers collecting diets using the tentacles around their mouth [39,40]. The tentacles activity commonly reflects food consumption in the sea cucumber *Cucumaria frondosa* [28]. Crawling behavior is the movement pattern that sea cucumbers crawl to a place where is suitable for their survival [30,31]. Significantly greater tentacles activity frequency and crawling frequency occurred in group M than those in group C, which probably leads to better body growth of *A. japonicus*. It has been well documented that juvenile *A. japonicus* lose their balance when they move to the silt or sand [41,42], because their ambulacral feet are adapted for attaching to a large enough surface [43]. The surface of the plastic box used in group M was smooth and covered with powdered diets, which probably provided an optimal shelter and habitat for *A. japonicus*, and thus improved their fitness-related behaviors. This indicates that the production efficiency of *A. japonicus* can be further improved in the existing nursery tanks. For example, adding artificial reefs to the bottom of nursery tanks could represent a promising candidate to promote the growth performance of *A. japonicus*.

4.3. Feeding Behavior and Body Size of S. intermedius

Sea urchin seeds (>10 mm of test diameter) are further used for longline culture and stock enhancement [12]. It takes about three months for juvenile *S. intermedius* (3–4 mm of test diameter) to become qualified seeds (10–20 mm of test diameter) by feeding fresh kelp in land-based nursery tanks [10]. The slow growth of *S. intermedius* along with the high cost of diets greatly hinders the development of sea urchin aquaculture. The present study found that group M showed a significantly larger body size compared with those in group U after 34 days, suggesting that the IMTA system displays a great potential for seed production (>10 mm of test diameter) in large quantities. A significantly higher rate of Aristotle's lantern reflex was consistently found in group M than that in group U in this study. Aristotle's lantern reflex, which refers to the process of grasping and jawing food using the teeth [44], is commonly used to represent the ability of food consumption in sea urchins [27]. For example, Hu et al. [24] found that superior Aristotle's lantern reflex led to significantly higher food consumption in sea urchins. Thus, it is probable that the improved feeding behavior of *S. intermedius* contributed to the better utilization of the powdered diets in this new IMTA system.

Furthermore, although fresh kelp can be used for *S. intermedius* culture, this approach is inefficient and not environmentally sustainable [45], because it displays a low level of energy protein and varies seasonally in nutrients [46,47]. Formulated feed (e.g., sea cucumber diet), which can be supplied stably with high nutrients [48,49], greatly improved somatic growth during the juvenile stages of *S. intermedius* in this study. Most importantly, *S. intermedius* can utilize wasted sea cucumber diets deposited in the bottom of tanks. Thus, there is no need for additional supplementation of fresh kelp in the culture period. This greatly decreases the cost in *S. intermedius* aquaculture.

4.4. Survival of S. intermedius

The present study found that mass mortality and morbidity of *S. intermedius* occurred in both groups M and U. However, group M showed significantly lower mortality and morbidity compared to those in group U. Mass mortality and morbidity of sea urchins are consistent with Lawrence et al. [10], in which 80–90% of *S. intermedius* died due to the increasing water temperature. Black-mouth disease, which is caused by opportunistic bacteria, such as the genus *Bacillus firmus* [50], is one of the most serious diseases threatening the survival of *S. intermedius* in aquaculture [24]. The optimum temperature for the growth of both the pathogenic bacteria *B. firmus* and sea urchins *S. intermedius* is ~15 °C [12,50], which consequently results in the disease outbreak during the production season of sea urchins. Our previous study documented that eliminating interactions in multi-layer cultures greatly improved the survival after disease challenge assays in cultured *S. intermedius* [22]. However, it remains unclear whether this approach is applicable in the IMTA system. Here, the present results indicate that segregation in multi-layer culture is essential for *S. intermedius* culture in the IMTA system. It greatly contributes to improving the survival of *S. intermedius* when disease outbreaks, although the potential differences were not evaluated in pathogenic bacteria between groups in the present study.

In addition, there were various seaweeds in the powdered diets of sea cucumbers [17]. Dietary supplementation with seaweed promotes the activities of immunocytes and immunologic factors [51], because seaweeds have antiviral and antimicrobial activities [52,53]. For example, dietary supplementation with seaweeds *Sargassum whitti* and *Ulva prolifera* facilitates the levels of lysozyme in fishes *Mugil cephalus* and *Scophthalmus maximus*, respectively [54,55]. Therefore, another explanation is that sea urchins being fed a sea cucumber diet is probably beneficial to their survival in the new IMTA, which further highlights the superiority of this new system.

Author Contributions: Conceptualization, F.H., H.W., C.Z. and D.Y.; methodology, F.H., H.W. and C.Z.; software, F.H. and G.W. ; formal analysis, F.H., H.W., R.T., J.G. and G.W.; investigation, F.H., H.W., R.T., J.G. and G.W.; data curation, F.H., H.W. and R.T.; writing—original draft preparation, F.H., H.W. and C.Z.; writing—review and editing, F.H., H.W. and C.Z.; visualization, F.H., H.W. and R.T.; supervision, D.Y. and C.Z.; project administration, C.Z.; funding acquisition, C.Z. All authors have read and agreed to the published version of the manuscript.

Funding: This research was funded by the National Key Research and Development Program of China (2018YFD0901604), a research project for marine economy development in Liaoning province (for Jun Ding), the National Natural Science Foundation of China (41506177), a grant Chinese Outstanding Talents in Agricultural Sciences (for Yaqing Chang) and a grant for innovative talents in universities in Liaoning Province (for Chong Zhao).

Institutional Review Board Statement: Not applicable.

Informed Consent Statement: Not applicable.

Data Availability Statement: The data presented in this study are available on request from the corresponding author.

Acknowledgments: We thank John Lawrence, Yaqing Chang and Konrad Wojnarowski for academic and editorial suggestions, and thank Mingfang Yang, Xiaomei Chi, Yongchao Li, Peng Ding, Yihai Qiao and Xiang Li for their assistance.

Conflicts of Interest: The authors declare no conflict of interest.

References

1. Chary, K.; Aubin, J.; Sadoul, B.; Fiandrino, A.; Covès, D.; Callier, M.D. Integrated multi-trophic aquaculture of red drum (*Sciaenops ocellatus*) and sea cucumber (*Holothuria scabra*): Assessing bioremediation and life-cycle impacts. *Aquaculture* **2020**, *516*, 734621. [CrossRef]
2. Neori, A.; Chopin, T.; Troell, M.; Buschmann, A.H.; Kraemer, G.P.; Halling, C.; Shpigel, M.; Yarish, C. Integrated aquaculture: Rationale, evolution and state of the art emphasizing seaweed biofiltration in modern mariculture. *Aquaculture* **2004**, *231*, 361–391. [CrossRef]
3. Troell, M.; Joyce, A.; Chopin, T.; Neori, A.; Buschmann, A.H.; Fang, J.G. Ecological engineering in aquaculture potential for Integrated Multi-Trophic Aquaculture (IMTA) in marine offshore systems. *Aquaculture* **2009**, *297*, 1–9. [CrossRef]
4. Yu, Y.; Sun, J.; Zhao, Z.; Ding, P.; Yang, M.; Hu, F.; Qiao, Y.; Wang, L.; Chang, Y.; Zhao, C. Effects of water temperature, age of feces, light intensity and shelter on the consumption of sea urchin feces by the sea cucumber *Apostichopus japonicus*. *Aquaculture* **2022**, *554*, 738134. [CrossRef]
5. Shpigel, M.; Shauli, L.; Odintsov, V.; Ben-Ezra, D.; Neori, A.; Guttman, L. The sea urchin, *Paracentrotus lividus*, in an Integrated Multi-Trophic Aquaculture (IMTA) system with fish (*Sparus aurata*) and seaweed (*Ulva lactuca*): Nitrogen partitioning and proportional configurations. *Aquaculture* **2018**, *490*, 260–269. [CrossRef]
6. Israel, D.; Lupatsch, I.; Angel, D.L. Testing the digestibility of seabream wastes in three candidates for integrated multi-trophic aquaculture: Grey mullet, sea urchin and sea cucumber. *Aquaculture* **2019**, *510*, 364–370. [CrossRef]
7. Neofitou, N.; Lolas, A.; Ballios, I.; Skordas, K.; Tziantziou, L.; Vafidis, D. Contribution of sea cucumber *Holothuria tubulosa* on organic load reduction from fish farming operation. *Aquaculture* **2019**, *501*, 97–103. [CrossRef]

8. Grosso, L.; Rakaj, A.; Fianchini, A.; Morroni, L.; Cataudella, S.; Scardi, M. Integrated Multi-Trophic Aquaculture (IMTA) system combining the sea urchin *Paracentrotus lividus*, as primary species, and the sea cucumber *Holothuria tubulosa* as extractive species. *Aquaculture* **2021**, *534*, 36268. [CrossRef]

9. Zhang, X.L. *China Fishery Statistical Yearbook 2019*; China Agriculture Press: Beijing, China, 2021. (In Chinese)

10. Lawrence, J.M.; Zhao, C.; Chang, Y.Q. Large-scale production of sea urchin (*Strongylocentrotus intermedius*) seed in a hatchery in China. *Aquac. Int.* **2019**, *27*, 1–7. [CrossRef]

11. Ru, X.S.; Zhang, L.B.; Li, X.N.; Liu, S.L.; Yang, H.S. Development strategies for the sea cucumber industry in China. *J. Oceanol. Limnol.* **2019**, *37*, 300–312. [CrossRef]

12. Chang, Y.Q.; Ding, J.; Song, J.; Yang, W. *Biology and Aquaculture of Sea Cucumbers and Sea Urchins*; China Ocean Press: Beijing, China, 2004. (In Chinese)

13. Su, Y.M.; Cai, X.X.; Sun, J.; Zhou, W. Effects of several kinds of feedstuff on growth and digestibility of Juvenile sea urchin *Strongylocentrotus intermedius*. *J. Dalian Fish. Univ.* **2008**, *23*, 242–246. (In Chinese with an English abstract)

14. De Ridder, C.; Lawrence, J.M. Food and feeding mechanisms: Echinoidea. In *Echinoderm Nutrition*; Jangoux, M., Lawrence, J.M., Eds.; Balkema Press: Rotterdam, The Netherlands, 1982; pp. 57–115.

15. Harrold, C.; Pearse, J.S. The ecological role of echinoderms in kelp forests. In *Echinoderm Studies*; Jangoux, M., Lawrence, J.M., Eds.; Balkema Press: Rotterdam, The Netherlands, 1987; pp. 137–233.

16. Hagen, N.T. Enlarged lantern size in similar-sized, sympatric, sibling species of *Strongylocentrotid* sea urchins: From phenotypic accommodation to functional adaptation for durophagy. *Mar. Biol.* **2008**, *153*, 907–924. [CrossRef]

17. Yuan, X.; Yang, H.; Zhou, Y.; Mao, Y.; Zhang, T.; Liu, Y. The influence of diets containing dried bivalve feces and/or powdered algae on growth and energy distribution in sea cucumber *Apostichopus japonicus* (Selenka) (Echinodermata: Holothuroidea). *Aquaculture* **2006**, *256*, 457–467. [CrossRef]

18. Liu, Y.; Dong, S.; Tian, X.; Wang, F.; Gao, Q. The effect of different macroalgae on the growth of sea cucumbers (*Apostichopus japonicus* Selenka). *Aquac. Res.* **2010**, *2*, e881–e885. [CrossRef]

19. Seo, J.Y.; Shin, I.S.; Lee, S.M. Effect of dietary inclusion of various plant ingredients as an alternative for *Sargassum thunbergii* on growth and body composition of juvenile sea cucumber *Apostichopus japonicaus*. *Aquac. Nutr.* **2011**, *17*, 549–556. [CrossRef]

20. Sun, J.; Yu, Y.; Zhao, Z.; Tian, R.; Li, X.; Chang, Y.; Zhao, C. Macroalgae and interspecific alarm cues regulate behavioral interactions between sea urchins and sea cucumbers. *Sci. Rep.* **2022**, *12*, 3971. [CrossRef]

21. Wang, J.Q.; Cheng, X.; Yang, Y.; Wang, N.B. Polyculture of juvenile sea urchin (*Strongylocentrotus intermedius*) with juvenile sea cucumber (*Apostichopus japonicus* Selenka). *J. Dalian Fish. Univ.* **2007**, *22*, 102–108. (In Chinese with an English abstract)

22. Hu, F.; Yang, M.; Chi, X.; Ding, P.; Sun, J.; Wang, H.; Yu, Y.; Chang, Y.; Zhao, C. Segregation in multi-layer culture avoids precocious puberty, improves thermal tolerance and decreases disease transmission in the juvenile sea urchin *Strongylocentrotus intermedius*: A new approach to longline culture. *Aquaculture* **2021**, *543*, 736956. [CrossRef]

23. Hu, F.; Chi, X.; Yang, M.; Ding, P.; Yin, D.; Ding, J.; Huang, X.; Luo, J.; Chang, Y.; Zhao, C. Effects of eliminating interactions in multi-layer culture on survival, food utilization and growth of small sea urchins *Strongylocentrotus intermedius* at high temperatures. *Sci. Rep.* **2021**, *11*, 15116. [CrossRef]

24. Hu, F.; Yang, M.; Ding, P.; Zhang, X.; Chen, Z.; Ding, J.; Chi, X.; Luo, J.; Zhao, C.; Chang, Y. Effects of the brown algae *Sargassum horneri* and *Saccharina japonica* on survival, growth and resistance of small sea urchins *Strongylocentrotus intermedius*. *Sci. Rep.* **2020**, *10*, 12495. [CrossRef]

25. Deng, H.; He, C.; Zhou, Z.; Liu, C.; Tan, K.; Wang, N.; Jiang, B.; Gao, X.; Liu, W. Isolation and pathogenicity of pathogens from skin ulceration disease and viscera ejection syndrome of the sea cucumber *Apostichopus japonicus*. *Aquaculture* **2009**, *287*, 18–27. [CrossRef]

26. Broeke, J.; Perez, J.; Pascau, J. *Image Processing with ImageJ*, 2nd ed.; Packt Publishing Ltd.: Birmingham, UK, 2015.

27. Ding, J.Y.; Zheng, D.F.; Sun, J.N.; Hu, F.Y.; Yu, Y.S.; Zhao, C.; Chang, Y.Q. Effects of water temperature on survival, behaviors and growth of the sea urchin *Mesocentrotus nudus*: New insights into the stock enhancement. *Aquaculture* **2020**, *519*, 734873. [CrossRef]

28. Holtz, E.; MacDonald, B.A. Feeding behavior of the sea cucumber *Cucumaria frondosa* (Echinodermata: Holothuroidea) in the laboratory and the field: Relationships between tentacle insertion rate, flow speed, and ingestion. *Mar. Biol.* **2009**, *156*, 1389–1398. [CrossRef]

29. Sun, J.M.; Zhang, L.B.; Pan, Y.; Lin, C.G.; Wang, F.; Kan, R.T.; Yang, H.S. Feeding behavior and digestive physiology in sea cucumber *Apostichopus japonicas*. *Physiol. Behav.* **2015**, *139*, 336–343. [CrossRef]

30. Navarro, P.; García-Sanz, S.; Barrio, J.; Tuya, F. Feeding and movement patterns of the sea cucumber *Holothuria sanctori*. *Mar. Biol.* **2013**, *160*, 2957–2966. [CrossRef]

31. Pan, Y.; Zhang, L.B.; Lin, C.G.; Sun, J.M.; Kan, R.T.; Yang, H.S. Influence of flow velocity on motor behavior of sea cucumber *Apostichopus japonicus*. *Physiol. Behav.* **2015**, *144*, 52–59. [CrossRef]

32. Lin, C.G. Effects of Four Physical Environment Factors on the Movement and Feeding Behavior of Sea Cucumber *Apostichopus japonicas* (Selenka). Master's Thesis, Institute of Oceanology, Chinese Academy of Sciences, Qingdao, China, 2014. (In Chinese)

33. Chen, J. Overview of sea cucumber farming and sea ranching practices in China. *SPC Beche-De-Mer Inf. Bull.* **2003**, *18*, 18–23.

34. Siikavuopio, S.I.; Dale, T.; Mortensen, A. The effects of stocking density on gonad growth, survival and feed intake of adult green sea urchin (*Strongylocentrotus droebachiensis*). *Aquaculture* **2007**, *262*, 78–85. [CrossRef]

35. Qi, S.B.; Zhang, W.J.; Jing, C.C.; Wang, H.F.; Zhao, S.; Zhou, M.; Chang, Y.Q. Long-term effects of stocking density on survival, growth performance and marketable production of the sea urchin *Strongylocentrotus intermedius*. *Aquac. Int.* **2016**, *24*, 1323–1339. [CrossRef]

36. Richardson, C.M.; Lawrence, J.M.; Watts, S.A. Factors leading to cannibalism in *Lytechinus variegatus* (Echinodermata: Echinoidia) held in intensive culture. *J. Exp. Mar. Biol. Ecol.* **2011**, *399*, 68–75. [CrossRef]

37. Mos, B.; Byrne, M.; Cowden, K.L.; Dworjanyn, S.A. Biogenic acidification drives density-dependent growth of a calcifying invertebrate in culture. *Mar. Biol.* **2015**, *162*, 1541–1558. [CrossRef]

38. Hu, F.; Ding, P.; Yu, Y.; Wen, B.; Cui, Z.; Yang, M.; Chi, X.; Sun, J.; Luo, J.; Sun, Z.; et al. Effects of artificial reefs on selectivity and behaviors of the sea cucumber *Apostichopus japonicas*: New insights into the pond culture. *Aquac. Rep.* **2021**, *21*, 100842. [CrossRef]

39. Roberts, D.; Moore, H. Tentacular diversity in deep-sea deposit-feeding holothurians: Implications for biodiversity in the deep sea. *Biodivers. Conserv.* **1997**, *6*, 1487–1505. [CrossRef]

40. Hudson, I.R.; Wigham, B.D.; Solan, M.; Rosenberg, R. Feeding behaviour of deep-sea dwelling holothurians: Inferences from a laboratory investigation of shallow fjordic specie. *J. Mar. Syst.* **2005**, *57*, 201–218. [CrossRef]

41. Zhou, S.; Shirley, T.C. Habitat and depth distribution of the red sea cucumber *Parastichopus californicus* in a Southeast Alaska bay. *Alsk. Fish. Res. Bull.* **1996**, *3*, 23–131.

42. Dissanayake, D.C.T.; Stefansson, G. Habitat preference of sea cucumbers: *Holothuria atra* and *Holothuria edulis* in the coastal waters of Sri Lanka. *J. Mar. Biol. Assoc. UK* **2012**, *92*, 581–590. [CrossRef]

43. Qiu, T.L.; Zhang, L.B.; Zhang, T.; Yang, H.S. Effects of mud substrate and water current on the behavioral characteristics and growth of the sea cucumber *Apostichopus japonicus* in the Yuehu lagoon of northern China. *Aquac. Int.* **2014**, *22*, 423–433. [CrossRef]

44. Agustina, Y. Effect of the covering behavior of the juvenile sea urchin *Strongylocentrotus intermedius* on pral characteristics and growth of the sea cucumber *Apostichopus japonicus* in the Yuredation by the spider crab *Pugettia quadriceps*. *Fish. Sci.* **2001**, *67*, 1181–1183.

45. Lawrence, J.M.; Lawrence, A.L.; McBride, S.; George, S.B.; Watts, S.A.; Plank, L.R. Developments in the use of prepared feeds in sea-urchin aquaculture. *J. World Aquac. Soc.* **2001**, *32*, 34–39.

46. Kelly, M.S. Survivorship and growth rates of hatchery-reared sea urchins. *Aquac. Int.* **2002**, *10*, 309–316. [CrossRef]

47. Schlosser, S.C.; Lupatsch, I.; Lawrence, J.M.; Lawrence, A.L.; Shpigel, M. Protein and energy digestibility and gonad development of the European sea urchin, *Paracentrotus lividus* (Lamarck), fed algal and prepared diets during spring and fall. *Aquac. Res.* **2005**, *36*, 972–982. [CrossRef]

48. Akiyama, T.; Unuma, T.; Yamamoto, T. Optimum protein level in a purified diet for young red sea urchin, *Pseudocentrotus depressus*. *Fish. Sci.* **2001**, *67*, 361–363. [CrossRef]

49. Kennedy, E.J.; Robinson, S.M.C.; Parson, G.J.; Castell, J.D. Effect of protein source and concentration on somatic growth of juvenile green sea urchins *Strongylocentrotus droebachiensis*. *J. World Aquac. Soc.* **2005**, *36*, 320–336. [CrossRef]

50. Li, T.W.; Xu, S.L.; Wang, R.B.; Xu, S.F.; Su, X.R. Preliminary studies on the black mouth disease of sea urchin, *Strongylocentrotus intermedius* (Strongylocentrotidae Echinoidea). *Mar. Sci.* **2000**, *24*, 41–43. (In Chinese with an English abstract)

51. Wang, C.; Hu, W.J.; Wang, L.S.; Qiao, H.J.; Wu, H.Y.; Xu, Z.G. Effects of dietary supplementation with *Sargassum horneri* meal on growth performance, body composition, and immune response of juvenile turbot. *J. Appl. Phycol.* **2019**, *31*, 771–778. [CrossRef]

52. Hemmingson, J.A.; Falshaw, R.; Furneaux, R.H.; Thompson, K. Structure and antiviral activity of the galactofucan sulfates extracted from *Undaria pinnatifida* (Phaeophyta). *J. Appl. Phycol.* **2006**, *18*, 185–193. [CrossRef]

53. Narasimhan, M.K.; Pavithra, S.K.; Krishnan, V.; Chandrasekaran, M. In vitro analysis of antioxidant, antimicrobial and antiproliferative activity of *Enteromorpha antenna*, *Enteromorpha linza* and *Gracilaria corticata* extracts. *Jundishapur J. Nat. Pharm. Prod.* **2013**, *8*, 151–159. [CrossRef]

54. Kanimozhi, S.; Krishnaveni, M.; Deivasigmani, B.; Rajasekar, T.; Priyadarshni, P. Immunomo-stimulation effects of *Sargassum whitti* on *Mugil cephalus* against *Pseudomonas fluorescence*. *Int. J. Curr. Microbiol. Appl. Sci.* **2013**, *2*, 93–103.

55. Guo, Z.; Yang, N.; Wang, Z.; Lu, Q.; Wang, R.; Lin, Y. Optimal dietary *Enteromorpha prolifera* in turbot (*Scophthalmus maximus*). *Fish. Sci.* **2015**, *34*, 423–427. (In Chinese)

Journal of
*Marine Science
and Engineering*

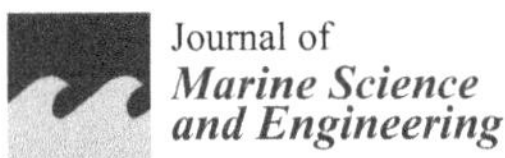

Article

A Na⁺/H⁺-Exchanger Gene from *Penaeus monodon*: Molecular Characterization and Expression Analysis under Ammonia Nitrogen Stress

Yundong Li [1,2,3,4], Shigui Jiang [1,2], Hongdi Fan [1], Qibin Yang [1,2], Song Jiang [1], Jianhua Huang [1], Lishi Yang [1], Wenwen Zhang [1], Xu Chen [2] and Falin Zhou [1,2,3,*]

[1] Key Laboratory of South China Sea Fishery Resources Exploitation and Utilization, Ministry of Agriculture and Rural Affairs, South China Sea Fisheries Research Institute, Chinese Academy of Fishery Sciences, Guangzhou 510300, China

[2] Key Laboratory of Efficient Utilization and Processing of Marine Fishery Resources of Hainan Province, Sanya Tropical Fisheries Research Institute, Sanya 572018, China

[3] Hainan Yazhou Bay Seed Laboratory, Sanya 572025, China

[4] Key Laboratory of Tropical Hydrobiology and Biotechnology of Hainan Province, Hainan Aquaculture Breeding Engineering Research Center, College of Marine Sciences, Hainan University, Haikou 570228, China

* Correspondence: zhoufalin@aliyun.com

Abstract: Na⁺/H⁺-exchanger (NHE) assumes a significant part in different particle transport in creatures. A clone of *Penaeus monodon* NHE cDNA was examined in this study (*PmNHE*), and its impact on high-concentration ammonia nitrogen stress was researched. The 877-amino acid (aa) protein was encoded by a full-length *PmNHE* cDNA that was 2788 base pairs (bp) long and had a 2643-bp open reading frame (ORF). The findings show that *PmNHE* was expressed in all of the *P. monodon* organs that were tested, including the intestine, muscle, hemolymph, heart, hepatopancreas, stomach, epidermis, gill, testis, and ovary, and the intestine and muscle were found to have the highest levels of *PmNHE* expression. The expression of *PmNHE* in the gill tissue of *P. monodon* was significantly up-regulated under high levels of ammonia nitrogen stress. The expression of *PmNHE* in the intestine of *P. monodon* under high-concentration ammonia nitrogen stress was significant. When exposed to high concentrations of ammonia nitrogen stress, *P. monodon* exhibited shorter survival times than the two control groups. Hence, it is suggested in the present study that *PmNHE* may have a significant impact on the environment with high levels of ammonia nitrogen.

Keywords: NHE; Penaeus monodon; ammonia nitrogen stress; intestine; gill; RNAi

Citation: Li, Y.; Jiang, S.; Fan, H.; Yang, Q.; Jiang, S.; Huang, J.; Yang, L.; Zhang, W.; Chen, X.; Zhou, F. A Na⁺/H⁺-Exchanger Gene from *Penaeus monodon*: Molecular Characterization and Expression Analysis under Ammonia Nitrogen Stress. *J. Mar. Sci. Eng.* **2022**, *10*, 1897. https://doi.org/10.3390/jmse10121897

Academic Editor: Ka Hou Chu

Received: 26 October 2022
Accepted: 2 December 2022
Published: 5 December 2022

1. Introduction

The apical Na⁺/H⁺-*exchanger* (NHE) can release ammonia nitrogen from the cytoplasm into the environment, so it has attracted the attention of researchers. In the human kidney, the proximal tubules secrete ammonia equivalent to 75–120% of total urinary ammonia, suggesting that NHE-3 plays a greater role in NH4⁺ excretion than Rh C Glycoprotein [1,2]. The study found that the NHE-3 may be the primary mechanism by which NH4⁺ is secreted from the apical plasma membrane of proximal tubules. NHE-3 is a member of the Na⁺/H⁺-*exchanger* family. NHE-3-mediated ammonia secretion may be due to the substitution of NH4+ for H+ at the cytosolic H+ binding site, activating Na⁺/NH4⁺ exchange activity. It has been shown that the NH4⁺ in the cytoplasm competes with the H⁺ in the cytoplasm to exchange with Na⁺ in the lumen, hence the Na⁺/NH4⁺ exchange [3,4].

Aquaculture systems are adversely affected by ammonia nitrogen, a significant pollutant. Research has been conducted to understand the mechanisms by which ammonia negatively affects aquatic animals [5,6]. The research on the mechanism of ammonia excretion of crustaceans found that inhibition of NHE activity in *Carcinus maenas*, *Cancer pagurus*, and *Astacus leptodactylus* with the nonspecific inhibitor amiloride blocked ammonia nitrogen excretion from gills [7,8]. Moreover, Towle et al. identified the NHE transport mode on the gills of the *Carcinus maenas* at the molecular level and believed that it may belong to a transport mode cation or proton binding side [9]. Furthermore, it showed that the expression level of NHE mRNA in the gills of *Eriocheir sinensis* and *Portunus trituberculatus* was decreased after 2 days of exposure to ammonia nitrogen.

Most studies focus on the acid–base balance and salinity stress [10–13]. Little research has been conducted on whether NHE responds to nitrogen stress induced by ammonia and its role in the aquaculture water environment with high ammonia nitrogen concentration. Therefore, in this study, the NHE gene of *P. monodon* was cloned for the first time by race technology, bioinformatics analysis and tissue expression analysis were carried out, and the mRNA expression trend of NHE after 96 h acute ammonia nitrogen stress was detected. Moreover, RNA interference technology was employed to explore the death curve of shrimp under high-concentration ammonia nitrogen stress after NHE silencing.

2. Materials and Methods

2.1. Experimental Animals

Examples of *P. monodon* were procured from the Experimental Base of South China Sea Fisheries Research Institute in Shenzhen. The shrimp with a body length of 12.2 ± 0.6 cm and a weight of 16.2 ± 1.1 g were used in experiments. They were refined in sifted circulated air through seawater at 28 ± 1 °C, salinity 25–28 psu, pH 7.8–8.2, and fed business feed.

2.2. Ammonia Nitrogen Stress

A preliminary trial was conducted with six ammonia nitrogen concentrations: 0, 20, 40, 60, 80, and 100 mg·L^{-1}. This allowed us to calculate the 96 h LC50 (medial lethal dose) [5,6]. There were 180 shrimp used in all. Thirty shrimp were housed in tanks each holding 3.5 L of saltwater with the desired amount of ammonia nitrogen. Every two hours during the preliminary experiment, mortality was noted. A linear regression equation was formulated to determine the median lethal (LC50-96 h) and safety concentrations (SC) [14,15]. It was found that the LC50 and SC concentrations were 50 and 5 mg·L^{-1}, respectively. To control the level of ammonia nitrogen, clean seawater was mixed with ammonium chloride (NH$_4$Cl). Three groups were employed in the experiment: pure seawater served as the control group, 50 mg·L^{-1} represented a high level of ammonia nitrogen concentration, and 5 mg·L^{-1} represented a low level of ammonia nitrogen concentration. Three parallel experiments were conducted in each experimental group. Every 2 h, shrimp survival in each experimental group was monitored and documented, and the dead shrimp were immediately removed from the bucket. After ammonia nitrogen stress, three shrimp were taken from each replicate at 0, 3, 6, 12, 24, 48, and 96 h. The shrimp's gills and intestinal tissues were rapidly removed and flash-frozen in liquid nitrogen.

2.3. RNA Extraction and Sequencing

Using the HiPure Fibrous RNA Plus Kit (Megan, Guangzhou, China), total RNA from ovarian tissue was extracted. The RNA products were checked for integrity using 1.2% agarose gel electrophoresis, and concentration and purity were checked using Nanodrop (Thermo Scientific NC2000, Waltham, MA, USA). Finally, the concentration was ≥ 1.2 μg/μL, the total amount was ≥ 60 μg as the sample quality standard, and the samples that met the requirements were used in the experiments.

2.4. PmNHE cDNA Cloning

We created the specific primers NHE-F and NHE-R (Table 1) based on Primer Premier 5.0 (RuiBiotech, Beijing, China). The 5 and 3 ends of *PmNHE* were obtained using a method developed by Clonetech in Japan. The following conditions were used in the PCR: 1 cycle at 94 °C for three minutes, 35 cycles at 94 °C for 30 s, 67 °C for 30 s, and 72 °C for 45 s, and a final cycle at 72 °C for ten minutes. After being sequenced and purified on a gel, the PCR products were analyzed. The product was purified according to the operating instructions of the purification kit (TIANGEN, Guangzhou, China).

Table 1. Primers and sequences used in the study.

Primer Name	Nucleotide Sequence (5$'$ $\rightarrow$ 3$'$)	Purpose
3$'$race GSP	TACCTCTGCTACCTCACGGCGGAACT	3$'$RACE
3$'$race NGSP	GTAGAGCTGAGAAGGAAGCCCTCGTA	3$'$RACE
5$'$race GSP	TCCGTGCTGCCCTAGAATCTCC	5$'$RACE
5$'$race NGSP	TCCGCCGTGAGGTAGCAGAGGT	5$'$RACE
qNHE-F	TGGCTTCACAGAGACCTT	RT-PCR
qNHE-R	CACGCATCCGAACAGAAT	RT-PCR
qEF-1a-F	AAGCCAGGTATGGTTGTCAACTTT	RT-PCR
qEF-1a-R	CGTGGTGCATCTCCACAGACT	RT-PCR
dsNHE-F	TAATACGACTCACTATAGGGGCACAAGGAGGTCTAATGGG	dsRNA
dsNHE-R	TAATACGACTCACTATAGGGGAGGTAGCAGAGGTACGCAAG	dsRNA
dsGFP-F	TAATACGACTCACTATAGGGATGGCTAGCAAAGGAGAAGAACTTT	dsRNA
dsGFP-R	TAATACGACTCACTATAGGGAACGGGAAAAGCATTGAACACCA	dsRNA

2.5. Bioinformatic Analysis

Using ExPASy ProtParam software, protein physicochemical properties were predicted (http://web.expasy.org/protparam/, accessed on 5 October 2021), and domain analysis was conducted using SMART4.0 (http://smart.embl-heidelberg.de/, accessed on 5 October 2021). NetPhos 2.0 (http://www.cbs.dtu.dk/services/NetPhos/, accessed on 5 October 2021) and NetNGlyc 1.0 Server (http://www.cbs.dtu.dk/services/NetNGlyc/, accessed on 5 October 2021) were used for predicting protein phosphorylation and glycosylation, respectively. The Clustal X software was used to align multiple sequences, and then, BioEdit and MEGA 6.0 software was used to construct phylogenetic trees.

Combining cDNA and RACE-PCR sequences was used to create the full-length *PmNHE* gene using DNA-man software version 10. The open reading frame (ORF) was located using ORF Finder (https://www.ncbi.nlm.nih.gov/orffinder/, accessed on 8 October 2021). EMBOSS (http://www.bioinformatics.nl/emboss-explorer/, accessed on 8 October 2021) was used to predict the amino acid (aa) sequence. The protein physicochemical properties and aa sequences were predicted using the ExPASy ProtParam software (http://web.expasy.org/protparam/, accessed on 8 October 2021). The SMART4.0 online program was used to analyze protein domains (http://smart.embl-heidelberg.de/, accessed on 10 October 2021). The protein phosphorylation sites were predicted using the NetPhos2.0 program (http://www.cbs.dtu.dk/services/NetPhos/, accessed on 10 October 2021), and the protein glycosylation sites were predicted using the NetNGlyc 1.0 Server (http://www.cbs.dtu.dk/services/NetNGlyc/, accessed on 10 October 2021). The Clustal X was used to align multiple sequences, and BioEdit and MEGA 6.0 were used to create a phylogenetic tree.

2.6. qRT-PCR Analysis of PmNHE mRNA Expression

The expression of *PmNHE* mRNA in various organs was discovered using qRT-PCR. As the reference gene, elongation factor 1α (EF1a) was utilized [16,17] (Table 1). *PmNHE* is compatible with the reaction condition for EF1a. The Roche Light Cycler® 480II was used to perform qRT-PCR with green fluorescence measurement. The relative CT method ($2^{-\Delta\Delta CT}$) was used to obtain the PCR data [13,15,16]. For each time point in the experiment, three individuals were sampled, and for RT-PCR, three technical replicates were made to ensure accuracy. A one-way ANOVA and Tukey's multiple range test were used for the statistical study (IBM, New York, USA). At $p < 0.05$, the differences were deemed to be significant. Data from the tests are displayed as mean SD (standard deviation).

2.7. RNA Interference

Using Primer Premier 5.0, the primers dsNHE-f, dsNHE-r, dsGFP-F, and dsGFP-R (Table 1) required for the synthesis of dsNHE were created. Ex Taq was used as a template to amplify the T7 promoter-containing DNA fragment. Clear and bright bands were obtained, agarose gel recovery was performed according to the instructions of the gel recovery kit, and the cDNA was stored at −20 °C with a concentration greater than 125 ng/μL and meeting the experimental requirements for later use. The synthesis of dsRNA was carried out according to the instructions of the T7 RiboMAXTMExpress RNAi System kit. Purification was performed to reserve the obtained dsRNA. Using the same procedure, green fluorescent protein (GFP) double-chain RNA and the pDGFP recombinant vector were produced.

In this experiment, dsRNA was injected at a ratio of 35 μg/g into *P. monodon* at a weight of (5.0 ± 1.0) g. There were three groups of injection tests: groups that received injections of PBS, dsGFP, and ds*PmNHE*. Before the injection, healthy shrimp were randomly taken from the molting interphase. Their intestines were set in RNAlater as 0 h samples for inspecting RNAi efficiency. The shrimp specimens were transferred to a bucket containing high-concentration ammonia nitrogen 24 h after being injected. Every three hours, dead shrimp from each group were recorded and collected.

Healthy shrimp were taken from each treatment group at 3, 6, 9, 12, 24, and 48 h following exposure to high-concentration ammonia nitrogen stress, and their intestines were collected and set in RNAlater. Ten shrimp were simultaneously injected with dsNHE and dsGFP and set in two separate buckets. In order to evaluate the efficacy of dsRNA interference, intestinal tissues were randomly collected after 24 and 48 h and set in RNAlater as samples. All of the RNA samples were mixed with RNAlater and kept at −80 °C.

3. Results

3.1. PmNHE Sequence Display and Bioinformatics Analysis

The full-length *PmNHE* cDNA was obtained by splicing the open reading frame and $5'/3'$ non-coding region sequence sequencing results. *PmNHE* was 2788 bp long (GenBank accession No. MT164534), including an ORF of 2643 bp, a $5'$-untranslated region (UTR) of 76 bp, and a $3'$ UTR of 78 bp (Figure 1). The ORF of the *PmNHE* gene encoded 877 amino acids with a molecular weight of 97.97 KDa and a theoretical isoelectric point of 6.54. Bioinformatics prediction analysis showed that *PmNHE* had a variety of functional sites, contained 23 phosphorylation sites (20 serine sites and 3 threonine sites), 6 glycosylation sites (highlighted in green font), and 12 transmembrane domains (highlighted with light blue line segments). The sequence contains a sodium/proton *exchanger* domain (highlighted by gray shading) at 78-490 aa. The poly A structure is used in the $3'$-UTR sequence (highlighted in italics).

The *PmNHE* (Genebank No. MT164534) start code (ATG) and termination code (TAA) are listed in a square box. The signal peptide sequence is highlighted in red font. The glycosylation site is highlighted in green font, and the Na^+/H^+-*exchanger* domain is expressed in gray. The Poly A structure is highlighted in italics.

3.2. Phylogenetic Tree Analysis and Multiple Sequence Alignment

In NCBI, the amino acid sequence of the *NHE* gene of *P. monodon* was compared with other species, and it was found that it had a high homology with the amino acid sequence of the NHE protein of *Penaeus vannamei*.

The NJ evolutionary tree of *PmNHE* and this protein of other species was constructed by MEGA6.0 (Figure 2). The results show that the relationship between invertebrates was the closest, and the relationship between *Penaeus vannamei* and *P. monodon* was the closest and clustered together, and they clustered into a branch with *Scylla olivacea* and *Hyalella azteca*.

```
   1 GAGTGAGTGTAGCTCGAGGTGTCACGAGGAAACTGGTGCTTCTCTCGTCAGCCGCCTCACCTACTCGCTTGAGTCGatgtggagagtatg   90
   1                                                                                M  W  R  V  W    5
  91 ggcgtcccgggcgtgtgtgttggtgtcgcttactctgtgggtgataggaaattcgtgggcggcggcgctagaggccagcacacacggcgg  180
   6  A  S  R  A  C  V  L  V  S  L  T  L  W  V  I  G  N  S  W  A  A  A  L  E  A  S  T  H  G  G   35
 181 cgagggcgcgggcggcgacggcaacgccagcgcccatgacgacggcaacggcagctgccacagcgaggagcacgagggcgggatccacct  270
  36  E  G  A  G  G  D  G  \  A  S  A  H  D  D  G  \  G  S  C  H  S  E  E  H  E  G  G  I  H  L   65
 271 cctgtccttccggtggaacgaggtgggcgtctactacaccgtcaccaccttcgtcatcgtcgccggcctctgcaaagtcgcattccatca  360
  66  L  S  F  R  W  N  E  V  G  V  Y  Y  T  V  T  T  F  V  I  V  A  G  L  C  K  V  A  F  H  Q   95
 361 gatccactggctctccaacaagatcccagagtcctgtgtcctgatcatcttgggcgtcttactgggcatcgtcgtcttcttcaccgttga  450
  96  I  H  W  L  S  N  K  I  P  E  S  C  V  L  I  I  L  G  V  L  L  G  I  V  V  F  F  T  V  D  125
 451 tgatggcaacggaacctcaacctgtagctacaacttcgaggttcctcactttacgtcagacaagttcttctttgttctactccccccat  540
 126  D  G  \  G  T  S  T  C  S  Y  N  F  E  V  P  H  F  T  S  D  K  F  F  F  V  L  L  P  P  I  155
 541 catcttagagtctgcatactccctccatgaccgcgccttctttgacaacttaggcacagtcctcgtgtttgccgtgattggaacgctgtt  630
 156  I  L  E  S  A  Y  S  L  H  D  R  A  F  F  D  N  L  G  T  V  L  V  F  A  V  I  G  T  L  F  185
 631 caatatcttcaccataggtccagctctgtatggtgttgcacaaggaggtctaatgggcgccattgctattggcttcacagagaccttagt  720
 186  N  I  F  T  I  G  P  A  L  Y  G  V  A  Q  G  G  L  M  G  A  I  A  I  G  F  T  E  T  L  V  215
 721 attttcatccctcatctcagccgtggaccccgtggcagtgttagccatttttccaggagcttggcgttaacaaagatctctatttccttgt  810
 216  F  S  S  L  I  S  A  V  D  P  V  A  V  L  A  I  F  Q  E  L  G  V  N  K  D  L  Y  F  L  V  245
 811 ctttggggaatctcttctcaatgacggcgtcaccatcgtcgtgtacaccaccttgacctcgttcgtcaccatggaggtcatctcggcagg  900
 246  F  G  E  S  L  L  N  D  G  V  T  I  V  V  Y  T  T  L  T  S  F  V  T  M  E  V  I  S  A  G  275
 901 tcaatatgccttagccgtcgcctccttcttcatcgtggtgtttgcggggcggtcatcggcattctgttcggatgcgtgacggcgctcat  990
 276  Q  Y  A  L  A  V  A  S  F  F  I  V  V  F  A  G  A  V  I  G  I  L  F  G  C  V  T  A  L  I  305
 991 caccaagtacacggctgaagtcagagtggtggagcccctggcactcctaggtcttgcgtacctctgctacctcacggcggaactctttca 1080
 306  T  K  Y  T  A  E  V  R  V  V  E  P  L  A  L  L  G  L  A  Y  L  C  Y  L  T  A  E  L  F  H  335
1081 tttctcaggaatcatcagtctcatcatgtgtggactcttccaggcaaattatgcctttcagaatatttcgcagaagtcttacacgtgtgt 1170
 336  F  S  G  I  I  S  L  I  M  C  G  L  F  Q  A  N  Y  A  F  Q  \  I  S  Q  K  S  Y  T  C  V  365
1171 caaatacttcactaagatggccagtgccacaagcgacactatcatcttcatgttcttgggaatggccttggtgagcaaagaccatatttg 1260
 366  K  Y  F  T  K  M  A  S  A  T  S  D  T  I  I  F  M  F  L  G  M  A  L  V  S  K  D  H  I  W  395
1261 gcaccctggcttcatcctgtggacaattgggttctgcttaatcttcagattcattggggtagctcttttaaccatagtcatgaatcatta 1350
 396  H  P  G  F  I  L  W  T  I  G  F  C  L  I  F  R  F  I  G  V  A  L  L  T  I  V  M  N  H  Y  425
1351 tcgaatgaagaagattggtctccaagaacaattcatcacagcgtatggggggcttaagaggagctgtggccttctctctggccaacatgtt 1440
 426  R  M  K  K  I  G  L  Q  E  Q  F  I  T  A  Y  G  G  L  R  G  A  V  A  F  S  L  A  N  M  L  455
1441 agatcaaacagttgacccgaggaggatatttatcactacaacgcttatggttatattattcacaggtttcatacagggaatatcaatcaa 1530
 456  D  Q  T  V  D  P  R  R  I  F  I  T  T  T  L  M  V  I  L  F  T  G  F  I  Q  G  I  S  I  K  485
1531 gcccctggtgaacctcctcagaatacagaagaagcgttcagaccataagaaactgaatgaagaaatcaatgacactgccatggaccacat 1620
 486  P  L  V  N  L  L  R  I  Q  K  K  R  S  D  H  K  K  L  N  E  E  I  \  D  T  A  M  D  H  I  515
1621 catggcgggcgtcgaggagattctagggcagcacggagacttctatcttcgggagctgattattcactacaatgacaaatatctcaagaa 1710
 516  M  A  G  V  E  E  I  L  G  Q  H  G  D  F  Y  L  R  E  L  I  I  H  Y  N  D  K  Y  L  K  K  545
1711 gtggtttgtcaagccttgtagtgaatccaagctcacgagactcttcgagaaaattgcaatttcagaacactatgctcatctctacggtcc 1800
 546  W  F  V  K  P  C  S  E  S  K  L  T  R  L  F  E  K  I  A  I  S  E  H  Y  A  H  L  Y  G  P  575
1801 agtggcgatgattgaagataaagtgaagcctttagtgacgaaatccgctgcgtcgaagagaattcgcacaatttccgtgtctccttcgca 1890
 576  V  A  M  I  E  D  K  V  K  P  L  V  T  K  S  A  A  S  K  R  I  R  T  I  S  V  S  P  S  H  605
1891 taatgatgaaatttttgctgcagcccaatttaagcttagaagacgaagaagacgatgaatttgaagaacaggagaacaaggttatccaacc 1980
 606  N  D  E  I  L  L  Q  P  \  L  S  L  E  D  E  E  D  D  E  F  E  E  Q  E  N  K  V  I  Q  P  635
1981 agctgaaccagtagagctgagaaggaagccctcgtacttacctataaggaggagtcgagtggcatcagaatccagtgtcaccgagccgca 2070
 636  A  E  P  V  E  L  R  R  K  P  S  Y  L  P  I  R  R  S  R  V  A  S  E  S  S  V  T  E  P  Q  665
2071 ggtacagcggtcacatgatgcacaagcactgcggaaagcattcagaaacaacccatacaataagctccattacaagtacaacccaaacct 2160
 666  V  Q  R  S  H  D  A  Q  A  L  R  K  A  F  R  N  N  P  Y  N  K  L  H  Y  K  Y  N  P  N  L  695
2161 cgtgggggaggaggaccaggagctagcggagcacctgcatcgccgacacctgaacgcccgccgcatgacgcgcctcgcatcctgttcccg 2250
 696  V  G  E  E  D  Q  E  L  A  E  H  L  H  R  R  H  L  N  A  R  R  M  T  R  L  A  S  C  S  R  725
2251 tcttcctgtatccaacattgtcgagtctttacgacctaaaacagaaataagttttgggcaccttgaagatggtgtcgcagagctggtgca 2340
 726  L  P  V  S  N  I  V  E  S  L  R  P  K  T  E  I  S  F  G  H  L  E  D  G  V  A  E  L  V  Q  755
2341 acgccatcatgaacgtcggagcaccatgtccttgcgcagacgtcctcatccattgcaagcctgcttctcttccctgaacctggaaatcc 2430
 756  R  H  H  E  R  R  S  T  M  S  L  R  R  R  P  H  P  L  Q  A  C  F  S  S  P  E  P  G  N  P  785
2431 catggcttcacctaccattctgaggggagcttcgggaggcctcagaaccccggtccagccttcctgacatcagagatcatgactcgcactc 2520
 786  M  A  S  P  T  I  L  R  E  L  R  E  A  S  E  P  R  S  S  L  P  D  I  R  D  H  D  S  H  S  815
2521 tcgcctgagtgctacagcctccgcgaggggagagcggcagcctttattcccagacaggtcaaagagtgaagggacagcaaaccctgatgt 2610
 816  R  L  S  A  T  A  S  A  R  G  E  R  Q  P  L  F  P  D  R  S  K  S  E  G  T  A  N  P  D  V  845
2611 aacataccagaatgaaaagttgaatgaagaagttcctatggttgctacatctactaacacgcaaccatcaagtccaaaagaaaccagatt 2700
 846  T  Y  Q  N  E  K  L  N  E  E  V  P  M  V  A  T  S  T  N  T  Q  P  S  S  P  K  E  T  R  F  875
2701 cagaaaatgaGGATGTGGTTTGATGAATTATCCATAATCTGGAAGCCCAAATATAGCATAAG*AAAAAAAAAAAAAAAAAAAAAAAAAAAA* 2788
 876  R  K  *                                                                                   877
```

Figure 1. Nucleotide and deduced amino acid sequence of *PmNHE*. (* *termination codon, tga*).

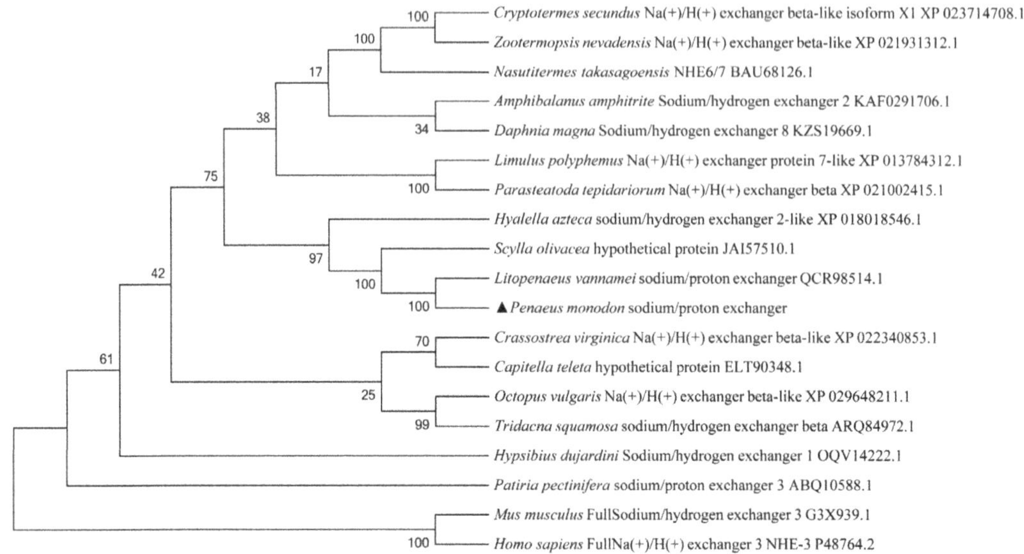

Figure 2. Phylogenetic analysis of *PmNHE*. (▲represents the *PmNHE* gene obtained in this study).

The results of amino acid sequence multiple alignment analysis (Figure 3) show that the amino acids of *P. monodon* NHE and other species have high homology, among which, *PmNHE* and *Penaeus vannamei* (qcr98514.1) have the highest similarity of 93.92%; the protein similarity with *Limulus polyphemus* (xp_013784312.1) was 42.71%; the protein similarity with *Portunus trituberculatus* (anv19765.1) was 35.77%.

3.3. Expression Analysis of PmNHE mRNA in Different Tissues

In order to study the expression of the *PmNHE* gene in different tissues, ten tissues including the muscle, hepatopancreas, gill, epidermis, intestine, heart, hemolymph, stomach, ovary, and testis were mainly detected (Figure 4). The order of expression from high to low is intestine > muscle > hemolymph > heart > hepatopancreas > stomach > epidermis > gill > testis > ovary, and the intestine and muscle had the highest levels of *PmNHE* mRNA expression.

3.4. Expression Changes of PmNHE under Acute Ammonia Nitrogen Stress

The gills and intestines of *P. monodon* were subjected to two concentrations of ammonia nitrogen, and RT-qPCR was used to detect changes in *PmNHE* expression. The expression levels in the intestine and gills varied in both the safe and half-lethal concentrations of ammonia nitrogen when compared to the control group (Figure 5).

In the gill tissue, the fluctuation range of *PmNHE* in the safe concentration was small, and with the increase in stress time, the expression of the *PmNHE* gene was not significantly different from that of the control group. Under the stress of half-lethal ammonia nitrogen concentration, the expression of *PmNHE* at 3, 12, 48, 72, and 96 h was up-regulated compared with the control group ($p < 0.01$). In the intestine, *PmNHE* showed a trend of first decreasing and then increasing in the safe concentration, which was higher than that of the control group at 6 h and lower than that of the control group at 24 h. The expression of *PmNHE* under the half-lethal concentration stress was inhibited.

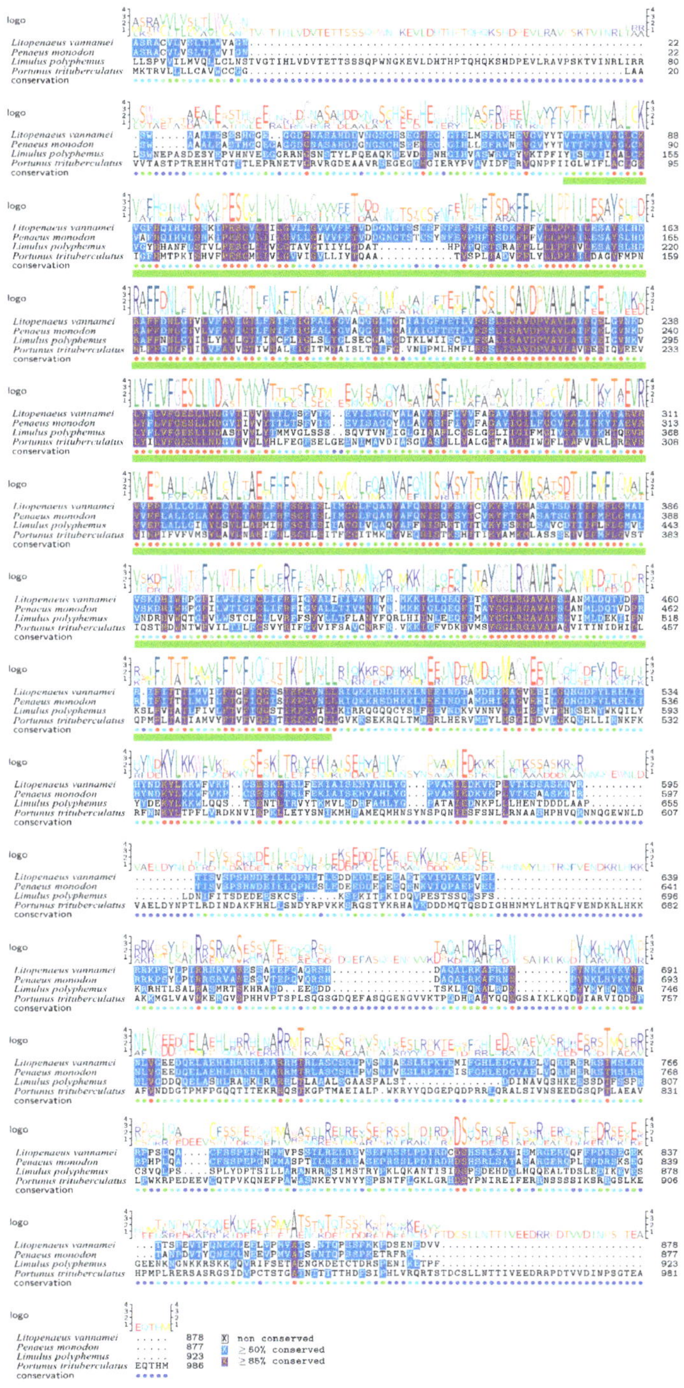

Figure 3. Multiple alignment of amino acid sequences of NHE gene from *Peenaeus monodon* and other species. Sequence ID: *Litopenaeus vannamei*, QCR98514.1; *Limulus Polyphemus*, XP_013784312.1; *Portunus trituberculatus*, ANV19765.1.

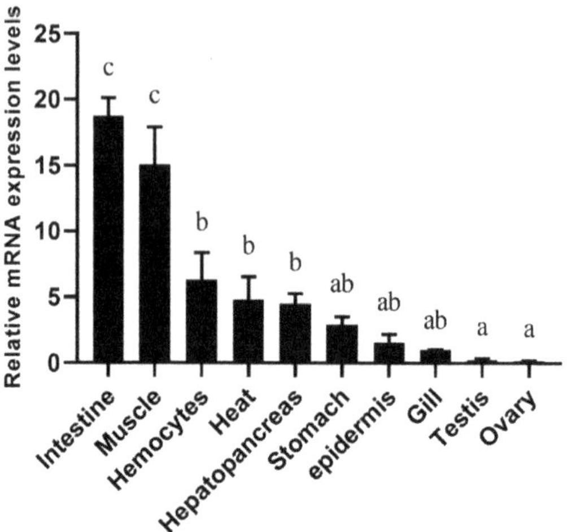

Figure 4. Relative expression of *PmNHE* in different tissues of *Penaeus monodon*. The values are x ± SD (*n* = 3). Bars with different lowercase letters indicate significant differences (*p* < 0.05).

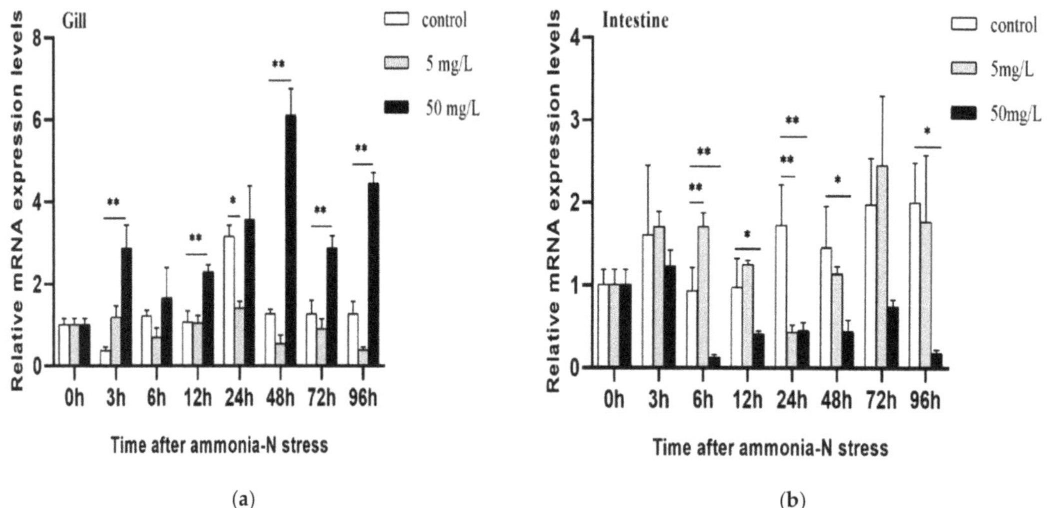

Figure 5. The mRNA expression levels of *PmNHE* in the gills and intestine at different time intervals of different-concentration ammonia nitrogen stress treatments. (**a**) Gill and (**b**) intestine under ammonia nitrogen stress. Data are shown as means ± SD (standard deviation) of three separate individuals. Significant differences are indicated with asterisks, * *p* < 0.05, ** *p* < 0.01.

3.5. Mortality in High-Concentration Ammonia Nitrogen Environment after PmNHE Silencing

Under high-concentration ammonia nitrogen, the expression of *PmNHE* was significantly up-regulated in gills and down-regulated in the intestine. It is suggested that *PmNHE* may be involved in the mechanism of response to high-concentration ammonia nitrogen stress. In order to explore the role of *PmNHE* in the process of high-concentration ammonia nitrogen stress, in this study, RNA interference experiments were performed by injecting ds*PmNHE* synthesized in vitro, and 24 h after ds*PmNHE* injection, acute high-concentration ammonia nitrogen stress was applied to *P. monodon*. The results show that with phosphate-buffered saline (PBS) and dsGFP as controls, the expression of *PmNHE* was lower than that of the control group at 24 h after injection (Figure 6), and there was always a difference between the ds*PmNHE* injection group and the two control groups during the subsequent acute ammonia nitrogen stress experiment ($p < 0.01$). It showed that the whole experimental process had good interference efficiency.

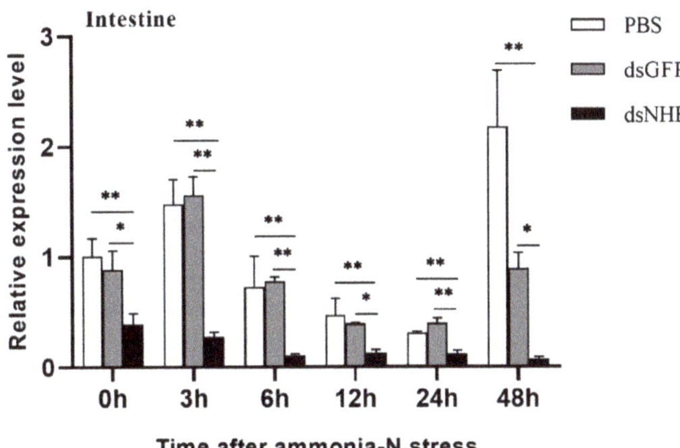

Figure 6. Silencing efficiency of *PmNHE* under ammonia-nitrogen stress. Data are shown as means ± SD (standard deviation) of three separate individuals. Significant differences are indicated with asterisks, * $p < 0.05$, ** $p < 0.01$.

The relative expression of *PmNHE* in the ds*PmNHE* and GFP groups at different time points under high-concentration ammonia-nitrogen stress was determined.

We compared the survival curves of the *PmNHE* silenced group and the control group under high levels of ammonia nitrogen stress to explore the relationship between the *PmNHE* gene and high-concentration ammonia nitrogen stress (Figure 7). The shrimp in the GFP group all died within 69 h, with an average survival time of 40.66 h, the shrimp in the PBS group all died within 69 h, with an average survival time of 42.32 h, the average survival time of the shrimp in the *PmNHE*-dsRNA group was 37.96 h. From the trend of the survival curve, it can be seen that the survival rate of shrimp in the silenced group was higher than that in the control group in the first 24 h under high-concentration ammonia nitrogen stress; after 36 h of stress, the death rate of shrimp in the *PmNHE*-silenced group began to increase, which was different from that of the two control groups.

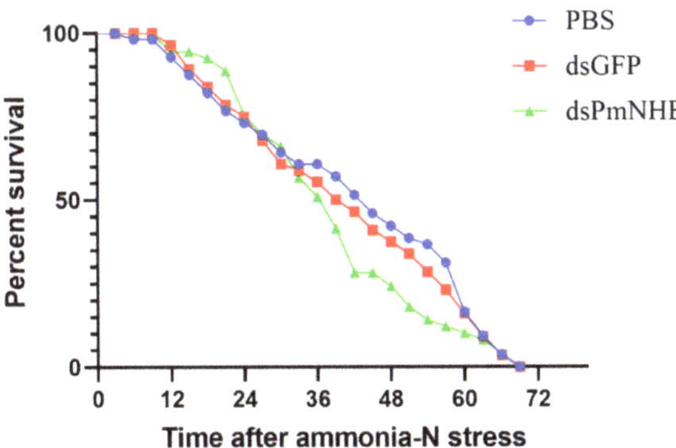

Figure 7. Mortality under ammonia-nitrogen stress after silencing of *PmNHE*.

ds*PmNHE*, GFP, and PBS group shrimp were simultaneously stressed under a higher concentration of ammonia nitrogen; the number of dead shrimp in each group was observed and recorded every 3 h.

4. Discussion

Numerous studies have shown that NHEs play an important role in the transport of various ions in animals [18]. In this research, the full-length cDNA encoding NHE was cloned from *P. monodon* for the first time. The amino acid sequence analysis of *PmNHE* by the SMART program found that *PmNHE* had 12 transmembrane domains, and the sequence contained a sodium/hydrogen ion exchanger domain (78–490 aa). Exploring this domain revealed that NHE transporters contained 10–12 transmembrane domains at the amino terminus and a larger cytoplasmic domain at the carboxy terminus. The transmembrane region M3–M12 had the same characteristics as other members of this family. The M6 and M7 transmembrane regions were highly conserved and therefore considered to be involved in the transport of sodium and hydrogen ions [19]. Phylogenetic tree analysis showed that the evolution relationship between *PmNHE* of *P. monodon* and NHE of *Penaeus vannamei* was the closest. The similarity between *PmNHE* and the NHE of *Litopenaeus vannamei* reached 93.92%, the similarity with the protein of *Limulus americanus* was 42.71%, and the similarity with the protein of *Portunus trituberculatus* was 35.77%; these data indicated that NHE sequences were species specific.

Quantitative analysis of various tissues showed that *PmNHE* was expressed in a variety of tissues, and the expression of *PmNHE* was the most abundant in the intestines of *P. monodon*, and the expression level in the muscle was second only to that of the intestinal tissues. It has been shown that the expression of NHE was the highest in the intestine of *Penaeus vannamei*, which was similar to our results [20]. Osmoregulation and ion regulation in crustaceans were mainly performed by gill tissues with multiple functions and the intestines, which was the excretory and digestive organ [21,22]. In addition, muscle has also been demonstrated to have an ion regulation function [23]. Therefore, we selected the gill and intestine as the research objects to explore their expression patterns under ammonia nitrogen stress.

The quantitative results show that the expression of *PmNHE* in the gill tissue of *P. monodon* was significantly up-regulated under a high concentration of ammonia nitrogen stress, which was similar to the result of the significant up-regulation of the NHE-2 gene in the gill tissue of *Oncorhynchus mykiss* when the concentration of ammonia nitrogen increased [24]. Other researchers found that it could block the excretion of ammonia nitrogen from their gills by inhibiting the NHE activity of *Carcinus maenas*, *Cancer pagurus*, and *Astacus leptodactylus* by non-specific inhibitor amiloride [7,8]. Therefore, we speculated that *PmNHE* also played a role in excreting ammonia in the gills of *P. monodon*, which may be in line with the "Na$^+$/NH4$^+$ exchange complex" model of ammonia excretion in freshwater fish. This model suggested that ammonia transport was carried out through Rh glycoproteins, and the hydration of CO_2 and H$^+$ by *Na$^+$/NH$_4^+$-exchange* (NHE-2) and *H$^+$-ATPase* led to the acidification of the gill boundary layer, thereby promoting ammonia transport [25]. Furthermore, we also conducted a quantitative analysis of the intestines, and the results show that the expression of *PmNHE* in the intestines of *P. monodon* under high levels of ammonia nitrogen stress was significantly inhibited compared with the control group. Silva et al. quantitatively analyzed the foregut and hindgut of *Monopterus albus* treated with high ammonia nitrogen and showed that the expression of NHE3 was increased in the foregut after 6 h of ammonia exposure, about 28 times that of the control group [25]. The expression of the NHE1 gene in the foregut and hindgut changed little over time. The expression of the NHE1 and NHE3 genes in the foregut and hindgut following chronic exposure was not different. They speculated that NHE3 in the foregut of *Monopterus albus* may have a significant impact on the active excretion of NH$_4^+$ under the high-concentration ammonia nitrogen environment. The high ammonia nitrogen environment made *PmNHE* up-regulated in the gill tissue of *P. monodon* and inhibited in the intestine. We speculated that it may be due to the differences between species; *PmNHE* in the gill tissue of *P. monodon* played an important role under ammonia nitrogen stress. To further explore the role of *PmNHE* in an environment with high levels of ammonia nitrogen, we compared the survival curves of the *PmNHE* silenced group and control group under high-concentration ammonia nitrogen stress. We found that after high levels of ammonia nitrogen stress, the death curves of shrimp in the GFP and PBS control groups were essentially consistent, and the average death time was similar. The average survival time of the shrimp in the *PmNHE*-dsRNA injection group was lower than that in the two control groups. This result indicates that *PmNHE* may have a significant impact on the high ammonia nitrogen environment.

5. Conclusions

In this research, the full-length *PmNHE* gene was cloned, the structure of the *PmNHE* gene was analyzed by bioinformatics, the expression changes of *PmNHE* in various tissues and in the gill and intestine of *P. monodon* under acute ammonia nitrogen stress were investigated, and the death curve under high levels of ammonia nitrogen stress was explored by RNAi technology. The results provide a basis for further exploring the role of the sodium/hydrogen ion transporter in *P. monodon* under ammonia nitrogen stress.

Author Contributions: S.J. (Shigui Jiang) and F.Z. conceived and designed the experiments. Y.L. performed the bioinformatics analysis and prepared the manuscript, the table, and the figures. H.F., J.H. and W.Z. conducted the experiment. L.Y., X.C., S.J. (Song Jiang) and Q.Y. collected the samples and performed sequencing. All authors have read and agreed to the published version of the manuscript.

Funding: This study was funded by the National Key R&D Program of China (2022YFD2401900); China Agriculture Research System (CARS-48); Hainan Yazhou Bay Seed Laboratory (Project of B21Y10701 and B21HJ0701); Research and Development Projects in Key Areas of Guangdong Province (2021B0202003); Central Public interest Scientific Institution Basal Research Fund, CAFS (2020TD30); Central Public Interest Scientific Institution Basal Research Fund, South China Sea Fisheries Research Institute, CAFS (2020ZD01,2021SD13); Hainan Provincial Natural Science Foundation of China (320QN359,322RC806,320LH008); Guangdong Basic and Applied Basic Research Foundation (2020A1515110200); Guangzhou Science and Technology Planning Project (202102020208); and Hainan Provincial Association for Science and Technology of Young Science and Technology Talents Innovation Plan Project (QCQTXM202206).

Institutional Review Board Statement: The use of all the shrimps in these experiments was approved by the Animal Care and Use Committee at the Chinese Academy of Fishery Sciences (CAFS), and we also applied the national and institutional guidelines for the care and use of laboratory animals at the CAFS.

Informed Consent Statement: Not applicable.

Data Availability Statement: Not applicable.

Conflicts of Interest: The authors declare no conflict of interest.

References

1. Hamm, L.L.; Nakhoul, N.; Hering-Smith, K.S. Acid-Base Homeostasis. *Clin. J. Am. Soc. Nephrol.* **2015**, *10*, 2232–2242. [CrossRef] [PubMed]
2. Weiner, I.D.; Verlander, J.W. Ammonia Transporters and Their Role in Acid-Base Balance. *Physiol. Rev.* **2017**, *97*, 465–494. [CrossRef] [PubMed]
3. Aronson, P.S.; A Suhm, M.; Nee, J. Interaction of external H^+ with the Na^+-H^+ exchanger in renal microvillus membrane vesicles. *J. Biol. Chem.* **1983**, *258*, 6767–6771. [CrossRef] [PubMed]
4. Kinsella, J.L.; Aronson, P.S. Interaction of NH_4^+ and Li+ with the renal microvillus membrane Na^+-H^+ exchanger. *Am. J. Physiol. Cell Physiol.* **1981**, *241*, C220–C226. [CrossRef] [PubMed]
5. Li, Y.D.; Zhou, F.L.; Huang, J.H.; Yang, L.S.; Jiang, S.; Yang, Q.B.; He, J.G.; Jiang, S.G. Transcriptome reveals involvement of immune defense, oxidative imbalance, and apoptosis in ammonia-stress response of the black tiger shrimp (*Penaeus monodon*). *Fish Shellfish. Immunol.* **2018**, *83*, 162–170. [CrossRef]

6. Li, Y.; Zhou, F.; Yang, Q.; Jiang, S.; Huang, J.; Yang, L.; Ma, Z.; Jiang, S. Single-Cell Sequencing Reveals Types of Hepatopancreatic Cells and Haemocytes in Black Tiger Shrimp (*Penaeus monodon*) and Their Molecular Responses to Ammonia Stress. *Front. Immunol.* **2022**, *13*, 883043. [CrossRef]
7. Lucu, Č.; Devescovi, M.; Siebers, D. Do amiloride and ouabain affect ammonia fluxes in perfused *Carcinus* gill epithelia? *J. Exp. Zool.* **1989**, *249*, 1–5. [CrossRef]
8. Weihrauch, D.; Becker, W.; Postel, U.; Luck-Kopp, S.; Siebers, D. Potential of active excretion of ammonia in three different haline species of crabs. *J. Comp. Physiol. B* **1999**, *169*, 25–37. [CrossRef]
9. Towle, D.W.; E Rushton, M.; Heidysch, D.; Magnani, J.J.; Rose, M.J.; Amstutz, A.; Jordan, M.K.; Shearer, D.W.; Wu, W.S. Sodium/proton antiporter in the euryhaline crab Carcinus maenas: Molecular cloning, expression and tissue distribution. *J. Exp. Biol.* **1997**, *200*, 1003–1014. [CrossRef]
10. Li, Y.D.; Zhou, F.L.; Tang, Y.P.; Huang, J.H.; Yang, L.S.; Jiang, S.; Yang, Q.B.; Jiang, S.G. Variation in bacterial communities among stress-sensitive and stress-tolerant black tiger shrimp (*Penaeus monodon*) individuals. *Aquac. Res.* **2020**, *52*, 2146–2159. [CrossRef]
11. Li, Y.D.; Zhou, F.L.; Huang, J.H.; Yang, L.S.; Jiang, S.; Yang, Q.B.; Jiang, S.G. Transcriptome and miRNA Profiles of Black Tiger Shrimp, Penaeus monodon, Under Different Salinity Conditions. *Front. Mar. Sci.* **2020**, *7*, 579381. [CrossRef]
12. Fan, H.; Li, Y.; Yang, Q.; Jiang, S.; Yang, L.; Huang, J.; Jiang, S.; Zhou, F. Isolation and characterization of a MAPKK gene from *Penaeus monodon* in response to bacterial infection and low-salinity challenge. *Aquac. Rep.* **2021**, *20*, 100671. [CrossRef]
13. Si, M.-R.; Li, Y.-D.; Jiang, S.-G.; Yang, Q.-B.; Jiang, S.; Yang, L.-S.; Huang, J.-H.; Chen, X.; Zhou, F.-L. A CSDE1/Unr gene from *Penaeus monodon*: Molecular characterization, expression and association with tolerance to low salt stress. *Aquaculture* **2022**, *561*, 738660. [CrossRef]
14. Li, Y.; Yang, Q.; Su, T.; Zhou, F.; Yang, L.; Huang, J. The toxicity of ammonia-N on *Penaeus monodon* and immune parameters. *J. Shanghai Ocean. Univ.* **2012**, *21*, 358–362. [CrossRef]
15. Zhou, K.; Zhou, F.; Huang, J.; Yang, Q.; Jiang, S.; Qiu, L.; Yang, L.; Zhu, C.; Jiang, S. Characterization and expression analysis of a chitinase gene (PmChi-4) from black tiger shrimp (*Penaeus monodon*) under pathogen infection and ambient ammonia nitrogen stress. *Fish Shellfish. Immunol.* **2017**, *62*, 31–40. [CrossRef] [PubMed]
16. Li, Y.; Zhou, F.; Fan, H.; Jiang, S.; Yang, Q.; Huang, J.; Yang, L.; Chen, X.; Zhang, W.; Jiang, S. Molecular Technology for Isolation and Characterization of *Mitogen-Activated Protein Kinase Kinase 4* from *Penaeus monodon*, and the Response to Bacterial Infection and Low-Salinity Challenge. *J. Mar. Sci. Eng.* **2022**, *10*, 1642. [CrossRef]
17. Si, M.-R.; Li, Y.-D.; Jiang, S.-G.; Yang, Q.-B.; Jiang, S.; Yang, L.-S.; Huang, J.-H.; Chen, X.; Zhou, F.-L. Identification of multifunctionality of the PmE74 gene and development of SNPs associated with low salt tolerance in *Penaeus monodon*. *Fish Shellfish Immunol.* **2022**, *128*, 7–18. [CrossRef]
18. Tseng, Y.; Yan, J.; Furukawa, F.; Hwang, P. Did Acidic Stress Resistance in Vertebrates Evolve as Na^+/H^+ Exchanger-Mediated Ammonia Excretion in Fish? *BioEssays* **2020**, *42*, e1900161. [CrossRef]
19. Dibrov, P.; Fliegel, L. Comparative molecular analysis of Na^+/H^+ exchangers: A unified model for Na^+/H^+ antiport? *FEBS Lett.* **1998**, *424*, 1–5. [CrossRef]
20. Li, H.; Ren, C.; Jiang, X.; Cheng, C.; Ruan, Y.; Zhang, X.; Huang, W.; Chen, T.; Hu, C.; Qiu, G.F. Na^+/H^+ exchanger (NHE) in Pacific white shrimp (*Litopenaeus vannamei*): Molecular cloning, transcriptional response to acidity stress, and physiological roles in pH homeostasis. *PLoS ONE* **2019**, *14*, e0212887. [CrossRef]
21. Freire, C.A.; Onken, H.; McNamara, J.C. A structure–function analysis of ion transport in crustacean gills and excretory organs. *Comp. Biochem. Physiol. Part A Mol. Integr. Physiol.* **2008**, *151*, 272–304. [CrossRef] [PubMed]
22. Chen, T.; Ren, C.; Wang, Y.; Gao, Y.; Wong, N.-K.; Zhang, L.; Hu, C. Crustacean cardioactive peptide (CCAP) of the Pacific white shrimp (*Litopenaeus vannamei*): Molecular characterization and its potential roles in osmoregulation and freshwater tolerance. *Aquaculture* **2016**, *451*, 405–412. [CrossRef]
23. Mcfarland, W.; Lee, B.D. Osmotic and ionic concentrations of penaeidean shrimps of the Texas coast. *Bull. Mar. Sci.* **1963**, *13*, 391–417.
24. Zimmer, A.M.; Nawata, C.M.; Wood, C.M. Physiological and molecular analysis of the interactive effects of feeding and high environmental ammonia on branchial ammonia excretion and Na^+ uptake in freshwater rainbow trout. *J. Comp. Physiol. B* **2010**, *180*, 1191–1204. [CrossRef]
25. Silva, J.M.; Coimbra, J.; Wilson, J. Weatherloach (*Misgurnus anguillicaudatus*) actively excretes ammonia through NHE. *Comp. Biochem. Physiol. Part A Mol. Integr. Physiol.* **2008**, *150*, S57. [CrossRef]

Journal of
*Marine Science
and Engineering*

Article

DNA Barcoding Is a Useful Tool for the Identification of the Family *Scaridae* in Hainan

Bo Liu [1,2], Yali Yan [1], Nan Zhang [1,3], Huayang Guo [1,3], Baosuo Liu [1,3], Jingwen Yang [1], Kecheng Zhu [1,3] and Dianchang Zhang [1,3,4,5,*]

[1] South China Sea Fisheries Research Institute, Chinese Academy of Fishery Sciences, Key Laboratory of South China Sea Fishery Resources Exploitation and Utilization, Ministry of Agriculture and Rural Affairs, Guangzhou 510300, China
[2] College of Fisheries and Life Science, Shanghai Ocean University, Shanghai 201306, China
[3] Southern Marine Science and Engineering Guangdong Laboratory (Guangzhou), Guangzhou 511458, China
[4] Tropical Aquaculture Research and Development Center, South China Sea Fisheries Research Institute, Chinese Academy of Fishery Sciences, Sanya 572018, China
[5] Sanya Tropical Fisheries Research Institute, Sanya 572018, China
* Correspondence: zhangdch@scsfri.ac.cn

Abstract: Species markers can be quickly and accurately assessed using DNA barcoding. We investigated samples from the parrotfish family *Scaridae* using DNA barcoding in Hainan. A total of 401 DNA barcodes were analyzed, including 51 new barcodes generated from fresh material, based on a 533 bp fragment of the cytochrome c oxidase subunit I (*CO* I) gene. There were 350 *CO* I barcode clusters that matched 43 species from the Barcode of Life Data Systems (BOLD) and GenBank databases. The results showed the following average nucleotide compositions for the complete dataset: adenine (A, 22.7%), thymine (T, 29.5%), cytosine (C, 29.5%), and guanine (G, 18.2%). The mean genetic distance between confamilial species was nearly 53-fold greater than that between individuals within the species. In the neighbor-joining tree of *CO* I sequences, *Chlorurus sordidus* and *C. spilurus* clustered together, and all other individuals clustered by species. Our results indicated that DNA barcoding could be used as an effective molecular tool for monitoring, protecting, and managing fisheries, and for elucidating taxonomic problem areas that require further investigation.

Keywords: *Scaridae*; DNA barcoding; cytochrome c oxidase subunit I (*CO* I)

Citation: Liu, B.; Yan, Y.; Zhang, N.; Guo, H.; Liu, B.; Yang, J.; Zhu, K.; Zhang, D. DNA Barcoding Is a Useful Tool for the Identification of the Family *Scaridae* in Hainan. *J. Mar. Sci. Eng.* **2022**, *10*, 1915. https://doi.org/10.3390/jmse10121915

Academic Editor: Francesco Tiralongo

Received: 11 November 2022
Accepted: 18 November 2022
Published: 6 December 2022

1. Introduction

Parrotfish (family *Scaridae*), belonging to Actinopterygii: *Perciformes*, are herbivorous fish that live in tropical and subtropical coral reefs and are relatively abundant in biomass [1]. They play a crucial role in coral reef ecosystems [2,3], and as consumers of benthic algae, directly affect the structure and composition of benthic communities, and positively affect coral survival and growth [4]. Parrotfish are also involved in calcium carbonate cycling in coral reefs [5,6], and decompose coral reef skeletons into sand-sized sediments [7,8]. They maintain a coral-dominated community structure by feeding on fast-growing algae and can also influence reef development and complexity by decomposing reef carbonates [2,9]. Thus, sediments produced through parrotfish activities are an important source of island construction and maintenance in atoll coral reef environments, particularly in the setting of current rising sea levels and changes in island morphology [10]. The Healthy Reef Initiative (http://www.healthyreefs.org/cms/; accessed on 1 June 2022) uses parrotfish biomass as one of the key indicators in their coral reef health reports [11]. They have beautiful body shapes and bright colors, which greatly enhance the aesthetic quality of coral reefs and improve their economic value through ecotourism [3]. They are used as important food resources, and their biomass has recently decreased because of increased fishing [12,13].

Parrotfish, which are widely distributed in tropical, subtropical, and temperate areas [14], include approximately 100 species, divided among 10 genera [15]. There are 33 species of *Scaridae* in Taiwan and 31 species in mainland China [16]. The fish are known to change sex (from female to male) and color (inpectoral fin color from brown to blue) during growth [12,17]. Additionally, parrotfish have a behavioral mode called "grouping crypsis" (http://fishdb.sinica.edu.tw/; accessed on 1 June 2022). Juvenile parrotfish are discolored because of their social behavior. When they swim together, regardless of whether they are of the same species, their body color responds to the majority, becoming consistently grayish-brown, or exhibiting a longitudinal gradient in these colors. Once alone, juvenile fish instantly display beautiful and bright colors that blend into the surrounding environment, and these color changes occur very quickly. These extremely variable characteristics lead to difficulties in the morphological identification of *Scaridae*.

In the last decade, research on parrotfish has largely focused on their role and the mechanisms of their effects on coral reef ecosystems [9,18]. These fish can be excavators, scrapers, or grazers [19]. With increasing concern about parrotfish populations, the focus has been on coral reef health [20]. Furthermore, some researchers have explored the feeding ecology, habitat, diet, and habitat shifts of parrotfish by examining their fatty acid concentrations, composition, and levels [17]. Therefore, based on the economic, development and conservation of parrotfish, it is necessary to accumulate basic information about parrotfish and develop appropriate conservation and management measures [21]. A study using otoliths to identify parrotfish found that small otoliths were most similar to large *Scarus oviceps*, and least similar to large *Hipposcarus longiceps* [22]. However, there are several obvious shortcomings in the identification of otoliths: (i) there is no complete database of otolith morphology; (ii) because of the wide variety of parrotfish, it is difficult to find enough variability in otolith shape to identify species; and (iii) polishing otoliths is time-consuming, and the loss is large. Much experience is required to accurately identify otoliths.

In the past two decades, molecular techniques have become a popular and critical method for identifying species and resolving taxonomic ambiguities [23]. Molecular methods are useful in elucidating phylogenetic relationships and evolutionary patterns for biological ecology where classical morphological methods are not applicable [24]. DNA barcoding methods have been used to complement or refine morphological species identifications [25,26]. Studies have shown that the identification and discrimination of DNA barcoding are accurate and rapid [27–29]. DNA barcoding has been used to identify species using cytochrome c oxidase subunit I (*CO* I) sequences [30]. Therefore, they have been widely used for species identification [31–34].

In this study, more parrotfish samples were investigated to further evaluate the effectiveness of DNA barcoding for distinguishing parrotfish. The objectives of this study were to examine the reliability of *CO* I as a DNA barcode in parrotfish gene composition, and to determine intra- and interspecific genetic distances, codon characteristics, and molecular phylogenetic trees. The DNA barcoding data generated can be used as an effective molecular tool to achieve better monitoring and conservation outcomes for the family *Scaridae*.

2. Material and Methods

2.1. Sample Collection

A total of 51 parrotfish were collected from Hainan and Sansha in China (collection information available in Table 1). All samples in this experiment were obtained with the assistance of local fishermen and buyers on 5 October 2018. Back muscles were collected from each sample and preserved in 95% ethanol prior to DNA extraction. All samples were preserved at the South China Sea Fisheries Research Institute, Chinese Academy of Fisheries Sciences, China.

Table 1. Voucher and sequence information for the 51 specimens. Process IDs are sequence numbers of voucher specimens in GenBank, and Voucher IDs are voucher numbers in the South China Sea Fisheries Research Institute. BOLD is a summary of identification based on the barcode sequence of each species obtained using the BOLD.

Genus	Species Studied	BOLD	Voucher ID	Process ID	Location
Calotomus	*Calotomus carolinus*	*Calotomus carolinus*	ss-18	MK765061	Sansha
Chlorurus	*Chlorurus japanensis*	*Chlorurus japanensis*	ss-67	MK765062	Sansha
	Chlorurus microrhinos	*Chlorurus microrhinos*	ss-53	MK765063	Sansha
	Chlorurus microrhinos	*Chlorurus microrhinos*	ss-54	MK765064	Sansha
	Chlorurus sordidus	*Chlorurus sordidus*	Ssfri-F0165-01	MK765065	Hainan
	Chlorurus sordidus	*Chlorurus sordidus*	Ssfri-F0165-02	MK765066	Hainan
	Chlorurus sordidus	*Chlorurus sordidus*	Ssfri-F0165-03	MK765067	Hainan
	Chlorurus sordidus	*Chlorurus sordidus*	Ssfri-F0165-04	MK765068	Hainan
	Chlorurus sordidus	*Chlorurus spilurus*	ss-16	MK765069	Sansha
	Chlorurus sordidus	*Chlorurus spilurus*	ss-44	MK765070	Sansha
	Chlorurus sordidus	*Chlorurus spilurus*	ss-45	MK765071	Sansha
Hipposcarus	*Hipposcarus longiceps*	*Hipposcarus longiceps*	ss-17	MK765072	Sansha
Scarus	*Scarus dimidiatus*	*Scarus dimidiatus*	ss-65	MK765073	Sansha
	Scarus chameleon	*Scarus chameleon*	Ssfri-F0158-02	MK765074	Hainan
	Scarus forsteni	*Scarus forsteni*	ss-48	MK765075	Sansha
	Scarus forsteni	*Scarus forsteni*	ss-57	MK765076	Sansha
	Scarus forsteni	*Scarus forsteni*	ss-66	MK765077	Sansha
	Scarus forsteni	*Scarus forsteni*	Ssfri-F0164-01	MK765078	Hainan
	Scarus forsteni	*Scarus forsteni*	Ssfri-F0164-02	MK765079	Hainan
	Scarus forsteni	*Scarus forsteni*	Ssfri-F0164-03	MK765080	Hainan
	Scarus forsteni	*Scarus forsteni*	Ssfri-F0164-04	MK765081	Hainan
	Scarus forsteni	*Scarus forsteni*	Ssfri-F0164-05	MK765082	Hainan
	Scarus forsteni	*Scarus forsteni*	Ssfri-F0164-06	MK765083	Hainan
	Scarus ghobban	*Scarus ghobban*	ss-14	MK765084	Sansha
	Scarus ghobban	*Scarus ghobban*	ss-15	MK765085	Sansha
	Scarus ghobban	*Scarus ghobban*	Ssfri-F0347-01	MK765086	Hainan
	Scarus ghobban	*Scarus ghobban*	Ssfri-F0347-02	MK765087	Hainan
	Scarus ghobban	*Scarus ghobban*	Ssfri-F0347-03	MK765088	Hainan
	Scarus ghobban	*Scarus ghobban*	Ssfri-F0347-04	MK765089	Hainan
	Scarus ghobban	*Scarus ghobban*	Ssfri-F0347-05	MK765090	Hainan
	Scarus niger	*Scarus niger*	ss-7	MK765091	Sansha
	Scarus niger	*Scarus niger*	ss-41	MK765092	Sansha
	Scarus niger	*Scarus niger*	ss-42	MK765093	Sansha
	Scarus oviceps	*Scarus oviceps*	ss-33	MK765094	Sansha
	Scarus oviceps	*Scarus oviceps*	ss-46	MK765095	Sansha
	Scarus psittacus	*Scarus psittacus*	Ssfri-F0171-01	MK765096	Hainan
	Scarus psittacus	*Scarus psittacus*	Ssfri-F0171-02	MK765097	Hainan
	Scarus globiceps	*Scarus rivulatus*	ss-93	MK765098	Sansha
	Scarus globiceps	*Scarus rivulatus*	Ssfri-F0161-03	MK765099	Hainan
	Scarus globiceps	*Scarus rivulatus*	Ssfri-F0161-04	MK7650100	Hainan
	Scarus globiceps	*Scarus rivulatus*	Ssfri-F0161-05	MK7650101	Hainan
	Scarus rivulatus	*Scarus rivulatus*	Ssfri-F0161-06	MK7650102	Hainan
	Scarus rivulatus	*Scarus rivulatus*	Ssfri-F0161-07	MK7650103	Hainan
	Scarus rivulatus	*Scarus rivulatus*	Ssfri-F0161-08	MK7650104	Hainan
	Scarus rivulatus	*Scarus rivulatus*	Ssfri-F0161-09	MK7650105	Hainan
	Scarus rivulatus	*Scarus rivulatus*	Ssfri-F0161-10	MK7650106	Hainan
	Scarus schlegeli	*Scarus schlegeli*	ss-20	MK7650107	Sansha
	Scarus schlegeli	*Scarus schlegeli*	ss-47	MK7650108	Sansha
	Scarus schlegeli	*Scarus schlegeli*	ss-51	MK7650109	Sansha
	Scarus schlegeli	*Scarus schlegeli*	ss-72	MK7650110	Sansha
	Scarus spinus	*Scarus spinus*	Ssfri-F0166-01	MK7650111	Hainan

2.2. DNA Extraction

DNA was extracted from the muscle samples using a HiPure Mollusc DNA Kit (Axygen Biosciences, San Francisco, CA, USA) according to the manufacturer's instructions. All DNA samples were stored at −20 °C, followed by polymerase chain reaction (PCR) amplification.

2.3. PCR and DNA Sequencing

Fragments of the mitochondrial *COI* gene were amplified using the following universal fish barcoding primers: forward fish-F 5′-TCRACYAAYCAYAAAGAYATYGGCAC-3′ and reverse fish-R 5′-ACTTCAGGGTGACCGAAGAATCAGAA-3′. The total volume and thermal cycle sequences of the PCR were performed as described previously [35]. The amplified PCR products were checked for optimal fragment sizes on 1.5% agarose gels. The PCR products with a single bright band were sent to Beijing RuiBiotec (Beijing, China) for sequencing in both directions. All sequences were loaded onto BOLD in the project.

2.4. Analysis of the Utility of the BOLD as an Identification Tool for Parrotfish

A total of 350 sequences belonging to 43 species of parrotfish with genus and species names assigned were found in a search on BOLD. In the case of these records, the number of barcoded index numbers (BIN) associated with each species was recorded. If a species was associated with a BIN, an assessment was conducted to determine whether the BIN was associated with any other species.

2.5. Data Analysis

SSR Hunter 1.3 software was used to edit the sequences, Clustal X 1.83 was used to align all the sequences, and redundant sequences at both ends were removed. Mega 7.0 was used to analyze the nucleotide composition, number of mutation sites and codon composition of all sequences. Based on the Kimura 2-parameter (K2P) model, inter and intraspecific genetic distances were calculated, molecular phylogenetic trees were constructed using the neighbor-joining (NJ) method, and their confidence was tested by 1000 repeated samplings.

2.6. Ethics Statement

All experiments in this study were conducted in accordance with the regulations and guidelines established by the Animal Care and Use Committee of the South China Sea Fisheries Research Institute of the Chinese Academy of Fishery Sciences (No. SCSFRI96-253).

3. Results

3.1. Sequence Characteristics of CO I Gene Fragment

A total of 401 *CO* I sequences were obtained, representing 44 species and 10 genera. All the sequences were trimmed to a consensus length of 533 bp. The mean nucleotide compositions for the complete data set were as follows: 22.7% adenine (A), 29.5% thymine (T), 29.5% cytosine (C), and 18.2% guanine (G). The highest percentage of G-C (55.69%) was detected in the first codon, whereas the lowest (42.96%) was detected in the second codon (Table 2). Within the 533-bp nucleotide sequences in the complete data set, there were conserved sites (327, 61.53%), variable sites (204, 38.27%), parsimony-informative sites (194, 36.40%), and singleton sites (10, 1.88%). Transitional pairs (si = 458) were present in greater numbers than transversional pairs (sv = 52). The ratio of si/sv (R) was 21.00 for the data set (Table 2).

Table 2. Sequence variation of the *CO* I gene and average nucleotide frequencies of *CO* I partial sequences of *Scaridae* (%).

Domain	ii	si	sv	R	T	C	A	G	Total
1		458.00	52.00	21.00	29.5	29.5	22.7	18.2	530.9
1st	174.00	4.00	0.00	9.40	19.0	27.5	25.3	28.2	178.0
2nd	175.00	2.00	0.00	5.89	42.1	28.8	15.0	14.1	178.0
3rd	109.00	46.00	20.00	2.23	27.5	32.2	27.9	12.3	175.0

Note: ii = Invariant pairs; si = Transitional pairs; sv = Transversional pairs; R = si/sv.

3.2. Genetic Distance between Species and within Species

Intraspecific K2P distances ranged from 0.000 to 0.015, and most intraspecific genetic distances were below 0.01. There were four species with intraspecific genetic distances between 0.01 and 0.02 (Figure 1). The mean intraspecific genetic distance was 0.003. Among the 44 species, *Scarus flavipectoralis* and *Nicholsina usta* had the greatest interspecific genetic distance of 0.248, while *Chlorurus sordidus* and *C. spilurus* had the lowest interspecific genetic distance (0.002). Most interspecific genetic distances were above 0.1. Overall, the mean interspecific genetic distance was 0.159, nearly 53 times higher than that among individuals within the same species (Supplementary Table S1).

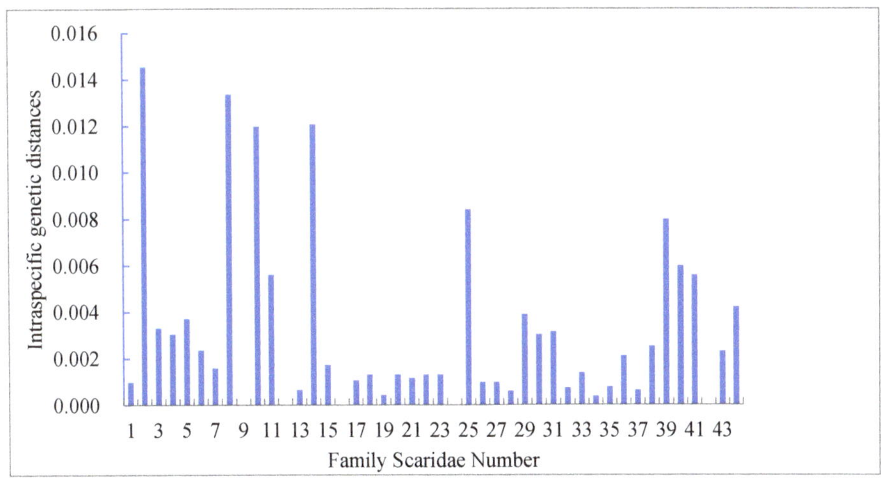

Figure 1. The intraspecific genetic distances of the family *Scaridae*. Note: The abscissa represents the species: 1. *Bolbometopon muricatum*, 2. *Cetoscarus ocellatus*, 3. *Hipposcarus longiceps*, 4. *Scarus iseri*, 5. *S. rubroviolaceus*, 6. *S. ghobban*, 7. *S. taeniopterus*, 8. *S. niger*, 9. *S. forsteni*, 10. *S. prasiognathos*, 11. *S. frenatus*, 12. *S. dimidiatus*, 13. *S. oviceps*, 14. *S. chameleon*, 15. *S. rivulatus*, 16. *S. globiceps*, 17. *S. quoyi*, 18. *S. flavipectoralis*, 19. *S. schlegeli*, 20. *S. fuscopurpureus*, 21. *S. psittacus*, 22. *S. pinus*, 23. *Chlorurus capistratoides*, 24. *C. japanensis*, 25. *C. bleekeri*, 26. *C. microrhinos*, 27. *C. frontalis*, 28. *C. spilurus*, 29. *C. sordidus*, 30. *Sparisoma radians*, 31. *S. aurofrenatum*, 32. *S. viride*, 33. *S. chrysopterum*, 34. *S. rubripinne*, 35. *S. rocha*, 36. *S. cretense*, 37. *S. atomarium*, 38. *Cryptotomus roseus*, 39. *Nicholsina denticulata*, 40. *N. usta*, 41. *Leptoscarus vaigiensis*, 42. *Calotlmus carolinus*, 43. *C. viridescens*, 44. *C. spinidens*.

3.3. Molecular Phylogenetic Tree

The NJ tree clustered *C. sordidus* and *C. spilurus* together, while the other individuals clustered by species (Figure 2). There were close relationships between *C. japanensis* and *C. capistratoides, C. carolinus*, and *C. viridescens; S. rivulatus* and *S. globiceps; S. rubroviolaceus* and *S. ghobban* and *S. schlegeli* and *S. ferrugineus*, which together formed a cohesive group with a moderately significant bootstrap value above 80%. Simultaneously, *Chlorurus, Cryp-*

totomus, Nicholsina, Leptoscarus, Hipposcarus, Bolbometopon, Sparisoma, Calotomus, Cetoscarus, and *Scarus* clustered into separate branches.

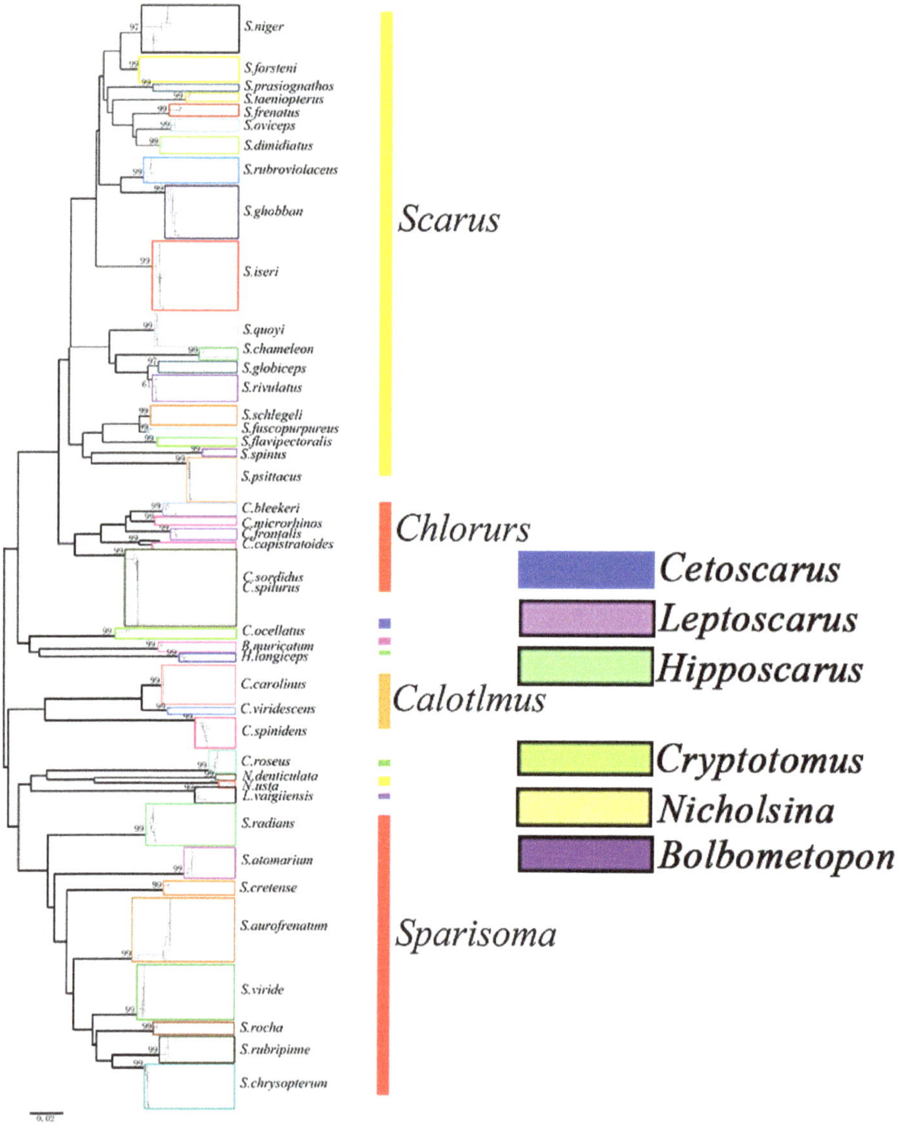

Figure 2. NJ tree resulting from analysis of the CO I gene for *Scaridae* species Bootstrap vaues Higher than 50% based on 1000 replicates are shown on the branches. The scale represents a genetic distance of 0.01 per million years.

3.4. New Country Records

Based on the samples collected, identified, and processed in this study, and a BLAST search in the BOLD and GenBank databases, three species, *C. carolinus*, *C. japanensis*, and *S. rivulatus*, are reported from mainland China for the first time. This increases the parrotfish diversity for mainland China species.

4. Discussion

CO I is commonly used as a barcode marker for animal species when the intraspecific K2P distance is below 1% and rarely exceeds 2% [36]. Hebert et al. suggested that the key point for the effective identification of species using *CO* I gene sequences is that the interspecific genetic distance must be greater than the intraspecific genetic distance, and the distances must differ by approximately 10 times [36,37]. The mean intraspecific genetic distance was 0.003 for the entire data set. The average interspecific genetic distance of the entire data set was 53 times that of average intraspecific genetic distance. The NJ tree clustered *C. sordidus* and *C. spilurus* together, and all other individuals clustered together by species, with high confidence. These results indicate that the *CO* I gene sequence can be used to effectively identify species in the family *Scaridae*.

On average, a G-C content of 47.75% was detected in the dataset. The base composition characteristics of the *CO* I gene were consistent with those reported for other teleosts, all of which had a GC content lower than AT content [38,39]. The first codon had the highest G + C content (48.3%), and the variation range was 53.39–56.74%. The second codon had the lowest G + C content (42.96%), and the variation range was 41.57–44.94%. The G + C content of the 3rd codon was 44.53%, ranging from 30.86–52.57%. In the CO I gene, si is often greater than sv, and the smaller the ratio R, the faster the evolutionary rate. In this study, the first and third codons had the largest and smallest R-values, respectively; the third codon had the fastest evolutionary rate, and the first codon the slowest. The possible explanation for this result might be that the variation range of G + C content is directly related to the evolution rate of codons. The larger the variation range of G + C content, the faster the evolution rate of codons.

The greatest genetic distance within the species was less than the smallest genetic distance between species, and a barcode gap was generated. Barcode gap is a key factor for the accurate identification of species from DNA barcodes. The intraspecific genetic distances of *C. sordidus* and *C. spirus* were 0.005 and 0.000, respectively. The interspecific genetic distance between the two species was 0.003. No barcode gaps were observed between the two species. In the NJ phylogenetic tree, *C. sordidus* and *C. spilurus* formed independent branches with a confidence level of 88. Therefore, the results of this study are identical to those of earlier studies, supporting the assertion that *C. spilurus* and *C. sordidus* are the same species [16].

The divergence between *Chlorurus* and *Scarus* was quite close, 6.0–7.4 mya, and the genera *Chloruus* and *Scarus* showed most of the variation after 3–5 million years [40]. However, Bellwood regarded the genera *Chlorurus* and *Scarus* as two distinct monophyletic lineages [41]. The topological structure reinforced the morphological diagnosis that these two genera belong to a monophyletic lineage, and together form a good clade [39]. Bayesian analysis was consistent with previous studies that provided strong support for confirming the identity of *Chlorurus* and *Scarus* [42]. In our study, *Chlorurus* and *Scarus* clustered into two independent branches, verifying the morphological diagnosis. In the phylogenetic tree, fish species of each genus formed an independent branch. Therefore, *CO* I is also suitable for identification at the genus level in the *Scaridae*.

The International Barcode of Life (iBOL, http://ibol.org/; accessed on 2 June 2022) is the global leader in DNA barcode work, determining species based on DNA barcodes, and sharing results freely [43]. Notably, the development of DNA barcode libraries is based on community efforts, and the use of the BOLD has led to DNA barcode technology being regarded as the standard for species recognition [38]. In BOLD, barcode sequences are stored and associated with other taxonomic data (voucher images, location data, etc.) to improve the accuracy of species recognition [44]. The BOLD has accelerated exchanges between countries worldwide, enabling global resources to be interoperable and species identification to be more standardized. BOLD is an accessible database for the analysis and search of DNA barcode data [45]. International life barcodes have several shortcomings: (i) Data sharing is not timely. Data provided by countries to the BLOD can be made public after only two years, because researchers hope to disclose the data only after analysis or article publication. (ii) According to the BLOD standards for data management, each DNA barcode must have complete voucher specimen information, acquisition information, and the original files of the sequenced peak map. However, many research groups and researchers from China cannot strictly follow these requirements and standards, and the quality of the data is greatly reduced. (iii) The data were not updated in time to match the genetic sequences of the same species, which were still identified by a previous name. Therefore, the entries were not unified. Although *C. sordidus* and *C. spilurus* belong to the same genus, their genetic distance is very small, and *C. spilurensis* syn. nov. is therefore synonymized with *C. sordidus* [16], and they cluster together in a molecular evolutionary tree. However, when the sequences are aligned in BOLD, they do not have the same species names.

5. Conclusions

By analyzing the sequences of the *CO* I gene of 401 parrotfish, we found that the average intraspecific genetic distance was 0.003, and the average interspecific genetic distance was 0.159, approximately 53 times the average intraspecific genetic distance. The NJ tree shows that *C. sordidus* and *C. spilurus* of blue-headed green parrotfish gather together, and individuals of other species grouped together, with high support, and different species can be effectively distinguished. The results showed that DNA barcode technology based on the *CO* I gene could be used to identify species of *Scaridae*.

Supplementary Materials: The following supporting information can be downloaded at: https://www.mdpi.com/article/10.3390/jmse10121915/s1, Table S1: Mean percent genetic distances between Scaridae species under the corrected K2P.

Author Contributions: Data curation, B.L. (Bo Liu).; formal analysis, N.Z.; investigation, H.G.; methodology, J.Y.; resources, Y.Y., N.Z., H.G., B.L. (Baosuo Liu) and K.Z.; validation, B.L. (Baosuo Liu); visualization, J.Y.; writing—original draft, B.L. (Bo Liu); writing—review & editing, D.Z. All authors have read and agreed to the published version of the manuscript.

Funding: This research was funded by the Financial Fund of Ministry of Agriculture and Rural affairs of China (NHYYSWZZZYKZX2020), National Marine Genetic Resource Center, and China-ASEAN Maritime Cooperation Fund.

Institutional Review Board Statement: All experiments in this study were approved by the Animal Care and Use Committee of South China Sea fisheries Research Institute, Chinese Academy of fishery Sciences (no. SCSFRI96-253) and performed according to the regulations and guidelines established by this committee. Written informed consent was obtained from the owners for the participation of their animals in this study.

Informed Consent Statement: Informed consent was obtained from all subjects involved in the study. Written informed con-sent has been obtained from the patients to publish this paper.

Data Availability Statement: All data generated or analyzed during this study are included in this published article.

Acknowledgments: This work was supported by the Financial Fund of Ministry of Agriculture and Rural affairs of China (NHYYSWZZZYKZX2020), National Marine Genetic Resource Center, and China-ASEAN Maritime Cooperation Fund.

Conflicts of Interest: No potential conflict of interest was reported by the authors.

References

1. Francini-Filho, R.B.; Moura, R.L. Evidence for spillover of reef fishes from a no-take marine reserve: An evaluation using the before-after control-impact (BACI) approach. *Fish. Res.* **2008**, *93*, 346–356. [CrossRef]
2. Bellwood, D.R.; Hoey, A.S.; Choat, J.H. Limited functional redundancy in high diversity systems: Resilience and ecosystem function on coral reefs. *Ecol. Lett.* **2003**, *6*, 281–285. [CrossRef]
3. Carlon, D.B.; LippÉ, C. Isolation and characterization of 17 new microsatellite markers for the ember parrotfish *Scarus rubroviolaceus*, and cross-amplification in four other parrotfish species. *Mol. Ecol. Notes* **2007**, *7*, 613–616. [CrossRef]

4.   Mumby, P.J.; Dahlgren, C.P.; Harbome, A.R.; Kappel, C.V.; Micheli, F.; Brumbaugh, D.R.; Holmes, K.E.; Mendes, J.M.; Broad, K.; Sanchirico, J.N.; et al. Fishing, Trophic Cascades, and the Process of Grazing on Coral Reefs. *Science* **2006**, *311*, 98. [CrossRef]
5.   Goatley, C.; Bellwood, D. Biologically mediated sediment fluxes on coral reefs: Sediment removal and off-reef transportation by the surgeonfish *Ctenochaetus striatus*. *Mar. Ecol. Prog. Ser.* **2010**, *415*, 237–245. [CrossRef]
6.   Glynn, P.; Manzello, D. Bioerosion and coral reef growth: A dynamic balance. In *Coral Reefs in the Anthropocene*; Springer: Dordrecht, The Netherlands, 2015; pp. 67–97. [CrossRef]
7.   Bellwood, D.R. Production and reworking of sediment by Parrotfish (family Scaridae) on the Great Barrier Reef, Australia. *Mar. Biol.* **1996**, *125*, 795–800. [CrossRef]
8.   Perry, C.T.; Kench, P.S.; O'Leary, M.J.; Morgan, K.; Januchowski-Hartley, F.A. Linking reef ecology to island building: Parrotfish identified as major producers of island-building sediment in the Maldives. *Geology* **2015**, *43*, 503–506. [CrossRef]
9.   Bellwood, D.R.; Hughes, T.P.; Folke, C.; Nystrom, M. Confronting the coral reef crisis. *Nature* **2004**, *429*, 827–833. [CrossRef] [PubMed]
10. Morgan, K.M.; Kench, P.S. Parrotfish erosion underpins reef growth, sand talus development and island building in the Maldives. *Sediment. Geol.* **2016**, *341*, 50–57. [CrossRef]
11. Vallès, H.; Oxenford, H.A. Simple family-level parrotfish indicators are robust to survey method. *Ecol. Indic.* **2018**, *85*, 244–252. [CrossRef]
12. Ebisawa, A.; Ohta, I.; Uehara, M.; Nakamura, H.; Kanashiro, K.; Yasui, R. Life history variables, annual change in sex ratios with age, and total mortality observed on commercial catch on Pacific steephead parrotfish, *Chlorurus microrhinos* in waters off the Okinawa Island, southwestern Japan. *Reg. Stud. Mar. Sci.* **2016**, *8*, 65–76. [CrossRef]
13. Edwards, C.B.; Friedlander, A.M.; Green, A.G.; Hardt, M.J.; Sala, E.; Sweatman, H.P.; Williams, I.D.; Zgliczynski, B.; Sandin, S.A.; Smith, J.E. Global assessment of the status of coral reef herbivorous fishes: Evidence for fishing effects. *Proc. Biol. Sci.* **2014**, *281*, 20131835. [CrossRef] [PubMed]
14. Carpenter, K.E.; Niem, V.H. (Eds.) FAO Species Identification Guide for Fishery Purposes. Bony Fishes Part 4 (Labridae to Latimeriidae), Estuarine Crocodiles, Sea Turtles, Sea Snakes and Marine Mammals. In *The Living Marine Resources of the Western Central Pacific*; FAO: Rome, Italy, 2001; Volume 6, pp. 468–3492. Available online: https://ci.nii.ac.jp/ncid/BA67023968 (accessed on 1 June 2022).
15. Nelson, J.S.; Grande, T.; Wilson, M.V.H. *Fishes of the World*; John Wiley & Sons: Hoboken, NJ, USA, 2016; Volume 15, pp. 429–430. [CrossRef]
16. Wu, H.L.; Shao, G.Z.; Lai, C.F.; Chong, D.H.; Lin, P.L. *Latin-Chinese Dictionary of Fish Names by Classification System*; China Ocean University Press: Qingdao, China, 2017; pp. 319–320. Available online: http://cpfd.cnki.com.cn/Article/CPFDTOTAL-ZGHI201 209001002.htm (accessed on 2 June 2022).
17. Arai, T.; Amalina, R.; Bachok, Z. Similarity in the feeding ecology of parrotfish (Scaridae) in coral reef habitats of the Malaysian South China Sea, as revealed by fatty acid signatures. *Biochem. Syst. Ecol.* **2015**, *59*, 85–90. [CrossRef]
18. Rotjan, R.D.; Dimond, J.L. Discriminating causes from consequences of persistent parrotfish corallivory. *J. Exp. Mar. Biol. Ecol.* **2010**, *390*, 188–195. [CrossRef]
19. Francini-Filho, R.; Moura, R.; Ferreira, C.M.; Coni, E.O.C. Live coral predation by Parrotfish (Perciformes: Scaridae) in the Abrolhos Bank, eastern Brazil, with comments on the classification of species into functional groups. *Neotrop. Ichthyol.* **2008**, *6*, 191–200. [CrossRef]
20. Adam, T.C.; Deron, E.B.; Ruttenberg, B.I.; Paddack, M.J. Herbivory and the Resilience of Caribbean Coral Reefs: Knowledge Gaps and Implications for Management. *Mar. Ecol. Prog. Ser.* **2021**, *520*, 1–20. [CrossRef]
21. Roos, N.C.; Carvalho, A.R.; Lopes, P.F.M.; Pennino, M.G. Modeling sensitive parrotfish (Labridae: Scarini) habitats along the Brazilian coast. *Mar. Environ. Res.* **2015**, *110*, 92–100. [CrossRef]
22. Jawad, L.A. A comparative morphological investigation of otoliths of six parrotfish species (Scaridae) from the Solomon Islands. *J. Fish Biol.* **2018**, *93*, 1046–1058. [CrossRef]
23. Barman, A.S.; Singh, M.; Pandey, P.K. DNA barcoding and genetic diversity analyses of fishes of Kaladan River of Indo-Myanmar biodiversity hotspot. *Mitochondrial DNA A* **2018**, *29*, 367–378. [CrossRef]
24. Ritchie, P.; Bargelloni, L.; Meyer, A.; Taylor, J.A.; Macdonald, J.A.; Lambert, D.M. Mitochondrial Phylogeny of Trematomid Fishes (Nototheniidae, Perciformes) and the Evolution of Antarctic Fish. *Mol. Phylogenet. Evol.* **1996**, *5*, 383–390. [CrossRef]
25. Kochzius, M.; Seidel, C.; Antoniou, A.; Botla, S.K.; Campo, D.; Cariani, A.; Vazquez, E.G.; Hauschild, J.; Hervet, C.; Hjorleifsdottir, S.; et al. Identifying Fishes through DNA Barcodes and Microarrays. *PLoS ONE* **2010**, *5*, e12620. [CrossRef]
26. Chen, W.T.; Ma, X.H.; Shen, Y.J.; Mao, Y.T.; He, S.P. The fish diversity in the upper reaches of the Salween River, Nujiang River, revealed by DNA barcoding. *Sci. Rep.* **2015**, *5*, 17437. [CrossRef] [PubMed]
27. Ma, H.Y.; Ma, C.Y.; Ma, L.B. Molecular identification of genus Scylla (Decapoda: Portunidae) based on DNA barcoding and polymerase chain reaction. *Biochem. Syst. Ecol.* **2012**, *41*, 41–47. [CrossRef]
28. Khedkar, G.D.; Jamdade, R.; Naik, S.; David, L.; Haymer, D. DNA barcodes for the fishes of the Narmada, one of India's longest rivers. *PLoS ONE* **2014**, *9*, e105490. [CrossRef]
29. Dhar, B.; Ghosh, S.K. Genetic assessment of ornamental fish species from North East India. *Gene* **2015**, *555*, 382–392. [CrossRef]
30. Hebert, P.D.N.; Cywinska, A.; Ball, S.L.; de Waard, J.R. Biological identification through DNA barcodes. *Proc. R. Soc. Lond.* **2003**, *270*, 313–321. [CrossRef]

31. Stein, E.D.; Martinez, M.C.; Stiles, S.; Miller, P.E.; Zakharov, E.V. Is DNA Barcoding Actually Cheaper and Faster than Traditional Morphological Methods: Results from a Survey of Freshwater Bioassessment Efforts in the United States. *PLoS ONE* **2014**, *9*, e95525. [CrossRef] [PubMed]

32. Sethusa, M.T.; Yessoufou, K.; der Bank, M.V.; der Bank, H.V.; Sethusa, M.T.; Millar, I.M.; Jacobs, A. DNA barcode efficacy for the identification of economically important scale insects (Hemiptera: Coccoidea) in South Africa. *Afr. Entomol.* **2014**, *22*, 257–266. [CrossRef]

33. Decru, E.; Moelants, T.; De Gelas, K.; Vreven, E.; Verheyen, E.; Snoeks, J. Taxonomic challenges in freshwater fishes: A mismatch between morphology and DNA barcoding in fish of the north-eastern part of the Congo basin. *Mol. Ecol. Resour.* **2016**, *16*, 342–352. [CrossRef] [PubMed]

34. Nigro, L.M.; Angel, M.V.; Blachowiak-Samolyk, K.; Hopcroft, R.R.; Bucklin, A. Identification, Discrimination, and Discovery of Species of Marine Planktonic Ostracods Using DNA Barcodes. *PLoS ONE* **2016**, *11*, e0146327. [CrossRef] [PubMed]

35. Liu, B.; Yang, J.W.; Liu, B.S.; Zhang, N.; Guo, L.; Guo, H.Y.; Zhang, D.C. Detection and identification of marine fish mislabeling in Guangzhou's supermarkets and sushi restaurants using DNA barcoding. *J. Food Sci.* **2022**, *87*, 2440–2449. [CrossRef]

36. Hebert, P.D.N.; Ratnasingham, S.; de Waard, J.R. Barcoding animal life: Cytochrome c oxidase subunit 1 divergences among closely related species. *Proc. Biol. Sci.* **2003**, *270*, 96–99. [CrossRef]

37. Zhang, Y.H.; Qin, G.; Zhang, H.X.; Wang, X.; Lin, Q. DNA barcoding reflects the diversity and variety of brooding traits of fish species in the family Syngnathidae along China's coast. *Fish. Res.* **2017**, *185*, 137–144. [CrossRef]

38. Chambers, E.A.; Hebert, P.D. Assessing DNA Barcodes for Species Identification in North American Reptiles and Amphibians in Natural History Collections. *PLoS ONE* **2016**, *11*, e0154363. [CrossRef] [PubMed]

39. Sun, L.Y.; Yang, T.Y.; Meng, W.; Yang, B.Q.; Zhang, T. Analysis of the mitochondrial genome characteristics and phylogenetic relationships of eight sciaenid fishes. *Mar. Sci.* **2017**, *41*, 48–54. Available online: https://oversea.cnki.net/HYKX201703007 (accessed on 2 June 2022).

40. Smith, L.L.; Fessler, J.L.; Alfaro, M.E.; Streelman, J.T.; Westneat, M.W. Phylogenetic relationships and the evolution of regulatory gene sequences in the Parrotfish. *Mol. Phylogenet. Evol.* **2008**, *49*, 136–152. [CrossRef] [PubMed]

41. Bellwood, D.R. A phylogenetic study of the parrotfish family Scaridae (Pisces: Labroidea), with a revision of genera. *Rec. Aust. Mus.* **1994**, *20*, 1–86. [CrossRef]

42. Choat, J.H.; Klanten, O.S.; Herwerden, L.V.; Robertson, D.R.; Clements, K.D. Patterns and processes in the evolutionary history of Parrotfish (Family Labridae). *Biol. J. Linn. Soc.* **2012**, *107*, 529–557. [CrossRef]

43. Ma, K.P. DNA barcode: From species to biome. *Biodivers. Sci.* **2015**, *23*, 279–280. [CrossRef]

44. Barco, A.; Raupach, M.J.; Laakmann, S.; Neumnn, H.; Knebelsberger, T. Identification of North Sea molluscs with DNA barcoding. *Mol. Ecol. Resour.* **2016**, *16*, 288–297. [CrossRef]

45. Zhang, J.B.; Hanner, R. DNA barcoding is a useful tool for the identification of marine fishes from Japan. *Biochem. Syst. Ecol.* **2011**, *39*, 31–42. [CrossRef]

Journal of
*Marine Science
and Engineering*

Article

Establishment and Application of Microsatellite Multiplex PCR System for *Cheilinus undulatus*

Fangcao Zhao [1,2,3], Liang Guo [1,3], Nan Zhang [1,3], Kecheng Zhu [1,3], Jingwen Yang [1,3], Baosuo Liu [1,3], Huayang Guo [1,3] and Dianchang Zhang [1,3,4,*]

1 Key Laboratory of South China Sea Fishery Resources Exploitation and Utilization, South China Sea Fisheries Research Institute, Chinese Academy of Fishery Sciences, Ministry of Agriculture and Rural Affairs, Guangzhou 510300, China
2 School of Fisheries and Life Science, Dalian Ocean University, Dalian 116023, China
3 Sanya Tropical Fisheries Research Institute, Sanya 572019, China
4 Guangdong Provincial Engineer Technology Research Center of Marine Biological Seed Industry, Guangzhou 510300, China
* Correspondence: zhangdch@scsfri.ac.cn

Abstract: *Cheilinus undulatus* is a valuable seawater economic fish with tender meat, fresh taste, and high nutritional value; however, its population is rapidly declining because of massive fishing and habitat destruction. Assessing changes in genetic diversity and inbreeding levels is a very valuable monitoring tool, and multiplex PCR has the advantages of being time-efficient and economical. Here, we selected 12 pairs of polymorphic microsatellite loci, developed two multiplex PCR amplification systems based on these microsatellites, and used them to examine 30 *C. undulatus* specimens. The number of alleles (Na) for the 12 microsatellite markers ranged from 2 to 8. The effective allele number (Ne) ranged from 1.724 to 4.592. The expected heterozygosity (He) and observed heterozygosity (Ho) ranged from 0.420 to 0.782 and 0.100 to 0.900, respectively. The polymorphic information content (PIC) ranged from 0.422 to 0.746, with a mean value of 0.557. 5 loci deviated from Hardy-Weinberg equilibrium (HWE, $p < 0.05$ after Bonferroni correction). The multiplex PCR amplification system established in this study provides a basis for germplasm identification, genetic diversity analysis, and assessment of the effects of accretion and release of *C. undulatus*.

Keywords: *Cheilinus undulatus*; microsatellite; multiplex PCR; genetic diversity

Citation: Zhao, F.; Guo, L.; Zhang, N.; Zhu, K.; Yang, J.; Liu, B.; Guo, H.; Zhang, D. Establishment and Application of Microsatellite Multiplex PCR System for *Cheilinus undulatus*. *J. Mar. Sci. Eng.* **2022**, *10*, 2000. https://doi.org/10.3390/jmse10122000

Academic Editor: Nguyen Hong Nguyen

Received: 21 October 2022
Accepted: 12 December 2022
Published: 15 December 2022

1. Introduction

Cheilinus undulatus, known as Maori, Napoleon, humphead wrasse, and so-mei [1], belongs to the order Perciformes [2]. The species is found in reefs and nearshore habitats with seagrass beds and mangroves distributed in tropical waters of the Pacific and Indian Oceans [3]. Its abundance is usually very low, and it feeds on mollusks, small fish, sea urchins, and crustaceans [4]. *C. undulatus* is one of the most valuable and expensive fish species in coral reefs [5], and the large coral triangle is the main distribution area [6]. Due to the heavy exploitation of the live reef fish trade (LRFT), it is classified as "vulnerable" in the IUCN 1996 Red Data Book [1]. International regulations treat *C. undulatus* as a wild fish that can be traded within a limited quota [7]. Human activities are a major cause of biodiversity decline, and marine animal extinctions began to accelerate in the 1970s, when fisheries harvesting peaked and began to linger. The marine animals under threat are mainly large animals at the top of the food chain. The populations of *C. undulatus*, a large fish in coral reefs, have rapidly declined because of heavy fishing and habitat destruction [8].

Fluctuations in population size can affect genetic diversity [9], and very small populations tend to cause inbreeding within the population [10], which can lead to a reduction in population fitness. Inbreeding decline is less pronounced in better environments but is readily apparent in harsh environments [11]. Current scientific studies have identified a dramatic decline in the size of *C. undulatus* populations [12]; however, the role of coral reef destruction and human fishing is not clear. Assessing changes in genetic diversity and inbreeding levels is a practical monitoring tool [13]. Current molecular markers used to monitor genetic diversity include single-nucleotide polymorphisms (SNPs) [14] and microsatellites, which are also called simple sequence repeats [15]. For monitoring specific populations, microsatellite markers have the advantages of abundant alleles at individual loci [16], low typing cost [17], and mature technology [18]. They have a wide range of applications in genetic mapping, population structure analysis, genetic diversity studies, and germplasm conservation studies [19–22]. Multiplex polymerase chain reaction (multiplex PCR) refers to the simultaneous amplification of multiple target sequences by adding two or more pairs of primers to the same PCR reaction system [23]. Multiplex PCR can increase the number of microsatellite markers detected in a single run, simplifying the test procedure and reducing the cost and amount of DNA used in the sample [24,25]. In this study, 12 polymorphic microsatellite markers were selected, and one 7-plex PCR amplification system and one 5-plex PCR amplification system were successfully constructed. Then, the two systems were used to examine the genetic diversity of 30 *C. undulatus* specimens. This is the first multiplex PCR amplification system for the *C. undulatus*, providing technical support for germplasm identification, genetic diversity analysis and assessment of the effects of accretion and release in this species.

2. Materials and Methods

2.1. Sample Collection

In the present study, 30 *C. undulatus* specimens from the South China Sea were studied. The fins were first cut and immediately stored in 95% ethanol, then replaced twice with 95% ethanol, and finally stored at $-20\,°\text{C}$ for backup.

2.2. Experimental Methods

DNA Extraction and Sequencing

Genomic DNA was extracted using a Marine Tissue Genomic DNA Extraction Kit (Mobio, Guangzhou, China), and its concentration was measured using a UV spectrophotometer after DNA extraction. The concentration was adjusted to 50 ng/µL and the DNA was stored at 4 °C. Samples from 12 randomly selected individuals were sequenced on the Illumina NovaSeq platform at the Institute of Bioinformatics (Beijing, China).

2.3. Primer Design

Multiplex PCR primers were designed using the MultiplexSSR pipeline, as described by Guo and Yang [26]. The resequencing data from 12 randomly selected individuals were processed using this pipeline. The amplicons had a minimum length of 80 bp, maximum length of 480 bp, and minimum spacing of 20 bp. The minimum length of repeat units, minimum number of genotyped individuals, minimum number of alleles, and minimum depth of genotypes were set to 3, 5, 5, and 1, respectively. The optimum annealing temperature was set at 60 °C.

2.4. Primer Selection

After Tandem repeat detection, a total of 13,264 SSRs were obtained. Then, after lobSTR processing, a total of 145 high quality SSRs and ranges were finally selected. Based on the predicted results of the SSRs, 12 pairs of specific primers were selected from the developed primer sequences, namely primer pairs Cun463, Cun378, Cun500, Cun626, Cun672, Cun586, Cun148, Cun752, Cun230, Cun27, Cun485, and Cun484 following the principle of high polymorphism at the loci. Each primer pair consisted of one forward primer and one reverse primer. The dosage ratio of the forward and reverse primers was 1:4 [27]. Furthermore, two fluorescently labeled universal primers M13 and PQE-F were included: the fluorescent marker that is compatible with the universal primer M13 is 5-FAM, and the fluorescent marker that is compatible with the universal primer PQE-F is 5-HEX [28]. The 12 primer pairs were divided into two groups: G1 group primers, including Cun463, Cun378, Cun500, Cun626, Cun672, Cun586, and Cun148, and G2 group primers, including primer pairs Cun752, Cun230, Cun27, Cun485, and Cun484. Primers were synthesized by Beijing Rui Bo Xing Ke Biotechnology Co. Site names and primer sequences are listed, and the generic primer names and primer sequence information are shown in Table 1.

Table 1. Information about site combination and primer concentration of Multiplex PCR1 and Multiplex PCR2.

Name	Number	Primer Pairs	Repetitive Units	Fragment Length Range (bp)	Primer Sequences	Primer Number
Multiplex PCR1	1	Cun463	CTAT	129–141	tgtaaaacgacggccagtcatgaaacaacccgtataccct	Cun463.F
					aatagccctgctccatacttca	Cun463.R
	2	Cun378	AGAT	155–159	ttgagaggatcgcatccatgtattgatcatgctttctgcc	Cun378.F
					taagtctgagccaatcgcatta	Cun378.R
	3	Cun500	GATA	297–313	tgtaaaacgacggccagtaacacaacacgcagcttagaga	Cun500.F
					aatggatcctttgagagcgata	Cun500.R
	4	Cun626	ATA	403–409	tgtaaaacgacggccagtctatttcctgcatgtctctccc	Cun626.F
					atggcccgtttatagacacaat	Cun626.R
	5	Cun672	TAGA	129–141	ttgagaggatcgcatccacacttcatctgtcccaccatta	Cun672.F
					ctccctcctgcttcactgtact	Cun672.R
	6	Cun586	AGAT	301–321	tgtaaaacgacggccagtcaagaattgacaaggtttccct	Cun586.F
					taactgcagtgatgaaccctgt	Cun586.R
	7	Cun148	ATCT	462–490	tgtaaaacgacggccagttgcaagagcattggtgatattc	Cun148.F
					aagggacaacaaggacactgtta	Cun148.R
	8	Generic primers			FAM-5′-tgtaaaacgacggccagt	M13
	9				HEX-5′-ttgagaggatcgcatcca	PQE-F
Multiplex PCR2	1	Cun752	ACAAA	157–167	tgtaaaacgacggccagttctggaagcacctcatgataga	Cun752.F
					ctttgactgcaaggtttcctct	Cun752.R
	2	Cun230	ATAG	252–272	tgtaaaacgacggccagtattaaacgcgctggttgttatt	Cun230.F
					accaccaactgtgtgaatgttt	Cun230.R
	3	Cun27	GATA	367–375	tgtaaaacgacggccagtctctgtgctctcttgtcattgg	Cun27.F
					gtcatgtcattgcatctcacct	Cun27.R
	4	Cun485	TGGA	404–412	ttgagaggatcgcatccacaggctaggaaggaagaaatca	Cun485.F
					tttatgcctctgtgacagcatt	Cun485.R
	5	Cun484	GATA	470–486	ttgagaggatcgcatccacatgtatactctgccaccctcca	Cun484.F
					agaagttgccaggaaattggta	Cun484.R
	6	Generic primers			FAM-5′-tgtaaaacgacggccagt	M13
	7				HEX-5′-ttgagaggatcgcatcca	PQE-F

2.5. Establishment of a Microsatellite Multiplex PCR Amplification System

The reaction volume for multiplex PCR in both G1 and G2 groups was 25 μL. Specific information is shown in Table 2. For PCR amplification, the amplification procedure used was as follows: 98 °C for 10 s, 57 °C for 40 s, 72 °C for 60 s, 35 cycles; 98 °C for 10 s, 53 °C for 40 s, 72 °C for 60 s, 15 cycles; and 72 °C for 30 min for extension [26]. After PCR, 5 μL was taken on an agarose gel to detect bright bands of the desired size. Gel electrophoresis was used to confirm that the primers amplified the target bands. The remaining samples were sent to a commercial company (Beijing Ruibio BioTech Co., Ltd., Beijing, China) for genotyping using an ABI 3730XL sequencer.

Table 2. Group G1 and Group G2 multiplex PCR reaction system.

G1 Group PCR System Reactants	Content (μL)	G2 Group PCR System Reactants	Content (μL)
Chun463.F (20 μM)	0.06	Cun752.F (10 μM)	0.06
Chun463.R (20 μM)	0.24	Cun752.R (10 μM)	0.24
Chun378.F (20 μM)	0.06	Cun230.F (20 μM)	0.06
Chun378.R (20 μM)	0.24	Cun230.R (20 μM)	0.24
Chun500.F (10 μM)	0.06	Cun27.F (10 μM)	0.06
Chun500.R (10 μM)	0.24	Cun27.R (10 μM)	0.24
Chun626.F (20 μM)	0.06	Cun485.F (10 μM)	0.06
Chun626.R (20 μM)	0.24	Cun485.R (10 μM)	0.24
Chun672.F (10 μM)	0.06	Cun484.F (10 μM)	0.06
Chun672.R (10 μM)	0.24	Cun484.R (10 μM)	0.24
Chun586.F (10 μM)	0.06	M13 (10 μM)	0.36
Chun586.R (10 μM)	0.24	PQE-F (10 μM)	0.36

Table 2. *Cont.*

G1 Group PCR System Reactants	Content (μL)	G2 Group PCR System Reactants	Content (μL)
Chun148.F (20 μM)	0.06	BSA (2 mg/mL)	0.45
Chun148.R (20 μM)	0.24	DNA (50 ng/μL)	2.0
M13 (10 μM)	0.36	Taq HS (Takara)	12.5
PQE-F (10 μM)	0.36	ddH2O	7.83
BSA (2 mg/mL)	0.45	Total	25.0
DNA (50 ng/μL)	2.0		
Taq HS (Takara)	12.5		
ddH2O	7.23		
Total	25.0		

2.6. Polymorphism Analysis

The amplification products were subjected to capillary electrophoresis using ABI3730XL. Peak types were converted to alleles using GeneMarker (version 2.2.2.0, SoftGenetics, Pennsylvania, USA) [29]. Genotype data were counted using GenAlex (version 6.5, ANU, Canberra, Australia) [30] and population genetics parameters were calculated including the number of individuals (N), number of different alleles (Na), number of effective alleles (Ne), observed heterozygosity (Ho), expected heterozygosity (He), unbiased expected heterozygosity(uHe), fixation index (F) and Hardy–Weinberg equilibrium (HW). The null allele frequency (F(null)) and polymorphism information content (PIC) were calculated using the software Cervus (version 3.0.7, Field Genetics, London, UK).

3. Results

3.1. Multiplex PCR System Establishment

The optimal amplification temperature for 12 microsatellite loci was first determined using gradient PCR amplification, and the results showed that the optimal annealing temperature was 53 °C. Two multiplex PCR systems, multiplex PCR1 and multiplex PCR2, were constructed using 12 pairs of microsatellites based on the annealing temperature and amplification length. Information on the combination of sites, primer concentrations used for multiplex PCR and information on the two universal primers and the fluorescent markers compatible with the universal primers is shown in Table 1. The results of PCR amplification and agarose gel electrophoresis of samples from 30 individuals showed that the PCR target bands were clear and did not overlap. The amplification efficiency of each locus was similar, and the amplified fragment sizes were as expected. The capillary electrophoresis diagram is shown in Figures 1 and 2.

3.2. Polymorphism Analysis

Thirty *C. undulatus* specimens were selected for PCR amplification, capillary electrophoresis, and genotyping of 12 loci in the two multiplex PCR systems, using fluorescently labeled primers. The genetic parameters were calculated and are presented in Table 3. The number of alleles of the 12 microsatellite markers ranged from 2 to 8. Among these, locus Cun148 had the highest number of alleles (8). The mean He was 0.594, and the mean Ho was 0.475. The null allele frequencies ranged from -1.104 to 0.716, with 2 loci having higher frequency invalid alleles (F(null) > 0.2), at Cun752 and Cun27. The polymorphic information content (PIC) ranged from 0.422 to 0.746, with a mean value of 0.557. A total of 9 loci were highly polymorphic (PIC $\geq$ 0.5) and 3 loci were moderately polymorphic (0.25 $\leq$ PIC < 0.5). Five loci deviated from Hardy–Weinberg equilibrium (HWE, $p < 0.05$ after Bonferroni correction) The above results show that the multiplex PCR method consisting of 12 primer pairs of microsatellites is stable and accurate in the population typing of *C. undulatus*. This can provide accurate and reliable genetic information for *C. undulatus* germplasm identification, family lineage management, and stocking effect evaluation.

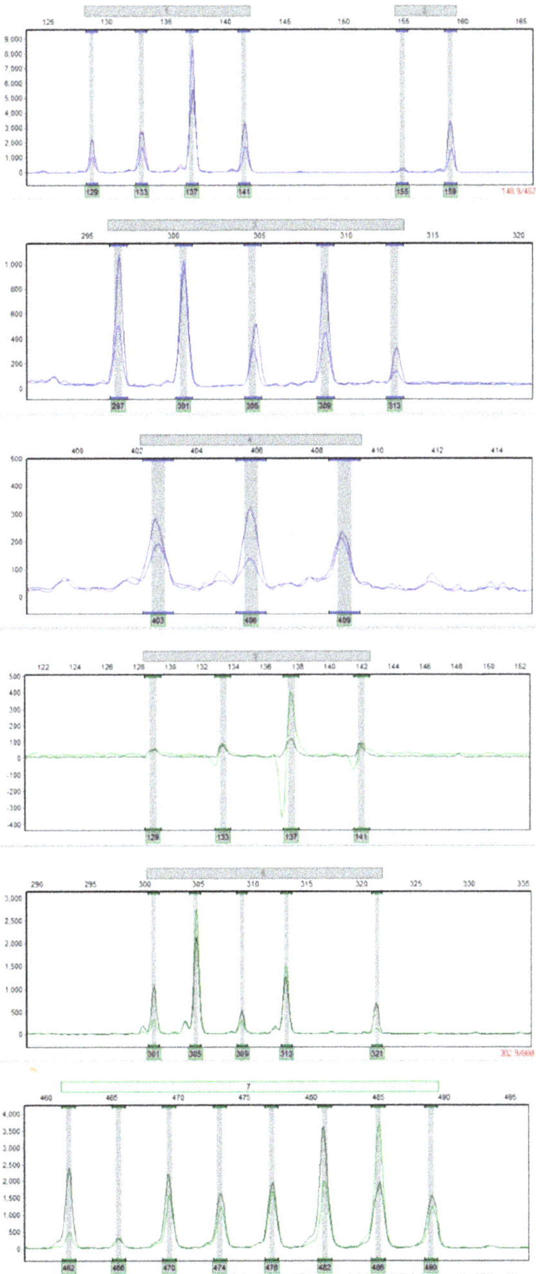

Figure 1. The capillary electrophoresis diagram of the seven primer pairs in group G1. (The 1 to 7 in this Figure correspond to alleles loci Cun463, Cun378, Cun500, Cun626, Cun672, Cun586 and Cun148, respectively).

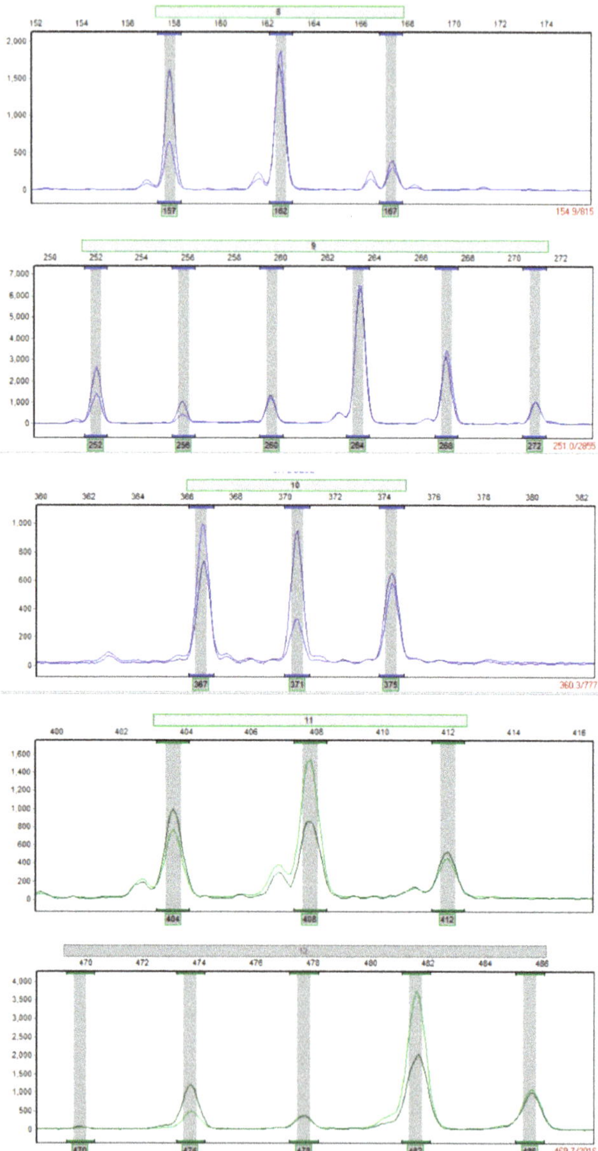

Figure 2. The capillary electrophoresis diagram of the five primer pairs in group G2. (The 8 to 12 in this Figure correspond to alleles loci Cun752, Cun230, Cun27, Cun485 and Cun484, respectively).

Table 3. Genetic parameters of 12 microsatellite loci in *C.undulatus*.

Locus	N	Na	Ne	Ho	He	uHe	F	F (Null)	PIC	HW
Cun463	30	4.000	1.835	0.367	0.455	0.463	0.194	0.125	0.422	ns
Cun378	30	2.000	1.724	0.267	0.420	0.427	0.365	0.189	0.460	ns
Cun500	30	5.000	3.455	0.667	0.711	0.723	0.062	0.039	0.671	**
Cun626	30	3.000	2.817	0.433	0.645	0.656	0.328	0.192	0.572	ns

Table 3. *Cont.*

Locus	N	Na	Ne	Ho	He	uHe	F	F (Null)	PIC	HW
Cun672	30	4.000	1.850	0.433	0.459	0.467	0.057	0.039	0.431	ns
Cun586	30	5.000	2.521	0.500	0.603	0.614	0.171	0.103	0.542	ns
Cun148	30	8.000	4.380	0.900	0.772	0.785	−0.166	−0.104	0.746	**
Cun752	30	3.000	2.203	0.300	0.546	0.555	0.451	0.256	0.486	**
Cun230	30	6.000	4.592	0.667	0.782	0.795	0.148	0.112	0.745	ns
Cun27	30	3.000	2.436	0.100	0.589	0.599	0.830	0.716	0.542	**
Cun485	30	3.000	2.332	0.500	0.571	0.581	0.125	0.071	0.536	ns
Cun484	30	5.000	2.320	0.567	0.569	0.579	0.004	−0.102	0.534	*
Mean	30	4.250	2.705	0.475	0.594	0.604	0.214	0.136	0.557	−

Note: ns = not significant ($p > 0.05$), *. Significant difference ($p < 0.05$), **. Extremely significant difference ($p < 0.01$).

4. Discussion

C. undulatus is one of the most valuable and expensive fish species in coral reefs [5], and its populations have experienced a dramatic decline [12]; however, the role of coral reef destruction and human fishing remains unclear. Assessing changes in genetic diversity and inbreeding levels is a very valuable monitoring tool [13]. For monitoring specific populations, microsatellites have several advantages and a wide range of applications. Multiplex PCR is a cost-effective and rapid method for obtaining accurate genetic information [31].

Here, we selected 12 polymorphic microsatellites, developed one 7-plex PCR amplification system and one 5-plex PCR amplification system, and used them to examine the genetic diversity of 30 *C. undulatus* specimens. Three main factors, primer combination, primer concentration, and annealing temperature, were optimized, establishing an accurate, rapid, and efficient microsatellite analysis technique for this species. We obtained a total of 12 alleles; 9 of the 12 loci were highly polymorphic, 3 were moderately polymorphic, and overall had a high level of polymorphism. After Hardy–Weinberg equilibrium validation, seven loci showed no significance and five loci deviated from Hardy–Weinberg equilibrium, which may be related to the limited sample size and the presence of invalid alleles. We further calculated the null allele frequencies and two loci had high null alleles. Therefore, it was further speculated that the five loci deviated from Hardy–Weinberg equilibrium mainly due to the limited sample size. The results showed that the multiplex amplification system was stable and reliable, and the loci were highly polymorphic. This is the first multiplex PCR amplification system for the *C. undulatus*.

Microsatellite multiplex PCR technology has been used for parentage analysis of *Cirrhinus molitorella* [32], and is also used commercially in large-scale selection for breeding [33]. It has also been developed and utilized as a powerful and low-cost parental assignment tool to support company-breeding programs [34]. Furthermore, it has been used to identify paternity assignments in grass carp [35] and in genetic diversity studies of *Portunus trituberculatus* [36]. Multiplex PCR has also been applied to detect hemolytic strains in fish and fishery products [37] and to assess *Lateolabrax japonicus* population genetics [38]. These studies illustrate the applicability of multiplex PCR, demonstrating its advantages in terms of performance, accuracy, experimental time, and experimental cost.

The small number of samples is a limitation in this study. More samples could improve the accuracy and reliability of the data. We can be able to take more samples and redesign primers to validate loci that have high frequency of invalid alleles and deviate from Hardy–Weinberg equilibrium to ensure the accuracy and reliability of this multiplex PCR system in the future, and more parental and offspring samples can be collected for parentage identification using the 12 microsatellite loci to elucidate genealogical information for the genetic improvement of *C. undulatus*, and to provide a basis for selective breeding.

5. Conclusions

In summary, two multiplex PCR amplification systems were constructed using 12 microsatellite markers to establish a quasi-rapid and efficient microsatellite analysis technique for *C. undulatus*. The constructed multiplex amplification systems were stable, and the loci were highly polymorphic, providing a basis for germplasm identification, genetic diversity analysis, and stocking effect evaluation of *C. undulatus*. This method can be used to select a set of primers with high polymorphism and stable PCR amplification from the developed microsatellite primers and establish an accurate, rapid, and efficient microsatellite analysis technique using multiplex PCR combination, which can provide technical support for population genetics and family lineage analysis of species.

Author Contributions: Conceptualization, D.Z.; methodology, F.Z. and L.G.; software, L.G.; validation, F.Z. and J.Y.; formal analysis, F.Z.; resources, N.Z. and K.Z.; data curation, F.Z.; writing—original draft preparation, F.Z.; visualization, F.Z.; supervision, H.G. and K.Z.; project administration, L.G. and B.L. All authors have read and agreed to the published version of the manuscript.

Funding: This study was funded by Central Public-Interest Scientific Institution Basal Research Fund, CAFS (2020GH06), and Financial Fund of the Ministry of Agriculture and Rural Affairs, P. R. of China (NHYYSWZZZYKZX2020), and National Marine Genetic Resource Center, and China-ASEAN Maritime Cooperation Fund.

Institutional Review Board Statement: All applicable international, national, and institutional guidelines for the care were followed by the authors.

Informed Consent Statement: Informed consent was obtained from all subjects involved in the study. Written informed consent has been obtained from the patients to publish this paper.

Data Availability Statement: All data generated or analyzed during this study are included in this published article.

Conflicts of Interest: The authors declare no conflict of interest.

References

1. Donaldson, T.J.; Sadovy, Y. Threatened Fishes of the World: Cheilinus Undulatus Rüppell, 1835 (Labridae). *Environ. Biol. Fishes* **2001**, *62*, 428. [CrossRef]
2. Hu, J.; Hou, X.; Yin, S.; Zhu, F.; Jia, Y.; Hu, Y. Genetic diversity of different geographical populations of Chelilinus undulatus revealed by microsatellite analysis. *Adv. Mar. Sci.* **2013**, *31*, 538–545.
3. Dorenbosch, M.; Grol, M.G.G.; Nagelkerken, I.; van der Velde, G. Seagrass Beds and Mangroves as Potential Nurseries for the Threatened Indo-Pacific Humphead Wrasse, Cheilinus Undulatus and Caribbean Rainbow Parrotfish, Scarus Guacamaia. *Biol. Conserv.* **2006**, *129*, 277–282. [CrossRef]
4. Ou, Y.; Li, J. Analysis and evaluation of nutrition composition of double-headed parrotfish Cheilinus undulates. *J. Trop. Oceanogr.* **2010**, *29*, 97–102.
5. Liu, D.; Wang, X.; Guo, H.; Zhang, X.; Zhang, M.; Tang, W. Chromosome-level genome assembly of the endangered humphead wrasse Cheilinus undulatus: Insight into the expansion of opsin genes in fishes. *Mol. Ecol. Resour.* **2021**, *21*, 2388–2406. [CrossRef]
6. Peng, Y.; Luo, J.; Yin, S.; Zhu, X.; Hu, J.; Liu, Z.; Zhu, F.; Qi, X.; Hu, Y. Screening and suitability analysis of microsatellite markers in Cheilinus undulatus. *Open J. Mar. Sci.* **2011**, *36*, 109–116.
7. Indriatmoko; Syam, A.R.; Syahputra, K. Control Region-Mitochondrial Partial DNA Analysis of Humphead Wrasse [Cheilinus Undulates (Ruppel, 1835)] from Anambas Islands, Indonesia. *Aquat. Procedia* **2016**, *7*, 125–131. [CrossRef]
8. Shen, Y.; Ma, K.; Yue, G.H. Status, Challenges and Trends of Aquaculture in Singapore. *Aquaculture* **2021**, *533*, 736210. [CrossRef]
9. Nakamura, H.; Teshima, K.; Tachida, H. Effects of Cyclic Changes in Population Size on Neutral Genetic Diversity. *Ecol. Evol.* **2018**, *8*, 9362–9371. [CrossRef]
10. Jangtarwan, K.; Kamsongkram, P.; Subpayakom, N.; Sillapaprayoon, S.; Muangmai, N.; Kongphoemph, A.; Wongsodchuen, A.; Intapan, S.; Chamchumroon, W.; Safoowong, M.; et al. Predictive Genetic Plan for a Captive Population of the Chinese Goral (Naemorhedus Griseus) and Prescriptive Action for Ex Situ and in Situ Conservation Management in Thailand. *PLoS ONE* **2020**, *15*, e0234064. [CrossRef]
11. Othman, O.E.M.; Payet-Duprat, N.; Harkat, S.; Laoun, A.; Maftah, A.; Lafri, M.; Silva, A.D. Sheep diversity of five Egyptian breeds: Genetic proximity revealed between desert breeds. *Small Rumin. Res.* **2016**, *144*, 346–352. [CrossRef]
12. Sill, S.R.; Dawson, T.P. Climate Change Impacts on the Ecological Dynamics of Two Coral Reef Species, the Humphead Wrasse (Cheilinus Undulatus) and Crown-of-Thorns Starfish (Ancanthaster Planci). *Ecol. Inf.* **2021**, *65*, 101399. [CrossRef]

13. Faria, J.; Pita, A.; Martins, G.M.; Ribeiro, P.A.; Hawkins, S.J.; Presa, P.; Neto, A.I. Inbreeding in the Exploited Limpet Patella Aspera across the Macaronesia Archipelagos (NE Atlantic): Implications for Conservation. *Fish. Res.* **2018**, *198*, 180–188. [CrossRef]

14. Wang, L.; Wu, Z.; Zou, C.; Lu, Y.; Yue, X.; Song, Z.; Yang, R.; You, F. Genetic Diversity and Signatures of Selection in the Mito-Gynogenetic Olive Flounder Paralichthys Olivaceus Revealed by Genome-Wide SNP Markers. *Aquaculture* **2022**, *553*, 738062. [CrossRef]

15. Machmoum, M.; Boujenane, I.; Azelhak, R.; Badaoui, B.; Petit, D.; Piro, M. Genetic Diversity and Population Structure of Arabian Horse Populations Using Microsatellite Markers. *J. Equine Vet. Sci.* **2020**, *93*, 103200. [CrossRef]

16. Xiong, L.; Wang, Q.; Qiu, G. Large-Scale Isolation of Microsatellites from Chinese Mitten Crab Eriocheir sinensis via a Solexa Genomic Survey. *Int. J. Mol. Sci.* **2012**, *13*, 16333–16345. [CrossRef]

17. De Deus Vidal, J., Jr.; Cortez, M.B.; Alves, F.M.; Koehler, S.; de Souza, A.P.; Koch, I. Development and Cross-Validation of Microsatellite Markers for Rauvolfia Weddeliana Müll.Arg. (Apocynaceae) Species Complex. *Braz. J. Bot.* **2018**, *41*, 681–686. [CrossRef]

18. Papetti, C.; Harms, L.; Jürgens, J.; Sandersfeld, T.; Koschnick, N.; Windisch, H.S.; Knust, R.; Pörtner, H.-O.; Lucassen, M. Microsatellite Markers for the Notothenioid Fish Lepidonotothen Nudifrons and Two Congeneric Species. *BMC Res. Notes* **2016**, *9*, 238. [CrossRef]

19. Li, J.Z.; Sjakste, T.G.; Röder, M.S.; Ganal, M.W. Development and Genetic Mapping of 127 New Microsatellite Markers in Barley. *Theor. Appl. Genet.* **2003**, *107*, 1021–1027. [CrossRef]

20. Chistiakov, D.A.; Hellemans, B.; Volckaert, F.A.M. Microsatellites and their genomic distribution, evolution, function and applications: A review with special reference to fish genetics. *Aquaculture* **2006**, *255*, 1–29. [CrossRef]

21. Gupta, P.; Varshney, R. The development and use of microsatellite markers for genetic analysis and plant breeding with emphasis on bread wheat. *Euphytica* **2000**, *113*, 163–185. [CrossRef]

22. Gu, Z.X.; Gou, T.J.; Xi, X.B. Applications of microsatellite markers in studies of genetics and breeding of fish. *Chin. J. Agric. Biotechnol.* **2006**, *3*, 83–87.

23. Chamberlain, J.S.; Gibbs, R.A.; Rainer, J.E.; Caskey, C.T. Deletion screening of the Duchenne muscular dystrophy locus via multiplex DNA amplification. *Nucleic. Acids. Res.* **1988**, *16*, 11141–11156. [CrossRef] [PubMed]

24. Bian, Y.; Liu, S.; Liu, Y.; Jia, Y.; Li, F.; Chi, M.; Zheng, J.; Cheng, S.; Gu, Z. Development of a Multiplex PCR Assay for Parentage Assignment of the Redclaw Crayfish (Cherax Quadricarinatus). *Aquaculture* **2022**, *550*, 737813. [CrossRef]

25. Lata, C.; Kumar, A.; Gangwar, O.P.; Prasad, P.; Adhikari, S.; Kumar, S.; Kulshreshtha, N.; Bhardwaj, S.C. Multiplex PCR Assay for the Detection of Lr24 and Lr68 in Salt Tolerant Wheat Genotypes. *Cereal Res. Commun.* **2022**, *50*, 1019–1027. [CrossRef]

26. Guo, L.; Yang, Q.; Yang, J.; Zhang, N.; Liu, B.; Zhu, K.; Guo, H.; Jiang, S.; Zhang, D. MultiplexSSR: A Pipeline for Developing Multiplex SSR-PCR Assays from Resequencing Data. *Ecol. Evol.* **2020**, *10*, 3055–3067. [CrossRef]

27. Sudo, R.; Miyao, M.; Uchino, T.; Yamada, Y.; Tsukamoto, K.; Sakamoto, T. Parentage Assignment of a Hormonally Induced Mass Spawning in Japanese Eel (Anguillla Japonica). *Aquaculture* **2018**, *484*, 317–321. [CrossRef]

28. Ge, C.; Cui, Y.; Jing, P.; Hong, X. An alternative suite of universal primers for genotyping in multiplex PCR. *PLoS ONE* **2014**, *9*, e92826. [CrossRef]

29. Holland, M.M.; Parson, W. GeneMarker®HID: A Reliable Software Tool for the Analysis of Forensic STR Data: GENE-MARKER®HID. *J. Forensic Sci.* **2011**, *56*, 29–35. [CrossRef]

30. Peakall, R.; Smouse, P. GenAlEx 6.5: Genetic analysis in Excel. Population genetic software for teaching and research—An update. *Bioinformatics* **2012**, *28*, 2537–2539. [CrossRef]

31. Jones, A.; Small, C.; Paczolt, K.; Ratterman, N. A practical guide to methods of parentage analysis. *Mol. Ecol. Resour.* **2010**, *10*, 6–30. [CrossRef]

32. Wang, Y.; Zhao, J.; Li, W.; Zhang, X.; Hong, X.; Zhu, X. Development of a Multiplex Microsatellite PCR Assay Based on Microsatellite Markers for the Mud Carp, Cirrhinus molitorella. *J. World Aquacult. Soc.* **2016**, *47*, 277–286. [CrossRef]

33. Domingos, J.; Carolyn, S.K.; Dean, R. Fate of genetic diversity within and between generations and implications for DNA parentage analysis in selective breeding of mass spawners: A case study of commercially farmed barramundi, Lates calcarifer. *Aquaculture* **2014**, *424*, 174–182. [CrossRef]

34. Vallecillos, A.; María-Dolores, E.; Villa, J.; Rueda, F.M.; Carrillo, J.; Ramis, G.; Soula, M.; Afonso, J.M.; Armero, E. Development of the First Microsatellite Multiplex PCR Panel for Meagre (Argyrosomus Regius), a Commercial Aquaculture Species. *Fishes* **2022**, *7*, 117. [CrossRef]

35. Fu, J.; Shen, Y.; Xu, X.; Chen, Y.; Li, D.; Li, J. Multiplex microsatellite PCR sets for parentage assignment of grass carp (Ctenopharyngodon idella). *Aquacult. Int.* **2013**, *21*, 1195–1207. [CrossRef]

36. Lee, H.J.; Lee, D.H.; Yoon, S.; Kim, D.H.; Kim, S.; Hyun, Y.S.; Min, G.; Chung, K. Characterization of 20 microsatellite loci by multiplex PCR in swimming crab, Portunus trituberculatus. *Genes Genom.* **2013**, *35*, 77–85. [CrossRef]

37. Hussain, I.A.; Jeyasekaran, G.; Shakila, R.J.; Raj, K.T.; Jeevithan, E. Detection of Hemolytic Strains of Aeromonas Hydrophila and A. Sobria along with Other Aeromonas Spp. from Fish and Fishery Products by Multiplex PCR. *J. Food Sci. Technol.* **2014**, *51*, 401–407. [CrossRef]

38. An, H.S.; Lee, J.W.; Kim, H.Y.; Kim, J.B.; Chang, D.S.; Park, J.Y.; Myeong, J.I.; An, C.M. Population Genetic Structure of the Sea Bass (Lateolabrax Japonicus) in Korea Based on Multiplex PCR Assays with 12 Polymorphic Microsatellite Markers. *Genes Genom.* **2014**, *36*, 247–259. [CrossRef]

Journal of
Marine Science and Engineering

Article

Daily Rhythmicity of Hepatic Rhythm, Lipid Metabolism and Immune Gene Expression of Mackerel Tuna (*Euthynnus affinis*) under Different Weather

Wenwen Wang [1,2,†], Jing Hu [1,2,†], Zhengyi Fu [1,2,3], Gang Yu [1,2] and Zhenhua Ma [1,2,3,*]

1   Tropical Aquaculture Research and Development Center, South China Sea Fisheries Research Institute, Chinese Academy of Fishery Sciences, Sanya 572018, China
2   Key Laboratory of Efficient Utilization and Processing of Marine Fishery Resources of Hainan Province, Sanya Tropical Fisheries Research Institute, Sanya 572018, China
3   College of Science and Engineering, Flinders University, Adelaide, SA 5001, Australia
*   Correspondence: zhenhua.ma@scsfri.ac.cn
†   These authors contributed equally to this work.

Abstract: In order to investigate the rhythmic changes in gene expression in the liver of mackerel tuna (*Euthynnus affinis*) under sunny and cloudy conditions, this experiment had four sampling times (6:00, 12:00, 18:00 and 24:00) set on sunny and cloudy days to determine the expression of their immune, metabolic and rhythmic genes. The results showed that daily rhythmicity was present within most of the rhythm genes (*CREB1, CLOCK, PER1, PER2, PER3, REVERBA, CRY2* and *BMAL1*), metabolic genes (*SIRT1* and *SREBP1*) and immune genes (*NF-kB1, MHC-I, ALT, IFNA3, ISY1, ARHGEF13, GCLM* and *GCLC*) in this study under the sunny and cloudy condition ($p < 0.05$). The expression levels of *CREB1, PER1, PER3, RORA, REVERBA, CRY1* and *BMAL1* within rhythm genes were significantly different ($p < 0.05$) in the same time point comparison between sunny and cloudy conditions at 6:00, 12:00, 18:00 and 24:00; metabolic genes had the expression levels of *LPL* at 6:00, 12:00, 18:00 and 24:00 in the same time point comparison ($p < 0.05$); immune genes only had significant differences in the expression levels of *IFNA3* at 6:00, 12:00, 18:00 and 24:00 ($p < 0.05$). This study has shown that rhythm, lipid metabolism and immune genes in the livers of mackerel tuna are affected by time and weather and show significant changes in expression.

Keywords: cosinor analysis; biological clock; circadian rhythm; gene expression

Citation: Wang, W.; Hu, J.; Fu, Z.; Yu, G.; Ma, Z. Daily Rhythmicity of Hepatic Rhythm, Lipid Metabolism and Immune Gene Expression of Mackerel Tuna (*Euthynnus affinis*) under Different Weather. *J. Mar. Sci. Eng.* **2022**, *10*, 2028. https://doi.org/10.3390/jmse10122028

Academic Editor: Nguyen Hong Nguyen

Received: 12 November 2022
Accepted: 16 December 2022
Published: 19 December 2022

1. Introduction

Due to the various changes caused by the Earth's rotation and autotransfer, all organisms have biological clocks that are autonomous endogenous timing mechanisms within the organism that regulate its adaptation to exogenous rhythmic changes in light, temperature and other environmental factors [1,2]. The daily rhythmicity of animal behavior, physiology, metabolism and immunity is controlled by biological clocks that are genetically synchronized with environmental cycles and can maintain a 24 h rhythm even in the absence of environmental cues [3]. The biorhythmic center that controls periodic changes in biological functions is called the biorhythmic pacemaker [4]. At the molecular level, biological rhythms are regulated through feedback loops formed at the level of highly conserved transcripts [5]. The biological clock consists of biological clock genes and transcription factors involved in the transcription–translation feedback loop, including *BMAL1, CLOCK,* period genes (*PER1/PER2/PER3*) and cryptochrome genes (*CRY1/CRY2*) [2,6]. These genes are not only expressed and function in the cells of the biological rhythm centers, but are also present in all tissues and cells of the organism. Therefore, transcription factors have an important role in regulating circadian rhythms [7].

The physiology, metabolism and immunity of most fish are regulated by the biological clock [8]. Additionally, the liver, the main metabolic organ for lipids, is controlled by circadian rhythms [9]. Some studies have found that circadian rhythms exist in reptiles and birds [10,11]. Mammalian lipid metabolism also follows a circadian rhythm [12,13]. A correlation between metabolic pathways and circadian rhythms has also been found in studies of mice rhythmically oscillating [14]. Rhythmic gene expression for lipid metabolism in Atlantic bluefin tuna found in fish studies [15]. In recent years, studies have been conducted on the genetic correlation between mammalian immunity and daily rhythmicity. Studies show that mammals develop innate immune effects and daily rhythmicity during the feeding of mammals exposed to microorganisms associated with food [16]. However, there are fewer studies on the immune system and circadian rhythms in fish [17].

The mackerel tuna (*Euthynnus affinis*) is a species of tunas known as the eastern little tuna, skipjack tuna or kawakawa [18]. The production of mackerel tuna is entirely dependent on fishing, and a large proportion of mackerel tuna is 12–32 cm juveniles since the main fishing methods are seining and trawling. Although the production of mackerel tuna is still increasing, the Catch-MSY model estimates that it is currently overfished and will continue to obey the overfishing trend [19]. Therefore, from the perspective of marine resource conservation, it is necessary to carry out captive breeding of mackerel tuna. Currently, captive breeding of mackerel tuna has been reported only in Japan [18]. Our research team realized the land-based recirculating water culture of mackerel tuna [20]. Although the survival rate and better artificial domestication are guaranteed in artificial culture, there are still many problems. The situation of an unsynchronized circadian rhythm system of aquatic animals with the existing environment caused by the change of relative time under artificial culture conditions may lead to a state of stress and undesirable consequences, such as slow growth and reduction of disease resistance, which is a great obstacle to the development of artificial culture [21]. In this study, the mRNA expression levels of immune, metabolic and rhythm genes in the liver of mackerel tuna were investigated by RT-qPCR. The study aimed to elucidate the daily rhythm expression of lipid metabolism genes, immunity genes and rhythm genes in the liver of the mackerel tuna. This study is essential for this species to be cultivated in captivity to maintain its population and provides basic information to ensure the healthy, green and sustainable development of the mackerel tuna farming industry.

2. Materials and Methods

2.1. Animal

Mackerel tuna (total length: 32.38 ± 4.71 cm, weight: 1163.12 ± 284.60 g) were acclimated for more than six months in indoor ponds (8×5 m) with a recirculating water culture system and a natural light condition at the Tropical Aquatic Research and Development Centre, South China Sea Fisheries Research Institute, Chinese Academy of Fishery Sciences, Hainan, China. Fish were fed ad libitum once a day at 09:00 h, and miscellaneous fresh fish (4×2 cm pieces) were used as feed. The water quality parameters were maintained at ammonia nitrogen < 0.1 mg L^{-1}, nitrite nitrogen < 0.02 mg L^{-1}, pH 7.8, dissolved oxygen > 7.0 mg L^{-1}, and salinity 33 psu.

2.2. Sample Collection

Samples were collected in two different weather conditions, sunny and cloudy. Four sampling times (6:00, 12:00, 18:00 and 24:00) were set for one daily cycle with three parallel ponds. The three fish were taken from each parallel pond. The environmental conditions at each sampling time are shown in Table 1. Fish were deprived of feed for 24 h before the sampling was conducted. Three fish were randomly selected for sampling at each time point. Samples were collected under dim red light at night. After anesthesia with eugenol (80 mg·L^{-1}), the body length and weight were measured, and the liver tissue was quickly collected, snap-frozen in liquid nitrogen and preserved at -80 °C until use.

Table 1. Environmental indicators.

		6:00	12:00	18:00	24:00
Sunny day	Water temperature (°C)	31.8 ± 0.13	33 ± 0.21	33.6 ± 0.17	32.6 ± 0.22
	Light intensity (Lx)	2.7 ± 0.02	1116 ± 0.12	913 ± 0.04	1 ± 0.01
	DO (mg·L)	7.76 ± 0.24	7.59 ± 0.19	7.6 ± 0.2	7.61 ± 0.25
Cloudy day	Water temperature (°C)	31.6 ± 0.11	33 ± 0.16	32.9 ± 0.2	32.6 ± 0.16
	Light intensity (Lx)	1.9 ± 0.01	698 ± 0.03	192 ± 0.05	1.4 ± 0.01
	DO (mg·L)	7.62 ± 0.17	7.54 ± 0.11	7.55 ± 0.13	7.58 ± 0.09

2.3. Total RNA Extraction and cDNA Synthesis

The tissue samples were homogenized in a centrifuge tube containing 1 mL of Trizol (Invitrogen, Carlsbad, CA, USA) and extracted total tissue RNA according to the instructions. The quality and quantity of total RNA were tested using agarose gel electrophoresis and a Micro UV spectrophotometer (Biotec Biotechnology Co., Ltd., Beijing, China). Reverse transcription was performed on 1 µg of total RNA using an EasyScript® All-in-One First-Strand cDNA Synthesis SuperMix (All-Strand Biotechnology Co., Ltd., Beijing, China).

2.4. Real-Time qPCR Analysis

Primers (Table 2) were designed by Primer Premier 5 [14] software based on the gene sequence of the mackerel tuna genome (Non-public data). GAPDH was used as the housekeeping gene. The qPCR was conducted with the Real-time qPCR analysis (Analytik Jena GmbH, Jena, Germany) using SYBR Green (Tiangen Biotech Co., Ltd., Beijing, China). The 20 µL of reaction, including 10 µL of 2× RealUniversal PreMix, 0.6 µL each of forward and reverse primers (10 µM), 2 µL of cDNA template and 6.8 µL of RNase-free ddH2O. Reaction conditions were: (1) Pre-denaturation at 95 °C for 15 min; (2) Amplification reaction, denaturation at 95 °C for 10 s, annealing at 56 °C for 20 s, extension at 72 °C for 30 s and 40 cycles. There were three repetitions of each test. The dissociation curves were analyzed to ensure only specific products were obtained with no formation of primer dimers in each reaction. At the end of the reaction, the relative expression level of the target gene was calculated using the $2^{-\Delta\Delta CT}$ method [16]. The reaction efficiency was 90–110%, and Pearson's coefficients of determination (R^2) were >0.97.

Table 2. Relevant primer information.

Gene	Full Names of Target Genes	Primer Sequence (5′-3′)	Amplicon Size/bp
GAPDH	Glyceraldehyde-3-phosphate dehydrogenase	F: ACACTCACTCCTCCATCTTTG R: TTGCTGTAGCCGAACTCAT	100
TRIM35	Tripartite motif-containing 35	F: GTCTGAAGAGCTGGTGGA R: TTACGAGGTGGTTTGTCC	85
NF-kB1	Nuclear factor of kappa light polypeptide gene enhancer in B-cells 1	F: CCCAAAGACTCCAGCATCA R: GCAGTTGTATCCCATCCTCAA	117
MHC-I	Major histocompatibility complex 1	F: GCCCTCCTGCTCCTTCTT R: GGTTTGCCTCCTCCAATT	83
ALT	Alanine aminotransferase	F: CAGGCTTACGGAGCAAAT R: TCGTGGTGGGATGAAGAT	100
IFNA3	Interferon a3	F: GGTCTGCGTCCCTGTATT R: AGCACTGTGACCCATTCG	102
ISY1	Splicing factor ISY1 homolog	F: GTTCGGATCAAAGAGTTGGG R: CAGACTGGCTGGTGTATAATGG	90
ARHGEF13	A-kinase anchor protein 13	F: CCGTCAACTTCTACAAGGA R: CGCACAATGCTGCTACTC	85

Table 2. *Cont.*

Gene	Full Names of Target Genes	Primer Sequence (5′-3′)	Amplicon Size/bp
SIRT1	Sirtuin 1	F: AGAAGAGGCTGCGGAAGT R: AGGCGTTTGCTGATTGGA	99
GCLM	Glutamate cysteine ligase modifier subunit	F: CTGAGCGACTGGTCTTCC R: CGTGATAGCGTCTGTTGG	93
GCLC	Glutamate-cysteine ligase catalytic subunit	F: CTGTTGAGAAGGGAGTGTC R: TGTTTCTGGTAAGGGTGC	87
GST	Glutathione s-transferase	F: CGCCAAGAAGAACAACAT R: TCTCGAAGAGCAGGGACT	117
LPL	Lipoprotein lipase	F: AGGATGCGACATACAGAACA R: GAAGAGGTGGATGGAACG	112
SREBP1	Sterol regulatory element binding transcription factor 1	F: GACTGACTTGACCGTGTTC R: CTCCTCCTCTTGTTCATCCT	112
CREB1	cAMP responsive element binding protein 1	F: TGCCCACTCCCATCTATC R: CTCCATCTGTGCCGTTATT	93
CLOCK	Clock circadian regulator	F: TGTGGACGACCTGGAGAC R: AGGAAACGGTAGTAGCAAG	84
PER1	Period 1	F: CCAAAGGCGGTTCAGTTA R: GAGGCTTCTTGTCTCCCAC	144
PER2	Period 2	F: TCTAATGGAGTCGTCAGGGAG R: AGCCGCTGGTTGAAGGAT	119
PER3	Period 3	F: TCATCGGACGGCATAAAG R: TGGGTGACTGGGAAATACTC	85
RORA	Homo sapiens RAR-related orphan receptor A	F: CTGGATAGGGTGGGTGGAA R: CGTTGGCCCGGATTAGAG	84
REV-ERBA	Nuclear receptor subfamily 1 group D member 1	F: CCTACAACCATCCCACAG R: ACCTTACATAGAAGCACCATA	88
CRY1	Cryptochrome 1	F: GTGGGCAGCCTCCTCTTA R: CCGTACTTGTCTCCGTGGTC	145
CRY2	Cryptochrome 2	F: CTACATGAAGCTCCGTAAGC R: CGGTCAAAGTTTGGGTTG	108
BMAL1	Brain and muscle Arnt-like 1	F: CGTCCAGTGGTAATGTCA R: CATGAGTGCTTCTCCTCC	176

2.5. Statistical Analysis

All data were expressed as mean ± standard deviation. The data were analyzed by SPSS 26.0 statistical software and plotted by Origin2021. The test data all conformed to Shapiro–Wilk and Chi-squared tests. A two-way ANOVA was used to test the interaction effects of different weather and times of the day and was performed using SPSS software. A one-way ANOVA was used for multiple comparisons between different sample time points on the same day, and an independent sample *t*-test was used to analyze the significant differences between the sunny and cloudy days at the same time points. The significant difference level was set as $p < 0.05$. Data were then fitted to a cosine wave to determine the presence of a significant daily rhythm. Raw data were analyzed using Acro circadian analysis programs (University of South Carolina, USA; http://www.circadian.org/softwar.html; accessed on 7 October 2022).

3. Results

3.1. Changes in Gene Expression Levels in the Mackerel Tuna Rhythm

Under the sunny condition, the expression levels of *CREB1, CLOCK, RORA, PER1, PER3* and *CRY1* were not significantly rhythmic in the liver (Table 3). The expression levels of *CREB1* and *CLOCK* were significantly higher at 18:00 than at the other time points ($p < 0.05$). The expression levels of *RORA, PER1* and *PER3* were significantly higher at 24:00 than at the other time points ($p < 0.05$; Figure 1a–f). The expression level of *CRY1* was significantly lower at 6:00 than at the remaining time points ($p < 0.05$, Figure 1g). The expression levels of *PER2, CRY2, REVERBA* and *BMAL1* were significantly rhythmic in the liver under the sunny condition (Table 3). The expression levels of *PER2* and *CRY2* were significantly higher at 24:00 than at the other time points ($p < 0.05$; Figure 1e,h; Table 3). The expression levels of *REVERBA* were significantly higher at 18:00 than at the other time points < 0.05; Figure 1i; Table 3). The expression levels of *BMAL1* were significantly higher at 24:00 than at the other time points ($p < 0.05$; Figure 1h; Table 3).

Table 3. Cosinor analysis board for rhythm gene expression under sunny and cloudy conditions.

Gene	Weather	Acro (*p*-Value)	Acrophase
CREB1	Sunny day	n.s.	
	Cloudy day	<0.01	12 ± 1.16
CLOCK	Sunny day	n.s.	
	Cloudy day	<0.005	6 ± 0.92
PER1	Sunny day	n.s.	
	Cloudy day	<0.001	12 ± 0.79
PER2	Sunny day	<0.001	0 ± 0.61
	Cloudy day	n.s.	
PER3	Sunny day	n.s.	
	Cloudy day	<0.05	0 ± 1.89
RORA	Sunny day	n.s.	
	Cloudy day	n.s.	
REVERBA	Sunny day	<0.001	18 ± 0.43
	Cloudy day	<0.001	12 ± 0.85
CRY1	Sunny day	n.s.	
	Cloudy day	n.s.	
CRY2	Sunny day	<0.001	0 ± 0.83
	Cloudy day	n.s.	
BMAL1	Sunny day	<0.001	12 ± 0.57
	Cloudy day	<0.005	6 ± 0.81

n.s. denotes statistical differences between the sampling points. Acrophases (circadian peak times) were calculated by a non-linear regression fit of a cosine function. Data are expressed as acrophase $\pm$ 95% confidence intervals.

Under the cloudy condition, the expression levels of *RORA, PER2, CRY1* and *CRY2* were not significantly rhythmic in the liver (Table 3). The expression levels of *CREB1, PER1, PER3* and *REVERBA* were significantly higher at 12:00 than the other three time points under the cloudy condition ($p < 0.05$). The expression levels of *CLOCK* and *BMAL1* were significantly higher at 6:00 than at the other time points ($p < 0.05$; Figure 1a,b,d,f,i,j; Table 3). The expression levels of *RORA, PER2, CRY1* and *CRY2* were significantly rhythmic in the liver under cloudy conditions (Table 3). The expression levels of *RORA, PER2* and *CRY2* were significantly higher at 12:00 than at the other three time points ($p < 0.05$). The expression level of *CRY1* was significantly higher at 6:00 than at the other three time points ($p < 0.05$; Figure 1c,e,g,h; Table 3).

A comparison of rhythm genes' expression levels at the same time point under different weather (sunny versus cloudy days) is shown in Figure 1. Expression levels of *CREB1, RORA, PER1, PER3, CRY1, REVERBA* and *BMAL1* at all the time points were significantly different between sunny and cloudy days at the same time point ($p < 0.05$; Figure 1a,c,d,f,g,i,j). Expression levels of *CLOCK* at 6:00, 18:00 and 24:00 were significantly different between sunny and cloudy days at the same time point ($p < 0.05$; Figure 1b).

The expression levels of *PER2* at 6:00, 12:00 and *PER2* at 6:00, 12:00 and 24:00 were significantly different between sunny and cloudy days at the same time point ($p < 0.05$; Figure 1e). The expression levels of *CRY2* at 12:00, 18:00 and 24:00 were significantly different between sunny and cloudy days at the same time point ($p < 0.05$; Figure 1h).

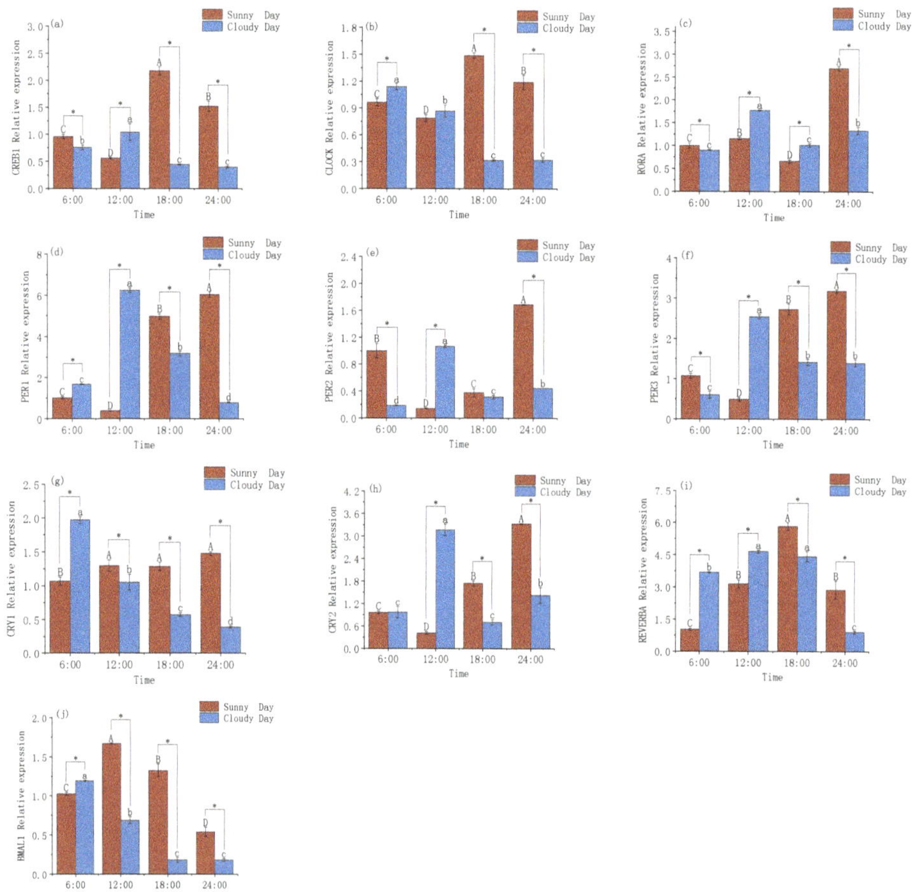

Figure 1. Expression of liver rhythm genes during 24 h in different weather in mackerel tuna. (**a**): *CREB1*; (**b**): *CLOCK*; (**c**): *RORA*; (**d**): *PER1*; (**e**): *PER2*; (**f**): *PER3*; (**g**): *CRY1*; (**h**): *CRY2*; (**i**): *REVERBA*; (**j**): *BMAL1*. Red in each graph represents sunny days, and blue represents cloudy days. The presence of different letters indicates significance by ANOVA and Tukey's tests ($p < 0.05$). * represents significant differences at the same time point ($p < 0.05$). Differences between those with different lowercase letters indicate significance ($p < 0.05$), while the opposite difference is not significant ($p > 0.05$); the same for the latter figure.

The results of the two-way analysis of different weather and time of day on mackerel tuna rhythm genes are shown in Table 4. The main effect of time and weather was significant ($p < 0.05$); there was a significant interaction between time and different weather on the level of rhythm gene expression in mackerel tuna ($p < 0.05$).

Table 4. Effects of light intensity and duration on the rhythm genes of mackerel tuna under different weather conditions.

Genes	Time	*p*-Value Weather	Interactions
CREB1	<0.001	<0.001	<0.001
CLOCK	<0.001	<0.001	<0.001
RORA	<0.001	<0.001	<0.001
PER1	<0.001	<0.001	<0.001
PER2	<0.001	<0.001	<0.001
PER3	<0.001	<0.001	<0.001
CRY1	<0.001	<0.001	<0.001
CRY2	<0.001	0.339	<0.001
REVERBA	<0.001	<0.001	<0.001
BMAL1	<0.001	0.026	<0.001

Results of the two-way ANOVA with SPSS for the measured factors. When interactions in the analysis are significant ($p < 0.001$), a between-group comparison and an independent samples t-test at the same time point are used (the same applies below).

3.2. Changes in the Expression Levels of Lipid Metabolism in Mackerel Tuna

Under the sunny condition, the expression levels of *SIRT1*, *GST* and *LPL* were not significantly rhythmic in the liver (Table 5). The expression levels of *SIRT1* and *GST* were significantly higher at 18:00 than the other three groups at different times in sunny conditions ($p < 0.05$).

Table 5. Cosinor analysis board for metabolic genes expression under the sunny and cloudy conditions.

Gene	Weather	Acro (*p*-Value)	Acrophase
SIRT1	Sunny day	n.s.	
	Cloudy day	<0.001	18 ± 0.59
GST	Sunny day	n.s.	
	Cloudy day	n.s.	
LPL	Sunny day	n.s.	
	Cloudy day	n.s.	
SREBP1	Sunny day	<0.05	0 ± 1.29
	Cloudy day	<0.001	18 ± 0.77

n.s. denotes statistical differences between the different sampling points. Acrophases (circadian peak times) were calculated by a non-linear regression fit of a cosine function. Data are expressed as acrophase $\pm$ 95% confidence intervals.

Under the cloudy condition, the expression levels of *GST* and *LPL* were not significantly rhythmic in the liver (Table 5). The expression levels of *GST* and *LPL* were significantly higher ($p < 0.05$) than the other three groups at 18:00 under different times in the cloudy condition (Figure 2b,c; Table 5). The expression levels of *SIRT1* and *SREBP1* were significantly rhythmic in the liver under overcast conditions (Table 5).

The expression levels of metabolic genes at the same time point under different weather on sunny and cloudy days were significantly different in all four groups ($p < 0.05$). No significant differences could be seen between *GST* at 6:00 and 18:00 in the same time point comparison; *GST* expression levels were significantly higher under the cloudy condition than under the sunny condition at 12:00 ($p < 0.05$), and *GST* expression levels were significantly higher under the sunny condition than under the cloudy condition at 24:00 ($p < 0.05$; Figure 2b). The expression levels of *LPL* were significantly higher under sunny light conditions at 6:00 and 24:00 than under the cloudy condition ($p < 0.05$); the expression levels of *LPL* were significantly higher under cloudy light conditions at 12:00 and 18:00 than under cloudy light conditions ($p < 0.05$) (Figure 2c). The expression level of SREBP1 at 12:00 under the cloudy condition was significantly higher than that of the sunny condition ($p < 0.05$); the expression level of *SREBP1* at 18:00 was not significantly different in the comparison at the same time point (Figure 2d).

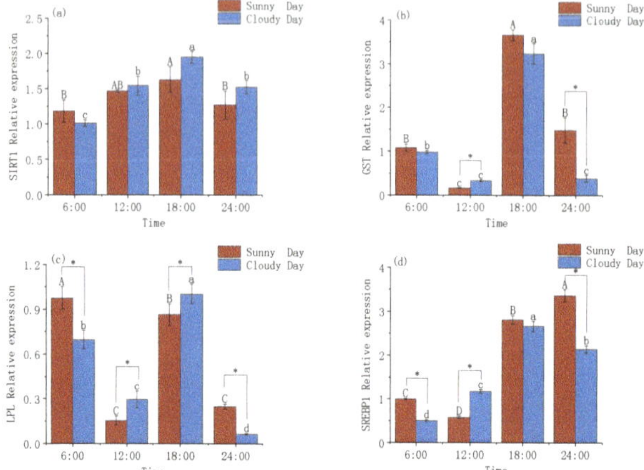

Figure 2. Expression of hepatic metabolic genes during 24 h in different weather in mackerel tuna. (**a**): *SIRT1*; (**b**): *GST*; (**c**): *LPL*; (**d**): *SREBP1*. Red in each graph represents sunny days, and blue represents cloudy days. The presence of different letters indicates significance by ANOVA and Tukey's tests ($p < 0.05$). * represents significant differences at the same time point ($p < 0.05$).

The results of the two-way analysis of different weather and time of day on metabolic genes in mackerel tuna are shown in Table 6. The main effect of time and weather was significant ($p < 0.05$); time and different weather had a significant interaction effect ($p < 0.05$) on metabolic gene expression levels in mackerel tuna.

Table 6. Effects of light intensity and duration on metabolic genes in mackerel tuna under different weather conditions.

| | | *p*-Value | |
Genes	Time	Weather	Interactions
SIRT1	<0.001	0.035	0.022
GST	<0.001	<0.001	<0.001
LPL	<0.001	0.007	<0.001
SREBP1	<0.001	<0.001	<0.001

3.3. Changes in the Expression Levels of Immune Genes in Mackerel Tuna

Expression levels of *TRIM35* under sunny conditions were not significantly rhythmic in the liver (Table 7). The expression level of *TRIM35* at 18:00 was significantly higher ($p < 0.05$) than the other three groups at different times in sunny conditions (Figure 3a; Table 7). Expression levels of *NF-kB1*, *MHC-I*, *ALT*, *IFNA3*, *ISY1*, *ARHGEF13* and *GCLC* were significantly rhythmic in the liver under sunny conditions (Table 7).

Table 7. Cosinor analysis board for immune genes expression under sunny and cloudy conditions.

Gene	Weather	Acro (*p*-value)	Acrophase
TRIM35	Sunny day	n.s.	
	Cloudy day	n.s.	
NF-KB1	Sunny day	<0.001	18 ± 0.5
	Cloudy day	<0.01	12 ± 1.13
MHC-I	Sunny day	<0.005	18 ± 1.03
	Cloudy day	n.s.	
ALT	Sunny day	<0.005	18 ± 1.07
	Cloudy day	<0.005	12 ± 1.03

Table 7. *Cont.*

Gene	Weather	Acro (*p*-value)	Acrophase
IFNA3	Sunny day	<0.001	18 ± 0.39
	Cloudy day	n.s.	
ISY1	Sunny day	<0.001	18 ± 0.79
	Cloudy day	<0.001	18 ± 0.64
ARHGEF13	Sunny day	<0.05	18 ± 1.43
	Cloudy day	<0.005	12 ± 1.03
GCLM	Sunny day	<0.05	12 ± 1.26
	Cloudy day	<0.001	12 ± 0.75
GCLC	Sunny day	n.s.	
	Cloudy day	<0.05	0 ± 1.36

n.s. denotes statistical differences between the different sampling points. Acrophases (circadian peak times) were calculated by a non-linear regression fit of a cosine function. Data are expressed as acrophase ± 95% confidence intervals.

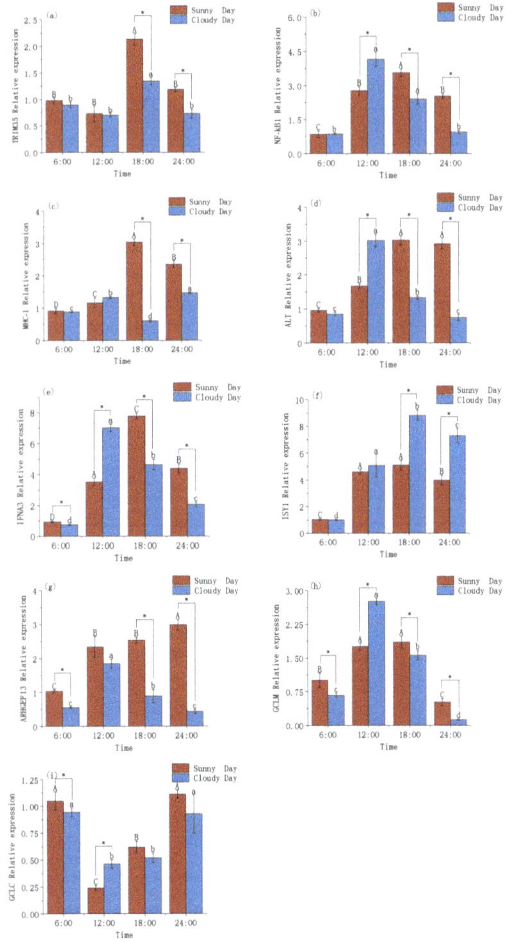

Figure 3. Expression of liver immune genes in different weather conditions in mackerel tuna over a 24 h period. (**a**): *TRIM35*; (**b**): *NF-kB1*; (**c**): *MHC-I*; (**d**): *ALT*; (**e**): *IFNA3*; (**f**): *ISY1*; (**g**): *ARHGEF13*; (**h**): *GCLM*; (**i**): *GCLC*. Red in each graph represents sunny days, and blue represents cloudy days. The presence of different letters indicates significance by ANOVA and Tukey's tests (*p* < 0.05). * represents significant differences at the same time point (*p* < 0.05).

Under the cloudy condition, the expression levels of *TRIM35*, *MHC-I* and *IFNA3* were not significantly rhythmic in the liver (Table 7). The expression levels of *TRIM35* at 18:00 were significantly higher than the other three time points at different times in the cloudy condition ($p < 0.05$), the expression levels of *MHC-I* at 24:00 were significantly higher than the other three time points in the cloudy condition ($p < 0.05$), and the expression levels of IFNA3 at 12:00 were significantly higher than the other three time points in the cloudy condition ($p < 0.05$; Figure 2a,b,e; Table 7). The expression levels of *NF-kB1*, *ALT*, *ISY1*, *ARHGEF13*, *GCLM* and *GCLC* were significantly rhythmic in the liver under overcast conditions (Table 7).

The expression levels of *TRIM35* at 6:00 and 12:00 were not significantly different between sunny and cloudy days at the same time point; the expression levels of *TRIM35* at 18:00 and 24:00 were significantly higher ($p < 0.05$) under the sunny condition than the under cloudy condition (Figure 3a). The expression levels of *NF-kB1* at 6:00 were not significantly different in the comparison at the same time point; *NF-kB1* expression levels were significantly higher under the cloudy condition at 12:00 than under the sunny conditions ($p < 0.05$); *NF-kB1* expression levels were significantly higher under the sunny condition at 18:00 and 24:00 than under the cloudy condition ($p < 0.05$) (Figure 3b). The expression levels of *MHC-I* at 6:00 and 12:00 were not significantly different in the same time point comparison; the expression levels of *MHC-I* at 18:00 and 24:00 under the sunny condition were significantly higher than those under the cloudy condition ($p < 0.05$) (Figure 3c). The expression levels of *ALT* at 6:00, 18:00 and 24:00 were significantly higher ($p < 0.05$) under the sunny condition than under the cloudy condition (Figure 3d). The expression levels of *ISY1* at 6:00 and 12:00 were not significantly different in the comparison at the same time point; the expression levels of *ISY1* at 18:00 and 24:00 under the cloudy condition were significantly higher than those under the sunny condition ($p < 0.05$; Figure 3f). The expression levels of *ARHGEF1* were significantly higher ($p < 0.05$) under the sunny condition than under the cloudy condition at 6:00, 18:00 and 24:00; the expression levels of *ARHGEF1* at 12:00 were not significantly different at the same time point comparison (Figure 3g). *GCLM* expression levels were significantly higher at 6:00, 18:00 and 12:00 in the same time point comparison between sunny and cloudy conditions ($p < 0.05$); the expression level of *GCLM* at 12:00 was significantly higher under cloudy conditions than under sunny conditions ($p < 0.05$; Figure 3h). The expression level of *GCLC* was significantly higher at 6:00 than in the same time point comparison between sunny and cloudy conditions ($p < 0.05$); The expression level of *GCLC* at 12:00 was significantly higher under cloudy conditions than under sunny conditions ($p < 0.05$; Figure 3i).

The results of the two-way analysis of different weather and time of day on immune genes in mackerel tuna are shown in Table 8. The main effects of time and weather were significant ($p < 0.05$), and there was a significant interaction between time and different weather on immune gene expression levels in mackerel tuna ($p < 0.05$).

Table 8. Effects of light intensity and duration on immune genes in mackerel tuna under different weather conditions.

Genes	Time	*p*-Value Weather	Interactions
TRIM35	<0.001	<0.001	<0.001
NF-kB1	<0.001	<0.001	<0.001
MHC-I	<0.001	<0.001	<0.001
ALT	<0.001	<0.001	<0.001
IFNA3	<0.001	0.022	<0.001
ISY1	<0.001	<0.001	<0.001
ARHGEF13	<0.001	<0.001	<0.001
GCLM	<0.001	<0.001	<0.001
GCLC	<0.001	0.205	0.002

4. Discussion

4.1. Rhythmic Gene Expression Patterns

In nature, in both animals and humans, there is a 24 h circadian rhythm called the biological clock [22]. *CREB1* (cyclic adenosine monophosphate response element binding protein 1) is a protein that regulates gene transcription and can participate in cycle regulation by regulating the expression of downstream target genes [23]. In this study, the daily rhythmicity of *CREB1* was only present under the cloudy condition, and the average *CREB1* gene expression was found to be higher under the sunny condition than under the cloudy condition. The light intensity may stimulate the *CREB1* gene expression in mackerel tuna, but the regulatory mechanism remains to be investigated.

The molecular mechanism of the biological clock is primarily the existence of a cell-autonomous transcriptional–translational feedback loop in which a pair of positive regulators (*CLOCK* and *BMAL1*) form a heterodimer that activates its transcription by binding to the negative regulators (*PER* and *CRY* promoters). Subsequently, the *PER* and *CRY* proteins bind to form a complex that enters the nucleus and acts on the *CLOCK* and *BMAL1* heterodimers, thereby feeding back to repress its own transcription [24]. *BMAL1* is thought to regulate the expression of rhythmic genes throughout the liver [25]. In this study, *PER2*, *CRY2* and *BMAL1* showed daily rhythmicity under the sunny condition and *CLOCK*, *PER1*, *PER3* and *BMAL1* showed daily rhythmicity under the cloudy condition. It has been shown that the expression levels of the rhythm genes *PER1*, *PER2*, *CRY1*, *CRY2* and *BMAL1* in the liver of the Atlantic bluefin tuna (*Thunnus thynnus*, L.) exhibit daily rhythmicity [26]. Similar to the results of the present study. However, in this study, PER1 showed daily rhythmicity only under cloudy conditions, while *PER2* and *CRY2* showed daily rhythmicity only under sunny conditions, which may be due to the different changes in light intensity under the cloudy and the sunny conditions, leading to an effect of light stimulation on the circadian system of fish. It has been shown that photoperiod has an effect on core biological clock gene expression in fish [27]. In this study, the expression levels of rhythm genes were at their highest under sunny conditions at 12:00, when light intensity was the strongest of the day, reflecting that light intensity at 12:00 under sunny conditions can cause up-regulation of the expression levels of rhythm genes in mackerel tuna to some extent.

4.2. Metabolic Gene Expression Patterns

The liver is the main organ of lipid metabolism in fish and plays a huge role in the body as a hub for fat transport, influencing the breakdown and absorption of nutrients and hormonal signaling [28]. Some hepatic metabolic pathways are driven by the circadian biological clock, resulting in a circadian rhythm, and disturbances in the biological clock can cause metabolic disorders in fish. *SIRT1* is involved in a wide range of glucose and lipid metabolism pathways, as well as in the regulation of gene transcription and cellular senescence through the deacetylation of several metabolism-controlling transcription factors [29]. Current research on the *SIRT1* gene focuses on humans, mice, livestock and poultry (pigs, sheep, chickens, etc.) [30]. Little research has been reported on the *SIRT1* gene in fish. It has been found that there is a close correlation between the expression level of the *SIRT1* gene and lipid metabolism in pigs [31]. In this study, there was no daily rhythm in the expression levels of the *SIRT1* gene under sunny conditions, while there was a daily rhythm in the expression levels of *SIRT1* under cloudy conditions. This may result from the involvement of *CLOCK* and *BMAL1* in the regulation. It has been shown that *SIRT1* interacts with *BMAL1*, *PER2* and *CRY1* in the liver of mice [32]. In rainbow trout [33], *SIRT1* interacts with *BMAL1*, *PER2*, *PER3*, *CLOCK* and *BMAL1*, with *CLOCK* and *BMAL1* controlling the rhythm of *SIRT1*, which is in agreement with the results of this study.

SREBP1 plays a key role in regulating lipid homeostasis, and *REVERBA* is a powerful transcriptional repressor that plays an important role in rhythms. In this study, *SREBP1* showed a clear daily rhythm under sunny conditions, and both *REVERBA* and *SREBP1* had a daily rhythm under cloudy conditions with similar trends, but *REVERBA* peaked at 12:00 and *SREBP1* peaked at 18:00 when the rhythm genes regulate *SREBP1* expression. This is probably due to the delayed expression levels of the *SREBP1* metabolic gene, resulting in a peak at 18:00, but *REVERBA* plays a regulatory role in *SREBP1* gene expression. It has been shown that *REVERBA* can regulate the expression of *SREBP1* in the salmon liver, that *SREBP1* exhibits daily rhythmicity, and that *REVERBA* has a regulatory effect on *SREBP1* [15], which is consistent with the results of this experimental study. In the present study, *SIRT1* and *SREBP1* showed an up-regulation trend both under sunny and cloudy conditions, and the expression level of *SIRT1* continued to increase with time. It has been shown that *LPL* is expressed in the liver of fish and can exhibit changes in lipid metabolism and deposition [34]. In this study, because of the high activity of mackerel tuna during the day under sunny or cloudy conditions, the mackerel tuna was active in the water at this time, prompting an increase in its metabolic levels and making its expression levels up-regulated during the day.

4.3. Immune Gene Expression Patterns

The TRIM family of proteins belongs to the RING family of E3 ubiquitin ligases. *NF-kB* is an important nuclear transcription factor involved in regulating the body's immune response, and abnormal regulation can lead to immune diseases, metabolic diseases, etc. The TRIM family of proteins is involved in the regulation of the NF-kB signaling pathway as E3 ubiquitin ligases [35–37]. Interferon (IFN) receptor proteins are a class of cytokines secreted by host cells that regulate the immune response. When a pathogen is present, interferon is usually released by the host cell, which is sensed by surrounding undisturbed cells and activates appropriate cellular defense mechanisms to eliminate the pathogen [38]. The composition, signaling pathways, function and evolutionary relationships of the *IFN* were extensively studied in fish many years ago [39–41]. It has been shown that *TRIM35* is a positive regulatory molecule in the natural immune signaling pathway [42]. In this study, *TRIM35* did not show daily rhythmicity under sunny conditions, but there were significant differences between groups. The expression level of *TRIM35* was up-regulated by sunny weather conditions. NF-kB1 showed significant daily rhythmicity in both sunny and cloudy conditions. However, under sunny conditions, the expression level of NF-kB1 showed an increasing trend and started to decrease by 24:00. *IFNA3* showed daily rhythmicity only in sunny conditions. It has been shown that the expression of EcTRIM21 significantly increased the IFN promoter activity and simultaneously increased the transcriptional levels of interferon-related molecules, with a positive regulatory effect [43]. This is consistent with the results of the present experimental study.

MHC-I (major histocompatibility complex) plays an important role in adaptive immunity in vertebrates, primarily recognizing intracellular antigens and triggering adaptive immunity. This gene is currently expressed in different fish species, such as the Japanese flounder (*Paralichthys olivaceus*) [44], turbot (*Scophthalmus maximus*) [45] and rainbow trout (*Oncorhynchus mykiss*) [46]. Interferons increase the expression levels of *MHC-I* class molecules on the cell surface, and an increase in *MHC-I* molecules on the surface of virus-infected cells contributes to the delivery of antigens to T cells, causing lysis of target cells. It was found that the MHC-I protein is extremely important in immune recognition in zebrafish by positively regulating IFN immunity and inflammatory responses [47]. In the present study, both *MHC-I* and *IFNA3* were expressed in daily rhythms under sunny conditions and were positively regulated. This is consistent with the present study results.

GCL (glutamate-cysteine ligase) is composed of different gene-edited *GCLC* (glutamate cysteine ligase catalytic subunit) and *GCLM* (glutamate cysteine ligase), with *GCLC* playing all catalytic roles and being subject to feedback inhibition by *GSH* and *GCLM* having regulatory functions [48]. It has been shown that altered single nucleotide polymorphisms in the *GCLC* and *GCLM* genes can regulate gene expression processes and thus participate in various disease processes [49]. In the present study, both *GCLM* and *GCLC* were rhythmic under overcast conditions, and expression levels were simultaneously downregulated at 12:00. It has been shown that the expression of *NF-kB1* and *GCLC* are mutually regulated [50]. In the present study, *NF-kB1* expression levels were down-regulated when *GCLC* expression levels were up-regulated, suggesting that the down-regulation of *GCLC* expression may be related to the inhibition of the signaling pathway.

5. Conclusions

In summary, the mRNA expression levels of rhythm and immune- and metabolism-related genes in the liver tissue of mackerel tuna were significantly changed under sunny and cloudy conditions, and some genes showed significant daily rhythmicity. Immune and metabolic genes can be regulated by rhythm genes and external factors through different signaling pathways and are jointly involved in regulating immunity and energy metabolism in mackerel tuna. This study provides practical implications for further understanding the regulation of lipid metabolism and immunity in mackerel tuna.

Author Contributions: Conceptualization, G.Y. and J.H.; Methodology, J.H.; Software, J.H.; Validation, J.H. and W.W.; Formal analysis, G.Y.; Investigation, Z.F.; Resources, Z.M.; Data curation, W.W.; Writing—original draft preparation, W.W.; Writing—review and editing, Z.M. and Z.F.; Visualization, J.H.; Supervision, G.Y.; Project administration, Z.M.; Funding acquisition, Z.M. All authors have read and agreed to the published version of the manuscript.

Funding: This work was supported by the Guangxi Innovation Driven Development Special Fund Project (grant no. Guike AA18242031), the Central Public-Interest Scientific Institution Basal Research Fund, CAFS (2020TD55), and the Central Public-Interest Scientific Institution Basal Research Fund South China Sea Fisheries Research Institute, CAFS (2021SD09).

Institutional Review Board Statement: The animal study protocol was approved by the Institutional Review Board (or Ethics Committee) of Animal Care and Use Committee of South China Sea fisheries Research Institute, Chinese Academy of Fishery Sciences (BIOL5312, 5 July 2021).

Informed Consent Statement: Not applicable.

Data Availability Statement: The original contributions presented in the study are included in the article. Further inquiries can be directed to the corresponding authors.

Conflicts of Interest: The authors declare no conflict of interest.

Ethics Statement: The experiment complied with the regulations and guidelines established by the Animal Care and Use Committee of the South China Sea fisheries Research Institute, Chinese Academy of Fishery Sciences.

References

1.   Feng, N.Y.; Bass, A.H. "Singing" Fish Rely on Circadian Rhythm and Melatonin for the Timing of Nocturnal Courtship Vocalization. *Curr. Biol.* **2016**, *19*, 2681–2689. [CrossRef] [PubMed]
2.   Hardin, P.E.; Panda, S. Circadian timekeeping and output mechanisms in animals. *Curr. Opin. Neurobiol.* **2013**, *23*, 724–731. [CrossRef] [PubMed]
3.   Mahman-Averbuch, H.; King, C.D. Disentangling the roles of circadian rhythms and sleep drive in experimental pain sensitivity. *Trends Neurosci.* **2022**, *11*, 796–797. [CrossRef] [PubMed]
4.   Patek, A.; Young, M.W.; Axelrod, S. Molecular mechanisms and physiological importance of circadian rhythms. *Nat. Rev. Mol. Cell Biol.* **2020**, *21*, 67–84. [CrossRef] [PubMed]
5.   Nakaya, M.; Wakamatsu, M.; Motegi, H.; Tanaka, A.; Sutherland, K.; Ishikawa, M.; Ozaki, M.; Shirato, H.; Hamada, K.; Hamada, T. A real-time measurement system for gene expression rhythms from deep tissue of freely moving mice under light-dark conditions. *Biochem. Biophys. Rep.* **2022**, *32*, 101344.
6.   Bertolucci, C.; Fazio, F.; Piccione, G. Daily rhythms of serum lipids in dogs: Influences of lighting and fasting cycles. *Comp. Med.* **2008**, *58*, 485–489.
7.   Takahashi, J.S.; Hee-Kyung, H.; Ko, C.H.; McDearmon, E.L. The genetics mammalian circadian order and disorder: Implications for physiology and disease. *Nat. Rev. Genet.* **2008**, *10*, 764–775. [CrossRef]
8.   Sato, F.; Kohsaka, A.; Bhawal, U.K.; Muragaki, Y. Potential Roles of Dec and Bmal1 Genes in Interconnecting Circadian Clock and Energy Metabolism. *Int. J. Mol. Sci.* **2018**, *3*, 781. [CrossRef]
9.   Lee, C.; Etchegaray, J.-P.; Cagampang, F.R.; Loudon, A.S.; Reppert, S.M. Posttranslational mechanisms regulate the mammalian circadian clock. *Cell* **2001**, *107*, 855–867. [CrossRef]
10. Panda, S.; Hogenesch, J.B.; Kay, S.A. Circadian rhythms from flies to human. *Nature* **2002**, *417*, 329–335. [CrossRef]

11. Chen, L.; Hu, C.; Lai, N.L.; Zhang, W.; Hua, J.; Lam, P.K.S.; Lam, J.C.W.; Zhou, B. Acute exposure to PBDEs at an environmentally realistic concentration causes abrupt changes in the gut microbiota and host health of zebrafish. *J. Environ.* **2018**, *240*, 17–26. [CrossRef]

12. Shibata, S.; Tominaga, K. Brain neuronal mechanisms of circadian rhythms in mammalians. *Yakuga Zasshi* **1991**, *111*, 270–283. [CrossRef] [PubMed]

13. Welsh, D.; Takahashi, J.; Kay, S. Suprachiasmatic nucleus: Cellautonomy and network properties. *Annu. Rev. Physiol.* **2010**, *72*, 551–577. [CrossRef] [PubMed]

14. Le Martelot, G.; Claudel, T.; Gatfield, D.; Schaad, O.; Kornmann, B.; Sasso, G.L.; Moschetta, A.; Schibler, U. REV-ERB alpha participates in circadian SREBP signalling and bile acid homeostasis. *PLoS Biol.* **2009**, *7*, e1000181. [CrossRef] [PubMed]

15. Eckel-Mahan, K.L.; Patel, V.R.; Mohney, R.P.; Vignola, K.S.; Baldi, P.; Sassone-Corsi, P. Coordination of the transcriptome and metabolome by the circadian clock. *Proc. Natl. Acad. Sci. USA* **2012**, *109*, 5541–5546. [CrossRef] [PubMed]

16. Sanchez, R.E.A.; Franck, K.; de la Iglesia, H.O. Slepp timing and the circadian clock in mammals: Past, present and the road ahead. *Semin. Cell Dev. Biol.* **2022**, *126*, 3–14. [CrossRef] [PubMed]

17. Brooks, J.F., II; Cassie, L.; Behrendt, K.; Ruhn, A.; Lee, S.; Raj, P.; Takahashi, J.S.; Hooper, L.V. The microbiota coordinates diurnal rhythms in innate immunity with the circadian clock. *Cell* **2021**, *16*, 4154–4167. [CrossRef]

18. Yazawa, R.; Takeuchi, Y.; Satoh, K.; Machida, Y.; Amezawa, K.; Kabeya, N.; Shimada, Y.; Yoshizaki, G. Eastern little tuna, Euthynnus affinis (Cantor, 1849) mature and reproduce within 1 year of rearing in land-based tanks. *Aquac. Res.* **2016**, *47*, 3800–3810. [CrossRef]

19. IOTC Secretariat. Assessment of Indian Ocean kawakawa (Euthynnus affinis) Using Data Poor Catch-Based Methods(2015). Available online: http://www.seafdec.org/documents/2020/02/pw-neritic-tuna_10-4.pdf (accessed on 7 November 2022).

20. Shengjie, Z.; Rui, Y.; Gang, Y.; Qiaer, W.; Zhenhua, M. Growth of Longtail Tuna *Thunnus tonggol* and Small Tuna *Euthymmus affinis* in Circulating Water Ponds. *Fish. Sci.* **2021**, *40*, 339–346.

21. Ceinos, R.M.; Chivite, M.; López-Patiño, M.A.; Naderi, F.; Soengas, J.L.; Foulkes, N.S.; Míguez, J.M. Differential circadian and light-driven rhythmicity of clock gene expression and behaviour in the turbot, Scophthalmus maximus. *PLoS ONE* **2019**, *14*, e0219153. [CrossRef]

22. Reeds, S.G. Sensory Systems, Perception, and Learning. Circadian Rhythms in Fish. In *Encyclopedia of Fish Physiology*; Academic Press: Cambridge, MA, USA, 2011; pp. 736–743.

23. Xiao, L.; Yong, D.; Yan-ling, Z. Research progress of CREB in angiogenesis, anti-apoptosis and lung cancer. *Tianjin Med.* **2019**, *47*, 431–435.

24. Husse, J.; Eichele, G.; Oster, H. Synchronization of the mammalian circadian timing system: Light can control peripheral clocks independently of the SCN clock: Alternate routes of entrainment optimize the alignment of the body's circadian clock network with external time. *Bioessays* **2015**, *37*, 1119–1128. [CrossRef] [PubMed]

25. Betancor, M.B.; Spragu, M.; Minghetti, M.; Migaud, H.; Tocher, D.R.; Davie, A. Daily Rhythms in Expression og Genes of Hepatic Lipid Metabolism in Atlantic Salmon (*Salmo salas* L.). *PLoS ONE* **2014**, *9*, e106739. [CrossRef] [PubMed]

26. Betancor, M.B.; Sprague, M.; Ortega, A.; de la Gándara, F.; Tocher, D.R.; Ruth, R.; Perkins, E.; Mourente, G. Central and peripheral clocks in Atlantic bluefin tuna (*Thunnus thynnus*, L.): Daily rhythmicity of hepatic lipid metabolism and digestive genes. *Aquaculture* **2020**, *523*, 735220. [CrossRef]

27. Huang, T.S.; Ruoff, P.; Fjelldal, P.G. Effect of continuous light on daily levels of plasma melatonin and cortisol and expression of clock genes in pineal gland, brain, and liver in atlantic salmon postsmolts. *Chronobiol. Int.* **2010**, *27*, 1715–1734. [CrossRef] [PubMed]

28. Mohamed, A.A.-R.; El-Houseiny, W.; EL-Murr, A.E.; Ebraheim, L.L.M.; Ahmed, A.I.; El-Hakim, Y.M.A. Effect of hexavalent chromium exposure on the liver and kidney tissues related to the expression of CYP450 and GST genes of Oreochromis niloticus fish: Role of curcumin supplemented diet. *Ecotoxicol. Environ. Saf.* **2020**, *188*, 109890. [CrossRef]

29. Brunet, A.; Sweeney, L.B.; Sturgill, J.F.; Chua, K.F.; Greer, P.L.; Lin, Y.; Tran, H.; Ross, S.E.; Mostoslavsky, R.; Cohen, H.Y.; et al. Stress-dependent regulation of FOXO transcription factors by the SIRT1 deacetylase. *Science* **2004**, *303*, 2011–2015. [CrossRef]

30. Finkel, T.; Deng, C.X.; Mostoslavsky, R. Recent progress in the biology and physiology of sirtuins. *Nature* **2009**, *460*, 587–591. [CrossRef]

31. Bai, L.; Pang, W.-J.; Yang, Y.-J.; Yang, G.-S. Modulation of Sirt1 by resveratrol and nicotinamide alters proliferation and differentiation of pig preadipocytes. *Mol. Cell. Biochem.* **2008**, *307*, 129–140. [CrossRef]

32. Asher, G.; Gatfield, D.; Stratmann, M.; Reinke, H.; Dibner, C.; Kreppel, F.; Mostoslavsky, R.; Alt, F.W.; Schibler, U. SIRT1 regulates circadian clock gene expression through PER2 deacetylation. *Cell* **2008**, *134*, 317–328. [CrossRef]

33. Naderi, F.; Míguez, J.M.; Soengas, J.L.; López-Patiño, M.A. SIRT1 mediates the effect of stress on hypothalamic clock genes and food intake regulators in rainbow trout, *Oncorhynchus mykiss*. *Comp. Biochem. Physiol. Part A Mol. Integr. Physiol.* **2019**, *235*, 102–111. [CrossRef] [PubMed]

34. Wang, L.; Kaneko, G.; Takahashi, S.I.; Watabe, S.; Ushio, H. Identification and gene expression profile analysis of a major type of lipoprotein lipase in adult madaka Oryzias latipes. *Fish. Sci.* **2015**, *81*, 163–173. [CrossRef]

35. Kentsis, A.; Borden, K.L. Construction Fish & Shellfish Immunology. of macromolecular assemblages in eukaryotic processes and their role in human disease: Linking EINGs together. *Curr. Protein Sci.* **2000**, *1*, 49–73.

36. Rajsbaum, R.; Garcia-Sastre, A.; Versteeg, G.A. TRIM munity: The roles of the TRIM E3-ubiquitin ligase family in innate antiviral immunity. *Mol. Biol.* **2014**, *426*, 1265–1284. [CrossRef] [PubMed]
37. Borden, K.L. RING fingers and B-boxes: Zinc-binding protein-protein interaction domains. *Biochem. Cell Biol.* **1998**, *76*, 351–358. [CrossRef]
38. Kong, Y.; Gao, C.; Du, X.; Zhao, J.; Li, M.; Shan, X.; Wang, G. Effects of single or conjoint administration of lactic acid bacteria as potential probiotics on growth, immune response and disease resistance of snakehead fish (*Channa argus*). *Fish Shellfish Immunol.* **2020**, *102*, 412–421. [CrossRef]
39. Xiao, X.; Zhu, W.T.; Zhang, Y.Q.; Liao, Z.; Wu, C.; Yang, C.; Zhang, Y.; Xiao, S.; Su, J. Broad-spectrum robust direct bactericidal activity of fish IFN ϕ1 reveals an antimicrobial peptide-like function for type I IFNs in vertebrates. *J. Immunol.* **2021**, *206*, 1337–1347. [CrossRef]
40. Gale, M., Jr. Interference with virus infection. *J. Immunol.* **2015**, *195*, 1909–1910. [CrossRef]
41. Zou, J.; Secombes, C.J. Teleost fish interferons and their role in immunity. *Dev. Comp. Immunol.* **2011**, *35*, 1376–1387. [CrossRef]
42. Langevin, C.; Levraud, J.; Boudinot, P. Fish antiviral tripartite motif (TRIM35) proteins. *Fish Shellfish Immunol.* **2019**, *86*, 724–733. [CrossRef]
43. Zheng, J.; Zhi, L.; Wang, W.; Ni, N.; Huang, Y.; Qin, Q.; Huang, X. Fish TRIM21 exhibits antiviral activity against grouper iridovirus and nodavirus infection. *Fish Shellfish Immunol.* **2022**, *127*, 956–964. [CrossRef] [PubMed]
44. Xu, T.; Chen, S. Genomic structure of DAA gene and polymorphism within MHC -DAA alleles in apanese flounder (*Paralichthys olivaceus*). *Hereditas* **2009**, *31*, 1020–1028. [PubMed]
45. Zhang, Y.; Chen, S. Full length cDNA cloning and tissue expression of major histocompatibility complex (MHC)IIa from turbot *Scophthalmus maximus. Oceanol. Limnol. Sin.* **2007**, *38*, 221–226.
46. Glamann, J. Complete coding sequence of rainbow trout Mhc II beta chain. *Scand. Immunol.* **1995**, *41*, 365–372. [CrossRef]
47. Chen, J.; Wang, L.; Huang, J.; Li, X.; Guan, L.; Wang, Q.; Yang, M.; Qin, Q. Functional analysis of a novel MHC-Iα genotype in orange-spotted grouper: Effects on Singapore grouper iridovirus (SGIV) replication and apoptosis. *Fish Shellfish Immunol.* **2022**, *121*, 487–497. [CrossRef] [PubMed]
48. Bansal, A.; Simon, M.C. Glutathione metabolism in cancer progression and treatment resistance. *Cell. Biol.* **2018**, *217*, 2291–2298. [CrossRef]
49. Zhang, S.; Wu, Y.; Liu, Z.; Tao, Q.; Huang, J.; Yang, W. Hepatic pathology of biliary atresia: A new comprehensive evaluation method using liver biopsy. *Turk. Gastroenterol.* **2016**, *27*, 257–263. [CrossRef]
50. Yang, H.; Magilnick, N.; Lee, C.; Kalmaz, D.; Ou, X.; Chan, J.Y.; Lu, S.C. Nrf1 and Nrf2 regulate rat glutamate-cysteine ligase catalytic subunit transcription indirectly via NF-kappaB and AP-1. *Mol. Cell. Biol.* **2005**, *25*, 5933–5946. [CrossRef]

Article

Aquaculture in an Offshore Ship: An On-Site Test of Large Yellow Croaker (*Larimichthys crocea*)

Youbin Yu [1,2], Wenyun Huang [1,2], Fei Yin [3], Huang Liu [1,*] and Mingchao Cui [1,2,*]

[1] Key Laboratory of Fishery Equipment and Engineering, Ministry of Agriculture and Rural Affairs, Fishery Machinery and Instrument Research Institute, Chinese Academy of Fishery Sciences, Shanghai 200092, China
[2] Qingdao National Laboratory for Marine Science and Technology, Qingdao 266237, China
[3] Key Laboratory of Applied Marine Biotechnology, Ministry of Education, State Key Laboratory for Managing Biotic and Chemical Threats to the Quality and Safety of Agro-Products, School of Marine Sciences, Ningbo University, Ningbo 315211, China
* Correspondence: liuhuang@fmiri.ac.cn (H.L.); cuimingchao@fmiri.ac.cn (M.C.)

Abstract: Ship aquaculture platforms are expected to become a meaningful way to expand offshore farming. The growth performance and nutritional composition of the large yellow croaker reared in an offshore ship aquaculture system and nearshore traditional cage system was evaluated in this study. The results showed that the aquaculture ship could effectively avoid the harsh environment such as typhoons and red tides. The test large yellow croaker adapted to the ship culture system in a short time. No serious stress events occurred during the whole rearing process. During the culture experimental period, the fish fed normally, and disease was controlled. The aquaculture ship has good environment conditions during breeding with a water temperature of 21.5–28.5 °C, salinity 20.7–31.8‰, pH 7.6–8.4, dissolved oxygen 7.2–12.8 mg/L, ammonia nitrogen < 0.08 mg/L, and the number of bacteria and vibrio in water were $1.2 * 10^3$–$1.6 * 10^3$ CFU/mL and $1.2 * 10^2$–$1.8 * 10^2$ CFU/mL, respectively. The survival, weight gain rate, and monthly weight gain of the large yellow croaker in the ship were 99.02%, 41.48%, and 67.52 g, respectively, which were significantly higher than those of cage culture. The crude protein content of the large yellow croaker raised in the ship was significantly higher than that in the cage group, and the crude fat content was significantly lower than that in the cage group. These results indicated that the growth performance and nutritional composition of the large yellow croaker reared in offshore ship were better than those of the fish in the cage. These findings enhanced our understanding of an offshore ship aquaculture model of large yellow croaker.

Keywords: aquaculture ship; traditional cage; large yellow croaker; growth performance; nutrient composition

Citation: Yu, Y.; Huang, W.; Yin, F.; Liu, H.; Cui, M. Aquaculture in an Offshore Ship: An On-Site Test of Large Yellow Croaker (*Larimichthys crocea*). *J. Mar. Sci. Eng.* **2023**, *11*, 101. https://doi.org/10.3390/jmse11010101

Academic Editor: Ka Hou Chu

Received: 2 December 2022
Revised: 26 December 2022
Accepted: 27 December 2022
Published: 4 January 2023

1. Introduction

With the increasing demand for high-quality protein, it was estimated that global fish production will increase by 32% by 2030 compared to 2018 [1,2]. However, in recent years, due to adverse factors such as limited space resources, environmental deterioration, excessive aquaculture density, and frequent disease outbreaks, the inland and nearshore aquaculture space has been seriously restricted [3,4]. To reduce the impact of aquaculture on land-based and nearshore areas, it was imperative to expand the aquaculture space. Enclosed aquaculture vessel was a new offshore aquaculture method in recent years. It was expected to be an important way to develop offshore aquaculture [5,6].

The development of offshore areas can be used as an important space for fish farming in the future. Offshore seas have the advantages of a wide area, good water quality, and few pollutions and pathogenic microorganisms [7]. Currently, many practical activities for offshore aquaculture have been carried out in China, Norway, Scotland, Korea, Japan, and other countries [8–11]. Studies have shown that offshore aquaculture can effectively improve the survival and growth of salmon and bluefin tuna and reduce the possibility of fish disease [12–14]. At the same time, offshore areas have relatively harsh environmental conditions, such as large wind waves and currents, which may damage the breeding equipment and fish health [15,16]. Therefore, offshore aquaculture needs to improve the requirements of technology, capital, operation, and other aspects.

As a new model of closed aquaculture, the relevant research data on the aquaculture ship was very scarce. It was reported that only Türkiye (Turkey) transformed a bulk carrier into an enclosed aquaculture vessel and carried out the breeding test of rainbow trout. The initial average body weight was 25.0 ± 2.7 g and reached 3.7 ± 0.4 kg after 11 months. The feed conversion ratio (FCR), specific growth rate (SGR), and breeding density were 1.1 ± 0.1, $1.51 \pm 0.3\%/d$, and 101 ± 2.1 kg/m^3, respectively [17]. Large yellow croaker is the main mariculture fish in China, and its production ranks first in mariculture all year-round. Due to its flavor, high flesh quality, and nutritional value, large yellow croaker has a large market space [18]. Large yellow croaker resources almost disappeared from the end of the last century to the current output of more than 250,000 tons, which mainly benefitted from the development of artificial aquaculture technology [19]. However, knowledge about the growth performance and nutritional composition of large yellow croaker that are reared in ships and cages for a short period is limited. Therefore, this experiment takes large yellow croaker as the target animal to explore whether the closed aquaculture system can be used in fish culture, determine the growth performance of large yellow croaker under this culture mode, and analyze the conditions that are suitable for large yellow croaker culture under this mode. In addition, if the platform can be used in offshore areas, the results of this study can establish a set of aquaculture technology standards, such as seed transfer and environmental control, and provide a scientific basis for the establishment of large yellow croaker and other fish.

2. Materials and Methods

2.1. Experimental Animals and Facilities

Traditional aquaculture cages (26°37′35″ N, 119°51′72″ E) were set in Sandu Bay, Xiapu County, Ningde City, Fujian Province, China. The closed aquaculture vessel of GuoXin No. 1 was transferred to the sea areas of Ningde in Fujian province, Zhoushan in Zhejiang province, and Qingdao in Shandong province to maintain the best growth conditions.

Large yellow croaker (340 ± 10 g) were obtained from Guoxin (Taizhou) aquaculture farm in Ningde City, Fujian Province. The test fish were incubated in an indoor hatchery and cultured in open cages in the inshore waters of the inner bay. Before the experiment, according to the initial density of 12.1 kg/m^3, 30,000 healthy large yellow croakers were randomly put into three 280 m^3 breeding tanks with 10,000 fish per tank. In addition, two traditional cages (10.5 m $*$ 7.0 m $*$ 6.0 m) were set up for the comparative test. The density of the traditional cage was subject to the actual production. The initial density of 5.06 kg/m^3 was used to put 6700 large yellow croaker of the same specification per cage. The large yellow croakers spent three weeks acclimatizing to the experimental environment before the experiment officially began. A period of respite, no feeding was performed for the first three days, and the feeding rate gradually increased to the normal level as time passed.

The mid-trial ship "Guoxin 101" was transformed from an offshore bulk carrier, with a total length of 96.8 m, a type of width of 16.6 m, a displacement of 5491 t (Figure 1A,B), a power of 2×662 kW, and a speed of 10.5 kn. It was equipped with three nearly quadrilateral breeding tanks with a volume of about 280 m^3 (Figure 1C,D; 8.8 m $*$ 7.8 m $*$ 5.15 m; the water level was 4.1 m). There was a movable awning on the upper opening. The marine aquaculture system was mainly composed of water circulation, oxygenation, illumination, sewage, and monitoring and alarm systems.

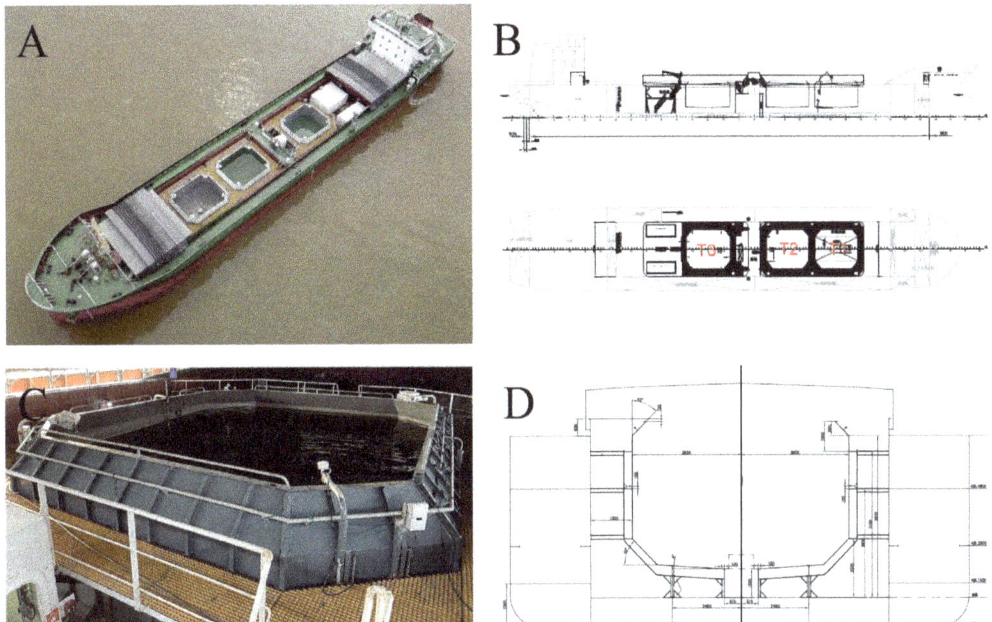

Figure 1. Gouxin No 101 closed aquaculture vessels. (**A,B**) represent overall photographs and schematic diagrams of the ship, respectively; T1, T2, and T3 are represented as the three different tanks in the ship; (**C,D**) are field pictures and schematic diagrams of the culture tank.

2.2. Experimental Procedures

In this experiment, the ship adopted flow-through systems where the aquaculture water was taken from 4 m underwater. The systems were carried out by water intake from the bottom of the ship and drainage from the starboard side. During the culturing period, the water change rate of the tank was maintained 16 times per day, and water flow was maintained at 0.2–0.4 m/s. The oxygen cone pressure method was adopted to dissolve pure oxygen into the breeding water to keep the dissolved oxygen above 8 mg/L, and no air was added in the breeding chamber. The light intensity was controlled at 300–500 Lux, and the photocycle was set to 12D: 12L from 5:30 am to 5:30 pm. In this study, the commercial special feed (main nutrients: crude protein $\geq$ 42%, crude fat $\geq$ 6%, crude fiber $\leq$ 5%, ash $\leq$ 18%, moisture $\leq$ 12%, total phosphorus $\geq$ 1.2%, lysine $\geq$ 2.2%) was the feed that was used for large yellow croakers. The fish were feed once in the morning and evening (6:00, 17:00). The daily standard feeding amount was 1.2–1.5% of the body weight of the fish and adjusted at any time according to the changes in water temperature, weather, and water quality, feeding behavior and residual bait situation, to reduce the residual bait amount as far as possible. Every two weeks, the underwater robot was used to clean the bottom of the tanks to ensure the cleanliness of the culturing environment. During ship culture, the noise pressure level of the tanks was recorded between 119–125 dB under mooring conditions and 132–134 dB under sailing conditions by using a hydrophone (C55, Cetacean Research Technology, Seattle, Washington, DC, USA). The planned experimental period was from June to August 2021. The caged group of large yellow croakers were kept according to cage culture rules for normal feeding.

2.3. Determination of Water Quality

During the trophic period, water temperature, salinity, dissolved oxygen, and pH were measured by an automated water quality analyzer (YSI-556, Beverly, MA, USA) once a day. The total ammonia nitrogen was measured once a week using a hash water quality detector (NA8000, Beverly, MA, USA).

The total number of bacteria and vibrio in the water of the tank in ship was measured by the plate counting method [20] and tested once a week. The TSA and TCBS (Sangon Biotech, Co., Ltd., Shanghai, China) was selected as medium. Water samples were collected in the culture tanks, diluted 100 times with sterile seawater, and then coated with 100 μL on the medium and placed at 28 °C for 24 h. Finally, the plate was observed and counted.

2.4. Experimental Sampling

For the period of the test, the number of dead fish and daily food intake of large yellow croaker of two groups per day were recorded to count the survival rate and the feed conversion ratio. A total of thirty fish were randomly collected from each tank in the ship every 15 days, and their body weight and body length were measured. At the same time, their visceral fat was dissected and weighed. After the experiment, large yellow croaker of two groups was starved for 24 h, thirty fish were randomly collected from each tank and cage to measure body weight, body length, and weight of visceral fat. Then, ten fish in each tank and cage were obtained with skinless muscle samples, which were placed at −80 °C for nutrient detection.

2.5. Muscle Chemical Analysis

According to AOAC method [21], the contents of total nitrogen, crude protein, moisture, and ash were analyzed. The moisture of the sample was weighed after desiccating to a constant weight at 105 °C in an oven (DHG-9240A, Keelrein Instrument Co., Ltd., Shanghai, China). The ash content was indicated after samples were incinerated in a muffle furnace (SX2-12-12N, Jinwen Instrument Co., Ltd., Nanjing, China) at 550 °C for 12 h. The crude protein was analyzed by single acid digestion in the Kjeldhal apparatus (Kelplus DISTYL-BS, Pelican Equipment Pvt. Ltd., Chennai, India). After extraction with chloroform-methanol (2:1, v/v), the crude lipids were measured following previously methods as indicated [22,23].

2.6. Data Calculation and Statistical Analyses

The parameters were determined based on the following calculations:

Survival rate (SR, %) = (final quantity of fish)/(initial quantity of fish) * 100%;
Weight gain rate (WGR, %) = (Wt − Wi)/Wi * 100%;
Feed conversion ratio (FCR, %) = feed intake/(Wt − Wi) * 100%;
Visceral index (VSI, %) = visceral weight/body weight * 100%;
Condition factor (CF, g/cm^3) = body wet weight/body length^3 * 100%;
Stocking density (SD, kg/m^3) = (quantity of fish * body wet weight)/volume of culture;

where Wi and Wt referred to the initial body weight and final body weight of large yellow croaker, respectively.

SPSS version 24.0 (IBM, Armonk, New York, NY, USA) was used for the statistical analyses. Data are expressed as the mean ± SD. Different groups were considered as significant at $p < 0.05$ by T-test and a one-way analysis of variance (ANOVA) and Duncan multiple comparative analysis. GraphPad Prism 8 was utilized to create plots of the data.

3. Results

3.1. Culturing Sites and Transfer Routes

The transfer route and anchorage seas of the ship Gouxin 101 and the site of the cage are shown in Figure 2. The ship could obtain the best culturing conditions by moving according to the demand, and the cage was fixed in the Sandu Bay. At the end of May, the temperature of water rose to more than 25 °C in Ningde Sea area in Fujian Province rapidly, and the aquaculture ship shifted from Sandu Bay to the Zhoushan Sea area in Zhejiang Province. Due to the occurrence of the red tide in anchorage seas of Lvhua Island in Zhoushan during the culturing, it was transferred to Peach Blossom Island for a short time. In mid-July, to avoid super typhoon "Yanhua", it moved to the waters off Qingdao, Shandong Province. While in Qingdao, due to short-term gale weather, the ship was temporarily transferred to the inner Bay of Jiaozhou Bay.

Figure 2. Location of experiment area in traditional aquaculture cages (marked with an orange star) and the anchorage location (marked with a red dot) and sailing route (marked with red lines) of the offshore aquaculture ship <http://www.gissky.net/map/ChinaZQYW.html, accessed on 8 November 2022>.

3.2. Parameters of Culture Environment

The daily average dissolved oxygen, temperature, salinity, pH, and weekly average total ammonia nitrogen of water in the ship and cage are recorded in Figure 3. During the experiment, the temperature of the cage increased steadily and kept at a relatively high level on the whole, ranging from 23.3 to 28.5 °C, 26.6 °C on average. The water temperature of the ship remained stable for the first 5 weeks, below the cage temperature. After the ship was moved to the sea area of Qingdao due to the typhoon, the temperature increased significantly and remained until the end of the test, similar to that of the cage group. The temperature of water in ship was maintained between 21.5 and 28.5 °C, with an average of 25.5 °C during culturing (Figure 3A). In the cage group, the salinity did not change significantly during the 8-week experiment, staying at 25–30‰, with an average of 29.4‰. The salinity of the ship increased steadily in the first 6 weeks from 20.7‰ to 29.1‰, which was lower than the cage during this period. It remained stable at around 30‰ for the next 2 weeks until the experiment was completed, when the salinity was similar to that of the cage group. The whole process was maintained between 20.7 and 31.8‰, with an average of 26.9‰ (Figure 3B). In addition, the level of dissolved oxygen on the ship was higher than that in the cage most of the time (Figure 3C). There was a similar pH between the two groups during the trial (Figure 3D). The ammonia nitrogen concentration of the two groups remained at a low level during the experiment, both of which were lower than 0.08 mg/L (Figure 3E). In this trial, environmental conditions in the ship group were harsh, including higher waves and stronger winds (Supplemental Data, Figure S1).

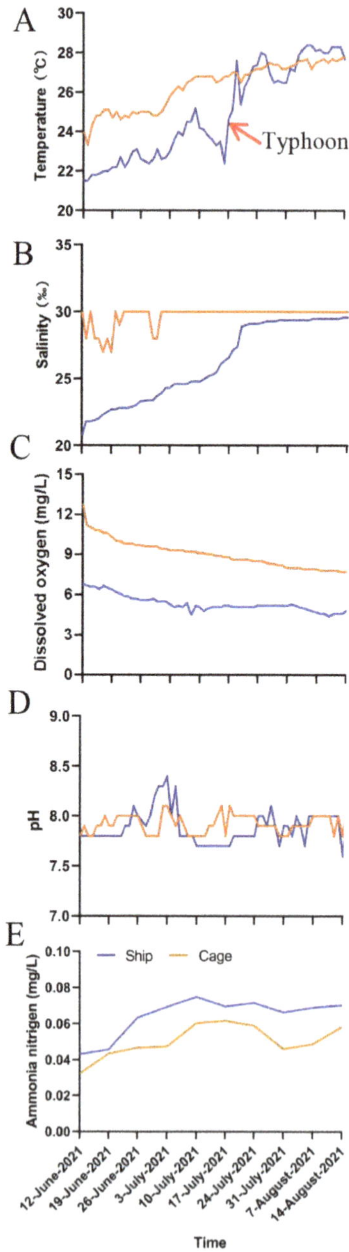

Figure 3. Comparison of water temperature (**A**), salinity (**B**), dissolved oxygen (**C**), pH (**D**), and ammonia nitrogen (**E**) in the ship and cages throughout the experiment period.

3.3. Total Number of Bacteria and Vibrio in Water

The total number of bacteria and vibrio in the tank of the ship were detected by the plate method, which remained stable at a low level and tended to increase slightly with the increase of breeding time. Still, there was no significant difference (Figure 4). The total number of bacteria was maintained at $1.2 * 10^3$–$1.6 * 10^3$ CFU/mL, and the total number of vibrio was maintained at $1.2 * 10^2$–$1.8 * 10^2$ CFU/mL.

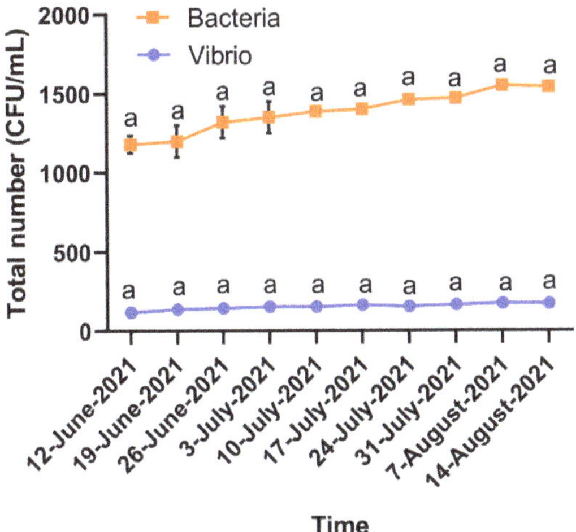

Figure 4. Changes in the total number of bacteria and vibrio in the fish tank of offshore aquaculture ship throughout the experiment period. Bars with different superscripts on the same microorganism are statistically different ($p < 0.05$, n = 3).

3.4. State of Behavior

Through daily observations, the behaviors of the large yellow croaker during swimming and feeding on the ship during the experiment were reported (Figure 5). Large yellow croaker can adapt to the ship environment, actively feed, and exhibit no irregular swimming behavior. Large yellow croaker were usually distributed in the 3–4 m water layer and swam in the reverse current. When feeding, all the large yellow croaker moved upstream to the surface of the water, in a relatively disorderly swimming.

3.5. Survival Rate and Growth Performance

Through almost two months of on-site culturing, the change in the survival of the large yellow croaker was well documented (Figure 6). The result showed that no large-scale death events and serious disease conditions of the fish occurred in the two groups. However, the mortality rate of large yellow croaker in the cage group was relatively high, with the highest single-day mortality rate of 0.23%, which was higher than the highest single-day mortality rate of 0.05% in the ship group.

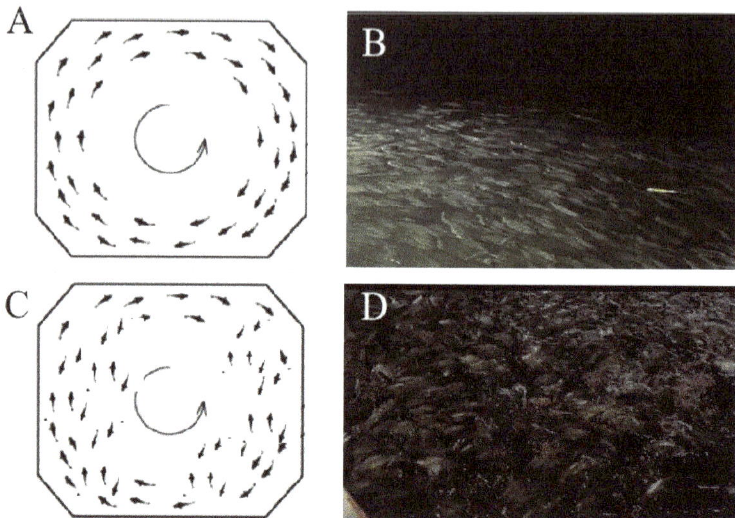

Figure 5. Swimming behavior of the large yellow croaker in the tank of the ship. (**A,B**) represent pictures and schematic images of the large yellow croaker colonies swimming against the current. (**C,D**) represent pictures and schematic diagrams of large yellow croakers swimming loosely while feeding, respectively.

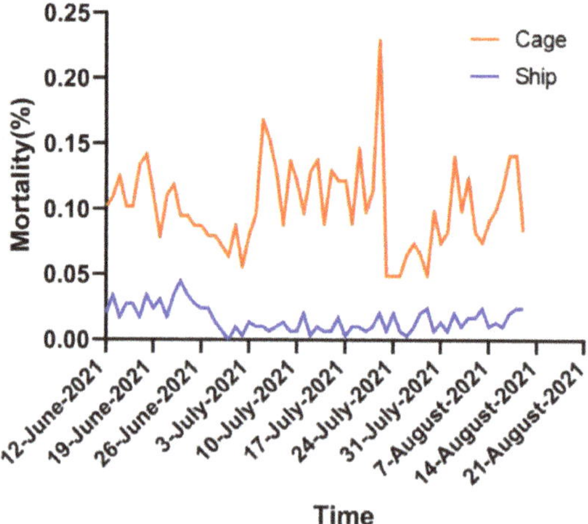

Figure 6. Daily mortality in ranched large yellow croaker in the nearshore cages (denoted in orange) and offshore ship (denoted in blue).

During the culture period, it was found that large yellow croakers in three tanks on the ship grew and developed well through the measured changes in body weight, body length, CF, and VSI (Figure 7). The body weight of large yellow croakers increased from 325.67 g to 460.71 g, with an average monthly weight gain of 67.52 g (Figure 7A). The body length of the large yellow croaker increased rapidly from 25.23 cm to 29.14 cm (Figure 7B). At the beginning of the experiment, the CF and VSI of the large yellow croaker were the highest. With the increase in culturing time, the VSI and CF decreased rapidly in the first 4 weeks and remained stable in the last few weeks (Figure 7C,D).

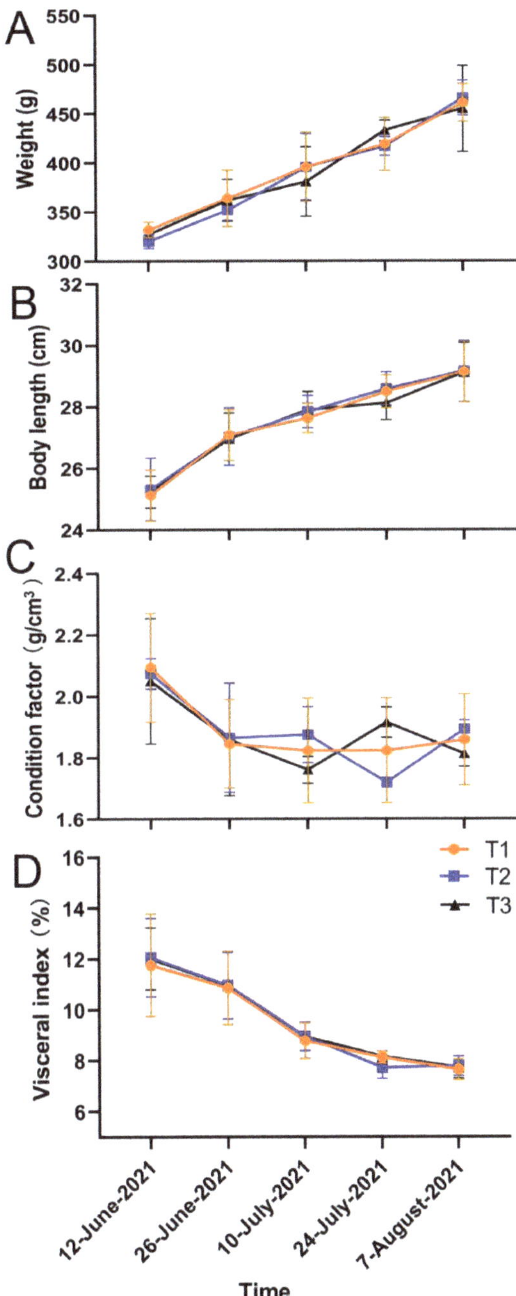

Figure 7. Body weight (**A**), body length (**B**), condition factor (**C**), and visceral index (**D**) in ranched large yellow croaker in the different tanks in an offshore aquaculture ship throughout the experiment period (n = 30); T1, T2, and T3 are represented as the three different tank.

At the end of the experiment, the survival rate (SR), weight gain rate (WGR), feed factor (FCR), condition factor (CF), and visceral index (VSI) of the large yellow croaker were significantly different under different aquaculture modes (Table 1). The SR, WGR, and CF of large yellow croakers in the ship group were significantly higher than in the cage group ($p < 0.05$). The FCR and VSI of the ship group were significantly lower than those of the cage group ($p < 0.05$). During 8 weeks of culturing, the density of the ship group increased rapidly from 12.1 kg/m^3 to 15.74 kg/m^3, while that of the cage group decreased to only 5.02 kg/m^3.

Table 1. Survival rate (SR, %), weight gain rate (WGR, %), feed conversion ratio (FCR, %), condition factor (CF, g/cm^3), visceral index(VSI, %), and stocking density (SD, kg/m^3) of the large yellow croaker that were reared under the different farming environments.

Group	SR (%)	WGR (%)	FCR (%)	CF (g/cm^3)	VSI (%)	SD (kg/m^3)
Cage	93.82 ± 1.96 [a]	9.11 ± 4.95 [a]	5.43 ± 2.89 [a]	1.32 ± 0.19 [a]	7.61 ± 0.21 [a]	5.02 ± 0.38
Ship	99.02 ± 0.20 [b]	41.48 ± 2.48 [b]	1.61 ± 1.06 [b]	1.85 ± 0.14 [b]	6.71 ± 0.15 [b]	15.74 ± 0.25

Note: Values represent the mean ± SD. Significant differences ($p < 0.05$, n = 30) among treatments were indicated by different letters.

3.6. Muscle Composition Analysis

The nutrient composition of the muscle of the large yellow croaker was significantly affected by the culture mode (Table 2). The contents of crude protein and ash in the ship group were significantly higher than those in the cage group ($p < 0.05$), and the ether extract and moisture contents were significantly lower than those of the cage group ($p < 0.05$). In conclusion, the muscle nutrient composition of the large yellow croaker in the ship group was better than in the cage group.

Table 2. Nutrient composition in the muscle tissues of the large yellow croaker that were reared under the different experimental groups.

Group	Mass Fraction (%)			
	Tissue Moisture	Ash	Crude Lipid	Crude Protein
Cage	76.76 ± 0.67 [a]	1.16 ± 0.03 [a]	7.53 ± 0.70 [a]	17.13 ± 0.40 [a]
Ship	71.14 ± 0.43 [b]	1.22 ± 0.03 [b]	5.26 ± 0.47 [b]	21.53 ± 1.29 [b]

Note: Values represent mean ± SD. Significant differences ($p < 0.05$, n = 10) among treatments were indicated by different letters.

4. Discussion

4.1. Comparison of Nearshore and Offshore Aquaculture

In China, mariculture is mainly concentrated nearshore and provides an important economic source for local fishermen [24]. However, with the general deterioration of the nearshore environment, the Chinese government has begun to reduce the area that is used for nearshore aquaculture [19]. On the other hand, there are nearly 70,000 km^2 of almost undeveloped offshore waters in China. The lack of shelter, such as islands in the offshore area, means more severe sea conditions, such as stronger winds, higher waves, and faster currents. This poses a greater challenge to the aquaculture system [25]. This study shows that the ship-platform can resist the more severe offshore sea conditions during the experiment without adverse effects on aquaculture and could ensure offshore aquaculture. In addition, the ship can avoid events such as typhoons, red tides, and other special disaster situations, through the way of autonomous navigation to escape. This also reduces the operational risk for producers. Open cage culture was vulnerable to changes in aquaculture environmental conditions that were caused by climate change, especially water temperature and dissolved oxygen, which have a significant impact on the production of cultured fish [26,27]. For example, due to the influence of high and low temperatures, the optimal growth time of large yellow croaker was only about half a year. In addition, the ship also realized the rapid growth of the large yellow croaker in the summer high-temperature period by transferring the culture in the sea area between Fujian, Zhejiang, and Shandong, which verified the feasibility of shortening the culture cycle. In conclusion, by taking advantage of the stability of the ship platform, the aquaculture ship can expand the space of offshore and move independently to avoid disaster weather and obtain an appropriate water temperature, thus increasing the stability of production.

Traditional nearshore cages use little mechanical equipment, the labor intensity of fishermen was large, and the risk of the marine operation was also high [28]. In contrast, although the cost of large machinery equipment was higher, it can reduce the unit cost of culture through large-scale operation and reduce the number of staff and work intensity. For example, the aquaculture ship uses mechanized feeding and sucking fish instead of manual operation, greatly reducing the use of labor. It was estimated that the production of large yellow croaker could reach 3000 tons per year by the 100,000-ton aquaculture ship, and the culture staff only needs 9–10 people.

4.2. Effects of Aquaculture Mode on Survival and Growth Performance of Large Yellow Croaker

Promoting the healthy growth of aquaculture animals is one of the most important objectives of aquaculture practitioners. Generally, it could be comprehensively evaluated through body weight indexes, body length indexes, key organ indexes, feed efficiency, etc. [29]. In this study, it was found that the large yellow croaker in the ship group gained more weight and died less during the test period. The survival and growth indexes were significantly better than those in the nearshore cage group at the end of the experiment. Similarly, Nicole et al. showed that compared with the traditional nearshore culture areas, the mortality rate of bluefin tuna in offshore culture areas was lower, and the prevalence of parasites was reduced by 30% [9]. Pang et al. showed that large, mechanized platforms offshore were more suitable for medium- and large-size abalone farming, and their growth rate was significantly faster than that of nearshore cages [30].

Previous studies have shown that the difference in water quality parameters that are caused by different farming modes was an important reason affecting the survival and growth of fish [31,32]. In this test, the water quality parameters of the ship and cage were the same except for temperature and dissolved oxygen. When large yellow croaker are exposed to high temperature for a long time, it causes an oxidative stress reaction, adversely affecting digestion, metabolism, and the immune system, and reduce its resistance to pathogens [33]. This study showed that the area of the cage was affected by high-temperature in summer, resulting in rising water temperature, which was higher than the optimal 22–26 °C water temperature for large yellow croaker for a long time [34]. However, the ship can keep the fish at optimum temperatures most of the time by moving. Therefore, in this study, high temperature may be an important reason for the high mortality and slow growth rate of the large yellow croaker in the cage group. In addition, studies have shown that very low dissolved oxygen (<2 mg/L) will cause large area death of large yellow crocks in a short time, and long-term low levels of dissolved oxygen will restrict aerobic metabolism, resulting in the slow growth of fish [35]. In this study, it was found that the water dissolved oxygen level was at a low level for a long time due to the high temperature around the cage. At the same time, the ship culture system could maintain the dissolved oxygen level above 8 mg/L. Therefore, low dissolved oxygen may also affect the survival and growth of cage large yellow croaker.

4.3. Effects of Aquaculture Mode on Nutrient Composition of Large Yellow Croaker

It was generally accepted that muscle quality was a complex set of characteristics, mainly including muscle sensory properties, nutritional value, and freshness [36,37]. Among them, the nutritional value was the most concern. The nutritional value of fish muscle was measured by various components, such as protein and fat, as well as ash and moisture content [38]. This study showed that the nutrient composition of the muscle of the large yellow croaker cultured in ship was better than that of the cage, mainly due to the higher protein and lower fat content. Due to the unique plasticity of skeletal muscle, it was prone to be seriously affected by exogenous factors. In the case of continuous swimming, important biochemical changes occurred in skeletal muscle, leading to changes in muscle nutrient composition [39,40]. There is a lot of research data showing that moderate exercise training can significantly improve the protein content of *Spinibarbus sinensis*, *Barbodes schwanenfeldi*, *Schizothorax prenanti*, and *Mylopharyngodon piceus*, and can also significantly reduce the lipid content of fish muscle [41–44]. It was also found in this study that large yellow croaker could keep swimming under stable flow field conditions (0.2–0.4 m/s), which was consistent with the phenomenon that was observed in the experiment. The cage was also located in the inner Bay Sea area, the water fluidity was poor, and difficult to form a continuous flow field. Therefore, flow velocity may be an important reason for the difference in muscle nutrient composition of the large yellow croaker. In addition, it was found in the experiment that the food intake of the large yellow croaker in the cage was low, and it failed to obtain enough nutrients to synthesize energy substances in the body, which may be another important reason affecting the nutrient composition of the large yellow croaker muscle.

5. Conclusions

In conclusion, we first found that the aquaculture ship was suitable for the breeding of the large yellow croaker, and its growth performance and nutrient composition were better than the cage. The result provides important basic support for the research and promotion of the aquaculture ship. However, the specific impact of environmental conditions on the large yellow croaker was unclear. Further research is necessary to evaluate the regulatory mechanisms of growth and provide a basis for the improvement of the management of fish in ships.

Supplementary Materials: The following supporting information can be downloaded at: https://www.mdpi.com/article/10.3390/jmse11010101/s1, Figure S1: Wave height (A) and wind scale (B) in the nearshore cages (denoted in orange) and offshore ship (denoted in blue) throughout the experiment period.

Author Contributions: Conceptualization, Y.Y., H.L. and M.C.; Data curation, W.H.; Funding acquisition, M.C.; Investigation, W.H.; Methodology, Y.Y., W.H. and H.L.; Project administration, H.L.; Software, M.C.; Validation, M.C.; Writing—original draft, Y.Y.; Writing—review and editing, F.Y., H.L. and M.C. All authors have read and agreed to the published version of the manuscript.

Funding: This research was funded by the National Key Research and Development Program of China (Grant No.2022YDF2401101), Program of Qingdao National Laboratory for Marine Science and Technology (Grant No.2021WHZZB1301), Key Research and Development Program of Shandong Province (Grant No.2021SFGC0701), and Central Public Interest Scientific Institution Basal Research Fund, YSFRI, Chinese Academy of Fisheries Science (Grant No.2021YJS005).

Institutional Review Board Statement: The animal study protocol was approved by the Ethics Committee of the Fishery Machinery and Instrument Research institute, Chinese Academy of Fishery Sciences (protocol code: YJS20210324NM and date of approval: 20210324).

Informed Consent Statement: Not applicable.

Data Availability Statement: Not applicable.

Acknowledgments: Thanks to Xuyang Jiang, Guangwei Meng, and Linlin Sun (Conson CSSC (Qingdao) Ocean Technology Co., Ltd., Qingdao, China) for help in the experiment. Special thanks to Shixian Huang for help in making the figures and tables.

Conflicts of Interest: The authors declare no conflict of interest.

References

1. FAO. *The State of World Fisheries and Aquaculture 2020: Sustainability in Action*; FAO: Rome, Italy, 2021.
2. World Bank. *Fish to 2030: Prospects for Fisheries and Aquaculture*; FAO: Rome, Italy, 2013.

3. Di Trapani, A.M.; Sgroi, F.; Testa, R.; Tudisca, S. Economic comparison between offshore and inshore aquaculture production systems of European sea bass in Italy. *Aquaculture* **2014**, *434*, 334–339. [CrossRef]

4. Davies, I.P.; Carranza, V.; Froehlich, H.E.; Gentry, R.R.; Kareiva, P.; Halpern, B.S. Governance of marine aquaculture: Pitfalls, potential, and pathways forward. *Mar. Pol.* **2019**, *104*, 29–36. [CrossRef]

5. Chu, Y.I.; Wang, C.M.; Park, J.C.; Lader, P.F. Review of cage and containment tank designs for offshore fish farming. *Aquaculture* **2020**, *519*, 734–752. [CrossRef]

6. Li, L.; Jiang, Z.; Ong, M.C.; Hu, W. Design optimization of mooring system: An application to a vessel-shaped offshore fish farm. *Eng. Struct.* **2019**, *197*, 109363. [CrossRef]

7. Lester, S.E.; Gentry, R.R.; Kappel, C.V.; White, C.; Gaines, S.D. Opinion: Offshore aquaculture in the United States: Untapped potential in need of smart policy. *Proc. Natl. Acad. Sci. USA* **2018**, *115*, 7162–7165. [CrossRef]

8. Kim, D.H.; Lipton, D.; Choi, J.Y. Analyzing the economic performance of the red sea bream *Pagrus major* offshore aquaculture production system in Korea. *Fish. Sci.* **2012**, *78*, 1337–1342. [CrossRef]

9. Aarset, B.; Carson, S.G.; Wiig, H.D.; Maren, I.E.; Marks, J. Lost in translation? Multiple discursive strategies and the interpretation of sustainability in the Norwegian salmon farming industry. *Food Ethics* **2020**, *5*, 1–21. [CrossRef]

10. Cao, L.; Wang, W.; Yang, Y.; Yang, C.T.; Yuan, Z.H.; Xiong, S.B.; Diana, J. Environmental impact of aquaculture and countermeasures to aquaculture pollution in China. *Environ. Sci. Pollut. Res. Int.* **2007**, *14*, 452–462. [PubMed]

11. Schupp, M.F.; Kafas, A.; Buck, B.H.; Krause, G.; Onyango, V.; Stelzenmueller, V.; Davies, I.; Scott, B.E. Fishing within offshore wind farms in the north sea: Stakeholder perspectives for multi-use from scotland and germany. *J. Environ. Manag.* **2021**, *279*, 111762. [CrossRef]

12. El-Thalji, I. Context analysis of offshore fish farming. *IOP Conf. Ser. Mater. Sci. Eng.* **2019**, *700*, 012065. [CrossRef]

13. Kirchhoff, N.T.; Rough, K.M.; Nowak, B.F. Moving cages further offshore: Effects on southern bluefin tuna, *T. maccoyii*, parasites, health and performance. *PLoS ONE* **2011**, *6*, e23705. [CrossRef] [PubMed]

14. Bjelland, H.V.; Føre, M.; Lader, P.; Kristiansen, D.; Holmen, I.M.; Fredheim, A.; Grøtli, E.I.; Fathi, D.E.; Oppedal, F.; Utne, I.B.; et al. *Exposed Aquaculture in Norway*; IEEE: New York, NY, USA, 2015; pp. 1–10.

15. Holmer, M. Environmental issues of fish farming in offshore waters: Perspectives, concerns and research needs. *Aquac. Environ. Interact.* **2010**, *1*, 57–70. [CrossRef]

16. Goseberg, N.; Chambers, M.D.; Heasman, K.; Fredriksson, D.; Fredheim, A.; Schlurmann, T. Technological approaches to long-line and cage-based aquaculture in open ocean environments. In *Aquaculture Perspective of Multi-Use Sites in the Open Ocean*; Springer: Berlin/Heidelberg, Germany, 2017; pp. 71–95.

17. Soner, B.; Volkan, K.; Ash, M.B. Floating fish farm unit. Is it an appropriate method for salmonid production. *Mar. Sci. Technol. Bull.* **2013**, *2*, 9–13.

18. Sun, P.; Bao, P.; Tang, B. Transcriptome analysis and discovery of genes involved in immune pathways in large yellow croaker (*Larimichthys crocea*) under high stocking density stress. *Fish Shellfish Immunol.* **2017**, *68*, 332–340. [CrossRef]

19. Gui, J.; Tang, Q.; Li, Z.; Liu, J.; Silva, D. *Aquaculture in China: Success Stories and Smodern Trends*; John Wiley & Sons: Hoboken, NJ, USA, 2018; pp. 297–308.

20. Li, Y.; Yang, C.Y.; Zheng, T.L. Bacterial survival modes and community characteristics in natural environment. *Chin. J. App. Environ. Biol.* **2013**, *19*, 553–560. [CrossRef]

21. AOAC. *Official Methods of Analysis of the Association of Official Analytical Chemists*, 17th ed.; Association of Official Analytical Chemists: Arlington, VA, USA, 2003.

22. Folch, J.; Lees, M.; Sloan Stanley, G.H. A simple method for the isolation and purification of total lipids from animal tissues. *J. Biol. Chem.* **1957**, *226*, 497–509. [CrossRef] [PubMed]

23. Barnes, H.; Black Stock, J. Estimation of lipids in marine animals and tissues detailed investigation of the Sulphophosphovanillin method for total lipids. *J. Exp. Mar. Biol. Ecol.* **1973**, *12*, 103–118. [CrossRef]

24. Chen, Y.; Huang, W.; Shan, X.; Chen, J.; Weng, H.; Yang, T.; Wang, H. Growth characteristics of cage-cultured large yellow croaker *Larimichthys crocea*. *Aquac. Rep.* **2020**, *16*, 100242. [CrossRef]

25. Soto, D.; Wurmann, C. Offshore aquaculture: A needed new frontier for farmed fish at aea. In *The Future of Ocean Governance and Capacity Development*; Brill Nijhoff: Santiago, Chile, 2019; pp. 379–384.

26. Huang, W.; Wang, N. Discussion on relationship between three kinds of parasitic disease outbreaks of large yellow croaker and the water temperature. *Heilongjiang Agricult. Sci.* **2012**, *12*, 71–73. (In Chinese)

27. Dan, X.M.; Yan, G.J.; Zhang, A.J.; Cao, Z.D.; Fu, S.J. Effects of stable and dielcycling hypoxia on hypoxia tolerance, postprandial metabolic response, and growth performance in juvenile qingbo (*Spinibarbus sinensis*). *Aquaculture* **2014**, *428*, 21–28. [CrossRef]

28. Klinger, D.H.; Levin, S.; Watson, J. The growth of finfish in global open-ocean aquaculture under climate change. *Proc. R. Soc. B Biol. Sci.* **2017**, *284*, 20170834. [CrossRef]

29. Du, Z.Y.; Turchini, G.M. Are we actually measuring growth?-An appeal to use a more comprehensive growth index system for advancing aquaculture research. *Rev. Aquac.* **2022**, *14*, 525–527. [CrossRef]

30. Pang, G.W.; Gao, X.L.; Hong, J.W.; Luo, X.; Wu, Y.S.; You, W.W.; Ke, C.H. Growing abalone on a novel offshore platform: An on-site test of the effects of stocking density and diet. *Aquaculture* **2022**, *549*, 737769. [CrossRef]

31. Saravanan, S.; Geurden, I.; Figureueiredo-Silva, A.C.; Nusantoro, S.; Kaushik, S.; Verreth, J.; Schrama, J.W. Oxygen consumption constraints food intake in fish fed diets varying in essential amino acid composition. *PLoS ONE* **2013**, *8*, e72757. [CrossRef] [PubMed]

32. Zhang, X.; Cai, X.; Liu, X.; Zhao, H.; Song, P.; Zhang, J. Proteomics of liver tissue of large yellow croaker (*Larimichthys crocea*) under high temperature stress. *J. Fish. China* **2021**, *45*, 862–870. (In Chinese)

33. Cai, X.; Zhang, J.; Lin, L.; Li, Y.; Liu, X.; Wang, Z. Study of a noninvasive detection method for the high-temperature stress response of the large yellow croaker (*Larimichthys crocea*). *Aquac. Rep.* **2020**, *18*, 100514. [CrossRef]

34. Dabruzzi, T.F.; Fangue, N.A.; Kadir, N.N.; Bennett, W.A. Thermal niche adaptations of common mudskipper (*Periophthalmus kalolo*) and barred mudskipper (*Periophthalmus argentilineatus*) in air and water. *J. Therm. Biol.* **2019**, *81*, 170–177. [CrossRef]

35. Ding, J.; Liu, C.; Luo, S.Y.; Zhang, Y.B.; Gao, X.M.; Wu, X.F.; Shen, W.L.; Zhu, J.Q. Transcriptome and physiology analysis identify key metabolic changes in the liver of the large yellow croaker (*Larimichthys crocea*) in response to acute hypoxia. *Ecotoxicol. Environ. Saf.* **2020**, *189*, 109957. [CrossRef]

36. Grigorakis, K. Compositional and organoleptic quality of farmed and wild gilthead sea bream (*Sparus aurata*) and sea bass (*Dicentrarchus labrax*) and factors affecting it: A review. *Aquaculture* **2007**, *272*, 55–75. [CrossRef]

37. Rasmussen, R.S. Quality of farmed salmonids with emphasis on proximate composition, yield and sensory characteristics. *Aquac. Res.* **2001**, *32*, 767–786. [CrossRef]

38. Zhu, Z.; Song, B.; Lin, X.; Xu, Z. Effect of sustained training on glycolysis and fatty acids oxidation in swimming muscles and liver in juvenile tinfoil barb *Barbonymus schwanenfeldii* (Bleeker, 1854). *Fish Physiol. Biochem.* **2016**, *42*, 1807–1817. [CrossRef] [PubMed]

39. Palstra, A.P.; Mes, D.; Kusters, K.; Roques, J.A.C.; Flik, G.; Kloet, K.; Blonk, R.J.W. Forced sustained swimming exercise at optimal speed enhances growth of juvenile yellowtail kingfish (*Seriola lalandi*). *Front. Physiol.* **2014**, *5*, 506. [CrossRef] [PubMed]

40. Palstra, A.P.; Planas, J.V. Fish under exercise. *Fish Physiol. Biochem.* **2011**, *37*, 259–272. [CrossRef] [PubMed]

41. Li, X.M.; Yu, L.J.; Wang, C.; Zeng, L.Q.; Cao, Z.D.; Fu, S.J.; Zhang, Y.G. The effect of aerobic exercise training on growth performance, digestive enzyme activities and postprandial metabolic response in juvenile qingbo (*Spinibarbus sinensis*). *Comp. Biochem. Physiol. A Mol. Integr. Physiol.* **2013**, *166*, 8–16. [CrossRef] [PubMed]

42. Song, B.; Lin, X.; Xu, Z. Effects of upstream exercise training on feeding efficiency, growth and nutritional components of juvenile tinfoil barbs (*Barbodes schwanenfeldi*). *J. Fish. China* **2012**, *36*, 106–114. [CrossRef]

43. Liu, G.; Wu, Y.; Qin, X.; Shi, X.; Wang, X. The effect of aerobic exercise training on growth performance, innate immune response and disease resistance in juvenile *Schizothorax prenanti*. *Aquaculture* **2018**, *486*, 18–25. [CrossRef]

44. Harimana, Y.; Tang, X.; Le, G.; Xing, X.; Zhang, K.; Sun, Y.; Li, Y.; Ma, S.; Karangwa, E.; Tuyishimire, M.A. Quality parameters of black carp (*Mylopharyngodon piceus*) raised in lotic and lentic freshwater systems. *LWT-Food Sci. Technol.* **2018**, *90*, 45–52. [CrossRef]

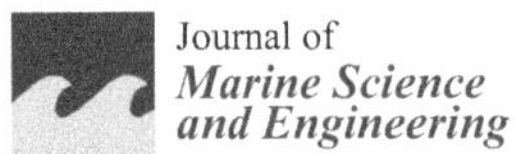

Journal of
Marine Science and Engineering

Article

Impact of Salinity Changes on the Antioxidation of Juvenile Yellowfin Tuna (*Thunnus albacares*)

Shengjie Zhou [1,2,†], Ninglu Zhang [1,2,†], Zhengyi Fu [1,2,3,†], Gang Yu [1,2], Zhenhua Ma [1,2,3,*] and Lei Zhao [4,*]

[1] Key Laboratory of Efficient Utilization and Processing of Marine Fishery Resources of Hainan Province, Sanya Tropical Fisheries Research Institute, Sanya 572018, China

[2] Tropical Aquaculture Research and Development Center, South China Sea Fisheries Research Institute, Chinese Academy of Fishery Science, Sanya 572018, China

[3] College of Science and Engineering, Flinders University, Adelaide 5001, Australia

[4] Yazhou Bay Agriculture and Aquaculture Co., Ltd., Sanya 572025, China

* Correspondence: zhenhua.ma@scsfri.ac.cn (Z.M.); leiforever@hotmail.com (L.Z.)

† These authors contributed equally to this work.

Abstract: To understand the impacts of salinity stress on the antioxidation of yellowfin tuna *Thunnus albacares*, 72 fishes (646.52 ± 66.32 g) were randomly divided into two treatments (32‰ and 29‰) and sampled at four time points (0 h, 12 h, 24 h, and 48 h). The salinity of the control group (32‰) was based on natural filtered seawater and the salinity of the stress group (29‰) was reduced by adding tap water with 24 h aeration to the natural filtered seawater. The superoxide dismutase (SOD), glutathione peroxidase (GSH-Px), and malondialdehyde (MDA) from liver, gill, and muscle tissues were used as the antioxidant indexes in this study. The results showed that the changes of SOD and GSH-Px in the gills were first not significantly different from the control group ($p > 0.05$) and finally significantly higher than the control group (SOD: 50.57%, GSH-Px: 195.95%, $p < 0.05$). SOD activity in fish liver was not significantly changed from 0 h to 48 h ($p > 0.05$), and was not significantly different between the stress group and control group ($p > 0.05$). With the increase in stress time, GSH-Px and MDA activities in the liver of juvenile yellowfin tuna increased first (GSH-Px: 113.42%, MDA: 137.45%) and then reduced (GSH-Px: −62.37%, MDA: −16.90%) to levels similar to the control group. The SOD activity in the white and red muscle of juvenile yellowfin tuna first decreased (white muscle: −27.51%, red muscle: −15.52%) and then increased (white muscle: 7.30%, red muscle: 3.70%) to the level of the control group. The activities of GSH-Px and MDA in white and red muscle increased first (white muscle GSH-Px: 81.96%, red muscle GSH-Px: 233.08%, white muscle MDA: 26.89%, red muscle MDA: 64.68%) and then decreased (white muscle GSH-Px: −48.03%, red muscle GSH-Px: −28.94%, white muscle MDA: −15.93%, red muscle MDA: −28.67%) to the level observed in the control group. The results from the present study indicate that low salinity may lead to changes in the antioxidant function of yellowfin tuna juveniles. In contrast, yellowfin tuna juveniles have strong adaptability to the salinity of 29‰. However, excessive stress may consume the body's reserves and reduce the body's resistance.

Keywords: low salinity stress; antioxidant index; superoxide dismutase; glutathione peroxidase; malondialdehyde

Citation: Zhou, S.; Zhang, N.; Fu, Z.; Yu, G.; Ma, Z.; Zhao, L. Impact of Salinity Changes on the Antioxidation of Juvenile Yellowfin Tuna (*Thunnus albacares*). *J. Mar. Sci. Eng.* **2023**, *11*, 132. https://doi.org/10.3390/jmse11010132

Academic Editor: Dariusz Kucharczyk

Received: 22 November 2022
Revised: 22 December 2022
Accepted: 3 January 2023
Published: 6 January 2023

1. Introduction

Thunnus albacares belongs to Perciformes, Scomberia, Scombrida, commonly known as yellowfin tuna [1]. Yellowfin tuna is an oceanic migratory fish that lives in the vast upper middle waters of tropical, subtropical, and temperate oceans. As one of the high economic value tuna species, yellowfin tuna grows quickly among the tuna species with high flesh quality [2]. It is a preferred species for offshore aquaculture [1,3,4]. Yellowfin tuna has been cultured in Mexico, Panama, and Indonesia [5]. Mexico has established a large number of yellowfin tuna culture bases, but the limited supply of wild fry has seriously restricted the culture industry of this species [6]. The artificial cultivation of yellowfin tuna in China is still at the initial stage. The in-depth marine aquaculture technology and variety development innovation team from the Chinese Academy of Fishery Sciences has achieved indoor circulating water and offshore deep-water cage culture of yellowfin tuna in Lingshui Li Autonomous County, Hainan Province, and carried out a series of studies on aquaculture biology and disease prevention and control [7,8]. Up to now, open published biological data on yellowfin tuna farming are relatively scarce, which restricts the development of its artificial farming. Yellowfin tuna is a migratory fish in the ocean, and its migration route is closely related to salinity [1], and the physical responses of yellowfin tuna to ambient salinity changes are still unclear.

As an ecological factor, ambient salinity has a series of physiological effects on fish. As reported in previous studies, salinity is an important factor regulating fish growth, metabolism, and various physiological activities [9]. Ambient salinity can also directly affect the promotion of aquaculture [10]. An inappropriate salinity range will affect the physiology and biochemistry, immunity, growth and development, feeding, and reproduction of fish, and fish will show different adaptive states under various salinity conditions [11]. Recent studies have shown that salinity changes can cause physiological stress reactions in fish, accompanied by the production of excessive reactive oxygen free radicals, which can lead to oxidative stress reactions that damage the antioxidant defense system of fish [12,13]. Excessive oxygen free radicals produced by oxidative stress will attack biological membranes, proteins, and nucleic acids, causing oxidative damage such as cytoplasmic efflux, enzyme inactivation, and genetic replication errors [14]. Therefore, such processes will disrupt the normal physiological and behavioral activities of fish. Salinity fluctuations exist in coastal areas, with possible causes including heavy rainfall and river injection, which pose a threat to cage- and land-based culture that relies on naturally filtered seawater. The low salinity of 28.5 ‰ has been observed in the coast of Hainan, China [15], but the optimum growth salinity of yellowfin tuna is 31.2~33.3 ‰ [16]. Such temporary environmental fluctuations may pose unknown challenges to yellowfin tuna.

The antioxidant system in fish can resist oxidative damage [17]. Superoxide dismutase (SOD) is an essential antioxidant enzyme in the antioxidant system [18]. SOD is the primary substance for scavenging free radicals in organisms. It decomposes superoxide anion free radicals (O^{-2}) through disproportionation. After scavenging O^{-2}, SOD generates H_2O_2 [19]. Glutathione peroxidase (GSH-Px) is a critical non-enzymatic component [20]. As an antioxidant, GSH-Px can participate in the elimination of toxic peroxides [21]. Malondialdehyde (MDA) is a lipid peroxide, which can indirectly reflect the degree of lipid peroxidation in tissue cells [22]. In the present study, SOD, GSH-Px, and MDA were used as the antioxidant indicators to evaluate the impacts of salinity on the antioxidant system of juvenile yellowfin tuna. The purpose of this study is to understand the physiological response of yellowfin tuna to acute low salt in the process of salinity adaptation, in order to solve the change of seawater salinity reduction in a short time in subgrade culture or cage culture. The results from the present study will add biometric data to the culture of yellowfin tuna.

2. Materials and Methods

2.1. Experimental Fish and Design

The juvenile yellowfin tuna were provided by Lingshui Research Station of Tropical Fisheries Research and Development Center, South China Sea Fisheries Research Institute, Chinese Academy of Fishery Sciences. The length and weight of the experimental fish were 28.97 ± 2.17 cm and 646.52 ± 66.32 g, respectively. A total of 72 fish were randomly distributed into six 5000 L fiberglass tanks with recirculating seawater systems for a 7-day acclimation. The fish were fed daily from 08:30 to 09:00. Fresh miscellaneous fish (4 cm × 2 cm pieces) were fed with 5–8% body weight daily by satiety. No feeding was conducted on the day before and during the experiment. During the experiment, all of the tanks were supplied with filtered seawater with a water exchange rate of 300% tank volume per day. The ambient salinity of 32‰ was used as the control with three replicates, and 29‰ was used as the stress group with three replicates (Figure 1). The salinity of the stress group was gradually adjusted by adding tap water with 24 h aeration to the natural filtered seawater at 1‰ per 1 h. When the salinity of the stress group reached 29‰, the experiment began. The experiment lasted 48 h. The photoperiod was maintained at 14:10 h (light:dark). The salinity, water temperature, DO, and pH were monitored using an HQ40d portable multi-parameter (HQ40d18, Hach, Loveland, CO, USA), and the nitrite nitrogen and ammonia nitrogen were measured by a nitrite nitrogen testing kit (Zhecheng Biotechnology, Beijing, China). During the experimental period, the water temperature was maintained at 29.5 ± 0.5 °C, DO was >7.50 mg·L^{-1}, pH was 7.93 ± 0.12, ammonia nitrogen was <0.1 mg·L^{-1}, and nitrite nitrogen <0.05 mg·L^{-1}.

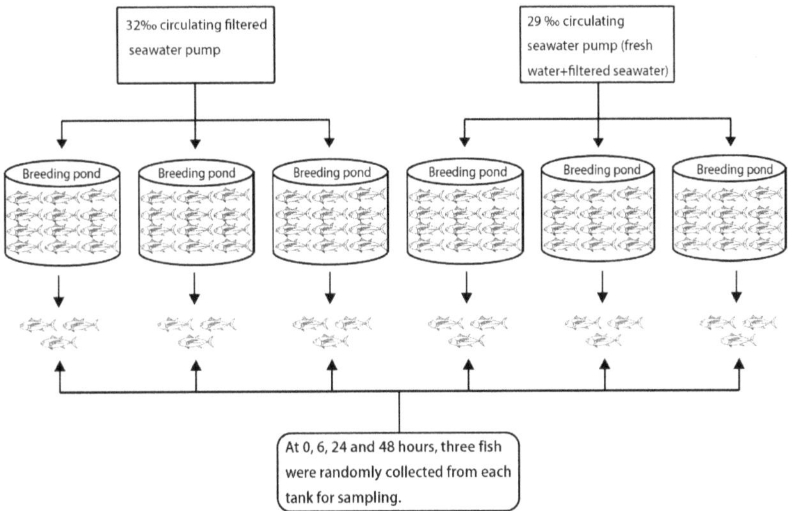

Figure 1. Experimental design of the effect of salinity change on the antioxidant capacity of juvenile yellowfin tuna (*Thunnus albacares*).

2.2. Analytical Method

At 0, 6, 24, and 48 h, three fish from each tank were randomly collected for enzyme activity measurement. The fish were anesthetized with overdose AQUI-S (100 mg L^{-1}, AQUI-S New Zealand Ltd., Lower Hutt, New Zealand) and dissected on an ice tray. The liver, gill, and muscle samples were weighed and immediately stored in liquid nitrogen. For each assay, an individual sample was partially thawed, weighed, and homogenized using a glass homogenizer on ice in 0.2 M NaCl. The suspensions were centrifuged at 2500 r·min^{-1} for 10 min at 2 °C. Then, the supernatant was incubated in the enzyme substrate and read on a spectrophotometer (Synergy H1, BioTek Instruments, Winooski, VT, USA). All of the measurements were conducted in triplicate.

The protein content was determined by the Coomassie Brilliant Blue method with bovine serum protein as the standard used the protein quantitative kit (catalog number A045-4, Nanjing, China), incubated at 37 °C for 30 min at 562 nm wavelength, and the protein concentration measured by microplate colorimetry. The SOD test kit (catalog number A001-3, Nanjing, China) was used to measure the activity of the SOD in the animal tissue samples. The activity of the SOD was determined by the xanthine oxidase method. The absorbance value was measured at the wavelength of 550 nm by colorimetry to calculate its activity. The activity unit was defined as follows: when the SOD inhibition rate reached 50% per milligram of tissue protein in 1 mL of the reaction solution, the corresponding amount of SOD was 1 SOD activity unit (U·mgprot^{-1}). The GSH-Px activity in tissues can be measured with a GSH-Px determination kit (catalog number A005-1, Nanjing, China). The GSH-Px activity was expressed by the consumption rate of GSH in the enzymatic reaction, while the more stable yellow substance formed by GSH and dithiodinitrobenzoic acid can be determined by colorimetry to calculate the GSH-Px activity. Through the colorimetric method, a 1 cm optical path cuvette was used at 412 nm wavelength, the distilled water was adjusted to zero, the absorbance value was measured, and its activity was calculated. The activity unit U represents that the GSH concentration in the reaction system is reduced by 1% per milligram of protein per minute by deducting non-enzymatic reaction $\mu mol \cdot L^{-1}$. The MDA determination kit (catalog number A003-1, Nanjing, China) was used to measure the content of MDA in the animal tissues. The MDA was condensed with thiobarbituric acid to form a red substance, and MDA was determined by colorimetry at 532 nm. The method according to Gan L as well as other methods were used with slight modifications [23,24].

2.3. Statistical Analysis

The measurement of each variable was expressed as a mean of three samples. SPSS 26.0 software (SPSS, Chicago, IL, USA) was used for the statistical analysis. All figures were drawn using the Origin 2019 software (OriginLab Corporation, Northampton, MA, USA), and the data were expressed as mean $\pm$ standard deviation (mean $\pm$ SD). At the same sampling time, an independent T-test was used to compare the difference between the control and the stress groups. The mean values of the specific activities of each enzyme in the stress group between the sampling times were compared with one-way ANOVA. The significant difference was set at $p < 0.05$.

3. Results

3.1. Change of Gill State of Juvenile Yellowfin Tuna under Salinity Stress

The SOD activity in the fish gills in the stress group increased with time from 0 to 48 h and reached the maximum value at 48 h (830.86 ± 80.66 U·mgprot^{-1}, Figure 2a). The SOD activity in the fish gills of the stress group was significantly higher than those in the control group at 24 h and 48 h ($p < 0.05$, Figure 2a). The GSH-Px activity in the stress group was not significantly different from 0 h to 24 h ($p > 0.05$), and reached the maximum value of 5.25 ± 1.01 U·mgprot^{-1} at 48 h ($p < 0.05$, Figure 2b). In the stress group, the MDA content in the gills of the juvenile yellowfin tuna was not significantly different from 0 h to 24 h ($p > 0.05$), and the highest value (2.87 ± 0.12 nmol · mgprot^{-1}) was observed at 48 h (Figure 2c).

3.2. Changes in Liver Status of Juvenile Yellowfin Tuna under Salinity Stress

The SOD activity in the liver of juvenile yellowfin tuna in the stress group fluctuated slightly from 0 h to 48 h with the extension of the stress time, and no significant differences were observed between the sampling times and between the control and stress groups in the same sampling time ($p > 0.05$, Figure 3a). The GSH-Px activity in the liver of fish in the control group and the experimental group experienced a change trend of first increasing (113.42%) and then decreasing (-19.69%), and reached the maximum value at 24 h (1.50 ± 0.27 U·mgprot^{-1}, Figure 3b). Before 24 h, the GSH-Px activity of the fish liver in the stress group was significantly higher than that observed in the control group ($p < 0.05$, Figure 3b), and the lowest GSH-Px activity in the fish liver was observed in the stress group at 48 h. The content of MDA in the liver of the fish in the stress group increased first (137.45%) and then decreased (-97.33%, Figure 3c). At 24 h, the content of MDA reached the highest value of 2.92 ± 0.23 nmol·mgprot^{-1}. There was a significant difference between the stress group at each time point ($p < 0.05$). The MDA content in the stress group at 6 and 48 h was significantly lower than in the control group ($p < 0.05$, Figure 3c).

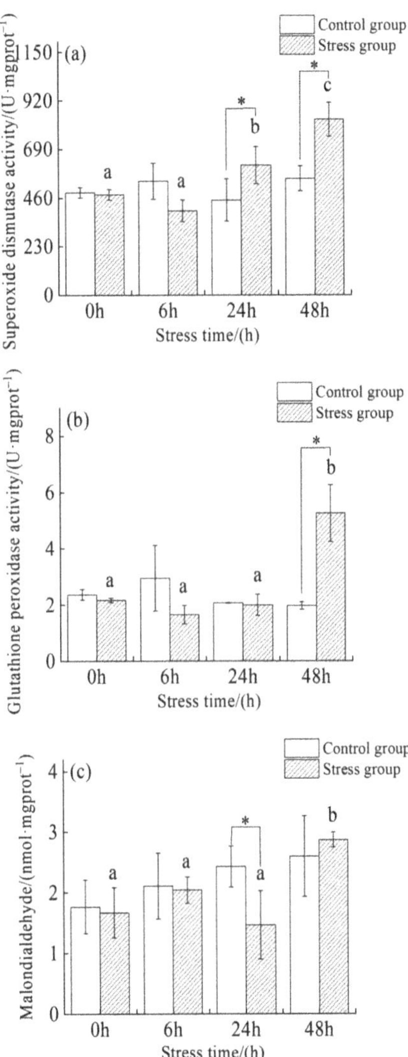

Figure 2. Effect of acute low salt stress on the antioxidation of gills of juvenile yellowfin fish (*Thunnus albacares*), (*n* = 9). (**a**), Superoxide dismutase activity; (**b**), Glutathione peroxidase activity; (**c**), Malondialdehyde. Different letters indicate that there is a significant difference in the same salinity group at different time points ($p < 0.05$). Different letters indicate the difference between the experimental group and the control group over time, and * indicates the difference between the experimental group and the control group.

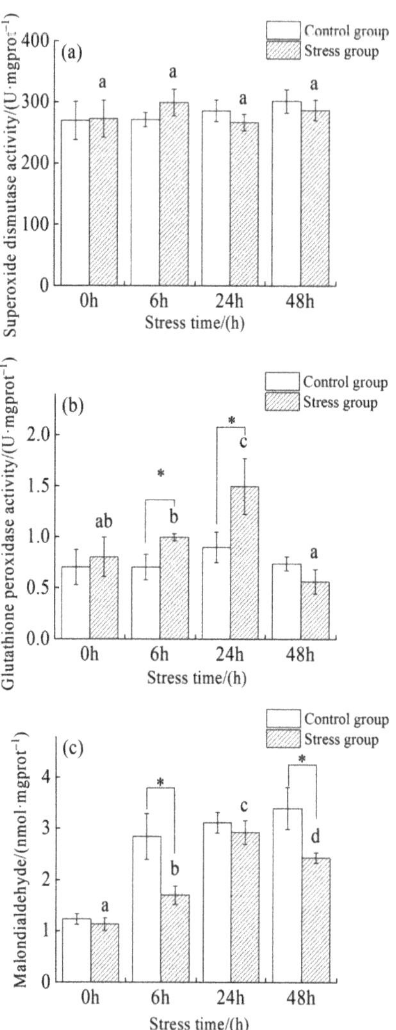

Figure 3. Effect of acute low salt stress on the antioxidation of liver of juvenile yellowfin fish (*Thunnus albacares*), (*n* = 9). (**a**), Superoxide dismutase activity; (**b**), Glutathione peroxidase activity; (**c**), Malondialdehyde. Different letters indicate the difference between the experimental group and the control group over time, and * indicates the difference between the experimental group and the control group.

3.3. Changes in Red Muscle Status of Juvenile Yellowfin Tuna under Salinity Stress

The SOD activity in the red muscle of fish was not significantly different at 0, 24, and 48 h in the stress group ($p > 0.05$), and the lowest value of 813.13 ± 39.81 U·mgprot^{-1} was observed at 6 h ($p < 0.05$, Figure 4a). The highest GSH-Px activity (15.71 ± 1.65 U·mgprot^{-1}) in the red muscle of the fish from the stress group was observed at 24 h, and the lowest value (3.35 ± 0.10 U·mgprot^{-1}) was recorded at 48 h ($p < 0.05$, Figure 4b). In the stress group, the highest MDA value of 3.74 ± 0.36 nmol·mgprot^{-1} was observed in the red muscle of fish at 24 h ($p < 0.05$, Figure 4c), while no significant difference was observed at other sampling times ($p > 0.05$). At 6 and 24 h, the MDA value of the fish in the stress group was significantly higher than those in the control group ($p < 0.05$).

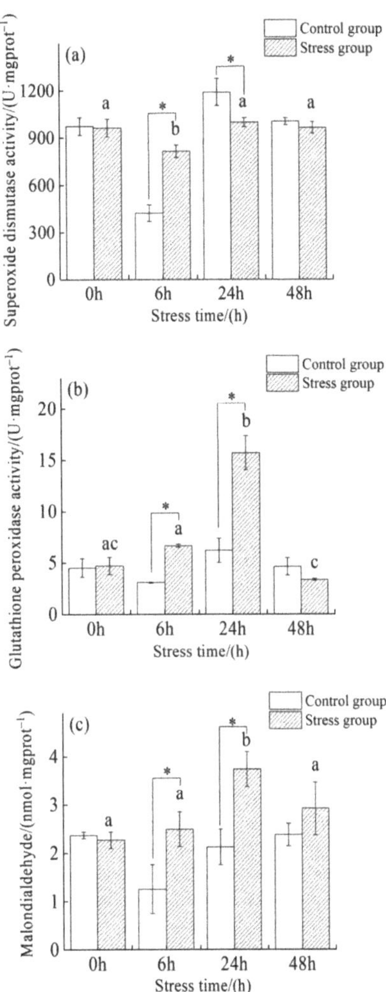

Figure 4. Effect of acute low salt stress on the antioxidation of red muscle of juvenile yellowfin fish (*Thunnus albacares*), (*n* = 9). (**a**), Superoxide dismutase activity; (**b**), Glutathione peroxidase activity; (**c**), Malondialdehyde. Different letters indicate the difference between the experimental group and the control group over time, and * indicates the difference between the experimental group and the control group.

3.4. Changes in White Muscle Status of Juvenile Yellowfin Tuna under Salinity Stress

The SOD activity of the white muscle of the juvenile yellowfin tuna in the stress group decreased (−28.56%) at 6 h ($p < 0.05$, Figure 5a) and gradually stabilized after 24 h ($p > 0.05$). The SOD activity in the white muscle of the fish at 6 and 48 h was significantly higher than that observed in the control group ($p < 0.05$). However, the SOD activity in the stress group was significantly lower than in the control group at 24 h ($p < 0.05$, Figure 5a). In the stress group, the highest GSH-Px activity (5.97 ± 0.14 U·mgprot^{-1}) in the white muscle of the fish was observed at 24 h, and the lowest GSH-Px activity (1.71 ± 0.16 U·mgprot^{-1}) was observed at 48 h ($p < 0.05$, Figure 5b). At 6 and 24 h, the GSH-Px activity of the fish from the stress group was significantly higher than that observed in the control group ($p < 0.05$, Figure 5b). At 48 h, the GSH-Px activity of the fish from the stress group was significantly lower than that recorded in the control group ($p < 0.05$). The MDA value in the white muscle of the fish was not significantly different between the sampling times and treatments ($p > 0.05$, Figure 5c).

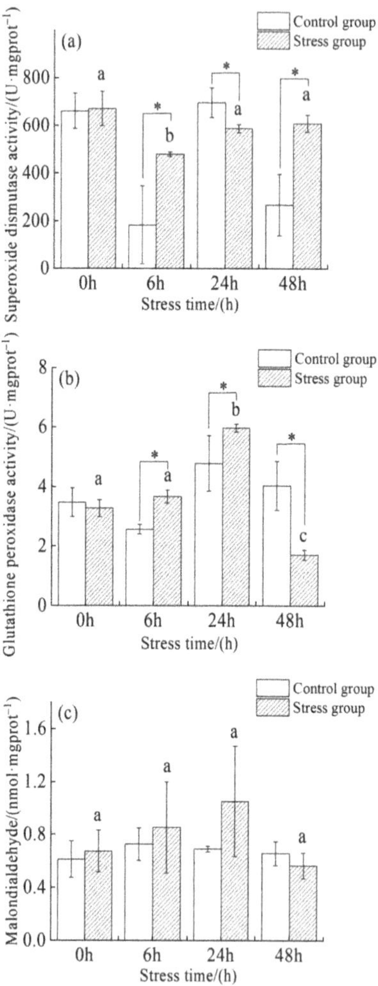

Figure 5. Effect of acute low salt stress on the antioxidation of white muscle of juvenile yellowfin fish (*Thunnus albacares*), ($n = 9$). (**a**), Superoxide dismutase activity; (**b**), Glutathione peroxidase activity; (**c**), Malondialdehyde. Different letters indicate the difference between the experimental group and the control group over time, and * indicates the difference between the experimental group and the control group.

4. Discussion

The gill is a key organ for osmoregulation in fish and plays an important role in ion regulation [25]. When salinity changes, gills not only need to carry out osmotic regulation in time, but also need to bear certain oxidative stress pressure [26]. In the present study, the change of salinity within 48 hours increased the antioxidant enzyme activity in the gills of the juvenile yellowfin tuna. The impacts of salinity on the activity of antioxidant enzymes in gills have been intensively studied. As SOD is the first part of antioxidant defense, it will encounter superoxide anion free radicals. During stress, SOD has a faster reaction speed and higher sensitivity, which is a good indicator for monitoring the aquatic ecosystem [27]. Ghanavatinasab et al. evaluated the impact of salinity on *Sparus flavipectus* when transferring fish from 20 to 5, 12, and 34. They found no significant differences in SOD activity between the groups after 14 days of the experiment [28]. The responses of SOD activity in fish to the salinity changes are species-dependent. In black porgy *Acanthopagrus schlegeli*, when environmental salinities decreased from 35 to 10 ppt, SOD activity of fish upregulated nearly twofold [29]. Moreover, the activity of SOD in *Scapharca broughtonii* showed similar responses after exposure to low environmental salinities [30]. It is speculated that when fish are subjected to salinity stress, different tissues in the fish may undergo various levels of oxidative stress and different resistance strategies. When the salinity dropped suddenly, the SOD in the gills of the juvenile yellowfin tuna participated in scavenging free radicals. Generally speaking, the H_2O_2 produced in most animals is mainly eliminated by GSH-Px [31]. At the same time, GSH-Px can also remove lipid peroxides such as fatty acids, and maintain the normal function of the cell membrane [32]. In the present study, the GSH-Px activities in stressed fish gills reached the highest level at 24 h and were reduced to similar levels to those observed in the control group. Such changes in the GSH-Px activities of fish gills may suggest that fish may fully recover from salinity shock after a 24 h adaption. In vivo, MDA is the final product of oxidation when free radicals act on lipid peroxidation, which can cause the cross-linking and polymerization of life macromolecules such as proteins and nucleic acids and has cytotoxicity [33]. At the same time, the MDA content also shows the degree of cell plasma membrane damage [34]. When the salinity decreases, the MDA in the gills of juvenile yellowfin tuna remains stable within a certain range in the whole process of antioxidation in a low-salt environment, which also proves that the cytoplasmic membrane is not damaged.

The activity of antioxidant enzymes in different tissues of the same fish is also different. The liver is a multifunctional organ that integrates metabolism, immunity, digestion, and other functions [25]. The liver is the tissue with more oxidation reactions, so the activity of antioxidant enzymes is higher [33]. The results of this study showed that the SOD activity in the liver of juvenile yellowfin tuna had no significant impact after the sudden drop in ambient salinity, which was contrary to the research results of *Epinephelus moora* [35], *Takifugu obscurus* [36], and *Oryzias melastigma* [37]. The cultivation salinity of *Dicentrarchus labrax* decreased from 37 to 5, and the SOD activity increased significantly after 12 h of stress [38,39]. The alternations in the SOD activity were relatively low, possibly due to the low response mechanism in the liver of migratory fish in the deep sea [40]. The increased content of GSH-Px and MDA in the first 24 h may relate to the elimination of excessive reactive oxygen free radicals in the body. Afterward, the content of GSH-Px and MDA in the fish liver decreased to a level similar to the control group, possibly as a result of the reaction between glutathione peroxidase and a variety of antioxidant enzymes [41,42]. In the present study, the adaptation processes of SOD, GSH-Px, and MDA were stable after 24 h in the liver, indicating that the antioxidant system in the fish liver had adapted to the salinity reduction. In contrast, however, antioxidant changes in the gills of the yellowfin tuna were different from those of the liver. The inhalation and excretion of fish gills are closely related to the gills. In the acute low-salt environment, the yellowfin tuna is required to maintain osmotic balance, and it also bears a certain amount of oxidative stress pressure [43]. The branchial vascular system is innervated and autonomously controlled by its nerves [44]. When the antioxidation data of their own tissues are stable, the gill tissue behaves differently from the liver and muscle in response to the oxidative stress caused by the sudden change of salinity. It is speculated that different tissues are subject to different degrees of oxidative stress, and there are also differences in the resistance strategies.

The red muscle and white muscle in fish are both part of the greater lateral muscle. As the yellowfin tuna needs to rely on the gills to obtain oxygen, the red muscle and white muscle are clearly distinguished. The red muscle has a high fat content, is rich in myoglobin and has a large amount of blood, and is dark red. The yellowfin tuna has strong endurance and developed red muscle. The white muscle contains no fat and is light white [26]. Muscle is widely distributed in the fish body and has a large proportion [26]. Determining antioxidant enzyme activity in fish muscle can also reflect the stress status of the fish. In *Acipenser schrencki,* the muscles are sensitive tissues to osmotic pressure perception and can respond positively when salinity changes [45]. A previous study has demonstrated that the SOD activity in *Rachycentron canadum* muscle increased with the decrease in salinity [46]. However, the responses of SOD activity to the salinity stress in the muscles of yellowfin tuna were different from previous studies [47]. The SOD activity in the muscle of the juvenile yellowfin tuna decreased first and then increased and tended to be stable at the end. Such responses may suggest the antioxidant system in fish muscle has fully adapted to the ambient salinity of 29‰ at 48 h.

The malondialdehyde content in the muscle increased with time, and the overall expression level of the yellowfin tuna red muscle was higher than that of the white muscle [46,47]. The higher SOD activity in the yellowfin tuna red muscle reflected the low potential mechanism in the antioxidant process of yellowfin tuna red muscle [40]. In the present study, the content of the total antioxidant capacity reached the highest value at 24 h and decreased at 48 h, which may be related to the adaptation process of muscle tissue to the salinity stress. However, there are few reports about the effect of salinity stress on antioxidant enzymes in the white and red muscles of fish, and this subject may be worthy of further investigation.

5. Conclusions

In this study, the juvenile yellowfin tuna were subjected to a 48 h acute low-salt stress experiment to evaluate the physical response of the fish. The results showed that, in terms of antioxidant levels, the juvenile yellowfin tuna could adapt to 29‰ within 48 h. Under salinity stress, the antioxidant status in the liver, gills, and muscles of the fish recovered to average levels at 48 h, indicating that the antioxidant system was able to clear the excess free radicals and oxidation intermediates and maintain the dynamic balance of the antioxidant system in the cells in this salinity changing range. This study helps to understand the physiological changes of yellowfin tuna under low salinity, and provides guidance for the artificial culture of yellowfin tuna.

Author Contributions: Conceptualization, S.Z. and G.Y.; methodology, S.Z.; software, S.Z. and Z.F.; validation, Z.F.; formal analysis, S.Z.; investigation, S.Z. and N.Z.; resources, Z.F. and L.Z.; data curation, N.Z.; writing—original draft preparation, N.Z.; writing—review and editing, Z.M. and G.Y.; visualization, N.Z. and G.Y.; supervision, Z.M. and L.Z.; project administration, Z.M.; funding acquisition, Z.M. and L.Z. All authors have read and agreed to the published version of the manuscript.

Funding: This work was supported by the Hainan Major Science and Technology Project (ZDKJ2021011); the Central Public-Interest Scientific Institution Basal Research Fund, CAFS (2020TD55), and the Central Public-Interest Scientific Institution Basal Research Fund South China Sea Fisheries Research Institute, CAFS (2021SD09).

Institutional Review Board Statement: The experiment complied with the regulations and guidelines established by the Animal Care and Use Committee of the South China Sea Fisheries Research Institute, Chinese Academy of Fishery Sciences. The approval number is 2020TD55, which was approved on 5 January 2020.

Informed Consent Statement: Not applicable.

Data Availability Statement: The original contributions presented in the study are included in the article. Further inquiries can be directed to the corresponding authors.

Conflicts of Interest: The authors declare no conflict of interest.

References

1. Bisby, F.; Roskov, Y.; Culham, A.; Orrell, T.; Nicolson, D.; Paglinawan, L.; Bailly, N.; Appeltans, W.; Kirk, P.; Bourgoin, T.; et al. (Eds.) *Species 2000 & ITIS Catalogue of Life, 2012 Annual Checklist*; Reading: London, UK, 2012.
2. Murua, H.; Rodriguez-Marin, E.; Neilson, J.D.; Farley, J.H.; Juan-Jordá, M.J. Fast versus slow growing tuna species: Age, growth, and implications for population dynamics and fisheries management. *Rev. Fish Biol. Fish.* **2017**, *27*, 733–773. [CrossRef]
3. Zhipan, T.; Fei, W.; Siquan, T.; Qiuyun, M. Stock assessment for Atlantic yellowfin tuna based on extended surplus production model considering life history. *Acta Oceanol. Sin.* **2022**, *41*, 41–51.
4. Weifeng, Z.; Huijuan, H.; Wei, F.; Shaofei, J. Impact of Abnormal Climatic Events on the CPUE of Yellowfin Tuna Fishing in the Central and Western Pacific. *Sustainability* **2022**, *14*, 1217.
5. Benetti, D.D.; Partridge, G.J.; Stieglitz, J. Overview on status and technological advances in tuna aquaculture around the world. In *Advances in Tuna Aquaculture*; Benetti, D.D., Partridge, G.J., Buentello, A., Eds.; Academic Press: Cambridge, MA, USA, 2016; pp. 1–19.
6. Cobcroft, J. Addressing causes of early mortality in hatchery produced southern bluefin tuna larvae. Chapter 9: The effects of in-tank lighting on the early behaviour and performance of yellowtail kingfish larvae. *Agric. Vet. Sci.* **2014**, 96232.
7. Nguyen, K.Q.; Phan, H.T. Length-length, Length-weight, and Weight-weight Relationships of Yellowfin (*Thunnus albacares*) and Bigeye (*Thunnus Obesus*) Tuna Collected from the Commercial Handlines Fisheries in the South China Sea. *Thalassas* **2022**, *38*, 911–917. [CrossRef]
8. Liu, H.; Fu, Z.; Zhou, S.; Hu, J.; Yang, R.; Yu, G.; Ma, Z. The Complete Mitochondrial Genome of *Pennella* sp. Parasitizing Thunnus alabacares. *Front. Cell. Infect. Microbiol.* **2022**, *12*, 945152. [CrossRef]
9. Katz, S.L.; Syme, D.A.; Shadwick, R.E. Enhanced power in yellowfin tuna. *Nature* **2001**, *410*, 770–771. [CrossRef]
10. Ern, R.; Huong DT, T.; Cong, N.V.; Bayley, M.; Wang, T. Effect of salinity on oxygen consumption in fishes: A review. *J. Fish Biol.* **2014**, *84*, 1210–1220. [CrossRef]
11. Livingstone, D.R. Contaminant-stimulated Reactive Oxygen Species Production and Oxidative Damage in Aquatic Organisms. *Mar. Pollut. Bull.* **2001**, *42*, 656–666. [CrossRef]
12. Wang, J.Q.; Lui, H.; Po, H.; Fan, L. Influence of salinity on food consumption, growth and energy conversion efficiency of common carp (*Cyprinus carpio*) fingerlings. *Aquaculture* **1997**, *148*, 115–124. [CrossRef]
13. Choi, C.Y.; An, K.W.; An, M.I. Molecular characterization and mRNA expression of glutathione peroxidase and glutathione S-transferase during osmotic stress in olive flounder (*Paralichthys olivaceus*). *Comp. Biochem. Physiol. Part A* **2008**, *149*, 330–337. [CrossRef] [PubMed]
14. Liu, Y.; Wang, W.-N.; Wang, A.-L.; Wang, J.-M.; Sun, R.-Y. Effects of dietary vitamin E supplementation on antioxidant enzyme activities in *Litopenaeus vannamei* (Boone, 1931) exposed to acute salinity changes. *Aquaculture* **2007**, *265*, 351–358. [CrossRef]
15. Jiang, Z.; Huang, X.; Zhang, J. Dynamics of nonstructural carbohydrates in seagrass Thalassia hemprichii and its response to shading. *Acta Oceanol. Sin.* **2013**, *32*, 61–67. [CrossRef]
16. Artetxe-Arratea, I.; Frailea, I.; Marsac, F.; Farley, J.H.; Rodriguez-Ezpeleta, N.; Davies, C.R.; Clear, N.P.; Grewed, P.; Murua, H. A review of the fisheries, life history and stock structure of tropical tuna (skipjack *Katsuwonus pelamis*, yellowfin *Thunnus albacares* and bigeye *Thunnus obesus*) in the Indian Ocean. *Adv. Mar. Biol.* **2020**, *88*, 39–89.
17. Lushchak, V.I.; Bagnyukova, T.V.; Lushchak, O.V.; Storey, J.M.; Storey, K.B. Hypoxia and recovery perturb free radical processes and antioxidant potential in common carp (*Cyprinus carpio*) tissues. *Int. J. Biochem. Cell Biol.* **2005**, *37*, 1319–1330. [CrossRef]
18. Canli, E.G.; Gülüzar Atli Eroglu, A.; Atli, G.; Canli, M. Effects of fish size on the response of antioxidant systems of *Oreochromis niloticus* following metal exposures. *Fish Physiol. Biochem.* **2014**, *40*, 1083–1091.
19. Vasseur, P.; Rodius, F.; Doyen, P.; Rodius, F. Molecular cloning and expression study of pi-class glutathione S-transferase (pi-GST) and selenium-dependent glutathione peroxidase (Se-*GSH-Px*) transcripts in the freshwater bivalve Dreissena polymorpha. *Comp. Biochem. Physiol. Part C: Toxicol. Pharmacol.* **2008**, *147*, 69–77.
20. Sinha, A.K.; Abd Elgawad, H.; Zinta, G.; Dasan, A.F.; Rasoloniriana, R.; Asard, H.; Blust, R.; De Boeck, G. Nutritional Status as the Key Modulator of Antioxidant Responses Induced by High Environmental Ammonia and Salinity Stress in European Sea Bass (*Dicentrarchus labrax*). *PLoS ONE* **2015**, *10*, e0135091. [CrossRef]
21. Tao, F.; Wei, Z.; Wan, H.; Rong, C. Effects of benzo(a)pyrene exposure on glutathione peroxidase activity in the liver of *Boleophthalmus pectinirostris*. *J. Fish. Sci. China* **2000**, *7*, 19–21.
22. Pipe, R.K.; Porte, C.; Livingstone, D.R. Antioxidant Enzymes Associated with the Blood Cells and Hemolymph of the Mussel Mytilus edulis. *Fish Shellfish. Immunol.* **1993**, *3*, 221–233. [CrossRef]
23. Klein, R.D.; Rosa, C.E.; Colares, E.P.; Robaldo, R.B.; Martinez, P.E.; Bianchini, A. Antioxidant defense system and oxidative status in Antarctic fishes: The sluggish rockcod Notothenia coriiceps versus the active marbled notothen *Notothenia rossii*. *J. Therm. Biol.* **2017**, *68 Pt A*, 119–127. [CrossRef]
24. Gan, L.; Xu, Z.X.; Ma, J.J.; Xu, C.; Wang, X.D.; Chen, K.; Li, E.C. Effects of salinity on growth, body composition, muscle fatty acid composition, and antioxidant status of juvenile Nile tilapia *Oreochromis niloticus* (Linnaeus, 1758). *J. Appl. Ichthyol.* **2016**, *32*, 372–374. [CrossRef]
25. Yokota, S.U. Regulation of the ion-transporting mitochondrion-rich cell during adaptation of teleost fishes to different salinities. *Zool. Sci.* **2001**, *18*, 1163–1174.

26. Saoud, I.P.; Kreydiyyeh, S.; Chalfoun, A.; Fakih, M. Influence of salinity on survival, growth, plasma osmolality and gill Na$^+$-K$^+$-ATPase activity in the rabbitfish *Siganus rivulatus*. *J. Exp. Mar. Biol. Ecol.* **2007**, *348*, 183–190. [CrossRef]
27. Zhang, H.; Zhang, X.; Dong, C.; Band, Z.; Li, L.; Yub, J.; Hu, Y.; Chen, C. Effects of ozone treatment on SOD activity and genes in postharvest cantaloupe. *RSC Adv.* **2020**, *10*, 17452–17460. [CrossRef] [PubMed]
28. Hahn, O.; Drews, L.F.; An, N.; Tatsuta, T.; Gkioni, L.; Hendrich, O.; Zhang, Q.; Langer, T.; Pletcher, S.; Wakelam, M.J.O.; et al. A nutritional memory impairs survival, transcriptional and metabolic response to dietary restriction in old mice. *Cold Spring Harb. Lab.* **2019**. [CrossRef]
29. An, M.I.; Choi, C.Y. Activity of antioxidant enzymes and physiological responses in ark shell, *Scapharca broughtonii*, exposed to thermal and osmotic stress: Effects on hemolymph and biochemical parameters. *Comp. Biochem. Physiol. Part B Biochem. Mol. Biol.* **2010**, *155*, 34–42. [CrossRef]
30. Martínez-Álvarez, R.M.; Morales, A.E.; Sanz, A. Antioxidant Defenses in Fish: Biotic and Abiotic Factors. *Rev. Fish Biol. Fish.* **2005**, *15*, 75–88. [CrossRef]
31. Zoysa, M.D.; Whang, I.; Lee, Y.; Lee, S.; Lee, J.S.; Lee, J. Transcriptional analysis of antioxidant and immune defense genes in disk abalone (*Haliotis discus* discus) during thermal, low-salinity and hypoxic stress. *Comp. Biochem. Physiol. Part B Biochem. Mol. Biol.* **2009**, *154*, 387–395. [CrossRef]
32. Wilhelm, D.; Giulivi, C.; Boveris, A. Antioxidant defenses in marine fish I.Teleosts. *Comp. Biochem. Physiol. Part C Comp. Pharmacol. Toxicol.* **1993**, *106*, 409–413.
33. Vinagre, C.; Madeira, D.; Narciso, L.; Cabral, H.N.; Diniz, M. Effect of temperature on oxidative stress in fish: Lipid peroxidation and catalase activity in the muscle of juvenile seabass, *Dicentrarchus labrax*. *Ecol. Indic.* **2012**, *23*, 274–279. [CrossRef]
34. Liao, Y.; Zhang, C.; Peng, S.; Gao, Q.; Shi, Z. Effects of salinity on activities of liver antioxidant enzymes and plasma lysozyme of *Epinehelus moara*. *J. Shanghai Ocean. Univ.* **2016**, *25*, 169–176.
35. Bian, P.J.; Qiu, C.G.; Shan-Liang, X.U.; Lin, S. Effects of salinity on growth, activity of non-specific immune and antioxidant enzymes in obscure puffer *Takifugu obscures*. *Acta Hydrobiol. Sin.* **2014**, *38*, 108–114.
36. Wang, R.P.; Dai, L.L.; Chen, Y.F. Effects of short-term temperature or salinity stress on feeding behavior and antioxidant of marine medaka(*oryzias melastigma*). *Oceanol. Et Limnol. Sin.* **2019**, *10*, 730853.
37. Mak, Y.L.; Jia, J.W.; Chan, W.H.; Murphy, M.B.; Lam, J.C.; Chan, L.L.; Lam, P.K. Simultaneous quantification of Pacific ciguatoxins in fish blood using liquid chromatography-tandem mass spectrometry. *Anal. Bioanal. Chem.* **2013**, *405*, 3331–3340. [CrossRef]
38. Kalinina, E.V.; Chernov, N.N.; Novichkova, M.D. Role of Glutathione, Glutathione Transferase, and Glutaredoxin in Regulation of Redox-Dependent Processes. *Biochemistry* **2014**, *79*, 1562–1583. [CrossRef] [PubMed]
39. Farrell, A.P. (Ed.) *Encyclopedia of Fish Physiology: From Genome to Environment*; Academic Press: London, UK, 2011.
40. Xiaohong, L. *Preliminary Study on the Effect of Water Cadmium Exposure on Liver Toxicity and Lipid Metabolism of Rare Gudgeon Crucian Carp*; Southwest University: Chongqing, China, 2016.
41. Zhao, F.; Zhuang, P.; Zhang, L.Z.; Huang, X.R.; Zhang, T.; Feng, G.P. Responses of antioxidases in different tissues of *Acipenser schrenckii* to increased salinity in water. *Mar. Fish. Res.* **2008**, *29*, 65–69.
42. Yang, J.; Chen, G.; Huang, J.S.; Zhang, J.D.; Shi, G.; Tang, B.G.; Zhou, H. Effects of Temperature and Salinity on the Growth and Activities of Antioxidant Enzymes of Cobia (*Rachycentron canadum*) Juveniles. *J. Guangdong Ocean. Univ.* **2007**, *4*, 24–29.
43. Davenport, J. Osmotic control in marine animals. *Symp. Soc. Exp. Biol.* **1985**, *39*, 207–244.
44. Michael G, J.; Giacomo, Z. Nervous control of the gills. *Acta Histochem.* **2009**, *111*, 207–216.
45. Ghanavatinasab, Y.; Salati, A.P.; Movahedinia, A.; Shahriari, A. Changes in Gill Antioxidant Status in *Acanthopagrus sheim* Exposed to Different Environmental Salinities. *Iran. J. Sci. Technology. Trans. A Sci.* **2018**, *43*, 1479–1483. [CrossRef]
46. Zhang, C.; Zhang, Y.; Gao, Q.; Shiming, P.E.N.G.; Zhaohong, S.H.I. Effect of low salinity stress on antioxidant function in liver of juvenile Nibea albiflora. *South China Fish. Sci.* **2015**, *11*, 59–64.
47. Garcia, D.; Lima, D.; Silva, D.; de Almeida, E.A. Decreased malondialdehyde levels in fish (*Astyanax altiparanae*) exposed to diesel: Evidence of metabolism by aldehyde dehydrogenase in the liver and excretion in water. *Ecotoxicol. Environ. Saf.* **2020**, *190*, 110107. [CrossRef] [PubMed]

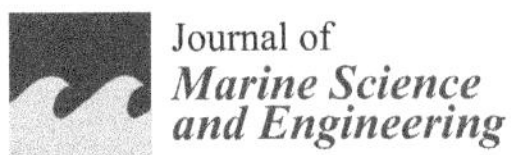

Journal of
*Marine Science
and Engineering*

Article

Pathology, Enzyme Activity and Immune Responses after *Cryptocaryon irritans* Infection of Golden Pompano *Trachinotus ovatus* (Linnaeus 1758)

Hua-Yang Guo [1,2,3,†], Wen-Fu Li [4,†], Ke-Cheng Zhu [1,2,3], Bao-Suo Liu [1,2,3], Nan Zhang [1,2,3], Bo Liu [1,2], Jing-Wen Yang [1,2] and Dian-Chang Zhang [1,2,3,*]

1   Key Laboratory of South China Sea Fishery Resources Exploitation and Utilization, Ministry of Agriculture and Rural Affairs, South China Sea Fisheries Research Institute, Chinese Academy of Fishery Sciences, Guangzhou 510300, China
2   Guangdong Provincial Engineer Technology Research Center of Marine Biological Seed Industry, Guangzhou 510300, China
3   Sanya Tropical Fisheries Research Institute, Sanya 572019, China
4   School of Life Sciences, Guangzhou University, Guangzhou 510006, China
*   Correspondence: zhangdch@scsfri.ac.cn; Tel.: +86-020-8910-8316; Fax: +86-02089022702
†   These authors contributed equally to this work.

Abstract: Golden pompano (*Trachinotus ovatus*) has become an economically important fish in China in the past decade. However, *Cryptocaryon irritans*, a parasitic ciliate, causes considerable economic losses in the mariculture of *T. ovatus*. To characterize the pathogenesis of *C. irritans* in *T. ovatus*, the pathological properties, immune-related enzyme activity and expression of the *NEMO* gene was analysed. The results from the histological sections showed that there was considerable metamorphosis and hyperplasia in the parasitized sites (skin) with leukocyte aggregation and mucous cell increases after *C. irritans* infection. Moreover, the activities of four enzymes, that is, alkaline phosphatase (AKP), acid phosphatase (ACP), superoxide dismutase (SOD) and lysozyme (LZM), were significantly increased in different tissues after *C. irritans* infection. Furthermore, the ORF of *T. ovatus* NF-kappa-B essential modulator (*ToNEMO*) measures 1650 bp, encoding 548 amino acids. The *ToNEMO* transcripts were universally expressed in all examined tissues, with higher levels being observed in the immune-relevant and central nervous tissues. The mRNA levels of *ToNEMO* after *C. irritans* infection were significantly increased in the gill, skin, liver, spleen and head kidney. These results suggested that *ToNEMO* might be involved in immune responses and helped to elucidate the physiological response after the *C. irritans* infection of fish.

Keywords: *Cryptocaryon irritans*; *Trachinotus ovatus*; histologic section; *NEMO*; enzyme activity

Citation: Guo, H.-Y.; Li, W.-F.; Zhu, K.-C.; Liu, B.-S.; Zhang, N.; Liu, B.; Yang, J.-W.; Zhang, D.-C. Pathology, Enzyme Activity and Immune Responses after *Cryptocaryon irritans* Infection of Golden Pompano *Trachinotus ovatus* (Linnaeus 1758). *J. Mar. Sci. Eng.* **2023**, *11*, 262. https://doi.org/10.3390/jmse11020262

Academic Editor: Dariusz Kucharczyk

Received: 8 November 2022
Revised: 7 January 2023
Accepted: 16 January 2023
Published: 23 January 2023

1. Introduction

Cryptocaryon irritans is an obligate ciliated parasite of fish. There are four stages in its life cycle: trophozoite, cyst precursor, cyst and larva [1]. *C. irritans* can cause severely fatal "white spot disease" in tropical or subtropical marine fish [1,2]. In recent years, as the density of marine aquaculture has increased, "white spot disease" has frequently broken out in the marine aquaculture areas of the South China Sea, which seriously threatens marine aquaculture fishes, especially *Trachinotus ovatus* [3]. However, there are still few studies on the histology of *T. ovatus* infection with *C. irritans* and its immune mechanism.

Moreover, *C. irritants* mainly invades the superficial epithelium of host fish, including the skin, gills and eyes, and it damages the physiological functions of these organs with pathological features of white spots on the fins, skin or gills. Fish may rub against the wall of the pool, swim irregularly, float on the surface or sink to the bottom or show lethargy or shortness of breath as signs of distress. Symptoms include pinhead-sized white nodules on the gills and body, mucus hyperproduction, skin discoloration, corneal cloudiness, ragged fins and pale gills. *C. irritans* infection results in the serious mortality of fish due to asphyxiation, osmotic imbalance and/or secondary bacterial infections. In fish stocks, mortality may increase rapidly within a few days, depending on the parasite strains, fish species and the water temperature [1].

In previous histological studies of *T. ovatus* challenged by *C. irritant*, their surface was directly mechanically injured, and the better-defined systemic immune tissue was indirectly damaged, with degeneration and necrosis [4]. The migration and aggregation of leukocytes, including lymphocytes, eosinophils, macrophages and neutrophils, was observed in gills, and IL-8 was significantly upregulated in skin [3].

Tumour-necrosis-factor-receptor-associated factor (TRAF) is a kind of intracellular junction protein previously discovered using yeast two-hybrid technology and glutathione transferase fusion technology [5]. TRAF participates in signal transduction pathways of Toll-like receptors (TLRs) and other families and regulates nuclear factor κB (NF-κB), which are important signal transduction proteins in cells [5,6]. NEMO is one of the most important components of the IKK (IκB kinase) complex, a key protein in the NF-κB pathway [7,8]. TRAF ubiquitination activation was shown to lead to the activation of the IκB kinase complex and the phosphorylation of NEMO (NF-kappa-B Essential Modulator) [5], which caused the released NF-κB to be transferred to the nucleus and initiate transcription of the TNF, Pro-IL-1b, IkBz, ATF3, Zc3 h12a and TTP genes [6]. The NF-κB pathway is essential for the immune response and inflammatory response to pathogens [9]. NEMO plays an essential role in the activation of the IKK kinase in the NF-κB signalling pathway [10]. NEMO is highly conserved. Structure prediction shows that it has an α-helix with three coiled coil regions, a leucine zipper domain and a C-terminal zinc finger structure [11].

The specific immunity of fish is lower than that of mammals, while the non-specific immunity of fish plays a major role in the body's immunity [12]. The antioxidant system of fish is the first line of defence in the innate immunity of fish. In addition to the intracellular activation of innate immune signalling pathways and the release of extracellular inflammatory factors, immune-related enzymes in the body are also important immune indicators when the body undergoes pathogen invasion. ACP plays a role in killing and digesting pathogens in immune responses. AKP is also a multi-functional enzyme involved in immune responses [13]. Superoxide dismutase (SOD) is the main antioxidant enzyme in the antioxidant system to remove excess active free radical ROS [14]. As a humoral immune factor, Lysozyme (LZM) activity can reflect the non-specific immune ability of fish to a certain extent [3,15].

Golden pompano is one of the most important farmed fish in the South China Sea [16]. Similar to other species, golden pompons are also affected by *C. irritans*. Although environmental studies have been reported [17–20], the infective characteristics and mechanisms of *C. irritans* in *T. ovatus* have not been adequately studied. In this study, the histopathology and gene expression of *NEMO* and immune-related enzyme activity were detected after *C. irritans* infection. The influence of the infection on the immune response of *T. ovatus* provides basic theoretical data and guidance for later studies of disease resistance breeding and blocking the effects of *C. irritans* on *T. ovatus*.

2. Materials and Methods

2.1. Ethics Statement

All trials in this study were authorized by the Animal Care and Use Committee of South China Sea Fisheries Research Institute, Chinese Academy of Fishery Sciences (No. SCSFRI96-253) and were performed according to the regulations and guidelines formulated by this committee.

2.2. C. irritans Challenge and Sampling

Healthy *T. ovatus* (body weight = 98 ± 15 g) were purchased from Linshui Marine Fish Farm in Hainan Province, China. Before the experiment, the fish were put into a circulating marine pool at 28 ± 2 °C with 25‰ salinity and acclimated for two weeks with commercial feed (Hengxin, Zhanjiang, China, crude protein > 37%, crude fat > 7%). The *C. irritans* strain used in this study was originally isolated from infected *T. ovatus* and artificially propagated in the laboratory according to the methods described in previous research [21].

Fifty healthy fish were randomly selected as the control group. Then, one hundred and twenty healthy fish were challenged with *C. irritans* at a dose of 600 theronts/fish. After infection for 0 h, 3 h, 6 h, 12 h, 1 d, 2 d and 3 d, five tissue samples (skin, gill, liver, spleen and head kidney) were collected from six challenged fish to examine enzyme activities and gene expression. Blood, brain, muscle, heart, gonad, spleen, kidney, liver, skin, gill, and intestine was detected via qRT-PCR. Before dissection, the fish were anaesthetized using MS222 (0.1 g/L; Sigma, Alcobendas, Spain). In addition, the second branchial arch and skins were taken and stored in 4% paraformaldehyde for tissue section samples. The same tissues from uninfected fish were regarded as a negative control at each time point. All samples were immediately frozen in liquid nitrogen and then stored at $-80\ ^\circ$C until use.

2.3. HE Staining of Tissue Sections

The tissue samples of the gills and skin stored in 4% paraformaldehyde were sent to Servicebio Biotechnology Co., Ltd. (Wuhan, China) for paraffin embedding, tissue sectioning and HE staining. The HE-stained sections were observed under a microscope (ZEISS Axio Scope A1, Jena, Germany) and photographed.

2.4. Determination of Immunity-Related Enzyme Activities

The acid phosphatase (ACP), alkaline phosphatase (AKP), superoxide dismutase (SOD) and lysozyme (LZM) activities and total protein content (TP) were measured in five tissues from infected fish (skin, gill, liver, spleen and head kidney). ACP, AKP, SOD and LZM enzyme activity and total protein content were measured with assay kits according to the protocol supplied by the manufacturer of the kits. Colorimetry was used to determine ACP and AKP activity, and enzyme activity that produced 1 mg of phenol by 1 g of tissue protein at 37 $^\circ$C in 15 min was defined as 1 U (U/g prot). To measure the SOD activity, 50 mL of supernatant was mixed with 1.3 mL of reaction solution containing. 75 mM nitroblue tetrazolium (NTB) and 20 mM riboflavin. The mixture was incubated at 37 for 40 min, and then, the appearance of NTB-diformazan was measured with a Mindray BS-420 automatic biochemical instrument (Shenzhen Mindray Biological Medical Electronics Co., Ltd., Shenzhen, China). The method of determining LZM activity has been described in a previous study [22].

2.5. Cloning of the NEMO Gene

The tissues were homogenized with a tissue homogenizer (JXFSTPRP-24, Shanghai Jingxin Industrial Development Co., Ltd., Shanghai, China) after adding magnetic beads. Total RNA (1 µg) was isolated from each tissue using the HiPure Universal RNA Kit (Magen, Guangzhou, China) and reverse transcribed into cDNA using the PrimeScript^{TM} RT Reagent Kit with gDNA Eraser (Perfect Real Time) (Takara, Japan). Moreover, the quantity and quality of the extracted RNA were detected using a NanoDrop 2000 spectrometer (Thermo Scientific, Waltham, MA, USA) and 1% agarose gels. The synthetic cDNA samples were stored at $-20\ ^\circ$C until use. The NEMO predicted sequences were obtained from *T. ovatus* genomic data [23]. To determine the accuracy of the encoding sequence of NEMO, gene-specific primers were designed (Table 1) based on the proposed sequence.

2.6. Bioinformatic Analysis of the NEMO Gene

To learn about the bioinformatic features of the *NEMO* gene, the Compute pl/Mw online tool on ExPASy (https://web.expasy.org/compute_pi/ (accessed on 11 December 2019)) was used to predict the molecular weight and isoelectric point of the protein encoded by *NEMO*. Showorf, an online tool from EMBOSS (http://www.bioinformatics.nl/emboss-explorer/ (accessed on 11 December 2019)), showed the open reading frame and predicted amino acid sequence. Other species' *NEMO* genes were searched for and downloaded using the NCBI database (https://www.ncbi.nlm.nih.gov/ (accessed on 11 December 2019)). The structure and domains of the *NEMO* gene of *T. ovatus* were predicted using the SMART online tool (http://smart.embl-heidelberg.de/ (accessed on 11 December 2019)). The amino acid sequences of the conserved domains of the *NEMO* gene of *T. ovatus* were analysed using the Clustal Omega online tool in EMBL (https://www.ebi.ac.uk/Tools/msa/clustalo/ (accessed on 11 December 2019)). The phylogenetic trees of NEMO were constructed using MEGA 7.0 software with the neighbour-joining (NJ) method.

Table 1. Information regarding the *NEMO* primers used in this study.

Primer	Sequences (5′→3′)
NEMO-ORF-R	ATGGTGCAGCCACAACCTG
NEMO-ORF-L	CTGTATACAGTCCATGA
NEMO-qRT-R	GGCTCACTGGAGACTGTT
NEMO-qRT-L	GAGGAAATCTGCCTTGTA
EF-1α-qRT-R	CCCCTTGGTCGTTTTGCC
EF-1α-qRT-L	GCCTTGGTTGTCTTTCCGCTA

2.7. Quantitative Real-Time PCR and Statistical Analysis

To determine the relative mRNA expression level of *T. ovatus NEMO* in normal tissues under normal conditions and after infection with *C. irritans*, qRT-PCR was used to analyse the expression pattern of *NEMO* in five infected tissues (skin, gill, liver, spleen and head kidney) from *T. ovatus*. Based on the *NEMO* cDNA sequences, specific primers for *NEMO* were designed with Primer 5.0 software, and the housekeeping gene *EF-1α* (elongation factor 1, alpha) was used as a reference in a previously reported study [24]. The qRT-PCR procedure was executed as previously described. To evaluate the relative expression of these genes, the $2^{-\Delta\Delta CT}$ method was used [25].

2.8. Statistical Analysis

The data are shown as the means ± SE (n = 3). The one-way analysis of variance (ANOVA) followed by Tukey's multiple comparisons test was used in the present study at a significance level of $p < 0.05$. Statistical analyses were implemented using SPSS 20.0 (SPSS, Chicago, IL, USA).

3. Result

3.1. Histological Analysis

HE staining of the gill is shown in Figure 1. In the control group, the gill filaments and gill lamellae of healthy *T. ovatus* were completed and arranged in a comb shape, and the mitochondrial-rich cells, blood cells and epithelial cells were clearly identified. After challenge with *C. irritans* for 3–6 h, round or elliptical trophonts floated in the water and then attached to the gill lamellae of *T. ovatus*, with a size of 20.36 ± 5.00 μm. Vesiculate- and horseshoe-shaped nuclei could be observed. After challenge for 12 h–1 d, the *C. irritans* trophont developed into an elliptical solid protomont with a size of 84.24 ± 4.00 μm; the gill lamellae were swollen, curved and hyperplastic. MRC and other structures were not observed. After challenge for 2–3 d, tomonts developed and asexually reproduced as theronts. Part of the tomonts split, and theronts were released into the water. Hyperplasia and curves were observed in the gill filaments, and a large number of tissues swelled, split, shed cells and exhibited tissue necrosis.

HE staining of the skin is shown in Figure 2. The structure of the epidermis and muscle layer of healthy *T. ovatus* was complete. After challenge for 3–6 h, there was no obvious change in the skin surface structure. After challenge for 12 h–2 d, the glandular parts of the epidermal layer thickened, and the mucous cells and secretion fluid increased. After challenge for 3 d, the skin tissue was destroyed, the epidermis layer was exposed, the cells fell off and tissue necrosis occurred.

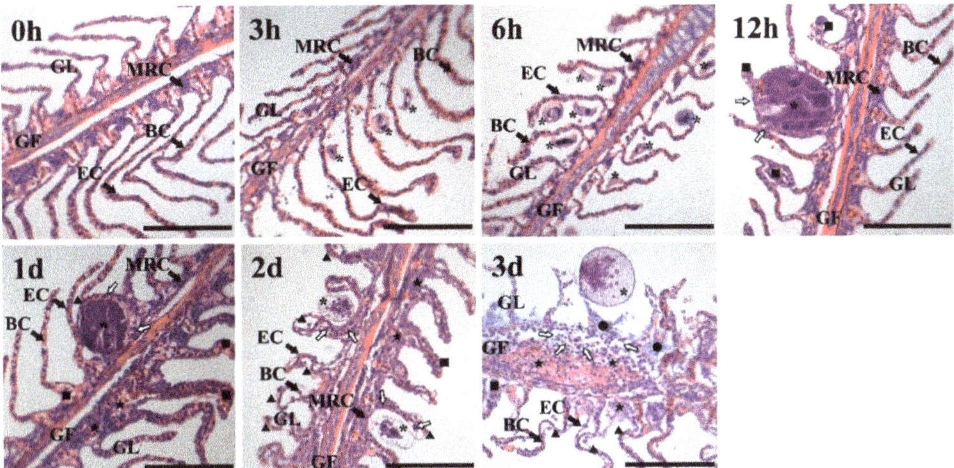

Figure 1. HE staining of the gill tissue of *T. ovatus* infected by *C. irritans*. Abbreviations: GF: gill filaments; GL: gill lamellae; EC: epithelial cell; MRC: mitochondrial-rich cell; BC: blood cell; ▲: split; ■: swell; ★: hyperplasia; •: necrosis. The white arrow shows white blood cells migrating and aggregating at the parasitic site. *: *C.irritans* (life cycle: 3–6 h: trophont; 12–24 h: protomont; 48 h: tomont; 72 h: theronts). Scale bar = 100 μm.

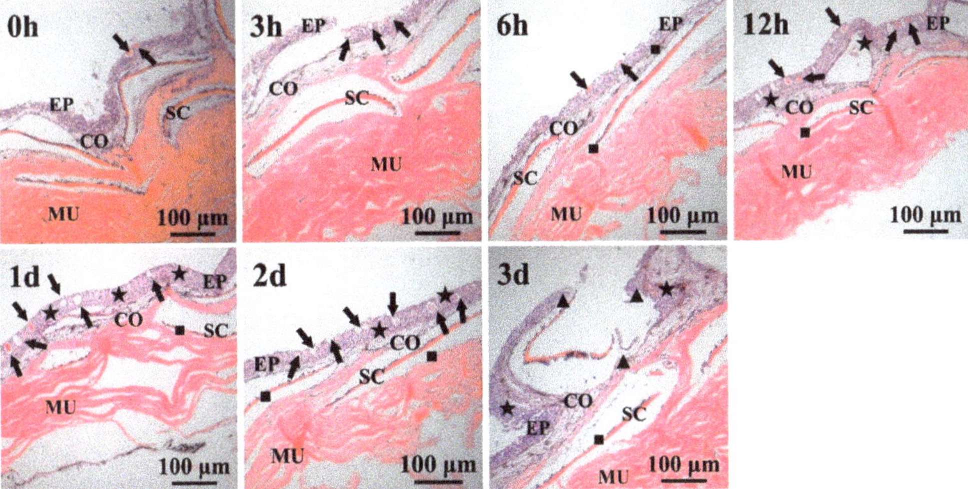

Figure 2. HE staining of the skin tissue of *T. ovatus* infected by *C. irritans*. Abbreviations: EP: epidermis layer; CO: dermis layer; MU: muscle layer; ▲: break; ■: flat; ★: thickening; the black arrow shows the distribution of mucous cells in the gland of the epidermis layer. Scale bar = 100 μm.

3.2. Immune-Related Enzyme Activity Analysis

To investigate the antioxidant activity and non-specific immune response of *T. ovatus* after infection with *C. irritans*, the enzyme activity of ACP, AKP, SOD and LZM in the defence against parasite infection was determined at local infection sites (skin and gills) and system immune tissues (liver, spleen and head kidney) (Figure 3).

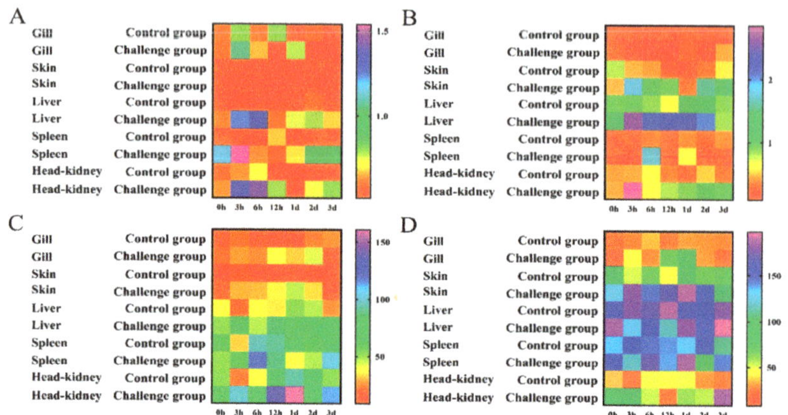

Figure 3. Enzyme activities of ACP (**A**), AKP (**B**), SOD (**C**) and LZM (**D**) were detected in gill, skin, liver, spleen and head kidney after infection with *C. irritans* (0 h, 3 h, 6 h, 12 h, 1 d, 2 d and 3 d). The heatmap was constructed using Graphpad Prism 5.0 software.

The results of the ACP activity assay are shown in Figure 3A. In the gill, ACP was upregulated to a peak at 6 h, then decreased and had a second peak at 1 d. In the skin, ACP was upregulated to its maximum (3.14-fold relative to the control) from 3 h to 6 h and then downregulated. In the liver, ACP was upregulated to a maximum (2.39-fold relative to the control) from 3 h to 6 h and then it decreased to normal levels. In the liver, ACP was upregulated at 6 h and then downregulated. In the head kidney, ACP was upregulated to a peak (1.97-fold relative to the control) from 3 h to 6 h and then downregulated to normal levels. ACP had the smallest change in the gill and the largest change in the head kidney. The ACP activity in the skin was the lowest, with an average of 0.15 (U/gprot), and the ACP activity in the spleen was the highest, with an average of 0.93 (U/gprot).

The results of the AKP activity assay are shown in Figure 3B. In the gill, AKP showed no significant changes between the challenge group and the control group. In the skin, AKP was upregulated to a peak (3.80-fold relative to the control); then, it decreased and had a second peak (4.65-fold relative to the control) at 2 d. In the liver, AKP was upregulated at 3 h and 6 h, and it reached a peak (3.55-fold relative to the control) at 12 h and then was maintained to 2 d. In the spleen, AKP had a peak (4.96-fold relative to the control) at 6 h. In the head kidney, AKP reached a peak (6.01-fold relative to the control) at 3 h. AKP had the smallest change in the gill and the largest change in the head kidney. The activity of AKP in the gills was the lowest, with an average of 0.24 (U/gprot), and the activity of AKP in the liver was the highest, with an average of 1.94 (U/gprot).

The results regarding SOD activity are shown in Figure 3C. In the gill, SOD was upregulated to a peak (2.98-fold relative to the control) from 3 h to 1 d. In the skin, SOD was upregulated to a peak (4.70-fold relative to the control) from 3 h to 1 d. In the liver, SOD was upregulated at 3 h and 12 h and then maintained at a certain level. In the spleen, SOD was upregulated at 6 h and 12 h, and then it decreased, with a second peak at 3 d. In the head kidney, SOD was upregulated to a maximum from 3 h to 1 d and then downregulated. SOD displayed the smallest change in the spleen and the largest change in the skin. The activity of SOD in the spleen was the lowest, with an average of 1.06 U/mg prot, and the activity of SOD in the skin was the highest, with an average of 2.52 U/mg prot.

The results of the LZM activity assays are shown in Figure 3D. In the gill, LZM was upregulated to a maximum (3.27-fold relative to the control) from 3 h to 1 d. In the skin, LZM reached a peak at 3 h and was then maintained at a certain level. In the liver, no significant change was observed in the challenge group. In the spleen, LZM had a peak (1.41-fold relative to the control) on day 1. In the head kidney, LZM was upregulated at 6 h and had a peak (8.49-fold relative to the control) on day 3. LZM displayed the smallest change in the liver and the largest change in the skin. The activity of LZM in the gills was the lowest, with an average of 43.30 U/mL, and the activity of LZM in the liver was the highest, with an average of 160.13 U/mL.

3.3. NEMO Bioinformatic Analysis

The *NEMO* ORF of *T. ovatus* was 1650 bp in length and encoded 548 amino acids (GenBank accession number: MW076540) (Figure 4). The predicted molecular weight of NEMO was 62.68 kDa, and the predicted isoelectric point was 5.52. It included a 68 amino acid Pfam NEMO domain (28–95 aa); a 222 amino acid coiled coil domain (138–359 aa); a 95 amino acid Pfam CC2-LZ (coiled coil region–leucine zipper domain) (365–459 aa) and a 19 amino acid ZnF C2H2 (zinc finger domain) (526–545 aa).

```
1    ATGGTGCAGCCACAACCTGATGGCCCAATGCAGTGGGACATGTCAGGGGAGGAGAGTGGAGGGGCCCTCAGGGTCCCACC  80
1    M  V  Q  P  Q  P  D  G  P  M  Q  W  D  M  S  G  E  E  S  G  G  A  L  R  V  P  P   27
81   AGAGCTGGCTGCCAATGAAGTGGTGACCAGGTTGCTGGGTGATAACCAGCAGCTCAGAGAGGCCTTGCGGCGCAGCAACC  160
28   E  L  A  A  N  E  V  V  T  R  L  L  G  D  N  Q  Q  L  R  E  A  L  R  R  S  N  L   54
161  TGGCCCTGCGTCAGCGCTGTGAGGAGATGGAGGGATGGCAGCGAAGGACAAGAGAGGAGAGAGAATTTCTCAGCTCTCGT  240
55   A  L  R  Q  R  C  E  E  M  E  G  W  Q  R  R  T  R  E  E  R  E  F  L  S  S  R      80
241  TTCCAGGAGGCCAGGGCCCTAGTGGAGAGGCTGGCCCAGGAGAACCACTCCTTACAGAGTCTGATCAGCGGTCCGTCCTC  320
81   F  Q  E  A  R  A  L  V  E  R  L  A  Q  E  N  H  S  L  Q  S  L  I  S  G  P  S  S   107
321  TAACCACTGCTGCAGCTCCAGCCAAACAGAAGACCTGCAGGGACGCCCTGCGAGGAACGGTCCGCTCGACGGACCGCAGC  400
108  N  H  C  C  S  S  S  Q  T  E  D  L  Q  G  R  P  A  R  N  G  P  L  D  G  P  Q  P   134
401  CCGTGGAAGGTGCCAACGAGTTCCTGCAGCTGCTGAAGAGCCATAAAGAAAAACTGGAGGAAGGGATGAGAGAGCTGAGG  480
135  V  E  G  A  N  E  F  L  Q  L  L  K  S  H  K  E  K  L  E  E  G  M  R  E  L  R      160
481  AGGAAGAACGAAGAGCTGGAGAGGGACAGAGATGAGGGAGAGAAAGAGAAAGAGCGGATGCGTCGCTGCATTGACCAGCT  560
161  R  K  N  E  E  L  E  R  D  R  D  E  G  E  K  E  K  E  R  M  R  R  C  I  D  Q  L   187
561  ACGAGCCAAACTGGCCCAGGCACAGGTAACTGAGGAGGTTGTTCAGCATCGCTCTGAGGCGCAACACTCCTCTGACTCTA  640
188  R  A  K  L  A  Q  A  Q  V  T  E  E  V  V  Q  H  R  S  E  A  Q  H  S  S  D  S  S   214
641  GCCTGGCTAAACTCACAGAGCAGCTTCAGGCCACACAGGGCAGGTATCGGGAGCTGGAGGAGAAGCTCGATTACCTGCAG  720
215  L  A  K  L  T  E  Q  L  Q  A  T  Q  G  R  Y  R  E  L  E  E  K  L  D  Y  L  Q      240
721  AAGAGTTCGGCTCAGCGTGACAGAACTGAAGCTCTGCTCAAACAAAAGGACAAGGACTGTGCTCAGCTGGCCAAAGACTG  800
241  K  S  S  A  Q  R  D  R  T  E  A  L  L  K  Q  K  D  K  D  C  A  Q  L  A  K  D  C   267
801  TGAGGCTCTGAAAGCTCAGGCAACTTCCCTGTTAGGAGAACTGAATGAGAGACAGAGCTGCCTGGAGAAAAGCGAGCATG  880
268  E  A  L  K  A  Q  A  T  S  L  L  G  E  L  N  E  R  Q  S  C  L  E  K  S  E  H  E   294
881  AACGCAAAATGTTGGAAGAAAAGCTGTGCAGTAAGATGAAGGCCCTGCAGGCAGCAGAGCGGGACCTGGAGCAGCAGAGG  960
295  R  K  M  L  E  E  K  L  C  S  K  M  K  A  L  Q  A  A  E  R  D  L  E  Q  Q  R      320
961  AAGCAGAATAATGTGGCCATGGACAAGCTGCTGCTGCAGACCCAGAGCCTGGAGCAGGCCCTGAAGACAGAGAGAAACGT  1040
321  K  Q  N  N  V  A  M  D  K  L  L  L  Q  T  Q  S  L  E  Q  A  L  K  T  E  R  N  V   347
1041 CGTCACAGAGGAGAAGAAAAAACTGACACAGCTGCAACACGCCTACACGTGTCTGTTTAGAGACTACGACTCTAAACTGA  1120
348  V  T  E  E  K  K  K  L  T  Q  L  Q  H  A  Y  T  C  L  F  R  D  Y  D  S  K  L  K   374
1121 AGAGTGAGCAGGGAGGAGATCTGTGCAGCCGTCTAGAGGAGGCAGAGCGAGCGTTGGCCCTGAAGCAGGATCTGATCGAT  1200
375  S  E  Q  G  G  D  L  C  S  R  L  E  E  A  E  R  A  L  A  L  K  Q  D  L  I  D      400
1201 AAACTGAAGGAGGAGGTGGAGCAGCAGAAAGGCTCACTGGAGACTGTTCCCGTCCTCGCTGCACAGTACAAGGCAGATTT  1280
401  K  L  K  E  E  V  E  Q  Q  K  G  S  L  E  T  V  P  V  L  A  A  Q  Y  K  A  D  F   427
1281 CCTCGCAGAGCGAGAAGCGAGAGAGAAGCTGAACCAGAAGAAGGAGGAGCTGCAGGATCAACTGACTCAGGCCCTGGCTG  1360
428  L  A  E  R  E  A  R  E  K  L  N  Q  K  K  E  E  L  Q  D  Q  L  T  Q  A  L  A  E   454
1361 AGATTGACAGGCTCAAACAGGAAGCCACAGCACGGGCACGTATGGAGCAGATGAAGATGAGGCACATGGAGGACTTTCCA  1440
455  I  D  R  L  K  Q  E  A  T  A  R  A  R  M  E  Q  M  K  M  R  H  M  E  D  F  P      480
1441 CCACGGGCTCCACTCATTCCTCCTCCACAAGGTCCAGCTTTCAACACTGTGCCCCCCGCCCCCTCATTCCGCAATCAGGG  1520
481  P  R  A  P  L  I  P  P  P  Q  G  P  A  F  N  T  V  P  P  A  P  S  F  R  N  Q  G   507
1521 TTTGGTTCCAGTCGGTGATCAAGGAGCAGCTGGAGCTGAGGATGTTCCAGTTTTATGTTGTCCTAAGTGCCAGTACCAGG  1600
508  L  V  P  V  G  D  Q  G  A  A  G  A  E  D  V  P  V  L  C  C  P  K  C  Q  Y  Q  A   534
1601 CTCCAGATATGGACACACTGCAGATACACGTCATGGACTGTATACAGTAA  1650
535  P  D  M  D  T  L  Q  I  H  V  M  D  C  I  Q  *   549
```

Figure 4. *NEMO* ORF sequence and predicted amino acid sequence analysis. NEMO domain is marked by light grey shadows. The low-complexity region is underlined. The CC2-LZ (coiled coil region–leucine zipper) domain is marked by dark grey shadows. The ZnF C2H2 (Zinc finger) domain is marked by a box. * stands for termination codon.

To study protein structure conservation, SMART was used to predict and align the NEMO protein structure of *T. ovatus* with other species (Figure 5A). Multiple alignment analysis of the NEMO amino acid sequence showed that NEMO has three conserved domains.

Multiple sequence alignment results revealed that the amino acid sequences of ToNEMO were highly conserved with the corresponding sequences of other species. To further study the homology of ToNEMO with other species, the amino acid sequence of ToNEMO was aligned with other species via MEGA 7.0 with the neighbour-joining (NJ) method

(Figure 5B). In teleosts, ToNEMO had the highest homology with *Seriola dumerili* NEMO and *Seriola lalandi dorsalis* NEMO isoform X3 and the lowest homology with *Acipenser ruthenus* NEMO. Compared with Mammalia, Lepidosauria, Aves and Amphibia, ToNEMO had the highest homology with Mammalia and the lowest homology with Amphibia.

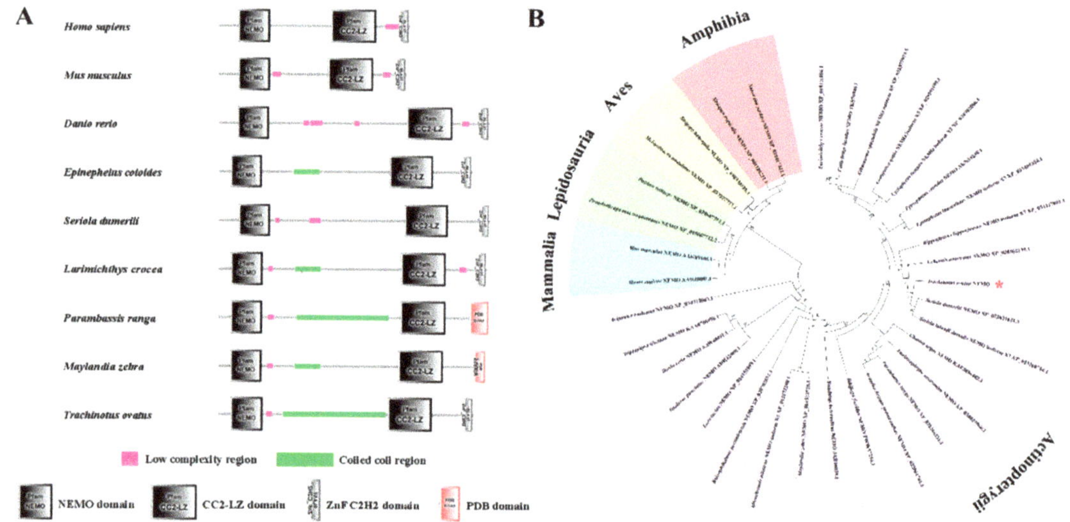

Figure 5. Protein structure (**A**) and phylogenetic tree (**B**) of comparing ToNEMO with other NEMO vertebrates. SMART was used to predict and aligned NEMO protein structure of *T. ovatus* and other species. MEGA 7.0 with neighbour-joining (NJ) method was used to structure phylogenetic tree. *T. ovatus* is marked by green shading (**A**) and red asterisk (**B**). The accession numbers are listed in Table S1.

3.4. NEMO Gene Expression Pattern Analysis

To study the relative mRNA level in various tissues, the constitutive expression of *NEMO* in blood, brain, muscle, heart, gonad, spleen, kidney, liver, skin, gill and intestine was detected via qRT-PCR (Figure 6A). The *NEMO* gene was highly expressed in the blood, brain and muscle, while low expression was observed in the intestine, gill and skin ($p < 0.05$).

To further study the expression pattern of *NEMO* in the defence against parasite infection, the mRNA levels of *NEMO* were determined in local infection sites (skin and gills) and systemic immune tissues (liver, spleen and head kidney) after *C. irritans* challenge (Figure 6B). The relative expression level of *NEMO* in gills was upregulated from 3 h to 12 h, and a peak (1.55-fold relative to the control) was observed at 12 h, and then, it was generally downregulated. In the skin, *NEMO* was upregulated to a peak from 3 h to 6 h, and then, it returned to normal levels. In the liver, *NEMO* was upregulated to its maximum (3.12-fold relative to the control) from 3 h to 12 h, then downregulated on 1 d, and it had a second peak (2.06-fold relative to the control) at 2 d. In the spleen, *NEMO* was upregulated at 6 h, 12 h, and 1 d and then returned to normal levels. In the head kidney, *NEMO* was upregulated to a peak from 6 h to 12 h and then generally downregulated to normal levels. According to these results, the difference in the relative expression of the *NEMO* gene in the gills and spleen was the smallest, and the difference in the relative expression in the liver was the largest. The highest peak of *NEMO* was observed at 6 h and 12 h after challenge with *C. irritans*.

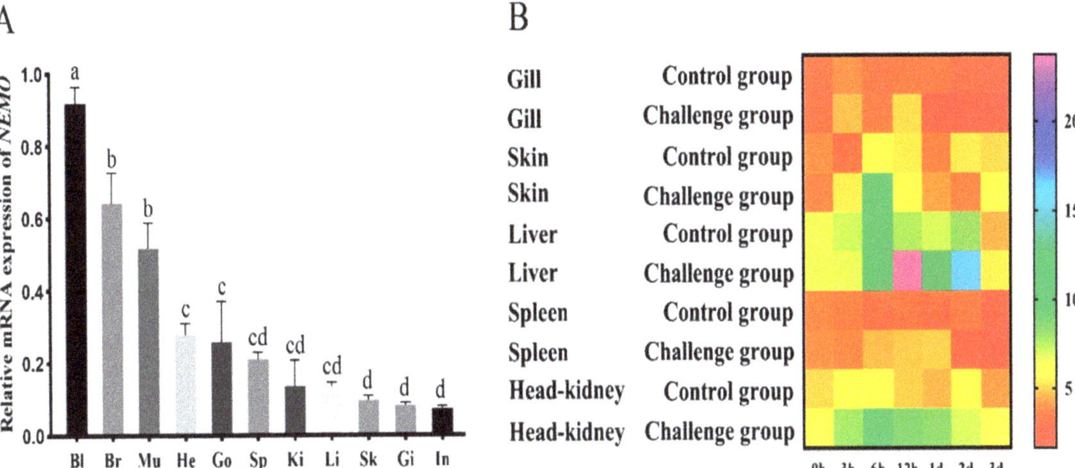

Figure 6. *ToNEMO* transcriptions in various tissues. (**A**) qRT-PCR was used to test relative *NEMO* mRNA levels in healthy fish, containing blood (Bl), brain (Br), muscle (Mu), heart (He), gonad (Go), spleen (Sp), kidney (Ki), liver (Li), skin (Sk), gill (Gi) and intestine (In). Significant differences at $p < 0.05$ are labelled with different letters, and mean $\pm$ SEM of each mRNA quantity is shown for each tissue tested. (**B**) Temporal mRNA expression analyses of *ToNEMO* in different tissues (gill, skin, liver, spleen and head kidney) after PBS (control), *C. irritans* challenges (0 h, 3 h, 6 h, 12 h, 1 d, 2 d and 3 d). EF-1a is used as the internal control to calibrate the cDNA templates for all the samples. The heatmap was constructed using Graphpad Prism 5.0 software.

4. Discussion

In the present study, *T. ovatus* was infected by stimulating *C. irritans*. HE staining was used to observe the life history of *C. irritans* and the histopathological characteristics of the infected *T. ovatus*. The enzyme activity of ACP, AKP, SOD and LZM was detected via colorimetry to study the histological and immune stress response of *T. ovatus* after challenge with *C. irritans*. Moreover, qRT-PCR was used to detect the relative expression levels of *NEMO* in response to *C. irritans* infection.

The infection of *T. ovatus* by *C. irritans* will result in direct mechanical damage to the parasitized tissue, internal tissue haemorrhage and necrosis, as well as the migration and proliferation of a large number of immune cells and the upregulation of chemokines [4,26]. Histopathological results showed hyperplasia in the gill filaments, cell shedding and tissue necrosis. The gill lamellae were swollen, separated and split seriously, and a large number of white blood cells had migrated and gathered in the parasitized part. Mucous cells in the skin had thickened, and mucous cells and secreted fluid quantity increased significantly. The skin was destroyed, the dermis layer was exposed, and cell shedding and tissue necrosis were observed. The pathological tissue morphology was similar to that in a previous report [4,26].

Immuno-related enzymes, such as ACP, AKP, SOD and LZM, are important immune indicators for aquaculture animals when they undergo pathogen invasion. ACP plays a role in killing and digesting pathogens in immune responses [13]. AKP is also a multifunctional enzyme involved in immune responses [13]. Research has shown that the serum ACP activity of snails (*Biomphalaria glabrata*) will increase significantly in response to the pathogen infection [27]. Studies have shown that when infected by acute viral necrosis virus (AVNV), the blood AKP level of *Chlmys frreri* was higher than that of the control group [28]. Increases in ACP and AKP activities indicate that defence against foreign materials is enhanced in *Apostichopus japonicus* [29]. According to these results, AKP activity in the head kidney and ACP activity in the skin were changed the most after

C. irritans challenge, suggesting that 6–12 h after infection, ACP and AKP may mainly participate in detoxification in these tissues.

Among antioxidant responses in fish, SOD production is a first line of defence against oxidative stress, converting superoxide anions to hydrogen peroxide and oxygen [30]. SOD showed significant upregulation in the immune tissues, suggesting that SOD may clear superoxide free radicals and enhance the antioxidant capacity of cells instead of directly acting on the pathogen at the site of *C. irritans* infection. LZM is also an important lysosomal enzyme that can lyse and digest pathogenic microorganisms [31]. The results of LZM showed significant upregulation in all five tissues. The upregulation of lysozymes on the skin may be due to the secretion of mucosal immune tissue as a barrier against the stimulation of cryptosporidium, and a high level of its expression in the immune system indicates the initiation of a non-specific immune response in the body.

Bioinformatic analysis showed that the NEMO domain, coiled coil region–leucine zipper domain and zinc finger domain were highly conserved in different species. It has been previously reported that orange-spotted grouper *TLR2* and its downstream pathway genes were significantly upregulated after challenge with *C. irritans* [32]. The challenge of *C. irritans* may activate the innate immune inflammatory response through the TLR2 pathway [6]. MyD88 and TIRAP are recruited by TLR2 and form the IRAK-TRAF6 complex. TRAF ubiquitination leads to the formation of the IKK complex via NEMO, IKKα and IKKβ, which results in the phosphorylation of IκB. Next, NF-κB is transferred to the nucleus to activate the target genes and induce the transcription of TNF, Pro-IL-1b, IkBz, ATF3, Zc3 h12a, TTP, etc. [6]. The NEMO domain is an important structure for recognizing and binding IKKβ [33]. The coiled coil domain ensures the integrity of the NEMO structure. Conservation among different species suggests that *T. ovatus* NEMO may play the same role in innate immunity as other species [34].

The infection of *T. ovatus* by *C. irritans* activates the innate immune signalling pathway of *TLR2*, and its downstream factors are significantly upregulated in infected sites and systemic immune tissues [35,36]. NEMO is a necessary regulator of NF-κB and plays a crucial role in the innate immunity response to pathogens, including bacteria and viruses. *NEMO* receives upstream signals from the TLR family [6].

The *NEMO* gene was upregulated in the skin, liver, spleen, head and kidney after infection, which is similar to the results found for the *TLR2* signalling pathway in orange-spotted groupers infected by *C. irritans* reported in a previous study [32]. After infection, *NEMO* in the gill, skin, spleen and head kidney reached a peak at 6–12 h after infection. Compared with *TLR2*, the time at which *NEMO* reached its peak was significantly delayed. qRT-PCR results suggested that the pathogen recognition reaction was initiated at 3–6 h, and the innate immune *NEMO* downstream pathway was upregulated to its peak at 6–12 h.

In conclusion, the histological sections showed that there was considerable metamorphosis and hyperplasia in the parasitized sites (skin) with leukocyte aggregation and mucous cell increases after *C. irritans* infection. Moreover, the activities of four enzymes were significantly increased in different tissues after *C. irritans* infection. The *ToNEMO* transcripts were universally expressed in all of the examined tissues, with higher levels being observed in the immune-relevant and central nervous tissues. The mRNA levels of *ToNEMO* after *C. irritans* infection were significantly increased in the gill, skin, liver, spleen and head kidney. This study could help deepen the understanding of the poisoning mechanism of *C. irritans* to golden pompano at the molecular level. It could provide some basic data for the study of the healthy culture and physiological function regulation of golden pompano.

Supplementary Materials: The following supporting information can be downloaded at: https://www.mdpi.com/article/10.3390/jmse11020262/s1, Table S1: Information for NEMO amino acid sequence alignment and constructing phylogenetic tree.

Author Contributions: K.-C.Z., D.-C.Z. conceived and designed the experiments; K.-C.Z. performed the experiments; B.-S.L. and N.Z. contributed to sample collection; H.-Y.G. and W.-F.L. analysed the data and wrote the paper; B.L., J.-W.Y. and D.-C.Z. assisted with writing and proofreading. All authors have read and agreed to the published version of the manuscript.

Funding: This study was supported by the National Natural Science Foundation of China (U20A2064), the Central Public-interest Scientific Institution Basal Research Fund CAFS (NO.2020TD29), the China Agriculture Research System (CARS-47) and the Guangdong Provincial Special Fund for Modern Agriculture Industry Technology Innovation Teams (2019KJ143).

Institutional Review Board Statement: All applicable international, national and institutional guidelines for the care were followed by the authors.

Informed Consent Statement: Informed consent was obtained from all subjects involved in the study. Written informed consent has been obtained from the patients to publish this paper.

Data Availability Statement: All data generated or analysed during this study are included in this published article.

Conflicts of Interest: The authors declare no conflict of interest.

References

1. Colorni, A.; Burgrss, P. *Cryptocaryon irritans* Brown 1951, the cause of 'white spot disease' in marine fish: An update. *Aquar. Sci. Conserv.* **1997**, *1*, 217–238. [CrossRef]
2. Zhong, Z.H.; Jiang, B.; Li, Z.C.; Li, S.Y.; Li, A.X. Quantification of parasite abundance: A novel method to evaluate anti-*Cryptocaryon irritans* efficacy. *Aquaculture* **2020**, *528*, 735482. [CrossRef]
3. Li, H.Y.; Wei, X.M.; Yang, J.L.; Zhang, R.R.; Zhang, Q.; Yang, J.M. The bacteriolytic mechanism of an invertebrate-type lysozyme from mollusk *Octopus ocellatus*. *Fish Shellfish Immunol.* **2019**, *93*, 232–239. [CrossRef] [PubMed]
4. Chen, F.Y.; Liang, W.W.; Chen, M.; Huang, T.; Li, L.P.; Lei, A.Y.; Wang, R.; Yang, X.M.; Luo, H.L.; Ouyang, X.H. Histopathology study on the *Cryptocaryon irritans* disease of *Trachinotus ovatus*. *J. Fish Res.* **2017**, *39*, 181–187. (In Chinese)
5. Braue, J.; Murugesan, V.; Holland, S.; Patel, N.; Naik, E.; Leiding, J.; Yacoub, A.T.; Prieto-Granada, C.N.; Greene, J.N. NF-kappaB Essential Modulator Deficiency Leading to Disseminated Cutaneous Atypical Mycobacteria. *Mediterr. J. Hematol. Infect. Dis.* **2015**, *7*, e2015010. [CrossRef] [PubMed]
6. Takeuchi, O.; Akira, S. Pattern recognition receptors and inflammation. *Cell* **2010**, *140*, 805–820. [CrossRef] [PubMed]
7. May, M.J.; D'Acquisto, D.; Madge, L.A.; Glochner, J.; Pober, J.S.; Ghosh, S. Selective Inhibition of NF-κB Activation by a Peptide That Blocks the Interaction of NEMO with the IκB Kinase Complex. *Science* **2000**, *289*, 1550–1554. [CrossRef]
8. Solt, L.A.; May, M.J. The IκB kinase complex: Master regulator of NF-κB signaling. *Immunol. Res.* **2008**, *42*, 3–18. [CrossRef] [PubMed]
9. Hayden, M.S.; Ghosh, S. NF-kappaB, the first quarter-century: Remarkable progress and outstanding questions. *Genes Dev.* **2012**, *26*, 203–234. [CrossRef]
10. Hayden, M.S.; Ghosh, S. Signaling to NF-kappaB. *Genes Dev.* **2004**, *18*, 2195–2224. [CrossRef]
11. Scheidereit, C. IkappaB kinase complexes: Gateways to NF-kappaB activation and transcription. *Oncogene* **2006**, *25*, 6685–6705. [CrossRef] [PubMed]
12. Yin, F.; Dan, X.M.; Sun, P.; Shi, Z.H.; Gao, Q.X.; Peng, S.M.; Li, A.X. Growth, feed intake and immune responses of orange-spotted grouper (*Epinephelus coioides*) exposed to low infectious doses of ectoparasite (*Cryptocaryon irritans*). *Fish Shellfish Immunol.* **2014**, *36*, 291–298. [CrossRef] [PubMed]
13. Kong, X.H.; Wang, S.P.; Jiang, H.X.; Nie, G.X.; Li, X.J. Responses of acid/alkaline phosphatase, lysozyme, and catalase activities and lipid peroxidation to mercury exposure during the embryonic development of goldfish *Carassius auratus*. *Aquat. Toxicol.* **2012**, *120*, 119–125. [CrossRef] [PubMed]
14. Kim, J.H.; Kim, S.K.; Hur, Y.B. Toxic effects of waterborne nitrite exposure on antioxidant responses, acetylcholinesterase inhibition, and immune responses in olive flounders, *Paralichthys olivaceus*, reared in bio-floc and seawater. *Fish Shellfish Immunol.* **2020**, *97*, 581–586. [CrossRef]
15. Zhang, C.N.; Zhang, J.L.; Liu, M.; Huang, M.X. Molecular cloning, expression and antibacterial activity of goose-type lysozyme gene in *Microptenus salmoides*. *Fish Shellfish Immunol.* **2018**, *82*, 9–16. [CrossRef] [PubMed]
16. Zhong, Z.H.; Guo, W.L.; Lei, Y.; Wang, F.; Wang, S.F.; Sun, Y.; Hu, W.T.; Zhou, Y.C. Antiparasitic efficacy of honokiol against *Cryptocaryon irritans* in pompano, *Trachinotus ovatus*. *Aquaculture* **2019**, *500*, 398–406. [CrossRef]
17. Watanabe, Y.; How, K.H.; Zenke, K.; Itoh, N.; Yoshinaga, T. Characterization of the proteases in the parasitic stage of *Cryptocaryon irritans*, and in vitro and in vivo effects of protease inhibitors on cryptocaryoniasis. *Aquaculture* **2019**, *512*, 734311. [CrossRef]
18. Yang, Q.; Zhu, K.C.; Guo, L.; Liu, B.S.; Guo, H.Y.; Zhang, N.; Yang, J.W.; Zhang, D.C. Molecular characterization of GRP94 and HSP90 alpha from *Trachinotus ovatus*, Linnaeus 1758 and their expression responses to various levels of stocking density stress and *Cryptocaryon irritans* infection. *Aquaculture* **2019**, *529*, 735601. [CrossRef]

19. Guo, L.; He, P.Y.; Zhu, K.C.; Guo, H.Y.; Liu, B.S.; Zhang, N.; Jiang, J.G.; Zhang, D.C. Functional identification of ToLAAO genes and polymorphism association analysis of *Cryptocaryon irritans* resistance in *Trachinotus ovatus*. *Aquac. Res.* **2021**, *53*, 208–220. [CrossRef]

20. Zhu, K.C.; Liu, J.; Liu, B.S.; Guo, H.Y.; Zhang, N.; Guo, L.; Jiang, S.G.; Zhang, D.C. Functional characterization of four ToRac genes and their association with anti-parasite traits in Trachinotus ovatus (Linnaeus, 1758). *Aquaculture* **2022**, *560*, 738514. [CrossRef]

21. Dan, X.M.; Li, A.X.; Lin, X.T.; Teng, N.; Zhu, X.Q. A standardized method to propagate *Cryptocaryon irritans* on a susceptible host pompano *Trachinotus ovatus*. *Aquaculture* **2006**, *258*, 127–133. [CrossRef]

22. Yin, F.; Sun, P.; Tang, B.J.; Dan, X.M.; Li, A.X. Immunological, ionic and biochemical responses in blood serum of the marine fish *Trachinotus ovatus* to poly-infection by *Cryptocaryon irritans*. *Exp. Parasitol.* **2015**, *154*, 113–117. [CrossRef] [PubMed]

23. Zhang, D.C.; Guo, L.; Guo, H.Y.; Zhu, K.C.; Li, S.Q.; Zhang, Y.; Zhang, N.; Liu, B.S.; Jiang, S.G.; Li, J.T. Chromosome-level genome assembly of golden pompano (*Trachinotus ovatus*) in the family Carangidae. *Sci. Data* **2019**, *6*, 216. [CrossRef] [PubMed]

24. Zhu, K.C.; Song, L.; Guo, H.Y.; Guo, L.; Zhang, N.; Liu, B.S.; Jiang, S.G.; Zhang, D.C. Elovl4a participates in LC-PUFA biosynthesis and is regulated by PPAR alpha beta in golden pompano *Trachinotus ovatus* (Linnaeus 1758). *Sci. Rep.-Uk.* **2019**, *9*, 4684. [CrossRef]

25. Livak, K.J.; Schmittgen, T.D. Analysis of relative gene expression data using real-time quantitative PCR and the 2(T)(-Delta Delta C) method. *Methods* **2001**, *25*, 402–408. [CrossRef] [PubMed]

26. Li, Y.W.; Jiang, B.; Dan, X.M.; Li, A.X. Advances in the research on mucosal immune response of fish against *Cryptocaryon irritans* infection. *J. Fish. China* **2019**, *43*, 156–167. (In Chinese)

27. Cheng, T.; Dougherty, W. Ultrastructural evidence for the destruction of *Schistosoma mansoni* sporocysts associated with elevated lysosomal enzyme levels in *Biomphalaria glabrata*. *J. Parasitol.* **1989**, *75*, 928–941. [CrossRef]

28. Xing, J.; Zhan, W.B.; Zhou, L. Endoenzymes associated with haemocyte types in the scallop (*Chlamys farreri*). *Fish Shellfish Immunol.* **2002**, *13*, 271–278. [CrossRef]

29. Huo, D.; Sun, L.; Ru, X.; Zhang, L.; Lin, C.; Liu, S.; Xin, X.; Yang, H. Impact of hypoxia stress on the physiological responses of sea cucumber *Apostichopus japonicus*: Respiration, digestion, immunity and oxidative damage. *Peer J.* **2018**, *6*, e4651. [CrossRef]

30. Sun, S.; Ge, X.; Zhu, J.; Xuan, F.; Jiang, X. Identification and mRNA expression of antioxidant enzyme genes associated with the oxidative stress response in the Wuchang bream (*Megalobrama amblycephala* Yih) in response to acute nitrite exposure. *Comp. Biochem. Physiol. C Toxicol. Pharmacol.* **2014**, *159*, 69–77. [CrossRef]

31. Sun, Y.; Ding, S.; He, M.; Liu, A.; Long, H.; Guo, W.; Cao, Z.; Xie, Z.; Zhou, Y. Construction and analysis of the immune effect of *Vibrio harveyi* subunit vaccine and DNA vaccine encoding TssJ antigen. *Fish Shellfish Immunol.* **2020**, *98*, 45–51. [CrossRef] [PubMed]

32. Li, Y.W.; Dan, X.M.; Zhang, T.W.; Luo, X.C.; Li, A.X. Immune-related genes expression profile in orange-spotted grouper during exposure to *Cryptocaryon irritans*. *Parasite Immunol.* **2011**, *33*, 679–987. [CrossRef] [PubMed]

33. Rushe, M.; Silvian, L.; Bixler, S.; Chen, L.L.; Cheung, A.; Bowes, S.; Cuervo, H.; Berkowitz, S.; Zheng, T.; Guckian, K.; et al. Structure of a NEMO/IKK-associating domain reveals architecture of the interaction site. *Structure* **2008**, *16*, 798–808. [CrossRef] [PubMed]

34. Barczewski, A.H.; Ragusa, M.J.; Mierke, D.F.; Pellegrini, M. The IKK-binding domain of NEMO is an irregular coiled coil with a dynamic binding interface. *Sci. Rep.* **2019**, *9*, 2950. [CrossRef] [PubMed]

35. Li, Y.W.; Luo, X.C.; Dan, X.M.; Huang, X.Z.; Qiao, W.; Zhong, Z.P.; Li, A.X. Orange-spotted grouper (*Epinephelus coioides*) TLR2, MyD88 and IL-1beta involved in anti-*Cryptocaryon irritans* response. *Fish Shellfish Immunol.* **2011**, *30*, 1230–1240. [CrossRef] [PubMed]

36. Zhu, K.C.; Liu, B.S.; Zhang, N.; Guo, H.Y.; Guo, L.; Jiang, S.G.; Zhang, D.C. Interferon regulatory factor 2 plays a positive role in interferon gamma expression in golden pompano, *Trachinotus ovatus* (Linnaeus 1758). *Fish Shellfish Immunol.* **2020**, *96*, 107–113. [CrossRef] [PubMed]

Journal of
Marine Science and Engineering

Article

The Impacts of Dietary Curcumin on Innate Immune Responses and Antioxidant Status in Greater Amberjack (*Seriola dumerili*) under Ammonia Stress

Chuanpeng Zhou [1,2], Zhong Huang [2,3], Shengjie Zhou [2,4], Jing Hu [2,4], Rui Yang [2,4], Jun Wang [1,2], Yun Wang [1,2], Wei Yu [2,3], Heizhao Lin [2,3,]* and Zhenhua Ma [2,4,]*

1   Key Laboratory of Aquatic Product Processing, Ministry of Agriculture and Rural Affairs, South China Sea Fisheries Research Institute, Chinese Academy of Fishery Sciences, Guangzhou 510300, China
2   Key Laboratory of Efficient Utilization and Processing of Marine Fishery Resources of Hainan Province, Sanya 512426, China
3   Shenzhen Base of South China Sea Fisheries Research Institute, Chinese Academy of Fishery Sciences, Shenzhen 518121, China
4   Tropical Aquaculture Research and Development Center, South China Sea Fisheries Research Institute, Chinese Academy of Fishery Sciences, Sanya 512426, China
*   Correspondence: linheizhao@163.com (H.L.); zhenhua.ma@hotmail.com (Z.M.)

Abstract: In this study, we investigated the effect of dietary curcumin on non-specific immune responses and antioxidative ability in *Seriola dumerili* under ammonia stress and post-recovery. Three diets were prepared to contain 0, 75, and 150 mg/kg of curcumin. A total of 225 greater amberjack (initial weight: 100.90 ± 0.03 g) were distributed into nine cylindrical tanks, constituting an experimental design with three treatments and three replicates. After 56 days of feeding, plasma, intestinal, and hepatic enzyme activities were evaluated. Then, an acute ammonia challenge experiment was conducted. Ten fish per tank were subjected to acute ammonia stress (total ammonia-N: 1000 mg/L) for eight minutes followed by six minutes of recovery. The results indicated that dietary curcumin significantly promoted intestinal and hepatic alkaline phosphatase (ALP) and acid phosphatase (ACP) levels as well as hepatic antioxidative enzymes such as superoxide dismutase (SOD), total antioxidant capacity (T-AOC), reduced glutathione (GSH), and glutathione peroxidase (GSH-Px) of greater amberjack. In addition, curcumin addition improved the activities of antioxidant enzymes, such as SOD, T-AOC, GSH, GSH-Px, and catalase (CAT), and reduced malondialdehyde (MDA) content in liver, spleen, head kidney, and brain tissues after post-recovery. The indexes related to immunity and antioxidant enzymes in the liver, gill, and spleen rose again to some extent, but they showed the worst recovery ability in the head kidney and brain tissue samples. These results indicate that dietary curcumin supplementation could increase non-specific immune responses, antioxidant ability, and enhance resistance to high ammonia stress in juvenile *S. dumerili*.

Keywords: curcumin; ammonia; enzyme activity; oxidative stress; *Seriola dumerili*

Citation: Zhou, C.; Huang, Z.; Zhou, S.; Hu, J.; Yang, R.; Wang, J.; Wang, Y.; Yu, W.; Lin, H.; Ma, Z. The Impacts of Dietary Curcumin on Innate Immune Responses and Antioxidant Status in Greater Amberjack (*Seriola dumerili*) under Ammonia Stress. *J. Mar. Sci. Eng.* **2023**, *11*, 300. https://doi.org/10.3390/jmse11020300

Academic Editor: Dariusz Kucharczyk

Received: 20 December 2022
Revised: 15 January 2023
Accepted: 17 January 2023
Published: 1 February 2023

1. Introduction

Whether in the wild or in an aquaculture environment, fish are often under stress (e.g., exposed to environmental contaminants, extreme conditions or water quality change, handling, transport, population density, or bacterial and viral invasion). These negative factors can lead to stress responses in fish species [1]. Between 60 and 80% of the nitrogenous waste excreted by fish species is ammonia. Ammonia is the final product of protein catabolism [2]. Ammonia includes ionized (NH_4^+) and unionized (NH_3) forms [3], and the toxic effects of ammonia is mainly attributed to NH_3, as it can diffuse freely along the concentration gradient to the gill membranes [2]. The accumulation of ammonia in water is a serious problem in today's high-density aquaculture environment. Ammonia concentrations in water can rise rapidly and lead to acute toxicity to aquatic organisms. Many studies have demonstrated that excessively high concentrations of ammonia are detrimental to fish health, leading to immunosuppression and high mortality [4] and damage to many organs (such as the gill, liver, etc.) of fish species [5].

Curcumin (*Rhizoma curcumae* Longae) is the main active ingredient in turmeric [6,7]. It has extensively pivotal functions such as reducing inflammation as well as antioxidant, antitumor, and anti-stress properties in mammalian and aquatic animals [8,9]. Previous studies have shown many kinds of functions of curcumin in fish and other aquatic animals. Numerous studies have demonstrated that dietary curcumin could improve antioxidant capacity and immune function in tilapia (*Oreochromis niloticus*) [10], rainbow trout (*Oncorhynchus mykis*) [11], Japanese Sea bass (*Lateolabrax japonicus*) [12], and largemouth bass (*Micropterus salmoides*) [13]. In addition, dietary curcumin can play a protective role against environmental stress via promotion of antioxidant enzyme activities such as GSH-Px, SOD, and CAT activities [14,15]. In another study on tilapia, supplementation of curcumin also markedly enhanced anti-oxidative status during exposure to natural challenging cold temperatures [16].

Greater amberjack *Seriola dumerili* is a marine pelagic fish species with a circumglobal distribution throughout warm and tropical waters (such as in Australia, New Zealand, Japan, China, the United States, and Chile) [17]. This species has been targeted for commercial farming in Japan, Australia, and the Mediterranean region because of its high growth rate, superior flesh quality, and high commercial value [17,18]. It was reported that the aquaculture production of amberjack in 2021 reached more than 20,000 tons in China [19]. In the future, the artificial culture of *S. dumerili* has great potential in China. With the rise of high-density intensive farming, *S. dumerili* cultured in marine cages face more stressors, such as temperature, salinity, pH, bacteria, parasites, etc., which can seriously affect the healthy development of *S. dumerili*. However, to date, limited studies have been published examining the effects of dietary curcumin on resistance to ammonia stress, particularly changes in the activities of enzymes associated with various tissues in this fish. Therefore, in the present study, we tried to determine the effects of curcumin on the innate immune response and hepatic histology of *S. dumerili*, and we aimed to study the impacts of acute ammonia exposure on non-specific immune parameters and antioxidant abilities more comprehensively. We also sought to investigate the ability of recovery after ammonia exposure, which will provide some insights for disease prevention and/or stress attenuation.

2. Materials and Methods

All the experiments were conducted in accordance with the Guidelines for the Care and Use of Laboratory Animals of South China Sea Fisheries Institute. This study was recognized by the Ethics Committee of South China Sea Fisheries Institute (No. 20200815).

2.1. Experimental Diets

Fishmeal, corn gluten meal, and soybean meal were used as dietary protein sources. Fish oil and lecithin were used as lipid sources. Contents of crude protein and crude lipid were 49.7% and 12.7%, respectively, and this formulation (Table 1) has been shown to be nutritionally adequate for the growth of *S. dumerili* [20]. Three diets were formulated to supplement with levels of curcumin (0, 75 mg/kg, and 150 mg/kg) (Purity, ≥98%, Xi'an Bluegrass Biotechnology Co., Ltd., Shaanxi, China), which were named as control group, 75 mg/kg group, and 150 mg/kg group, respectively. The experimental diets were made by blending all the components well with oil and then adding distilled water until a stiff dough was produced. The pellets with 1 mm diameter were wet-extruded by a pelletizer (Institute of Chemical Engineering, South China University of Technology, Guangzhou, China) and then air-dried (26 °C, 3 days), put in plastic self-sealing bags, and stored at −20 °C until use.

Table 1. Ingredients and proximate composition of experimental diets.

Ingredient	(% of Dry Matter)
Fish meal	60
Corn gluten meal	8
Soybean meal	10
Corn starch	8
Microcrystalline cellulose	2
Fish oil	7
Lecithin	1
Vitamin mixture [1]	0.5
Mineral mixture [2]	0.5
Choline chloride	0.5
Betaine	0.5
Carboxyl-methyl cellulose	2
Total	100
Proximate composition	
Dry matter	87.9
Crude protein	49.7
Crude lipid	12.7
Crude ash	10.7
Gross energy (kJ/g)	18.5

[1,2] The specific components of vitamin premix and mineral premix refer to our previous literature [21].

2.2. Fish and Animal Husbandry

The feeding experiment was conducted in the Sanya Basement of the South China Sea Fisheries Research Institute, CAFS. Juvenile *S. dumerili* were bought in a commercial fish farm. Before the feeding experiment, fish were taken to 450 L cylindrical fiberglass tanks and fed with control diet to domesticate them for 14 days. The fish were fasted for 24 h before grouping. A total of 225 fish with uniform size (100.90 ± 0.03 g) were randomly divided into nine tanks and raised for 56 days. Fish were artificially fed twice a day (at 8:00 and 16:00) until apparent satiation. During the period, water quality parameters were as follows: temperature 29.0 ± 1.0 °C, dissolved oxygen >6.0 mg/L, pH 8.2–8.4, salinity 33.0 ± 0.04‰, and ammonia <0.01 mg/L, respectively. The photoperiod was the natural solar cycle throughout the whole process.

2.3. Acute Ammonia Challenge Experiment

After eight weeks of the feeding trial, the fish were starved for 24 h. Ten fish of similar specification were collected from each tank and subjected to an acute ammonia stress (total ammonia-N: 1000 mg/L) test in cylindrical tanks (100-L) according to the methods described in Zhang et al. [2,22]. The water temperature ranged from 28 °C to 30 °C, the flow rate was 2.2 L/min, and the DO was >6 mg/L. Ammonium chloride (NH_4Cl) was used as an ammonia source and added to required final ammonia contents. The fish were sampled at eight minutes, and the rest were taken to aerated water in a tank for six minutes of post-exposure recovery.

2.4. Sample Collection and Chemical Analyses

At the end of the rearing experiment, the fish were sampled from three fish per tank at 0, 8, and 14 min (8 + 6 min; they were taken after 6 min of post-exposure recovery). The fish were quickly netted and anesthetized with diluted eugenol (1:10,000; Shanghai Reagent Corp., China), and then three fish were randomly selected from each tank, and blood was collected through the caudal vein with 2 mL heparinized syringes. After collection, blood was centrifuged (3000× g at 4 °C for 10 min) to obtain plasma (stored at −80 °C for use). The liver, intestine, brain, gill, spleen, and head kidney samples were stored at −80 °C until enzyme activities were determined. The liver was taken and then placed in a fixed solution (4% paraformaldehyde) with 10 mL centrifuge tubes for morphological observation.

2.5. Blood Biochemical Parameter Measurements

Plasma alkaline phosphatase (ALP) and acid phosphatase (ACP) activities were measured by ROCHE-P800 automatic biochemical analyzer (Roche, Basel, Switzerland) according to a standard kit method for each assay.

2.6. Tissues' Enzyme Activity Measurements

The liver, intestine, brain, gill, spleen, and head kidney samples were homogenized in cold phosphate buffer (diluted at 1:10) (phosphate buffer: 0.064 M at pH 6.4). Then the homogenate was centrifuged for 20 min (4 °C, $3000 \times g$) and the supernatant was taken to determine alkaline phosphatase (ALP), acid phosphatase (ACP), superoxide dismutase (SOD), catalase (CAT), total antioxidant capacity (T-AOC), malondialdehyde (MDA), reduced glutathione (GSH), glutathione peroxidase (GSH-Px), amylase, lipase, and trypsase levels according to the protocols provided by commercial kits (Jiancheng Institute of Biotechnology, Nanjing, China). ALP, ACP, SOD, CAT, T-AOC, MDA, GSH, GSH-Px, amylase, lipase, and trypsin activities were measured by a colorimetric method using wavelengths of 520 nm, 520 nm, 450 nm, 405 nm, 520 nm, 532 nm, 405 nm, and 412 nm, respectively. The Folin method was employed to measure the above six tissues' protein content (Lowry et al., 1951).

2.7. Statistical Analysis

The data in this study were tested for homogeneity of variances with the Levene test. All the data were subjected to one-way analysis of variance (ANOVA). When overall differences were significant, the Duncan's test was employed to compare means between treatments. The level of significant difference was $p < 0.05$. Statistical analysis was performed using the SPSS 25.0, and the results were expressed as mean $\pm$ SEM (standard error of the mean).

3. Results

3.1. Effect of Dietary Curcumin on Plasma ALP and ACP Activities of Greater Amberjack (Seriola dumerili)

The results of ALP were not significantly affected by dietary curcumin level ($p > 0.05$) (Table 2). Compared with the control group (0 mg/kg curcumin), the ACP was significantly improved in curcumin-addition groups ($p < 0.05$). There were no significant differences in ACP between curcumin treatment groups ($p > 0.05$).

Table 2. Plasma ALP and ACP of greater amberjack (*Seriola dumerili*) fed with diets containing different amounts of curcumin.

Dietary Curcumin (mg/kg)	0	75	150
ALP (king's unit/100 mL)	2.3 ± 0.15	2.17 ± 0.04	2.10 ± 0.10
ACP (king's unit/100 mL)	1.45 ± 0.02 [b]	2.67 ± 0.20 [a]	2.45 ± 0.09 [a]

The values are average $\pm$ standard error of three replications (n = 9). There was a significant difference in the average value of different superscript letters in the same row ($p < 0.05$).

3.2. Effect of Dietary Curcumin on Intestinal ALP and ACP Activities of Greater Amberjack (Seriola dumerili)

The results of intestinal ALP and ACP are shown in Table 3. The ALP and ACP levels in curcumin-addition groups were dramatically higher than that in the control treatment ($p < 0.05$). No statistical difference in ALP levels were observed between 75 mg/kg and 150 mg/kg of dietary curcumin groups ($p > 0.05$).

Table 3. Intestinal enzyme activities of greater amberjack (*S. dumerili*) fed with diets containing different amounts of curcumin.

Dietary Curcumin (mg/kg)	0	75	150
ALP (king's unit/gprot)	28,222 ± 2008 [b]	41,407 ± 2055 [a]	42,254 ± 3654 [a]
ACP (king's unit/gprot)	8710 ± 183 [b]	11,693 ± 419 [a]	11,330 ± 562 [a]

The values are average ± standard error of three replications (n = 9). There was a significant difference in the average value of different superscript letters in the same row ($p < 0.05$).

3.3. Effect of Dietary Curcumin on Hepatic Enzyme Activities of Greater Amberjack (Seriola dumerili)

Hepatic ALP levels in curcumin-addition groups were dramatically higher than those in the control treatment ($p < 0.05$) (Table 4). However, there were no significant differences in ALP between curcumin treatment groups. With the increase of dietary curcumin, the ACP levels were significantly promoted ($p < 0.05$). The ACP level in the 150 mg/kg group was significantly higher than those in control group and 75 mg/kg curcumin group ($p < 0.05$). Hepatic SOD and GPX activities in the group of 75 mg/kg dietary curcumin were dramatically higher than that in the control treatment ($p < 0.05$). However, there were no statistical differences in SOD and GPX between the control and 150 mg/kg dietary curcumin groups, respectively ($p > 0.05$). The T-AOC level in the 150 mg/kg dietary curcumin group was higher than that in the 75 mg/kg dietary curcumin group ($p < 0.05$), but it was not significantly different from that in the control group ($p > 0.05$). As for GSH, with the increase in dietary curcumin, it showed a trend of increasing and then decreasing. The GSH level in the 75 mg/kg dietary curcumin group was higher than that in the 150 mg/kg dietary curcumin group ($p < 0.05$), but it was not significantly different from that in control group ($p > 0.05$). However, with the increase of dietary curcumin, the CAT level showed a trend of increasing and then decreasing, but there was no statistical difference ($p > 0.05$). Inversely, the MDA level exhibited an opposite tendency, which was declining and then increasing.

Table 4. Hepatic enzyme activities of greater amberjack (*S. dumerili*) fed with diets containing different amounts of curcumin.

Dietary Curcumin (mg/kg)	0	75	150
ALP (king's unit/gprot)	16.62 ± 1.99 [b]	55.71 ± 4.07 [a]	52.25 ± 2.38 [a]
ACP (king's unit/gprot)	78.27 ± 4.48 [b]	90.44 ± 6.99 [b]	108.08 ± 1.91 [a]
SOD (U/mgprot)	6598 ± 772 [b]	8793 ± 768 [a]	5765 ± 204 [b]
CAT (U/mgprot)	172.89 ± 14.5	178.66 ± 4.88	160.87 ± 6.24
T-AOC (mM)	0.93 ± 0.02 [ab]	0.9 ± 0.01 [b]	0.98 ± 0.01 [a]
MDA (nmol/mgprot)	1.25 ± 0.11	1.16 ± 0.08	1.28 ± 0.05
GSH (μmol/gprot)	21.41 ± 1.37 [ab]	24.16 ± 1.05 [a]	18.12 ± 0.51 [b]
GPX (IU/mgprot)	9.19 ± 2.01 [b]	20.84 ± 1.06 [a]	11.29 ± 0.66 [b]

The values are average ± standard error of three replications (n = 9). There was a significant difference in the average value of different superscript letters in the same row ($p < 0.05$).

3.4. Intestinal Enzyme Activities in Greater Amberjack (S. dumerili) in Response to Acute Ammonia Exposure and Post-Recovery

The ALP level in the 75 mg/kg group after ammonia exposure was dramatically higher than that in the control treatment ($p < 0.05$) (Figure 1A). The ALP level was significantly increased with increasing dietary curcumin addition, reaching a maximum in the 150 mg/kg group after post-recovery ($p < 0.05$). Intestinal ALP levels in curcumin-addition groups after post-recovery were dramatically higher than that in the control treatment ($p < 0.05$) (Figure 1A). The intestinal ALP level in the control group after post-recovery was dramatically lower than that in the counterpart group after ammonia exposure ($p < 0.05$). Inversely, the ALP level in the 75 mg/kg group after post-recovery was significantly increased compared with the level after ammonia exposure ($p < 0.05$).

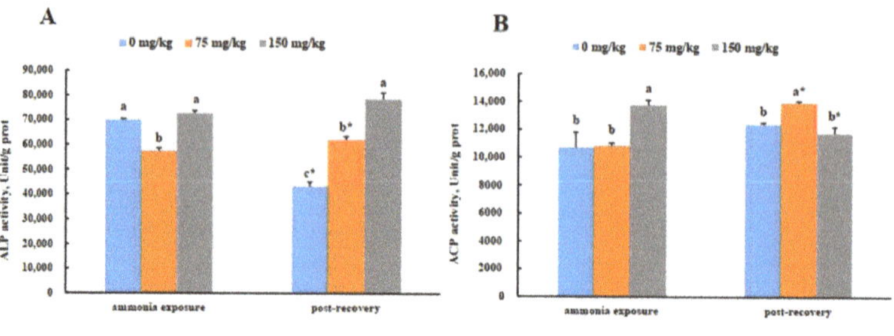

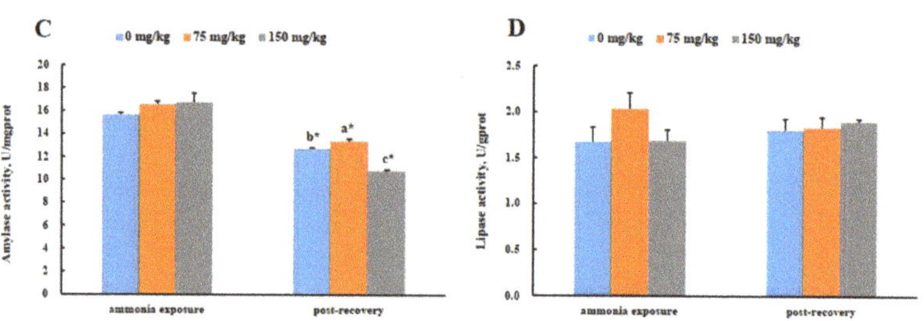

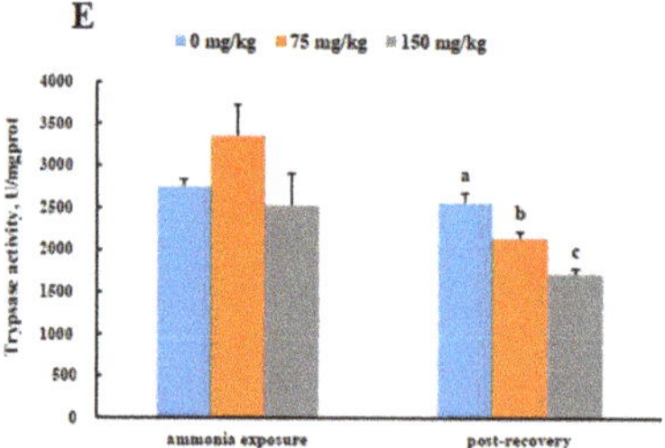

Figure 1. Intestinal ALP, ACP, amylase, lipase, and trypsase levels in greater amberjack (*S. dumerili*) fed dietary curcumin (**A**–**E**) and the response to acute ammonia exposure and post-recovery. Data are expressed as average ± standard error of the mean (SEM) (n = 9). The significant differences ($p < 0.05$) between values obtained in ammonia exposure and its post-recovery stage were determined using *t*-tests and are indicated by asterisks. Different letters indicate significant differences ($p < 0.05$) among different groups by Duncan's multi-range tests.

The ACP level in the 150 mg/kg group after ammonia exposure was dramatically higher than that in the control treatment ($p < 0.05$) (Figure 1B). The ACP level in the 75 mg/kg group was significantly higher than those in control and 150 mg/kg groups after post-recovery ($p < 0.05$). Compared with counterpart groups after ammonia exposure, the ACP level in the 75 mg/kg group after post-recovery was significantly increased ($p < 0.05$), while the ACP level in the 150 mg/kg group after post-recovery was significantly decreased ($p < 0.05$).

Compared to the control group after post-recovery, the amylase activity in the 75 mg/kg group after post-recovery was significantly increased ($p < 0.05$), while it was dramatically decreased in the 150 mg/kg group after post-recovery ($p < 0.05$) (Figure 1C). The amylase activities in the control, 75, and 150 mg/kg groups after post-recovery were significantly decreased compared with the respective groups after ammonia exposure ($p < 0.05$).

Trypsase activity decreased significantly with increasing dietary curcumin addition, reaching a minimum in the 150 mg/kg group after post-recovery ($p < 0.05$) (Figure 1E). The trypsase activities in the 75 and 150 mg/kg groups after post-recovery were statistically lower than that in the control treatment ($p < 0.05$).

3.5. Hepatic Enzyme Activities in Greater Amberjack (S. dumerili) in Response to Acute Ammonia Exposure and Post-Recovery

The hepatic ALP level was significantly increased with increasing dietary curcumin addition, reaching a maximum in the 150 mg/kg group after ammonia exposure ($p < 0.05$) (Figure 2A). Hepatic ALP levels in the 75 and 150 mg/kg groups after ammonia exposure were dramatically higher than that in the control treatment ($p < 0.05$). Hepatic ALP levels in curcumin-addition groups after post-recovery were dramatically higher than that in the control treatment ($p < 0.05$). The ALP level in the 150 mg/kg group after post-recovery was dramatically lower than that in the 75 mg/kg group after post-recovery ($p < 0.05$). Hepatic ALP levels in the control group and curcumin-addition groups after post-recovery were dramatically higher than those in the respective groups after ammonia exposure ($p < 0.05$).

Hepatic ACP levels were significantly decreased with increasing dietary curcumin addition, reaching a minimum in the 150 mg/kg group after ammonia exposure ($p < 0.05$) (Figure 2B). The ACP levels in the 75 and 150 mg/kg groups after ammonia exposure were dramatically lower than that in the control treatment ($p < 0.05$). Similar results were also observed after post-recovery. The hepatic ACP levels in curcumin-addition groups after post-recovery were dramatically lower than that in the control treatment ($p < 0.05$). The hepatic ACP levels in the control, 75, and 150 mg/kg groups after post-recovery were dramatically higher than those in the respective groups after ammonia exposure ($p < 0.05$).

Hepatic SOD levels were significantly decreased with increasing dietary curcumin addition, reaching a minimum in the 150 mg/kg group after ammonia exposure ($p < 0.05$) (Figure 2C). The SOD levels in the 75 and 150 mg/kg groups after ammonia exposure were dramatically lower than that in the control treatment ($p < 0.05$). The hepatic SOD levels in curcumin-addition groups after post-recovery were dramatically higher than that in the control treatment ($p < 0.05$). Compared with those three groups after ammonia exposure, the hepatic SOD levels in counterpart groups after post-recovery were statistically increased ($p < 0.05$).

CAT levels in the 75 and 150 mg/kg groups after ammonia exposure were dramatically lower than that in the control treatment ($p < 0.05$) (Figure 2D). Compared with the control group after ammonia exposure, the hepatic CAT level in the control group after post-recovery was significantly increased ($p < 0.05$).

Hepatic T-AOC levels in curcumin-addition groups after ammonia exposure were dramatically lower than that in the control treatment ($p < 0.05$) (Figure 2E). Compared with the control group after post-recovery, the hepatic T-AOC levels in curcumin-addition groups after post-recovery were significantly increased ($p < 0.05$). The hepatic T-AOC level in the 75 mg/kg group after post-recovery was dramatically raised compared with the higher curcumin-addition group (150 mg/kg) after post-recovery ($p < 0.05$). Compared with 75 and 150 mg/kg groups after ammonia exposure, the hepatic T-AOC level in 75 and 150 mg/kg groups after post-recovery was significantly increased ($p < 0.05$), respectively.

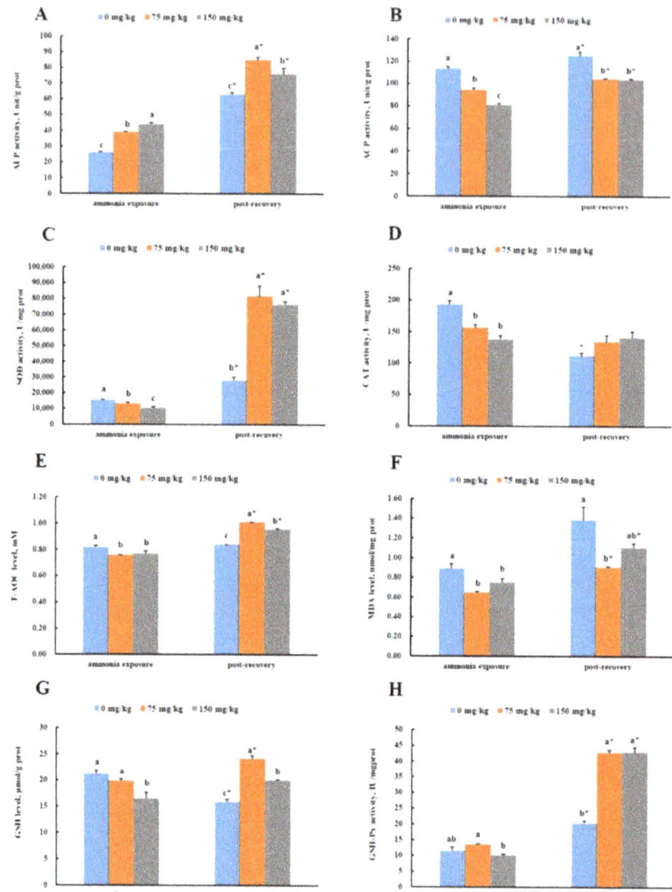

Figure 2. Hepatic ALP, ACP, SOD, CAT, T-AOC, MDA, GSH，and GSH-PX levels in greater amberjack (*S. dumerili*) fed dietary curcumin (**A–H**) and the response to acute ammonia exposure and post-recovery. Data are expressed as average ± standard error of the mean (SEM) (n = 9). The significant differences (*p* < 0.05) between values obtained in ammonia exposure and its post-recovery stage were determined using *t*-tests and are indicated by asterisks. Different letters indicate significant differences (*p* < 0.05) among different groups as assessed by Duncan's multi-range tests.

Hepatic MDA levels in curcumin-addition groups after ammonia exposure were dramatically lower than that in the control treatment (*p* < 0.05) (Figure 2F). The hepatic MDA level in the 75 mg/kg group after post-recovery was dramatically lower than that in the control treatment (*p* < 0.05). Compared with those three groups after ammonia exposure, the hepatic MDA levels in the control, 75, and 150 mg/kg groups after post-recovery were significantly increased (*p* < 0.05).

The hepatic GSH level in the 150 mg/kg group after ammonia exposure was dramatically lower than that in the control and 75 mg/kg treatments (*p* < 0.05) (Figure 2G). The hepatic GSH levels in the 75 and 150 mg/kg groups after post-recovery were dramatically higher than that in the control treatment after post-recovery (*p* < 0.05). The hepatic GSH level in the 75 mg/kg group after post-recovery was significantly higher than that in the 150 mg/kg groups after post-recovery (*p* < 0.05). The hepatic T-AOC levels in the control group after post-recovery was dramatically lower than that in the control treatment after ammonia exposure (*p* < 0.05). The hepatic T-AOC levels in the 75 mg/kg group after post-recovery was dramatically higher than that in counterpart treatment after ammonia exposure.

The hepatic GSH-Px level in the 75 mg/kg group after ammonia exposure was higher than that in the 150 mg/kg group after ammonia exposure ($p < 0.05$) (Figure 2H). No significant difference in hepatic GSH-Px level were observed between the control group and curcumin groups after ammonia exposure ($p < 0.05$). The hepatic GSH-Px levels in curcumin-addition groups after post-recovery were dramatically higher than that in the control treatment ($p < 0.05$). However, no statistical difference in hepatic GSH-Px levels was obtained between the 75 and 150 mg/kg groups after post-recovery ($p > 0.05$). Compared with those three groups after ammonia exposure, the hepatic GSH-Px levels in control, 75, and 150 mg/kg groups after post-recovery were statistically increased ($p < 0.05$).

3.6. Gill Enzyme Activities in Greater Amberjack (S. dumerili) in Response to Acute Ammonia Exposure and Post-Recovery

The gill ALP level in the 150 mg/kg group after ammonia exposure was dramatically lower than that in the control and 75 mg/kg treatments ($p < 0.05$) (Figure 3A). The gill ALP levels in the 75 and 150 mg/kg groups after post-recovery were dramatically lower than that in the control treatment ($p < 0.05$). Gill ALP levels in the control group and curcumin-addition groups after post-recovery were dramatically higher than those in the respective groups after ammonia exposure ($p < 0.05$).

Gill ACP levels were significantly decreased with increasing dietary curcumin addition, with a minimum in the 75 mg/kg group after ammonia exposure ($p < 0.05$) (Figure 3B). The gill ACP levels in the 75 and 150 mg/kg groups after ammonia exposure were dramatically lower than that in the control treatment ($p < 0.05$). Similarly, these results were also observed after post-recovery. The gill ACP levels in curcumin-addition groups after post-recovery were dramatically lower than that in the control treatment ($p < 0.05$). Compared with the control group after ammonia exposure, the gill ACP level in the control group after post-recovery was statistically increased ($p < 0.05$).

The gill SOD level was statistically increased with increasing dietary curcumin addition, reaching a maximum in the 150 mg/kg group after post-recovery ($p < 0.05$) (Figure 3C). Compared with the control group, gill SOD levels in the 75 and 150 mg/kg groups after post-recovery were statistically increased ($p < 0.05$). Gill SOD levels in the control and 75 mg/kg groups after post-recovery was dramatically lower than that in the control and 75 mg/kg treatments ($p < 0.05$), respectively.

The gill CAT level was significantly increased with increasing dietary curcumin addition, reaching a maximum in the 150 mg/kg group after ammonia exposure ($p < 0.05$) (Figure 3D). The gill CAT level in the 75 mg/kg group was significantly higher than those in the control and 150 mg/kg groups after post-recovery ($p < 0.05$). The gill CAT level in the 75 mg/kg groups after post-recovery was dramatically higher than that in the counterpart group after ammonia exposure ($p < 0.05$).

The gill T-AOC level in the 150 mg/kg group after ammonia exposure was dramatically higher than that in the control and 75 mg/kg treatments ($p < 0.05$) (Figure 3E). Gill T-AOC levels in the curcumin-addition groups after post-recovery were dramatically higher than that in the control treatment ($p < 0.05$). Compared with those two groups after ammonia exposure, the gill T-AOC level was significantly decreased in the control group after post-recovery, while it increased in the 75 mg/kg curcumin group after post-recovery ($p < 0.05$).

The gill MDA level in the 150 mg/kg group after ammonia exposure was dramatically lower than that in the control and 75 mg/kg treatments ($p < 0.05$) (Figure 3F). The gill MDA level in the 150 mg/kg groups after post-recovery was dramatically higher than that in the control and 75 mg/kg treatments ($p < 0.05$). Compared with the 150 mg/kg group after ammonia exposure, the gill MDA level was significantly increased in the 150 mg/kg group after post-recovery ($p < 0.05$).

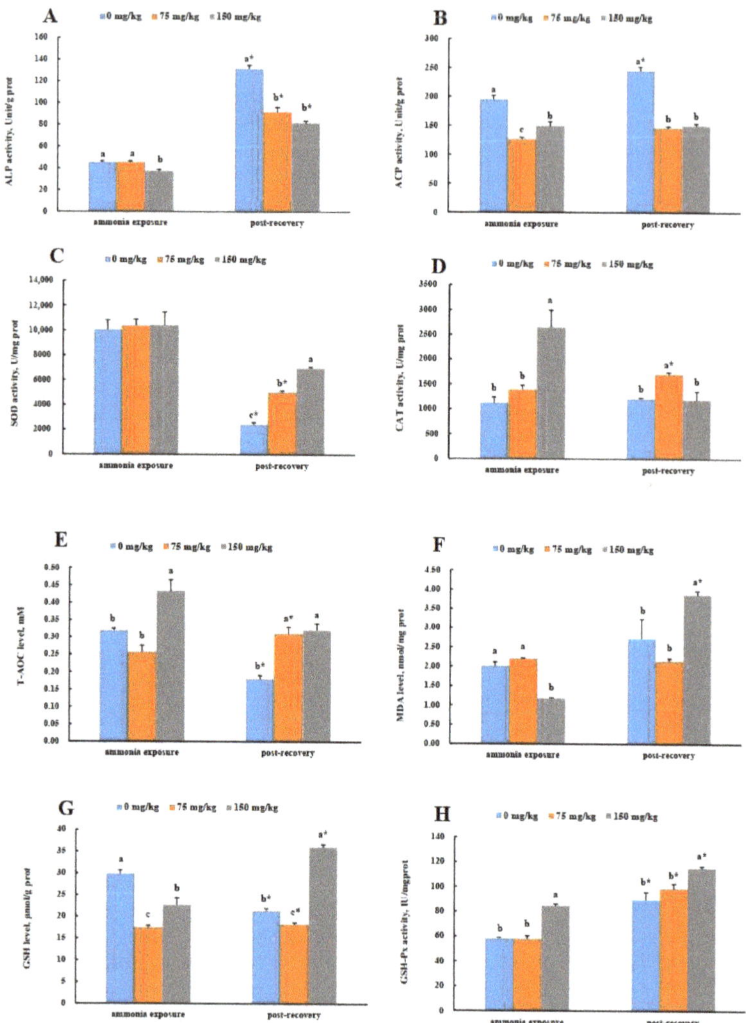

Figure 3. Gill ALP, ACP, SOD, CAT, T-AOC, MDA, GSH，and GSH-PX levels in greater amberjack (*S. dumerili*) fed dietary curcumin (**A–H**) and the response to acute ammonia exposure and post-recovery. Data are expressed as average ± standard error of the mean (SEM) (n = 9). The significant differences ($p < 0.05$) between values obtained in ammonia exposure and its post-recovery stage were determined using *t*-tests and are indicated by asterisks. Different letters indicate significant differences ($p < 0.05$) among different groups as assessed by Duncan's multi-range tests.

Gill GSH levels were significantly decreased with increasing dietary curcumin addition, with a minimum in the 75 mg/kg group after ammonia exposure ($p < 0.05$) (Figure 3G). Gill GSH levels in the curcumin-addition groups after ammonia exposure were dramatically lower than that in the control treatment ($p < 0.05$). Compared with the control group after post-recovery, gill GSH levels were statistically decreased in the 75 mg/kg group after post-recovery ($p < 0.05$), while they were significantly increased in the 150 mg/kg group after post-recovery ($p < 0.05$). Compared with those three groups after ammonia exposure, the gill GSH level was significantly decreased in the control group after post-recovery ($p < 0.05$), while it was increased in the 75 and 150 mg/kg group after post-recovery ($p < 0.05$).

The gill GSH-Px level was significantly increased with increasing dietary curcumin addition, reaching a maximum in the 150 mg/kg group after ammonia exposure ($p < 0.05$) (Figure 3H). The gill GSH-Px level was significantly increased with increasing dietary curcumin addition, reaching a maximum in the 150 mg/kg group after post-recovery ($p < 0.05$). Compared with those three groups after ammonia exposure, gill GSH-Px levels in the control, 75, and 150 mg/kg groups after post-recovery were noticeably increased ($p < 0.05$).

3.7. Spleen Enzyme Activities in Greater Amberjack (S. dumerili) in Response to Acute Ammonia Exposure and Post-Recovery

The spleen ALP level in the 150 mg/kg group after post-recovery were dramatically lower than those in the control and 75 mg/kg treatments ($p < 0.05$) (Figure 4A). Compared with the control and 75 mg/kg groups after ammonia exposure, spleen ALP levels in the control and 75 mg/kg groups after post-recovery were significantly increased ($p < 0.05$), respectively.

The spleen ACP level was significantly increased with increasing dietary curcumin addition, reaching a maximum in the 150 mg/kg group after ammonia exposure ($p < 0.05$) (Figure 4B). ACP levels in the curcumin-addition groups after ammonia exposure were dramatically higher than that in the control treatment ($p < 0.05$). The spleen ACP level in the 75 mg/kg group after post-recovery was dramatically higher than that in the control treatment ($p < 0.05$), while it was dramatically lower in the 150 mg/kg group after post-recovery was obtained ($p < 0.05$). Compared with those three groups after ammonia exposure, spleen ACP levels were statistically increased in the control and 75 mg/kg groups after post-recovery ($p < 0.05$), while it was decreased in the 150 mg/kg group after post-recovery ($p < 0.05$).

Spleen SOD levels in the curcumin-addition group after ammonia exposure were dramatically higher than that in the control treatment ($p < 0.05$) (Figure 4C). Spleen SOD levels in the 75 mg/kg groups were significantly higher than that in the 150 mg/kg group after ammonia exposure ($p < 0.05$). Spleen SOD levels in curcumin-addition groups after post-recovery were dramatically higher than that in the control treatment ($p < 0.05$). Compared with those two groups after ammonia exposure, spleen SOD levels in the control group and 75 mg/kg groups after post-recovery were significantly decreased ($p < 0.05$).

The spleen CAT level in the 150 mg/kg group after ammonia exposure was dramatically higher than that in the control treatment ($p < 0.05$) (Figure 4D). The spleen CAT level in the 75 mg/kg group after post-recovery was dramatically higher than that in the control treatment ($p < 0.05$). The spleen CAT level in the 150 mg/kg group after post-recovery was dramatically lower than that in the counterpart group after ammonia exposure ($p < 0.05$).

The spleen T-AOC level was significantly increased with increasing dietary curcumin addition, reaching a maximum in the 150 mg/kg group after ammonia exposure ($p < 0.05$) (Figure 4E). Compared with the control group, spleen T-AOC levels were significantly increased in the 75 and 150 mg/kg groups after ammonia exposure ($p < 0.05$). Similar results of spleen T-AOC were obtained after post-recovery. The spleen T-AOC levels were significantly increased with increasing dietary curcumin addition, reaching a maximum in the 150 mg/kg group after post-recovery ($p < 0.05$). Spleen T-AOC levels in curcumin-addition groups after post-recovery were dramatically higher than that in the control treatment ($p < 0.05$). Compared with the 75 mg/kg group after ammonia exposure, the spleen CAT level was significantly decreased in the counterpart group after post-recovery ($p < 0.05$).

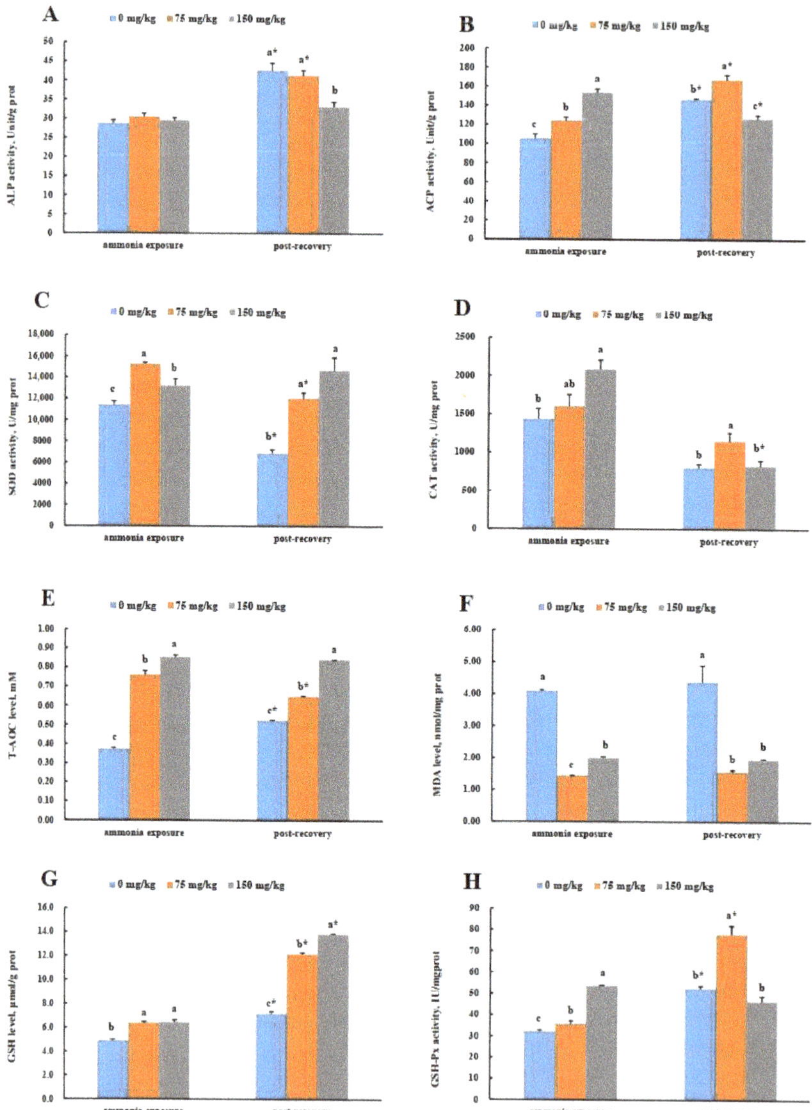

Figure 4. Spleen ALP, ACP, SOD, CAT, T-AOC, MDA, GSH，and GSH-PX levels in greater amberjack (*S. dumerili*) fed dietary curcumin (**A–H**) and the response to acute ammonia exposure and post-recovery. Data are expressed as average ± standard error of the mean (SEM) (n = 9). The significant differences (*p* < 0.05) between values obtained in ammonia exposure and its post-recovery stage were determined using *t*-tests and are indicated by asterisks. Different letters indicate significant differences (*p* < 0.05) among different groups as assessed by Duncan's multi-range tests.

Spleen MDA levels in curcumin-addition groups after ammonia exposure were dramatically lower than that in the control treatment (*p* < 0.05) (Figure 4F). The spleen MDA level in the 75 mg/kg group after ammonia exposure was statistically lower than that in the 150 mg/kg group after ammonia exposure (*p* < 0.05). Spleen MDA levels in curcumin-addition groups after post-recovery were dramatically lower than that in the control treatment ($p < 0.05$).

Spleen GSH levels in curcumin-addition groups after ammonia exposure were dramatically higher than that in the control treatment ($p < 0.05$) (Figure 4G). Spleen GSH levels were dramatically increased with increasing dietary curcumin addition, reaching a maximum in the 150 mg/kg group after post-recovery ($p < 0.05$). Spleen GSH levels in curcumin-addition groups after post-recovery were dramatically higher than that in the control treatment ($p < 0.05$). Compared with those three groups after ammonia exposure, spleen GSH levels in the control, 75, and 150 mg/kg groups after post-recovery were significantly increased ($p < 0.05$).

Spleen GSH-Px levels were statistically increased with increasing dietary curcumin addition, reaching a maximum in the 150 mg/kg group after ammonia exposure ($p < 0.05$) (Figure 4H). Compared with the control group, spleen GSH-Px levels were significantly increased in the 75 and 150 mg/kg groups after ammonia exposure ($p < 0.05$). The spleen GSH-Px level in the 75 mg/kg group after post-recovery was dramatically higher than that in the control treatment ($p < 0.05$). Spleen GSH-Px levels in the control and 75 mg/kg groups after post-recovery were dramatically higher than those in the respective groups after ammonia exposure ($p < 0.05$).

3.8. Head Kidney Enzyme Activities in Greater Amberjack (S. dumerili) in Response to Acute Ammonia Exposure and Post-Recovery

Head kidney ALP levels were significantly increased with increasing dietary curcumin addition, reaching a maximum in the 150 mg/kg group after ammonia exposure ($p < 0.05$) (Figure 5A). Compared with the control group, head kidney ALP levels were significantly increased in the 75 and 150 mg/kg groups after ammonia exposure ($p < 0.05$). Compared with the control group, head kidney ALP levels in the 75 and 150 mg/kg groups after post-recovery were statistically increased ($p < 0.05$). Head kidney ALP levels in the control and 150 mg/kg groups after post-recovery were dramatically higher than those in the respective treatments after ammonia exposure ($p < 0.05$).

The head kidney ACP level was significantly increased with increasing dietary curcumin addition, reaching a maximum in the 150 mg/kg group after post-recovery ($p < 0.05$) (Figure 5B).

The head kidney SOD level was significantly increased with increasing dietary curcumin addition, reaching a maximum in the 150 mg/kg group after ammonia exposure ($p < 0.05$) (Figure 5C). Head kidney SOD levels in the 75 and 150 mg/kg groups after post-recovery were dramatically higher than that in the control treatment ($p < 0.05$). The head kidney SOD level in the 75 mg/kg group after post-recovery was significantly higher than that in the 150 mg/kg groups after post-recovery ($p < 0.05$). Head kidney SOD levels in the control and 150 mg/kg groups after post-recovery were dramatically higher than those in the respective groups after ammonia exposure ($p < 0.05$).

The head kidney CAT level was significantly increased with increasing dietary curcumin addition, reaching a maximum in the 150 mg/kg group after ammonia exposure ($p < 0.05$) (Figure 5D). Compared with the 150 mg/kg group after ammonia exposure, the head kidney CAT level in the 150 mg/kg group after post-recovery was dramatically lower than that in the counterpart group after ammonia exposure ($p < 0.05$).

The head kidney T-AOC level in the 75 mg/kg group after ammonia exposure was dramatically higher than that in the control treatment ($p < 0.05$) (Figure 5E). Head kidney T-AOC levels were significantly increased with increasing dietary curcumin addition, reaching a maximum in the 150 mg/kg group after post-recovery ($p < 0.05$). Head kidney T-AOC levels in curcumin-addition groups after post-recovery were dramatically higher than that in the control treatment ($p < 0.05$). The head kidney T-AOC level in the 75 and 150 mg/kg groups after post-recovery was dramatically higher than those in the respective groups after ammonia exposure ($p < 0.05$).

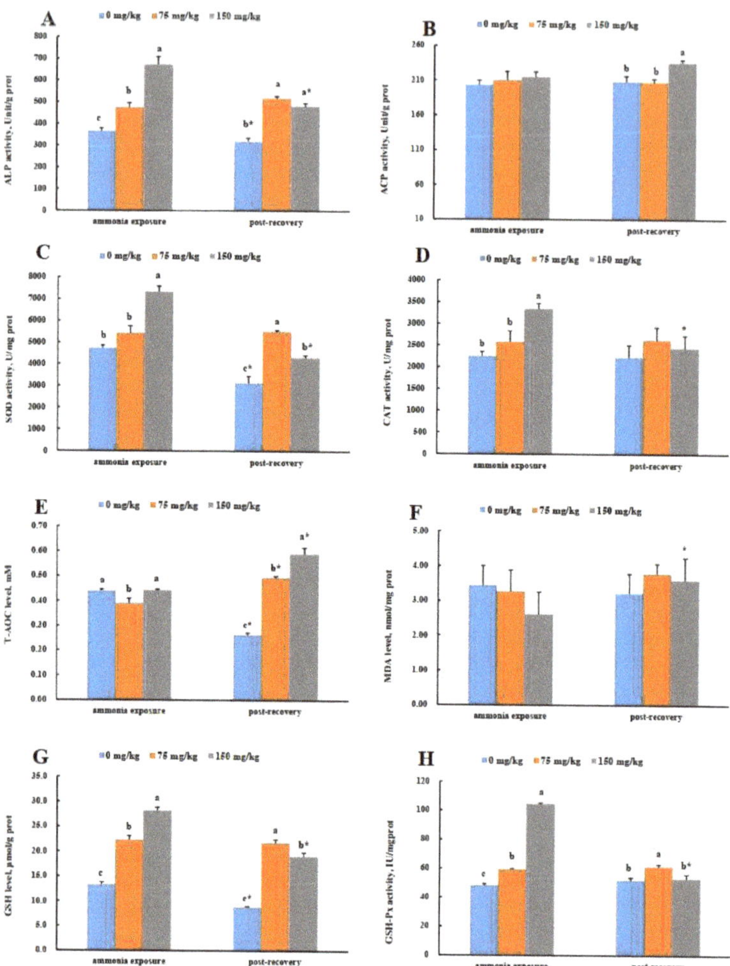

Figure 5. Head kidney ALP, ACP, SOD, CAT, T-AOC, MDA, GSH，and GSH-PX levels in greater amberjack (*S. dumerili*) fed dietary curcumin (**A–H**) and the response to acute ammonia exposure and post-recovery. Data are expressed as average ± standard error of the mean (SEM) (n = 9). The significant differences ($p < 0.05$) between values obtained in ammonia exposure and its post-recovery stage were determined using *t*-tests and are indicated by asterisks. Different letters indicate significant differences ($p < 0.05$) among different groups as assessed by Duncan's multi-range tests.

Compared with the 150 mg/kg group after ammonia exposure, the head kidney MDA level in the counterpart group after post-recovery was significantly increased ($p < 0.05$) (Figure 5F).

Head kidney GSH levels were significantly increased with increasing dietary curcumin addition, reaching a maximum in the 150 mg/kg group after ammonia exposure ($p < 0.05$) (Figure 5G). Compared with the control group, head kidney GSH levels were significantly increased in the 75 and 150 mg/kg groups after ammonia exposure ($p < 0.05$). Head kidney GSH levels in curcumin-addition groups after post-recovery were dramatically higher than that in the control treatment ($p < 0.05$). The head kidney GSH level in the 75 mg/kg group after post-recovery was dramatically increased compared with the 150 mg/kg groups after post-recovery ($p < 0.05$). Head kidney GSH levels in control and

150 mg/kg groups after post-recovery were dramatically lower than those in the respective groups after ammonia exposure ($p < 0.05$).

Head kidney GSH-Px levels were significantly increased with increasing dietary curcumin addition, reaching a maximum in the 150 mg/kg group after ammonia exposure ($p < 0.05$) (Figure 5H). Head kidney GSH levels in the 75 and 150 mg/kg groups after ammonia exposure were dramatically higher than that in the control treatment ($p < 0.05$). The head kidney GSH-Px level in the 75 mg/kg group after post-recovery was dramatically higher than that in the control treatment ($p < 0.05$). Compared with the control and 75 mg/kg groups after ammonia exposure, head kidney GSH-Px levels in the counterpart group after post-recovery were significantly increased ($p < 0.05$).

3.9. Brain Enzyme Activities in Greater Amberjack (S. dumerili) in Response to Acute Ammonia Exposure and Post-Recovery

The brain ALP level in the 150 mg/kg group after post-recovery was dramatically lower than that in the control treatment ($p < 0.05$) (Figure 6A). Compared with groups after ammonia exposure, brain ALP levels in the respective groups after post-recovery were noticeably declined ($p < 0.05$). Brain ACP levels were also observed among all groups after post-recovery ($p > 0.05$) (Figure 6B).

The brain SOD level was significantly increased with increasing dietary curcumin addition, reaching a maximum in the 150 mg/kg group after ammonia exposure ($p < 0.05$) (Figure 6C). The brain SOD level in the 150 mg/kg group after ammonia exposure was dramatically higher than that in other treatments ($p < 0.05$). Compared with the control group after post-recovery, the brain SOD level was significantly increased in the 150 mg/kg group after post-recovery ($p < 0.05$). Compared with the 75 and 150 mg/kg groups after ammonia exposure, brain SOD levels in the respective groups after post-recovery were significantly increased ($p < 0.05$).

The brain T-AOC level in the 75 mg/kg group after post-recovery was dramatically higher than that in the control treatment ($p < 0.05$) (Figure 6E). Compared with counterpart groups after ammonia exposure, the brain T-AOC level in the control group after post-recovery was significantly decreased ($p < 0.05$), while the brain T-AOC level in the 75 mg/kg group after post-recovery was significantly increased ($p < 0.05$).

Brain MDA levels in curcumin-addition groups after ammonia exposure were dramatically lower than that in the control treatment ($p < 0.05$) (Figure 6F). The brain MDA level in the 75 mg/kg group after ammonia exposure was significantly lower than that in the 150 mg/kg group after ammonia exposure ($p < 0.05$). Brain MDA levels in curcumin-addition groups after post-recovery were dramatically lower than that in the control treatment ($p < 0.05$). The brain MDA level in the control and 150 mg/kg groups after post-recovery was dramatically lower ($p < 0.05$), while it was higher in the 75 mg/kg group after post-recovery than those in counterpart groups after ammonia exposure ($p < 0.05$).

Brain GSH levels were significantly increased with increasing dietary curcumin addition, reaching a maximum in the 150 mg/kg group after post-recovery ($p < 0.05$) (Figure 6G). Brain GSH levels in curcumin-addition groups after post-recovery were dramatically higher than that in the control treatment ($p < 0.05$). Compared with the 75 mg/kg group after ammonia exposure, the brain GSH level in the counterpart group after post-recovery was statistically decreased ($p < 0.05$).

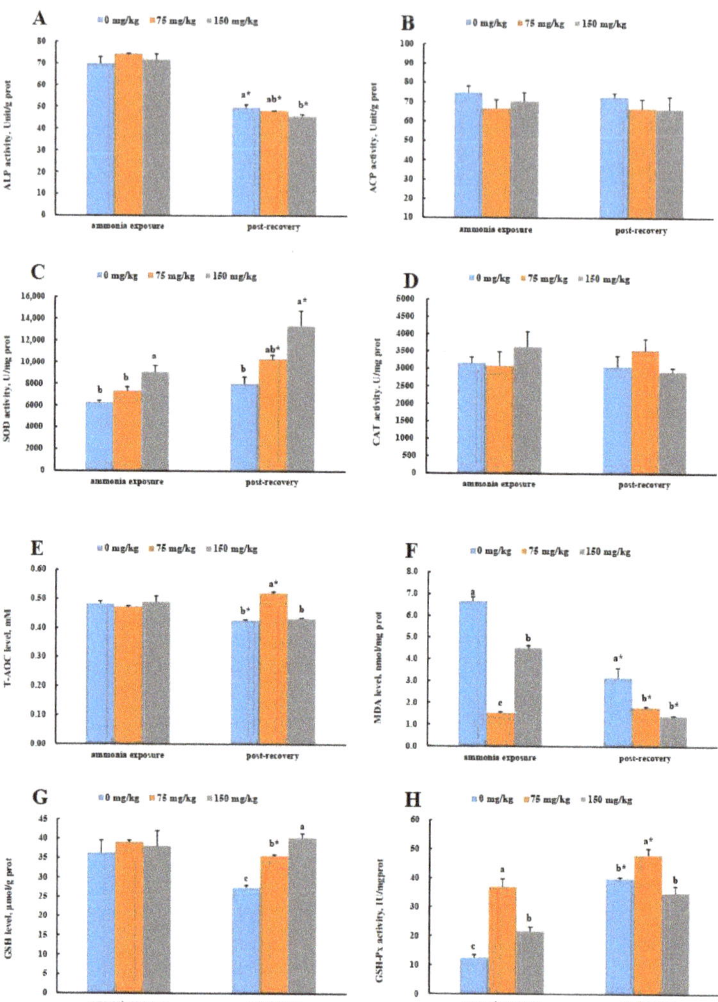

Figure 6. Brain ALP, ACP, SOD, CAT, T-AOC, MDA, GSH，and GSH-PX levels in greater amberjack (*S. dumerili*) fed dietary curcumin (**A–H**) and the response to acute ammonia exposure and post-recovery. Data are expressed as average ± standard error of the mean (SEM) (n = 9). The significant differences ($p < 0.05$) between values obtained in ammonia exposure and its post-recovery stage were indicated determined using *t*-tests and are indicated by asterisks. Different letters indicate significant differences ($p < 0.05$) among different groups as assessed by Duncan's multi-range tests.

Brain GSH-Px levels in curcumin-addition groups after ammonia exposure were dramatically higher than that in the control treatment ($p < 0.05$) (Figure 6H). The brain GSH-Px level in the 75 mg/kg group after ammonia exposure was significantly higher than that in the 150 mg/kg groups after ammonia exposure ($p < 0.05$). The brain GSH-Px level in the 75 mg/kg group after post-recovery was dramatically higher than that in the control treatment ($p < 0.05$). Compared with the control and 75 mg/kg groups after ammonia exposure, brain GSH-Px levels in counterpart groups after post-recovery were statistically increased ($p < 0.05$).

3.10. Effect of Dietary Curcumin on Hepatic Histological Changes of Greater Amberjack (Seriola dumerili)

Compared to the control group, liver tissues in the curcumin-addition groups had more intact morphology and showed no signs of cell swelling or increased vacuolation (Figure 7A–C).

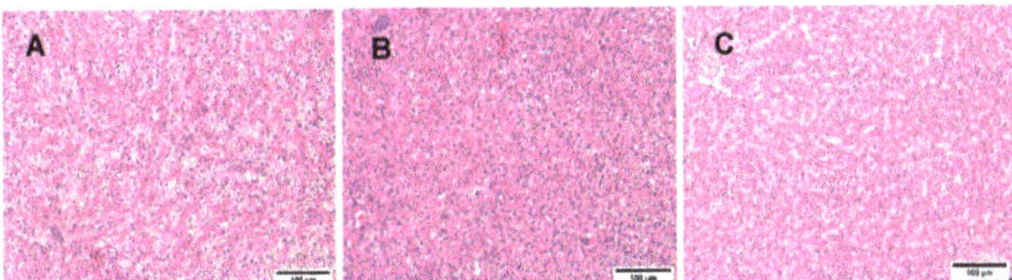

Figure 7. The microscopic hepatic structure of greater amberjack (*S. dumerili*) fed with diets containing different amounts of curcumin for eight weeks. (**A**) represents 0 mg/kg group; (**B**) represents 75 mg/kg group; and (**C**) represents 150 mg/kg group.

4. Discussion

Phosphatases, as part of the immune system, play a crucial role in protecting animals. Based on their catalytic optimum pH properties, phosphatases include two groups: alkaline phosphatases (ALPs) and acid phosphatases (ACPs) [23]. Phosphatases are pivotal lysosomal enzymes, and they are involved in humoral immune responses and degradation of exogenous nutrients, such as proteins, lipids, and carbohydrates [24]. ALPs are a superfamily of metalloenzymes. These enzymes exist in a wide range of organisms from bacteria to humans. ALPs can play an important catalytic role in the hydrolysis of phosphate monoesters to inorganic phosphate and the corresponding alcohols, phenols, and sugars [25]. ACP is a natural enzyme which exists widely in animals and plants. It plays an important role in many important immune defenses [26]. In acidic environments, ACP can play an immune role. It degrades or destroys pathogens by recognizing and catalyzing the hydrolysis of phosphate bonds of exogenous substances[27]. In the current study, ALP levels in the intestine and liver were significantly improved in fish fed with dietary curcumin compared with the control group. Similarly, the results of this study demonstrated that dietary curcumin promoted ACP levels in plasma, the intestine, and the liver of greater amberjack. All the results suggested that curcumin could enhance innate immune response in fish species, and curcumin has been considered to be an effective immunomodulator in both animals and humans [9].

The strength of the antioxidant capacity of the body defense system is closely related to health in fish [28,29]. Oxidative stress can lead to the generation of ROS and free radicals in fish [29,30]. In addition, a variety of antioxidant factors (e.g., SOD, T-AOC, GSH, and GSH-Px) are required to maintain immune system of fish [31]. Antioxidant capacity comprises both enzymatic and non-enzymatic antioxidant activities. Antioxidant enzymes include SOD, T-AOC, and GSH-Px, which form the first line of the organism's enzymatic defense mechanism against free radicals. T-AOC capacity is a comprehensive reflection of redox status in many aspects, such as the body, the antioxidant system of enzymes, and so on [28]. GSH is the primary antioxidant in the human body to counteract damage caused by free radicals, which is a contributing factor to aging and disease. In general, antioxidant defense in fish depends in some way on nutritional factors [32]. Curcumin is a natural ingredient that can also enhance the activities of antioxidative enzymes [33]. Curcumin exerts antioxidant effects mainly due to the presence of polyphenolic compounds as its main component [9,34]. Many plant extracts (also called herbal extracts) contain such substances, thus exerting antioxidant effects directly or indirectly [34,35]. These polyphenols can suppress oxidative stress and then prevent the catalytic mechanism of cytochrome P450CYP, and they can eventually neutralize ROS reagents [34]. The results of this study showed that dietary curcumin dramatically promoted the levels of hepatic SOD, T-AOC, GSH, and GSH-Px in greater amberjack. Similarly, previous studies showed that dietary herbal extract can upregulate antioxidant-related gene expression in fish [34,36]. The activities of glutathione peroxidase and superoxide dismutase were increased in Siberian sturgeon (*Acipenser baerii*) treated with different levels of barberry fruit (BF) extract [37]. Wang et al. (2020) also found an increase in the activity of SOD in black sea bream (*Acanthopagrus schlegelii*) fed berberine [38]. However, a study on rainbow trout showed that the expression of antioxidant-related genes was downregulated by dandelion flower (DF) extract [34]. Similar results were also found in gilthead seabream (*Sparus aurata*) fed with palm fruit extracts [39]. The reason for this contradiction may be due to a time-lag effect between transcription and translation [40]. Further studies are needed to determine the reason for the changes in the expression of antioxidant genes and/or antioxidant enzyme activities in response to dietary levels of plant extract. Meanwhile, in the present study, the hepatic histomorphology in the curcumin group showed more intact morphology and no signs of cell swelling or increased vacuolation, indicating that curcumin addition plays a beneficial role in liver health. This result could also confirm that curcumin has hypolipidemic effects and may prevent the accumulation of fatty acids in liver cells that may result from metabolic imbalances and nonalcoholic steatohepatitis [41].

Previous studies have shown that the specific activities of immune-related enzymes in aquatic animals under ammonia stress were significantly reduced [42–44], and high concentrations of ammonia have a serious impact on the catalytic action of antioxidant enzymes and the stability of cell membranes in aquatic animals and thus disrupt the osmotic balance [45]. In this study, curcumin addition improved the activities of antioxidant enzymes and MDA content in the liver, spleen, head kidney, and brain tissues after post-recovery. This suggests that curcumin attenuated oxidative stress caused by ammonia stress via increasing antioxidant enzyme activities and decreasing the content of MDA. This result coincided with previous studies on Japanese Sea bass [12] and largemouth bass [13]. In this study, curcumin addition increased ALP levels in the liver and head kidney, but it decreased ALP levels in the gills, spleen, and brain. Meanwhile, curcumin addition increased ACP levels in the intestine and head kidney, but it decreased ACP levels in the liver and gills. This may be due to the fact that enzyme activity is influenced by a variety of factors. Environmental conditions such as pH, temperature, salt concentration, substrate concentration, activators, and inhibitors may alter the spatial structure of the enzyme, thus changing its rate of activity and/or ability to bind substrates [46]. This inconsistency requires further research to elucidate.

In aquatic environments, ammonia accumulation could result in acute toxicity to aquatic animals. In the process of resisting environmental toxic and harmful substances, the organs of aquatic animals (e.g., the gill, liver, spleen, etc.) play a pivotal role [5]. In this study, the fish showed the ability to regulate themselves. Specifically, the indexes associated with immunity and antioxidant enzymes in the liver, gill, and spleen rose again to some extent, but they showed the worst recovery ability in head kidney and brain tissue. Similar studies can be found in juvenile blunt snout bream [2]. In addition, compared with other groups, the 75 mg/kg curcumin group generally significantly increased the recovery level of related enzymes in these tissues. This suggested that adding 75 mg/kg of curcumin can enhance the antioxidant abilities and promote the overall health of the fish.

5. Conclusions

In conclusion, ammonia stress leads to metabolic changes and induces oxidative stress in juvenile *S. dumerili*. Reduced antioxidant enzyme activities revealed that exposure to high concentrations of ammonia may induce more severe oxidative stress. Recovery of antioxidant enzyme activities revealed that post-exposure recovery may attenuate oxidative stress to some extent. Dietary curcumin supplementation could increase non-specific immune responses, antioxidant ability, and enhance resistance to high ammonia stress in juvenile *S. dumerili*. The outcomes of this study will facilitate future research on the effects of ammonia toxicity in juvenile *S. dumerili* and provide technical support for prevention of disease and alleviation of stress.

Author Contributions: C.Z. contributed to the original idea. H.L. and Z.M. contributed with experimental design. Z.H., J.W., Y.W., W.Y., S.Z., J.H., R.Y. and C.Z. conducted the experiment and analyzed all the samples. All 10 authors contributed to the draft and final manuscript. All authors have read and agreed to the published version of the manuscript.

Funding: This study was granted by the Central Public-interest Scientific Institution Basal Research Fund, CAFS (2022XT0404; 2020TD55); Central Public-interest Scientific Institution Basal Research Fund, South China Sea Fisheries Research Institute, CAFS (2021XK02); special funds for science technology innovation and industrial development of Shenzhen Dapeng New District (Grand No. PT202101-26); National Natural Science Foundation of China (32172984); the Fund of Southern Marine Science and Engineering Guangdong Laboratory (Zhanjiang) (ZJW-2019-06).

Institutional Review Board Statement: This study was conducted in strict accordance with the recommendation of the Ethics Committee of South China Sea Fisheries Institute. No protected species were used during the experiment.

Informed Consent Statement: Not applicable (Present study did not involve humans). Written informed consent has been obtained from all the patients to publish this paper.

Data Availability Statement: Data is contained within the article.

Conflicts of Interest: The authors declare no conflict of interest.

References

1. Wan, J.; Ge, X.; Liu, B.; Xie, J.; Cui, S.; Zhou, M.; Xia, S.; Chen, R. Effect of Dietary Vitamin C on Non-Specific Immunity and MRNA Expression of Three Heat Shock Proteins (HSPs) in Juvenile Megalobrama Amblycephala under PH Stress. *Aquaculture* **2014**, *434*, 325–333. [CrossRef]
2. Zhang, W.; Sun, S.; Ge, X.; Xia, S.; Zhu, J.; Miao, L.; Lin, Y.; Liang, H.; Pan, W.; Su, Y.; et al. Acute Effects of Ammonia Exposure on the Plasma and Haematological Parameters and Histological Structure of the Juvenile Blunt Snout Bream, Megalobrama Amblycephala, and Post-Exposure Recovery. *Aquac. Res.* **2018**, *49*, 1008–1019. [CrossRef]
3. Wajsbrot, N.; Gasith, A.; Diamant, A.; Popper, D.M. Chronic Toxicity of Ammonia to Juvenile Gilthead Seabream Sparus Aurata and Related Histopathological Effects. *J. Fish Biol.* **1993**, *42*, 321–328. [CrossRef]
4. Li, M.; Yu, N.; Qin, J.G.; Li, E.; Du, Z.; Chen, L. Effects of Ammonia Stress, Dietary Linseed Oil and Edwardsiella Ictaluri Challenge on Juvenile Darkbarbel Catfish Pelteobagrus Vachelli. *Fish Shellfish. Immunol* **2014**, *38*, 158–165. [CrossRef] [PubMed]
5. Benli, A.C.K.; Köksal, G.; Ozkul, A. Sublethal Ammonia Exposure of Nile Tilapia (Oreochromis Niloticus L.): Effects on Gill, Liver and Kidney Histology. *Chemosphere* **2008**, *72*, 1355–1358. [CrossRef] [PubMed]
6. Ammon, H.P.; Safayhi, H.; Mack, T.; Sabieraj, J. Mechanism of Antiinflammatory Actions of Curcumine and Boswellic Acids. *J. Ethnopharmacol.* **1993**, *38*, 113–119. [CrossRef]
7. Nabavi, S.F.; Daglia, M.; Moghaddam, A.H.; Habtemariam, S.; Nabavi, S.M. Curcumin and Liver Disease: From Chemistry to Medicine. *Compr. Rev. Food Sci. Food Saf.* **2014**, *13*, 62–77. [CrossRef]
8. Sharma, O.P. Antioxidant Activity of Curcumin and Related Compounds. *Biochem. Pharmacol.* **1976**, *25*, 1811–1812. [CrossRef]
9. Alagawany, M.; Farag, M.R.; Abdelnour, S.A.; Dawood, M.A.O.; Elnesr, S.S.; Dhama, K. Curcumin and Its Different Forms: A Review on Fish Nutrition. *Aquaculture* **2021**, *532*, 736030. [CrossRef]
10. Mahmoud, H.K.; Al-Sagheer, A.A.; Reda, F.M.; Mahgoub, S.A.; Ayyat, M.S. Dietary Curcumin Supplement Influence on Growth, Immunity, Antioxidant Status, and Resistance to Aeromonas Hydrophila in Oreochromis Niloticus. *Aquaculture* **2017**, *475*, 16–23. [CrossRef]
11. Yonar, M.E. Chlorpyrifos-Induced Biochemical Changes in Cyprinus Carpio: Ameliorative Effect of Curcumin. *Ecotoxicol. Environ. Saf.* **2018**, *151*, 49–54. [CrossRef] [PubMed]
12. Wu, J. Effects of Ammonia,Nitrite and Curcumin on Growth Metabolism and Disease-Resistant Genes in Japanese Sea Bass, Lateolabrax Japonicus. Master's Thesis, Guangdong Ocean University, Zhanjiang, China, 2020.
13. Zhao, L.; Tang, G.; Xiong, C.; Han, S.; Yang, C.; He, K.; Liu, Q.; Luo, J.; Luo, W.; Wang, Y.; et al. Chronic Chlorpyrifos Exposure Induces Oxidative Stress, Apoptosis and Immune Dysfunction in Largemouth Bass (Micropterus Salmoides). *Environ. Pollut.* **2021**, *282*, 117010. [CrossRef] [PubMed]
14. Sahin, K.; Orhan, C.; Yazlak, H.; Tuzcu, M.; Sahin, N. Lycopene Improves Activation of Antioxidant System and Nrf2/HO-1 Pathway of Muscle in Rainbow Trout (Oncorhynchus Mykiss) with Different Stocking Densities. *Aquaculture* **2014**, *430*, 133–138. [CrossRef]
15. Zhang, Y. Effects of Curcumin on Growth and Liver-Protection in Common Carp, Cyprinus Carpio. *Pak. J. Zool.* **2020**, *53*, 1211–1220. [CrossRef]
16. El-abd, H.; Abd El-latif, A.; Shaheen, A. Effect of Curcumin on Growth Performance and Antioxidant Stress Status of Nile Tilapia (Oreochromis Niloticus). *Iran. J. Fish. Sci.* **2021**, *20*, 1234–1246. [CrossRef]

Journal of
Marine Science and Engineering

Article

Effects of Water Velocity on Growth, Physiology and Intestinal Structure of Coral Trout (*Plectropomus leopardus*)

Zhenjia Qian [1,2], Jincheng Xu [2], Andong Liu [2], Jianjun Shan [2], Chenglin Zhang [2,*] and Huang Liu [2,*]

1   College of Fisheries and Life Science, Shanghai Ocean University, Shanghai 201306, China
2   Fishery Machinery and Instrument Research Institute, Chinese Academy of Fishery Sciences, Shanghai 200092, China
*   Correspondence: zhangchenglin@fmiri.ac.cn (C.Z.); liuhuang@fmiri.ac.cn (H.L.)

Abstract: This study aimed to investigate the effects of different water velocities on the growth performance, blood physiology, and digestive capacity of coral trout (*Plectropomus leopardus*) in a Recirculating aquaculture system (RAS). One hundred and twenty healthy, uniformly sized coral trout (body mass (92.01 ± 8.04) g; body length (15.40 ± 0.65) cm) were randomly assigned to three flow velocity groups (1 bl/s, 2 bl/s, and 2.5 bl/s) and one control group (0 bl/s). The results show that the weight gain rate (WGR) and specific growth rate (SGR) of coral trout in the 2.5 bl/s water flow velocity group were significantly lower than those in the control group and 1 bl/s water flow velocity group ($p < 0.05$), while their feed coefficient (FC) values were significantly higher than those of the control group and 1 bl/s water flow velocity group ($p < 0.05$). The blood glucose (GLU) concentration of coral trout in the 2 bl/s water flow velocity group and the 2.5 bl/s water flow velocity group significantly decreased compared to those in the control group ($p < 0.05$), while the lactic acid (LD) concentration increased. As the cortisol (COR) concentration and lipase (LPS) enzyme activity of coral trout did not significantly change ($p > 0.05$), the α- AMS enzyme activity significantly decreased ($p < 0.05$). Under 2.5 bl/s water flow velocity, the intestinal structure of coral trout changed, and the number of goblet cells decreased. High-water flow velocities affect the physiological homeostasis and intestinal digestion of coral trout, resulting in a decrease in their growth performance, indicating that coral trout is more sensitive to high-water flow velocities. In actual RAS aquaculture, the flow rate should be controlled within 1 bl/s.

Keywords: *Plectropomus leopardus*; water flow velocity; growth performance; blood biochemistry; intestinal digestibility

Citation: Qian, Z.; Xu, J.; Liu, A.; Shan, J.; Zhang, C.; Liu, H. Effects of Water Velocity on Growth, Physiology and Intestinal Structure of Coral Trout (*Plectropomus leopardus*). *J. Mar. Sci. Eng.* **2023**, *11*, 862. https://doi.org/10.3390/jmse11040862

Academic Editor: Gabriella Caruso

Received: 6 March 2023
Revised: 17 April 2023
Accepted: 17 April 2023
Published: 19 April 2023

1. Introduction

Plectropomus leopardus, commonly known as coral trout, is one of the most important species cultured in China, with such advantages as delicious meat, bright color, and rich nutritional value [1]. The limitations of the marine fishing industry and its high commercial value have promoted the development of the artificial breeding of coral trout [2]. China is a late adopter in the captive breeding of coral trout, and the production in this country is far from sufficient to meet the growing demand. There are currently two mainstream farming methods for coral trout, flowing water culturing, and cage culturing [3]. Flowing water culturing entails a serious waste of water and does not meet today's interests in water and energy conservation. Cage culturing may lead to waves and typhoons, which can cause some pollution to offshore waters. Recirculating aquaculture system (RAS) is developed for intensive aquaculture, which can save water resources and land resources, protect the environment, and help regulate the aquaculture water quality to provide the best aquaculture environment for fish [4]. The fish cultured under RAS have significant advantages in terms of food quality and safety. Coral trout is a warm-water reef fish that prefers to live near coral reefs with low flow velocities [5]. However, the influence of water flow velocity is often ignored in the artificial cultivation of coral trout.

As an environmental factor, water velocity also significantly affects the growth performance, morphological characteristics, and physiological changes of fish [6]. Bl/s is the body length per second. The tolerable water flow velocity of fish is related to their body length [7]. A previous study has reported that high-velocity and high-intensity continuous exercise training (2.0 bl/s) will reduce the growth rate and food conversion rate of juvenile tinfoil barbs (*Barbodes schwanenfeldi*) (15.12 ± 1.35 cm) [8]. Conversely, largemouth bass (*Micropterus salmoides*) (4.50 ± 0.36 cm) can show improved growth performance, digestion, and immunity at a high-water flow velocity (4.0 bl/s) [9]. *Salmo salar* (19.26 ± 0.08 cm) can obtain the best growth performance at 1 bl/s water flow velocity [7]. *Scophthalmus maximus* juvenile (161.9 ± 8 g) shows a reduced feed coefficient and improved specific growth rate at medium water flow velocity [10]. However, there are few reports on the appropriate water flow velocity for breeding coral trout. To optimize the aquaculture facilities and improve the aquaculture environment, it is essential to determine the appropriate water flow velocity for this species.

In this study, the growth performance, blood physiological changes, intestinal digestive enzyme activity, and digestive system structure of fish at different water flow velocities were used as indicators to explore the appropriate water flow velocities for coral trout in RAS. The results from the present study will provide a theoretical basis for the circulation of water culturing of coral trout and promote the healthy and sustainable development of aquaculture.

2. Materials and Methods

2.1. Experimental Fish and Design

A batch of juvenile coral trout was obtained from Delin Chengxin Aquaculture Co., Ltd. (Lingshui, China) with an initial body weight of 92.01 ± 8.04 g and a body length of 15.40 ± 0.65 cm. The culture experiments were carried out in a recirculating water culture tank at the pilot plant of the Institute of Fisheries Machinery and Instruments of the Chinese Academy of Fisheries Sciences (Figure 1). The system consists of a culture tank, a push-flow water pump, a wet and dry separation filter, a moving bed, a UV sterilization device, and a reservoir. The mobile bed has an oxygenating and aeration device to ensure the presence of sufficient dissolved oxygen in the experimental setup. Each tank is equipped with a push-flow water pump, and the intended water flow velocity is provided by a pump (Pond Spring, ABB-200), pushing the water at a speed of 0.5–7.5 bl/s (10–120 cm/s) in the tank.

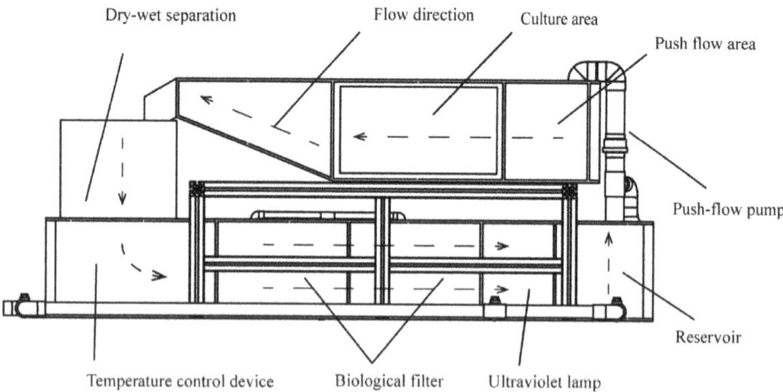

Figure 1. Circulating water culture tank.

The experiment was divided into three water flow velocity groups, with water flow velocities of 1 bl/s, 2 bl/s, and 2.5 bl/s. In addition, a control group was set up with a water flow velocity of 0 bl/s. Each group contained 3 tanks, and the experimental fish were evenly allocated to each tank. There were 10 fish in each tank and a total of 120 fish in the experiment. After one week of acclimatization, the experiment started and then lasted for

42 d. During the experimental period, the fish were fed at 2% of their body mass daily. The fish were fed three times daily at 8:00, 13:00, and 18:00, and the amount of food and residual bait was recorded. Body weight and length measurements were conducted on the 7th, 14th, 21st, 28th, 35th, and 42nd days of the experimental period to adjust the water flow velocity and feeding rate. A total of 6 fish (2 fish per tank) were collected from the control group and the different water flow velocity groups. They were weighed with water and put back into the original tank after weighing. We ensured that the feeding amount was always maintained at 2% body weight and that the water flow velocity was maintained at 1 bl/s, 2 bl/s, and 2.5 bl/s. The water temperature was maintained at $26.5 \pm 1.5\,^\circ$C, salinity 22–24‰, pH 8.07–8.31, dissolved oxygen 8.08 ± 0.26 mg/L, ammonia nitrogen 0.07 ± 0.05 mg/L, and nitrite 0.18 ± 0.12 mg/L.

2.2. Sample Collection and Processing

2.2.1. Sample Collection

On day 42, the experimental fish were sampled for blood samples, and intestinal and liver tissue. Before sampling, the experimental fish were starved for 24 h, and a total of 6 fish were collected from each experimental group (2 per tank). They were put into MS-222 water with a concentration of 120 mg/L for rapid anesthesia, and their body length was measured and weighed. First, blood was collected from the caudal vertebral vein of the experimental fish, and then the fish's body was dissected. The visceral mass was taken out and weighed along with the liver. The intestines were separated from the visceral mass, and the foregut, midgut, and hindgut were collected and transferred to a 10% concentration of formalin for preservation, which was used for the observation of intestinal tissue sections. The rest of the intestines were placed directly into a cryopreservation tube and stored in liquid nitrogen.

2.2.2. Sample Handling

After all the samples were collected, the blood samples were centrifuged at 3500 rpm for 10 min at $4\,^\circ$C using a cryogenic freezing centrifuge, and the upper serum layer was packed into lyophilized tubes and stored in a $-80\,^\circ$C refrigerator. The intestinal tissues were stored in a $-80\,^\circ$C refrigerator.

After thawing, the intestinal tissues were accurately weighed and added to 9 times the volume of saline at a ratio of mass (g): volume (mL) = 1:9, which was then homogenized at $4\,^\circ$C using a freezing homogenizer (MYD-48) to produce a 10-fold dilution of homogenate. The supernatant was then centrifuged at 2500 r/min for 10 min at $4\,^\circ$C, after which it was used to determine the activity of intestinal digestive enzymes.

2.3. Index Measurement

2.3.1. Growth Performance Parameters' Measurements

Feed coefficient (FC), weight gain rate (WGR, %), specific growth rate (SGR, %), condition factor (CF, %), hepatosomatic index (HSI, %), and viscerosomatic index (VSI, %) were calculated as follows:

$$FC = W_f / (W_t - W_0), \tag{1}$$

$$WGR = (W_t - W_0)/W_0 \times 100\%, \tag{2}$$

$$SGR = (\ln W_t - \ln W_0)/t \times 100\%, \tag{3}$$

$$CF = W_t / L^3 \times 100\%, \tag{4}$$

$$HSI = W_l / W_t \times 100\%, \tag{5}$$

$$VSI = W_v / W_t \times 100\%. \tag{6}$$

W_f is the total feeding of the fish on day t at the time of measurement, W_t is the mass of the fish measured at the time of the final sampling (g), W_0 is the mass of the fish measured at the time of the initial sampling (g), t is the number of days of breeding (d), L is the body length of the fish at the time of measurement (cm), W_l is the mass of the fish's liver at the time of sampling (g), and W_v is the mass of the fish's visceral mass at the time of sampling (g).

2.3.2. Blood Biochemical Parameters Measurements

Blood glucose (GLU) concentrations were measured via the glucose oxidase method, blood lactic acid (LD) concentrations were measured via the colorimetric method, and blood cortisol (COR) was measured via the double antibody sandwich method with an enzyme-linked immunoassay kit. All samples were tested according to the operating instructions of the kit (Jiancheng Institute of Biological Engineering, Nanjing, China).

2.3.3. Intestinal Digestive Enzyme Activity Measurements

Protein concentrations in intestinal tissues were determined by the Komas Brilliant Blue method. Intestinal α-amylase (α-AMS) and lipase (LPS) activities were determined via the enzyme starch colorimetric method and enzyme colorimetric method, respectively. All samples were assayed according to the operating instructions of the kit (Jiancheng Bioengineering Institute, Nanjing, China).

2.3.4. Fixation and Observation of Intestinal Section

The foregut, midgut, and hindgut of the experimental fish were fixed in 10% formalin for 24 h, dehydrated in 70%, 80%, 90, 95%, and anhydrous alcohol in that order, transplanted into xylene, embedded in paraffin, sectioned (5 μm) with a microtome (Leica RM2235, Wetzlar, Germany), then stained with hematoxylin and eosin, and sealed with neutral gum. A microscope (Leica DM1000, Wetzlar, Germany) was used to observe the intestinal tissues (magnification of intestinal tissues was 10). Image J software was used to measure the intestinal muscular thickness (MS), intestinal villus length (VL), and villus thickness (VT).

2.4. Statistical Analysis

Excel 2019 software was used for data processing. All statistical analyses were performed with SPSS 26.0 software. All figures were drawn using the Prism GraphPad 9 software. The data obtained in the text were subjected to one-way ANOVA followed by Duncan's multiple comparison analysis to determine if there were any significant differences ($p < 0.05$), and the experimental results are expressed as mean $\pm$ standard deviation (mean $\pm$ SD). Assumptions of normality, equality of variances, and outliers were confirmed prior to statistical analysis.

3. Results

3.1. Growth Performance

The statistical results show significant differences in fish growth performance among the experimental groups ($p < 0.05$, Table 1). The WGR and SGR of the coral trout in the 2.5 bl/s water flow velocity group were significantly lower than those in the other groups ($p < 0.05$), showing also significantly higher FC values than the control and 1 bl/s water flow velocity groups, the lowest feed conversion rate, and the lowest weight gain. The VSI of fish in the 2 bl/s water flow velocity group was the lowest, significantly lower than that of the control group ($p < 0.05$) and not significantly among the water flow velocity groups ($p > 0.05$). There was no significant difference ($p > 0.05$) among the water flow velocity groups as regards the CF and HSI of coral trout.

Table 1. Effect of water flow velocity on growth performance of coral trout.

Item	Control	1 bl/s Water Flow Velocity	2 bl/s Water Flow Velocity	2.5 bl/s Water Flow Velocity
FC	1.17 ± 0.14 [b]	1.14 ± 0.12 [b]	1.36 ± 0.21 [ab]	1.48 ± 0.23 [a]
WGR (%)	47.54 ± 5.63 [a]	47.73 ± 5.55 [a]	40.44 ± 5.63 [ab]	37.92 ± 5.57 [b]
SGR (%)	0.93 ± 0.09 [a]	0.93 ± 0.09 [a]	0.81 ± 0.10 [ab]	0.76 ± 0.10 [b]
CF (%)	2.53 ± 0.20	2.68 ± 0.05	2.63 ± 0.10	2.51 ± 0.08
HIS (%)	1.16 ± 0.16	1.22 ± 0.07	1.33 ± 0.18	1.24 ± 0.01
VSI (%)	5.06 ± 0.48 [a]	4.70 ± 0.42 [ab]	4.25 ± 0.13 [b]	4.40 ± 0.41 [ab]

The values are the mean ± standard deviation of four replications (n = 6). Values with different letters imply a significant difference among the groups ($p < 0.05$), and no letters means no significant difference ($p > 0.05$).

3.2. Blood Biochemistry

Water flow velocity significantly affected the blood biochemistry of the fish (Figure 2). The GLU concentrations of coral trout were significantly higher in the control and 1 bl/s groups than in the 2 bl/s and 2.5 bl/s water flow velocity groups ($p < 0.05$), and LD concentrations were lowest in the 1 bl/s water flow velocity group, which were significantly lower than in the 2.5 bl/s group ($p < 0.05$), with no significant difference from the control and 2 bl/s groups ($p > 0.05$). There was no significant difference in blood COR concentration of coral trout among the water flow velocity groups ($p > 0.05$) and no significant effect of water velocity on coral trout stress levels ($p > 0.05$).

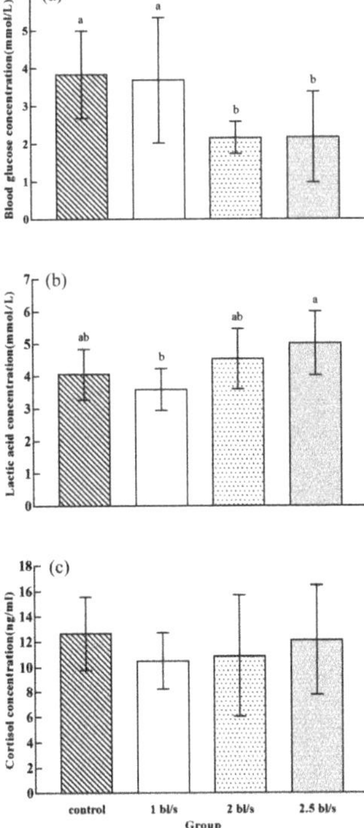

Figure 2. Effect of water flow velocity on blood biochemistry of fish (*Plectropomus leopardus*), (n = 6). (**a**) Blood glucose concentration; (**b**) lactic acid concentration; (**c**) cortisol concentration. Different letters indicate significant differences among groups ($p < 0.05$). No letter means no significant difference among groups ($p > 0.05$).

3.3. Intestinal Digestive Enzyme Activity

Figure 3a shows that the intestinal LPS activity of coral trout was not affected by water flow velocity ($p > 0.05$) and was maintained at normal levels in all groups. Intestinal α-AMS activity was significantly ($p < 0.05$) affected by water flow velocity (Figure 3b), with the values in the control and 1 bl/s water flow velocity groups both significantly higher than in the 2 bl/s and 2.5 bl/s water flow velocity groups ($p < 0.05$), with no significant difference among the control and 1 bl/s water flow velocity groups ($p > 0.05$). Water flow velocities above 2 bl/s reduced the digestive capacity of the fish intestine and affected growth and development.

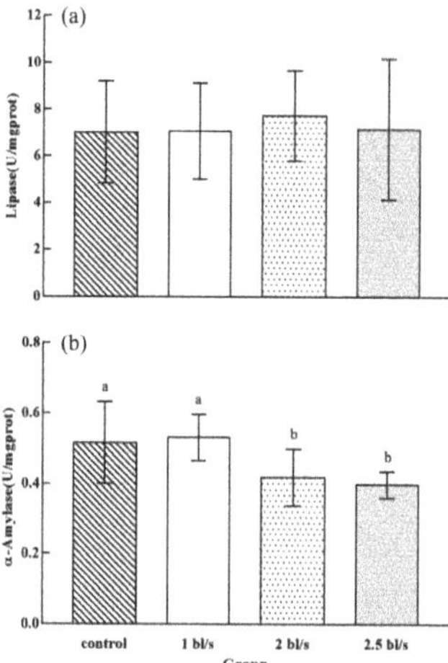

Figure 3. Effect of water flow velocity on the activity of intestinal digestive enzymes in fish (*Plectropomus leopardus*), ($n = 6$). (**a**) Lipase activity; (**b**) α-amylase activity. Different letters indicate significant differences among groups ($p < 0.05$). No letter means no significant difference among groups ($p > 0.05$).

3.4. Intestinal Structure

3.4.1. Foregut

Table 2 shows that the lengths of foregut VL of coral trout in the different water flow velocity groups were significantly different ($p < 0.05$) and that the 1 bl/s water flow velocity group was significantly shorter than those of the control group and other water flow velocity groups ($p < 0.05$). The value of MS in the 2.5 bl/s water flow velocity group was significantly lower than in the 2 bl/s water flow velocity group ($p < 0.05$). The VT of coral trout in the 2 bl/s velocity group was the highest—significantly higher than in the control and other velocity groups ($p < 0.05$). The villous tissue structure of coral trout in each water flow velocity group was complete, but the number of goblet cells in the medium–high-water flow velocity group (2 bl/s and 2.5 bl/s) was reduced compared with the control group (Figure 4).

Table 2. Effect of water flow velocity on the index of foregut tissue of coral trout.

Item	Control	1 bl/s Water Flow Velocity	2 bl/s Water Flow Velocity	2.5 bl/s Water Flow Velocity
VL (μm)	749.83 ± 21.17 [a]	565.72 ± 25.18 [b]	832.54 ± 59.57 [a]	764.4 ± 64.33 [a]
MS (μm)	198.72 ± 7.92a [b]	202.85 ± 38.01 [ab]	227.49 ± 20.69 [a]	190.8 ± 21.5 [b]
VT (μm)	60.72 ± 7.28 [b]	54.54 ± 7.65 [b]	84.31 ± 12.74 [a]	68.58 ± 7.12 [b]

The values are average ± standard deviation of four replications ($n = 6$). Values of different letters mean significant difference among the groups ($p < 0.05$), and those with no letters mean no significant difference ($p > 0.05$).

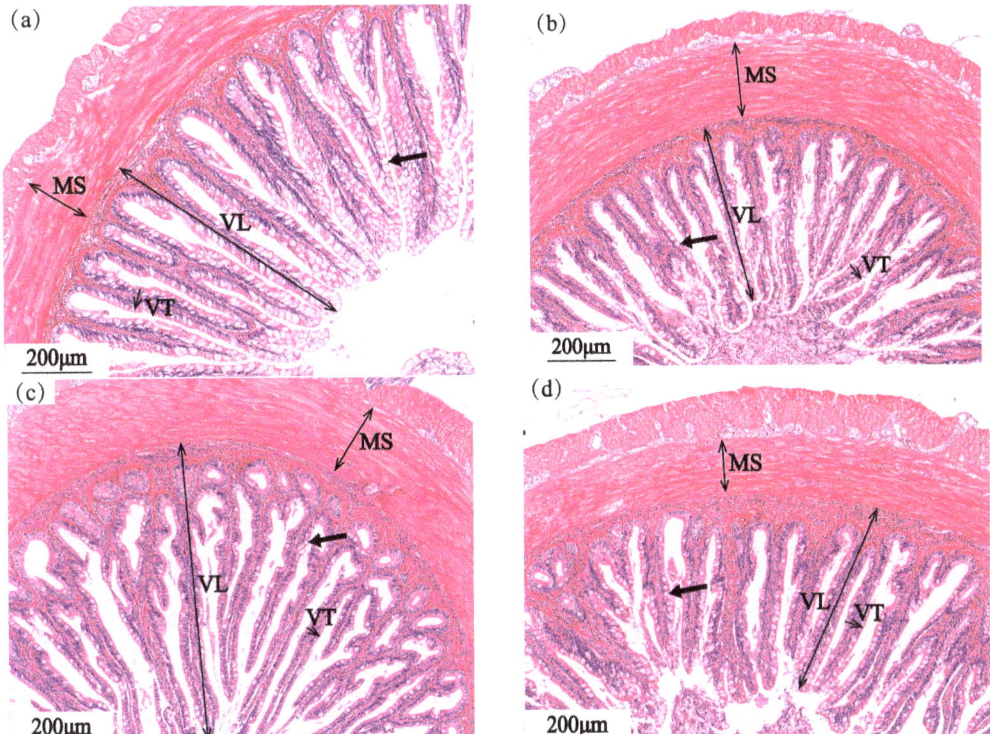

Figure 4. The effect of water flow velocity on the foregut structure of coral trout. (**a**) Hydrostatic control group: intestinal tissue was normal, muscular thickness (MS), villus length (VL), and villus thickness (VT) were uniform, and there were many goblet cells (thick arrows); (**b**) 1 bl/s water flow velocity group: muscular thickness (MS), villus length (VL), and villus thickness (VT) showed no significant changes compared with the control group, while goblet cells (thick arrows) were smaller and their number was relatively reduced; (**c**) 2 bl/s water flow velocity group: the muscle layer was thinner than in the control, while the number of goblet cells (thick arrows) was relatively reduced and their size was smaller; (**d**) 2.5 bl/s water flow velocity group: the muscle layer was relatively thin, while the number of goblet cells (thick arrows) in the contrast photograph was relatively small, and the morphological distribution was uneven.

3.4.2. Midgut

Table 3 shows the index values of the midgut tissues of coral trout in different water flow velocity groups, indicating significant differences among the groups. There was no significant difference in coral trout's VL among the 1 bl/s water flow velocity group and the control group ($p > 0.05$), but it was significantly higher than in other water flow velocity groups ($p < 0.05$). Compared with the control group and other velocity groups, the MS value of coral trout in the 2 bl/s velocity group was the highest ($p < 0.05$), and those in the 1 bl/s velocity group and 2 bl/s velocity group were significantly higher than in the control group and 2.5 bl/s velocity group ($p < 0.05$). The VT of coral trout was the widest in the 1 bl/s velocity group and the narrowest in the 2 bl/s velocity group. The numbers and sizes of goblet cells decreased with the increase in water flow velocity (Figure 5).

Table 3. Effect of water flow velocity on the index of midgut tissue in coral trout.

Item	Control	1 bl/s Water Flow Velocity	2 bl/s Water Flow Velocity	2.5 bl/s Water Flow Velocity
VL (μm)	498.13 ± 31.35 [ab]	523.97 ± 50.27 [a]	433.67 ± 27.69 [b]	443.2 ± 47.22 [b]
MS (μm)	110.88 ± 12.76 [c]	160.6 ± 16.83 [b]	235.57 ± 28.63 [a]	129.26 ± 6.27 [c]
VT (μm)	54.51 ± 6.64 [ab]	59.45 ± 7.71 [a]	48.77 ± 5.88 [b]	57.81 ± 3.43 [ab]

The values are average ± standard deviation of four replications ($n = 6$). Values with different letters show significant differences among the groups ($p < 0.05$), and those with no letters show no significant difference ($p > 0.05$).

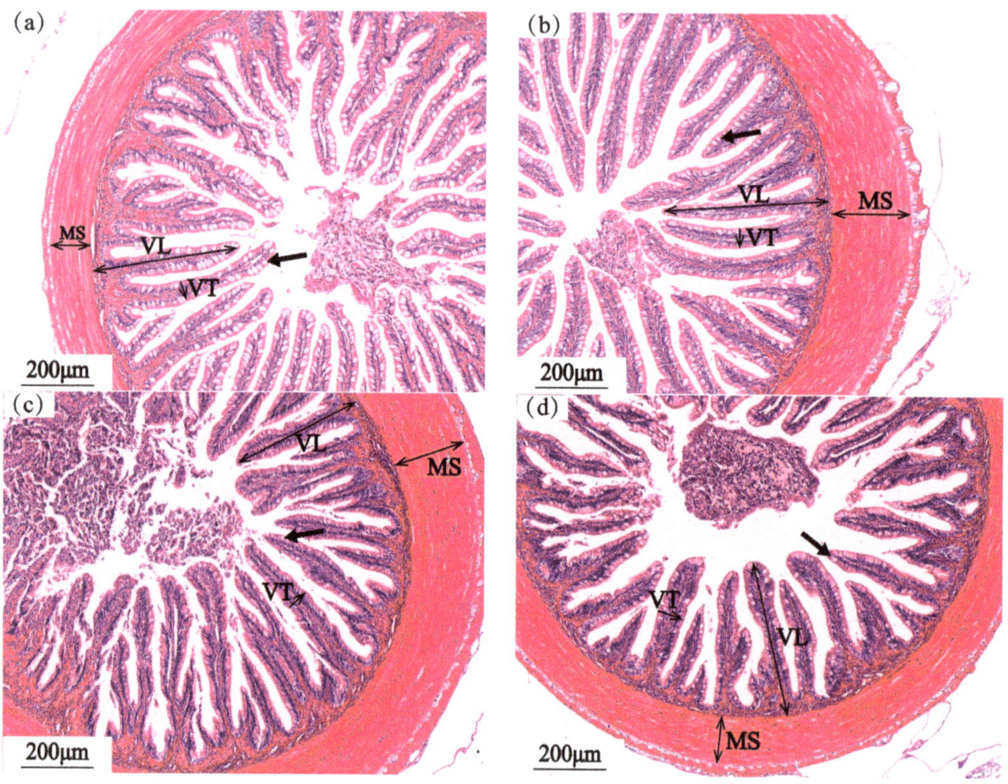

Figure 5. The effect of water flow velocity on the midgut structure of coral trout. (**a**) Hydrostatic control group: muscular thickness (MS), villus length (VL), and villus thickness (VT) were uniform, and there were many goblet cells (thick arrows); (**b**) 1 bl/s water flow velocity group: compared with the control group, the muscular layer was thicker, the villus length (VL) was longer, the villus thickness (VT) was not significantly different, and the number of goblet cells (thick arrows) was greater, but the size was relatively smaller; (**c**) 2 bl/s water flow velocity group: the muscular layer was thicker than in the control, the numbers and sizes of goblet cells (thick arrows) were significantly reduced, and some columnar epithelial cells in the mucosa layer were separated from the mucosa layer; (**d**) 2.5 bl/s water flow velocity group: muscular thickness (MS) was uniform, the number of goblet cells (thick arrows) was relatively small and the size was small.

3.4.3. Hindgut

In the hindgut tissue, the VL and MS of coral trout were significantly higher in the 1 bl/s water flow velocity group than in the control and other water flow velocity groups ($p < 0.05$). In terms of both MS and VT, the MS of coral trout in the 2.5 bl/s flow velocity group was the lowest, and there was no significant difference in VT among the 1 bl/s water flow velocity group and the control group ($p > 0.05$, Table 4). At 2 bl/s water flow velocity, the goblet cells showed normal levels, whereas at 2.5 bl/s water flow velocity, the goblet cells were fragmented and unevenly distributed, with significantly reduced numbers compared to all other groups (Figure 6).

Table 4. Effect of water flow velocity on the index of hindgut tissue of coral trout.

Item	Control	1 bl/s Water Flow Velocity	2 bl/s Water Flow Velocity	2.5 bl/s Water Flow Velocity
VL (μm)	628.24 ± 43.82 [b]	838.13 ± 47.86 [a]	627.75 ± 68.6 [b]	684.7 ± 56.29 [b]
MS (μm)	185.87 ± 13.17 [b]	216.57 ± 25.83 [a]	178.45 ± 22.08 [b]	171.56 ± 7.89 [b]
VT (μm)	67.8 ± 7.26 [ab]	54.01 ± 8.76 [b]	64.15 ± 9.47 [ab]	69.99 ± 11.34 [a]

The values are average ± standard deviation of four replications ($n = 6$). Values with different letters show significant difference among the groups ($p < 0.05$), and those with no letters showed no significant difference ($p > 0.05$).

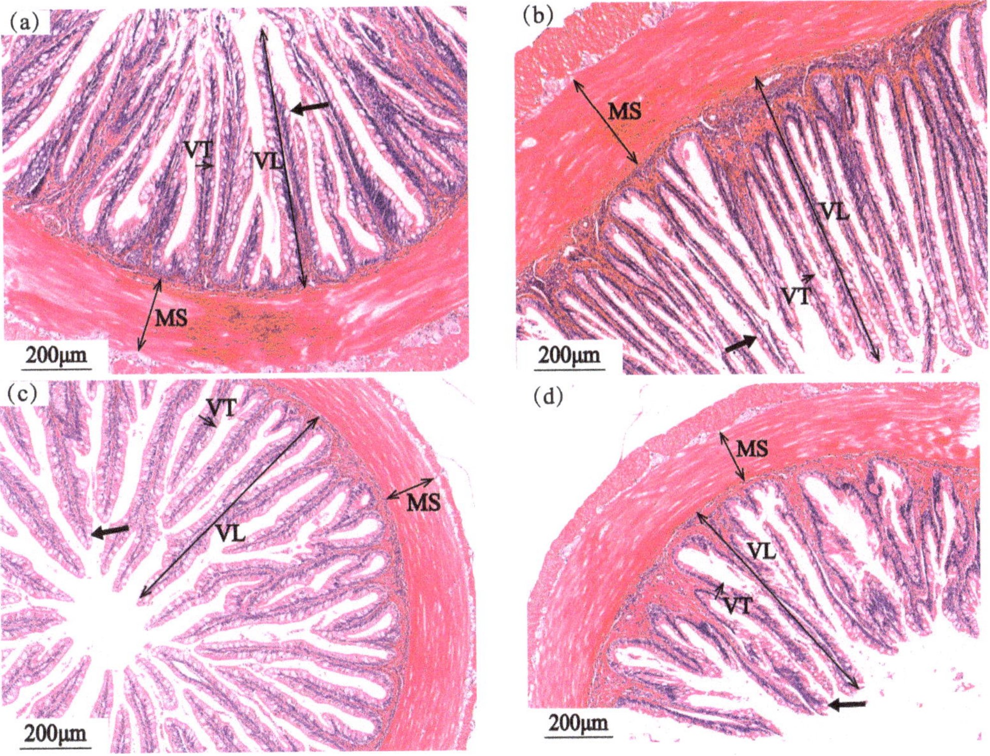

Figure 6. The effect of water flow velocity on the hindgut structure of coral trout. (**a**) Hydrostatic control group: muscular thickness (MS), villus length (VL), and villus thickness (VT) were uniform, and there were more goblet cells (thick arrows); (**b**) 1 bl/s water flow velocity group: the muscle layer was thicker than in the control, the villus was the longest, and the goblet cells (thick arrows) were relatively large; (**c**) 2 bl/s water flow velocity group: the muscle layer was the thinnest, and the goblet cells (thick arrows) were evenly distributed and numerous; (**d**) 2.5 bl/s water flow velocity group: some of the epithelial cells at the end of the villus fell off, and the goblet cells (thick arrows) were broken and unevenly distributed in size. The muscle layer was thicker, and the number of goblet cells (thick arrows) was significantly reduced compared with other groups.

4. Discussion

The different water flow velocities manifested significant differences in the growth performances of the fish. In a study by Wei [11], the highest SGR and WGR values were found in the 1 bl/s water flow velocity group of *Epinephelus coioides* (42.54 ± 0.62 g), and 2 bl/s water flow velocity negatively affected fish growth. Turbot (*Scophthalmus maximus*) (average body length 20.10 cm) achieved higher SGR values by swimming and exercising at 0.9 bl/s water flow velocity [12]. This is consistent with the results obtained in the present study, where the 1 bl/s water flow velocity group had the highest WGR and SGR values and the fastest growth. The desire to feed was enhanced when the current exercised the fish [13]. The SGR value of coral trout tended to decrease when the water flow velocity exceeded 1 bl/s, reaching a minimum at 2.5 bl/s water flow velocity. Goldfish (*Carassius auratus*) (10.1 ± 0.2 cm) that received high-water flow velocity exercise showed reduced feeding compared to the control [14]. Fish are capable of spontaneous swimming exercises at suitable water flow velocities, which helps to improve the efficiency of energy metabolism and accelerate their growth. At high-water flow velocities, the fish are forced to swim, and they fatigue faster, thus affecting their growth. At a high-water flow velocity (4 bl/s), *Schizothorax prenanti* (average body length 9.70 cm) allocated more energy to swimming and showed reduced growth performance [15].

It has been suggested that there are significant differences in the morphological characteristics of fish at different water flow velocities [16]. In the present experiment, however, there was no significant difference in the fatness of the fish among the groups, with the highest CF at 1 bl/s water flow velocity. The HSI value was also not significantly different, but the VSI ratio was, with the VSI in each water flow velocity group being smaller than that in the hydrostatic control group, and the VSI in the 2 bl/s water flow velocity group being the smallest. The reason for this is the difference in the thickness and mass of the fat envelope between the organs. Liu [17] pointed out that, as the water flow velocity increased, *Micropterus Salmoides* (8.13 ± 0.82 cm) broke down body fat to provide energy for maintained swimming movement, and the appropriate water flow velocity (5.0 bl/s) could reduce the excessive accumulation of fat in the fish, enhance the absorption of nutrients, and improve the quality of fish meat. However, when the water flow velocity exceeds the tolerance limit, the fish become fatigued, and no longer swims.

GLU metabolism is an important physiological metabolic process in animals. GLU can provide energy and carbon sources for the growth and metabolism of the body and meet the basic needs of the body's life activities [18]. In this experiment, the GLU concentration of the water flow velocity group was lower than that of the control group, and those in the 2 bl/s and 2.5 bl/s water flow velocity groups were significantly different from those in the control group, far below the average level. Under transport stress, the GLU concentration of silver pomfret (*Pampus argenteus*) increased significantly, and the breakdown of liver glycogen and muscle glycogen in these fish provided energy to ensure that the fish withstood the stress [19]. Different from the results of this experiment, the GLU concentration of black porgy (*Acanthopagrus schlegelii*) (6.75 ± 0.03 cm) increased at a high-water flow velocity (4.0 bl/s) [20]. The reason for this difference may be related to the swimming ability and living habits of fish. In another study, the GLU concentration of largemouth bass decreased with increasing water flow velocity [17]. Under acute hyperosmotic stress, the Chinese mitten crab (*Eriocheir sinensis*) undergoes corresponding adaptive physiological and biochemical changes, preferentially using GLU for energy supply to regulate osmotic pressure, and subsequently using nutrients, such as proteins [21]. At a high-water flow velocity (2.0 bl/s), tinfoil barbs (15.1 ± 1.35 cm) require more GLU to provide energy, which takes priority over protein and fat [22]. When GLU supply is inadequate, the organism derives additional energy from the breakdown of fat and protein to meet the needs of forced swimming to the detriment of nutrient accumulation in fish, and experimental fish undergo greater energy expenditure at high-water flow velocities ($\geq$2 bl/s), the specific regulatory and metabolic mechanisms of which require further study.

LD is commonly considered to be the main metabolic product produced by muscles after exercise, and that which cannot be processed in the muscles diffuses and is released into the hemolymph and blood for absorption and use by other tissues [23]. LD concentrations in the blood of black porgy (19.37 ± 1.88 cm) increased with exercise time, and fatigue was positively correlated with LD concentration [24]. In the present study, LD concentration in the blood showed a trend of decreasing and then increasing as the water flow velocity increased. A 1 bl/s water flow velocity correlated with the lowest LD concentration. At water flow velocities above 2 bl/s, the coral trout swimming activity was shown to involve anaerobic exercise, with higher LD concentrations than in the hydrostatic control group, thus requiring additional energy to maintain it. Swimming at high-water flow velocities leads to more energy expenditure, which is consistent with the findings in grass carp (*Ctenopharyngodon Idella*) (16.83 ± 0.96 cm) [25].

Cortisol is an important hormone secreted by the fish's body in response to external stimuli and is often used as an indicator of the physiological response to stress in fish. When the external environment causes stress in fish, it triggers the release of corticotrophin-releasing hormone (CRH) by the hypothalamus, which stimulates the secretion of adreno-cortico-tropic-hormone (ACTH) from the anterior pituitary gland of the hypothalamus, which is transmitted to the interrenal tissues to aid in the production of corticosteroids. The degree of stress in fish is related to the duration and intensity of the stress [26]. Atlantic salmon (19.26 ± 0.08 cm) subjected to 24 weeks of exercise at different water flow velocities (0.5 to 2.5 bl/s) showed similar levels of COR across treatment groups [7]. The research reported that GLU and COR are usually elevated for an initial period following stress, and then gradually decrease until normal levels are restored due to the negative feedback mechanisms arising from homeostasis and adaptation of the hypothalamic–pituitary–interrenal axis [27]. In another study, chronic high-flow stress (4.0 bl/s) also had no significant effect on COR levels in the blood of largemouth bass (4.50 ± 0.36 cm) [28]. Under crowding stress, a trend was observed of increasing and then decreasing COR concentrations in the whole bodies of *Ancherythroculter nigrocauda* (2.71 ± 0.31 cm) [29]. In the present study, there were no significant differences in COR concentrations under different water flow velocity conditions, and the experimental fish showed a good adaptive capacity. In conclusion, although the fish species, stressors, and modes of action were different, no significant differences were seen in COR concentrations among the treatment groups as the time was extended, indicating that the fish had an excellent adaptive capacity to water flow velocity.

The intestine is an essential digestive organ for fish, and the level of digestive enzyme activity directly affects the digestion and absorption of bait in this fish. The protease, LPS, and AMS activities of Jiffy Tilapia (*Oreochromis niloticus*) (15.1 ± 0.21 cm) were significantly increased under prolonged water flow [30]. The AMS activity of juvenile *Rhynchocypris lagowskii* (2.21 ± 0.07 g) showed a trend of increasing and then decreasing with increasing water flow velocity, which is similar to the results of this experiment [31]. A 1 bl/s water flow velocity correlated with the highest AMS activity in the intestine of *Oreochromis niloticus*, and the α-AMS activities of the 2 bl/s and 2.5 bl/s water flow velocity groups were significantly lower than those of the control and 1 bl/s groups. The high-water flow velocity group showed a poor ability to absorb starch-containing feeds, reducing feeding motivation. The lower GLU concentration observed in the high-water flow velocity group in the present study may be partly due to the lower activity of α-AMS at this water flow velocity, which affects the digestion and absorption of feed. LPS was not affected by water flow velocity, and there were no significant differences among the groups.

In the intestine, the secretion ability of mucous cells increases from front to back, and the mucous digestion ability of the rectum is the strongest in the intestine, which is closely related to the physiological function of the rectum [32]. An increase in VL in the intestine can increase the contact area with food, thus expanding the absorption area and improving the digestive capacity of the intestine [33]. The intestinal VL of experimental fish in the 1 bl/s water flow velocity group was significantly lower than that of other groups, possibly due to the fact that stimulation via water flow velocity weakens the digestive ability of the foregut, enhancing the digestive ability of the midgut and hindgut, and enabling them to secrete more goblet cells to promote food digestion. The specific causes need further experimental investigation. In this experiment, the VL, MS, and VT values in the midgut of each group were lower than those in the foregut and hindgut, and the relative absorption capacity of the midgut was poor. Therefore, the absorption of nutrients by the coral trout is mainly concentrated in the foregut and hindgut. Under stimulation by water flow velocity, the MS is increased to a certain extent, which can increase the elasticity of the intestine and promote food peristalsis. At 1 bl/s water flow velocity, the structural parameters of the foregut were slightly lower than those of the control group and other water flow velocity groups; further, both the midgut and hindgut were superior to those of other groups. In general, the intestinal structure of the 1 bl/s water flow velocity group was better than that of other water flow velocity groups, and its digestive enzyme activity was also higher than that of other water flow velocity groups. Medium and high-water flow velocities (2 bl/s and 2.5 bl/s) have a specific adverse effect on the shape and quantity of goblet cells, even causing cell breakage and other conditions, and thus have an inhibitory effect on digestive and immune functions. This is also in line with the lower digestive enzyme activity in these groups compared with the control.

In this experiment, the SGR and WGR values of coral trout showed high consistency with intestinal digestive enzyme activity. At a water flow velocity of 2.5 bl/s, the digestive enzyme activity of coral trout is weakened, resulting in a decrease in growth performance, while the GLU content also reflects nutritional deficiencies from a lateral perspective. At this water flow velocity, there was no significant impact on VT, VL, or MS. However, the reductions in the number of goblet cells and the structural changes are the main reasons for the reduced digestive capacity that was observed.

5. Conclusions

The results show that under water velocity stress, the growth indicators, serum biochemical indicators, intestinal digestive enzyme activity, and intestinal structure of coral trout have changed. High-water flow velocity ($\geq$2 bl/s) stress can lead to a decrease in GLU levels and an increase in LD content in fish, disrupting their normal physiological homeostasis. The high-water flow velocity has an inhibitory effect on the α-AMS activity of the coral trout and reduces the number of goblet cells, altering the intestinal structure and weakening the digestive capacity of the gut, ultimately leading to a reduction in SGR and WGR. This study shows that coral trout are sensitive to water flow velocity, and high flow velocity can exacerbate physiological disorders, weaken digestive enzyme activity, and alter the intestinal structure in this species. Therefore, in actual RAS aquaculture, the water flow speed should be controlled within 1 bl/s, and the impact of high-water flow velocity stress on this species' physiology and intestinal digestion should be avoided as much as possible to achieve a better aquaculture environment.

Author Contributions: Conceptualization, C.Z. and H.L.; methodology, H.L.; software, A.L. and J.S.; validation, Z.Q., J.X. and A.L.; formal analysis, Z.Q.; investigation, H.L.; resources, H.L.; data curation, Z.Q.; writing—original draft preparation, Z.Q.; writing—review and editing, C.Z. and H.L.; visualization, Z.Q.; supervision, J.X. and C.Z.; project administration, H.L.; funding acquisition, H.L. All authors have read and agreed to the published version of the manuscript.

Funding: The research is funded by Hainan Province Science and Technology Special Fund (ZDYF2022XDNY349); High-tech Ship Research Project of the Ministry of Industry and Information Technology, "Engineering Development and Key System Development of Movable Breeding Boat" (GXH (2019) No. 360).

Institutional Review Board Statement: The experiment conforms to the regulations and guidelines formulated by the Animal Protection and Use Committee of the Fishery Machinery and Instrument Research Institute, Chinese Academy of Fishery Sciences. The approval number is FMIRI-AWE-2023-001, and approval was given on 30 September 2022.

Informed Consent Statement: Not applicable.

Data Availability Statement: The original contributions presented in the study are included in the article. Further inquiries can be directed to the corresponding authors.

Acknowledgments: The authors would like to thank Zhenhua Ma and He Zhang for their support and help in the experiment.

Conflicts of Interest: The authors declare no conflict of interest.

References

1. Zhu, X.W.; Hao, R.J.; Zhang, J.P.; Tian, C.X.; Hong, Y.C.; Zhu, C.H.; Li, G.L. Improved growth performance, digestive ability, antioxidant capacity, immunity and *Vibrio harveyi* resistance in coral trout (*Plectropomus leopardus*) with dietary vitamin C. *Aquac. Rep.* **2022**, *24*, 101111. [CrossRef]
2. Sun, Z.J.; Xia, S.D.; Feng, S.M.; Zhang, Z.K.; Rahman, M.M.; Rajkumar, M.; Jiang, S.G. Effects of water temperature on survival, growth, digestive enzyme activities, and body composition of the leopard coral grouper *Plectropomus leopardus*. *Fish. Sci.* **2015**, *81*, 107–112. [CrossRef]
3. Lin, L. Effects of Environmental Factor on Growth and Blood Biochemical Indexes of *Plectropomus leopardus*. Master's Thesis, Tianjin Agricultural University, Tianjin, China, 2016.
4. Heinen, J.; Hankins, J.; Adler, P. Water quality and waste production in a recirculating trout-culture system with feeding of a higher-energy or a lower-energy diet. *Aquac. Res.* **1996**, *27*, 699–710. [CrossRef]
5. Deng, C.; Chen, S.L.; Ye, H.Z.; Qi, X.Z.; Luo, J. Analysis of Pigment and Enzyme Levels of *Plectropomus leopardus* with Body Color Difference. *Life Sci. Res.* **2020**, *24*, 15–20.
6. Davison, W. The effects of exercise training on teleost fish, a review of recent literature. *Comp. Biochem. Physiol. Part A Physiol.* **1997**, *117*, 67–75. [CrossRef]
7. Timmerhaus, G.; Lazado, C.C.; Cabillon, N.A.R.; Reiten, B.K.M.; Johansen, L.-H. The optimum velocity for Atlantic salmon post-smolts in RAS is a compromise between muscle growth and fish welfare. *Aquaculture* **2021**, *532*, 736076. [CrossRef]
8. Song, B.L.; Lin, X.T.; Xu, Z.N. Effects of upstream exercise training on feeding efficiency, growth and nutritional components of juvenile tinfoil barbs (*Barbodes schwanenfeldi*). *J. Fish. China* **2012**, *36*, 106–114. [CrossRef]
9. Chen, Z.L.; Ye, Z.Y.; Ji, M.D.; Zhou, F.; Ding, X.Y.; Zhu, S.M.; Zhao, J. Effects of flow velocity on growth and physiology of juvenile largemouth bass (*Micropterus salmoides*) in recirculating aquaculture systems. *Aquac. Res.* **2021**, *52*, 3093–3100. [CrossRef]
10. Li, X.Q. Study on Water Velocity and Polyculture in Recirculating Aquaculture System. Master's Thesis, Qingdao University of Technology, Qingdao, China, 2018.
11. Wei, X.L.; Yu, S.N.; Yang, Y.; Xiao, Y.Y.; Liu, Y.; Wei, F.S.; Tang, W.Q.; Li, C.H. Effects of exercise intensity on growth, blood innate immunity, hepatic antioxidant capacity, and HSPs70 mRNA expression of *Epinephelus coioides*. *J. Fish. Sci. China* **2017**, *24*, 1055–1064. [CrossRef]
12. Li, X.; Ji, L.Q.; Wu, L.L.; Gao, X.L.; Li, X.Q.; Li, J.; Liu, Y. Effect of flow velocity on the growth, stress and immune responses of turbot (*Scophthalmus maximus*) in recirculating aquaculture systems. *Fish Shellfish Immunol.* **2019**, *86*, 1169–1176. [CrossRef]
13. Grannell, A.; De Vito, G.; Murphy, J.C.; le Roux, C.W. The influence of skeletal muscle on appetite regulation. *Expert Rev. Endocrinol. Metab.* **2019**, *14*, 267–282. [CrossRef]
14. Nadermann, N.; Volkoff, H. Effects of short-term exercise on food intake and the expression of appetite-regulating factors in goldfish. *Peptides* **2020**, *123*, 170182. [CrossRef]
15. Liu, G.Y.; Wu, Y.J.; Qin, X.H.; Shi, X.T.; Wang, X.L. The effect of aerobic exercise training on growth performance, innate immune response and disease resistance in juvenile *Schizothorax prenanti*. *Aquac. Int.* **2018**, *486*, 18–25. [CrossRef]
16. Yan, G.J.; Cao, Z.D.; Peng, J.L.; Fu, S.J. The Effects of Exercise Training on the Morphological Parameter of Juvenile Common Carp. *J. Chongqing Norm. Univ. (Nat. Sci.)* **2011**, *28*, 18–21.
17. Liu, M.; Lin, Y.J.; Lian, Q.P.; Ni, M.; Gu, Z.M. Different water flow rates on the growth performance, antioxidant capacity, energy metabolism and tissue structure of *micropterus salmoides* under an in-pond recirculating aquaculture system. *Acta Hydrobiol. Sin.* **2023**, *47*, 25–36.
18. Ma, C.X.; Zhao, J.L.; Zeng, M.D.; Song, Y.D.; Zhou, Y.H. Effects of feeding and live refeeding on glucose metabolism in Mandarinfish *Siniperca chuatsi*. *Fish. Sci.* **2021**, *40*, 818–825.
19. Peng, S.M.; Shi, Z.H.; Li, J.; Yin, F.; Sun, P.; Wang, J.G. Effect of transportation stress on serum cortisol, glucose, tissue glycogen and lactate of juvenile silver pomfret (*Pampus argenteus*). *J. Fish. China* **2011**, *35*, 831–837. [CrossRef]
20. Yu, S.N.; Wei, X.L.; Wei, F.S.; Xiao, Y.Y.; Li, C.H. Effects of different exercise intensity on growth and serum and liver antioxidant indices of sparus macrocephalus. *Acta Hydrobiol. Sin.* **2018**, *42*, 255–263.
21. Wang, Y.R.; Li, E.C.; Chen, L.Q.; Wang, X.D.; Zhang, F.Y.; Gao, L.J.; Long, L.N. Effect of acute salinity stress on soluble protein, hemocyanin, haemolymph glucose and hepatopancreas glycogen of *eriocheir sinensis*. *Acta Hydrobiol. Sin.* **2012**, *36*, 1056–1062. [CrossRef]
22. Song, B.L. Effects of Water Current on Swimming Activity, Growth and Ecophysiological Aspect of Young *Barbodes Schwanenfeldi*. Ph.D. Thesis, Jinan University, Guangzhou, China, 2008.

23. Robles-Romo, A.; Zenteno-Savín, T.; Racotta, I.S. Bioenergetic status and oxidative stress during escape response until exhaustion in whiteleg shrimp *Litopenaeus vannamei. J. Exp. Mar. Biol. Ecol.* **2016**, *478*, 16–23. [CrossRef]
24. Chai, R.Y.; Yin, H.; Huo, R.M.; Wang, H.Y.; Huang, L.; Wang, P. Sustained Swimming on the Endurance Time and Physiological Metabolism of Acanthopagrus Schlegeli and Sciaenops Ocellatus. 2022. Available online: http://ssswxb.ihb.ac.cn/article/doi/10.7541/2023.2022.0186 (accessed on 31 August 2022).
25. Zhu, P.J. Effects of Different Water Velocities on Oxygen Consumption Rate, Morphological and Energy Metabolism of Grass Carp (*Ctenopharyngodon idella*). Master's Thesis, Huazhong Agricultural University, Wuhan, China, 2022.
26. Sun, G.X.; Li, M.; Wang, J.; Liu, Y. Effects of flow rate on growth performance and welfare of juvenile turbot (*Scophthalmus maximus* L.) in recirculating aquaculture systems. *Aquac. Res.* **2016**, *47*, 1341–1352. [CrossRef]
27. Zahedi, S.; Mirvaghefi, A.; Rafati, M.; Mehrpoosh, M. Cadmium accumulation and biochemical parameters in juvenile *Persian sturgeon, Acipenser persicus*, upon sublethal cadmium exposure. *Comp. Clin. Pathol.* **2013**, *22*, 805–813. [CrossRef]
28. Chen, Z.L. Effect of Flow Velocity on Largemouth Bass (*Micropterus salmoides*) Juveniles in Recirculating Aquaculture Systems. Master's Thesis, Zhejiang University, Hangzhou, China, 2021.
29. Li, P.; Chen, J.; Yu, D.H.; Li, Q.; Wang, G.Y.; Wei, H.J.; Sun, R.L.; Wang, S.R.; Sun, Y.H. Effects of transportation density and time on cortisol, lactate and glycogen of *ancherythroculter nigrocauda. Acta Hydrobiol. Sin.* **2020**, *44*, 415–422.
30. Zhao, L.Q.; Song, B.L. Effects of flow velocity on the behavior and digestive enzymes of juvenile tilapia. *Hebei Fish.* **2017**, *279*, 21–23.
31. Xu, Y.Q. Effects of Flow Rate on Growth, Nonspecific Immunity and Fatty Acid Composition in Juvenile *Phoxinus Lagowskii Dybowskii*. Master's Thesis, Dalian Ocean University, Dalian, China, 2020.
32. Wang, Y.B.; Zhang, J.; Li, X.M. Types and distribution of mucous cells in the digestive tract of *Plectropomus leopardus. Mar. Fish.* **2016**, *38*, 478–486.
33. Uni, Z.; Noy, Y.; Sklan, D. Posthatch changes in morphology and function of the small intestines in heavy-and light-strain chicks. *Poult. Sci.* **1995**, *74*, 1622–1629. [CrossRef] [PubMed]

Journal of
*Marine Science
and Engineering*

Article

Artificial Reefs Reduce Morbidity and Mortality of Small Cultured Sea Cucumbers *Apostichopus japonicus* at High Temperature

Huiyan Wang [†], Guo Wu [†], Fangyuan Hu, Ruihuan Tian, Jun Ding, Yaqing Chang, Yanming Su * and Chong Zhao *

Key Laboratory of Mariculture and Stock Enhancement in North China's Sea, Ministry of Agriculture and Rural Affairs, Dalian Ocean University, Dalian 116023, China
* Correspondence: suym@dlou.edu.cn (Y.S.); chongzhao@dlou.edu.cn (C.Z.)
† These authors contributed equally to this work.

Abstract: Summer mortality and morbidity are serious environment-related problems in cultured sea cucumbers (*Apostichopus japonicus*). Air exposure probably worsens the impact of high temperature on cultured sea cucumbers. In this present study, two laboratory experiments were designed to investigate the effects of artificial reefs on mortality, morbidity, crawling, feeding, and adhesion behaviors of small sea cucumbers (~1 g of wet body weight) after air exposure and disease outbreaks at 25 °C, respectively. Significantly lower mortality and morbidity occurred in the group with artificial reefs compared with those in the group without artificial reefs in the two experiments. This present study found that the stressed sea cucumbers cultured inside artificial reefs showed a significantly higher adhesion index, feeding behavior, and crawling frequency than those cultured without artificial reefs. In disease challenge assays, small sea cucumbers cultured inside the artificial reefs showed a significantly higher adhesion index and crawling frequency than those cultured without artificial reefs at 25 °C. Feeding, crawling, and adhesion behaviors of sea cucumbers cultured outside artificial reefs were not significantly different from those cultured without artificial reefs. The experimental results indicate that sea cucumbers with good fitness-related behaviors may be less affected by the disease and more likely to move into the crevices of artificial reefs. Fitness-related behaviors were poor in sea cucumbers cultured outside artificial reefs, so we considered them as affected individuals. Thus, artificial reefs provide a place to reduce the physical contact between unaffected and diseased/affected individuals, showing a potential to reduce disease transmission. Our present study establishes a cost-effective approach to increasing the survival of small sea cucumbers in seed production at high temperatures.

Keywords: sea cucumbers; morbidity; mortality; artificial reef; high temperature

Citation: Wang, H.; Wu, G.; Hu, F.; Tian, R.; Ding, J.; Chang, Y.; Su, Y.; Zhao, C. Artificial Reefs Reduce Morbidity and Mortality of Small Cultured Sea Cucumbers *Apostichopus japonicus* at High Temperature. *J. Mar. Sci. Eng.* **2023**, *11*, 948. https://doi.org/10.3390/jmse11050948

Academic Editor: Jean-Claude Dauvin

Received: 8 March 2023
Revised: 23 April 2023
Accepted: 26 April 2023
Published: 28 April 2023

1. Introduction

The sea cucumber (*Apostichopus japonicus*) is popular internationally because of its high nutritional and medicinal value [1]. High market demand stimulated the development of sea cucumber aquaculture [2,3]. The annual production of sea cucumber aquaculture was 222,707 tons in 2022 in China, which was 13.3% higher than that in 2021 [4]. Seed production is a common process in aquaculture, in which small sea cucumbers (~1 g of wet body weight) are intensively cultured in land-based factories until they reach the size available for the following pond culture and stock enhancement [5]. During this process, sea cucumbers are commonly challenged by high temperatures in summer [5]. Sea cucumbers are most suitable for survival in the temperature range between 10 and 21 °C, and the physiology and growth of sea cucumbers are greatly affected when water temperature is higher than 23 °C [6,7]. High temperature (>25 °C) damages the immune system and gut microbiota [6,7] and, consequently, causes mass mortality and morbidity of sea cucumbers in summer. It is very expensive to decrease the water temperature in seed production in summer. Therefore, it is essential to establish a cost-effective method to reduce the mortality and morbidity of sea cucumbers at high temperatures.

Changing seawater and removing feces are necessary in seed production. Thus, sea cucumbers are inevitably exposed to the air during this process. Air exposure leads to tissue damage and complicated oxidative stress [8,9], which increases the risk of death and disease in sea cucumbers [10]. Sea cucumbers are more severely affected when high temperature and air exposure happen together [11], which is probably responsible for the high mortality and morbidity of sea cucumbers in seed production in summer. In addition, skin ulcer syndrome (SUS) is one of the most common and severe diseases that occur to sea cucumbers in aquaculture, resulting in high mortality due to the high infectivity [12]. High temperature decreases the immunity of sea cucumbers and increases the transmission of pathogenic bacteria [13]. Therefore, sea cucumbers are more susceptible to disease in summer. Diseased sea cucumbers probably spread the disease to other individuals and increase the frequency of disease outbreaks [14,15], resulting in mass mortality and morbidity in aquaculture [16]. Therefore, it is of primary importance to reduce the mortality and morbidity of sea cucumbers after air exposure and disease outbreaks.

Although there is no obvious symptom (e.g., skin ulceration) in the early stages of the disease, sea cucumbers show decreased feeding, movement, and adhesion [16,17]. Thus, fitness-related behaviors are helpful for determining whether sea cucumbers are affected. Decreasing disease transmission is key to reducing mortality and morbidity during the disease outbreak. Our previous study suggests that separation effectively decreases the disease transmission of the sea urchin *Strongylocentrotus intermedius* [18]. However, it is costly and difficult to separate diseased/affected from unaffected individuals in aquaculture. We attempted to reduce mortality and morbidity by reducing the frequency that unaffected individuals have contact with diseased/affected ones [19]. Sea cucumbers are naturally attracted to artificial reefs [20], and they hide into crevices formed by rocks or reefs at high temperatures [21]. Diseased and affected individuals probably struggle to move to crevices because of their poor movement ability [16]. Tian et al. [11] found that artificial reefs improve fitness-related behaviors of stressed sea cucumbers at high temperatures. It is necessary to explore the key role of artificial reefs, which is beneficial to reduce the mortality and morbidity of sea cucumbers at high temperatures. Thus, we hypothesized that artificial reefs reduce mortality and morbidity by reducing physical contact between diseased/affected and unaffected sea cucumbers in seed production at high temperatures.

The main aim of the present study is to investigate whether artificial reefs improve sea cucumber survival in summer, as detailed below: (1) whether artificial reefs reduce the morbidity and mortality of sea cucumbers exposed to the air at high temperatures; (2) whether artificial reefs reduce the morbidity and mortality of sea cucumbers under a disease outbreak at high temperatures; and (3) whether artificial reefs improve the fitness-related behaviors of sea cucumbers exposed to both air exposure and disease outbreak at high temperatures.

2. Materials and Methods

2.1. Experimental Animals

Small green *A. japonicus* (~1 g of wet body weight) were transported from Yinhaima Aquatic Products Co., Ltd. (121°54′ E, 39°38′ N) to the Key Laboratory of Mariculture and Stock Enhancement in the North China's Sea and Ministry of Agriculture and Rural Affairs at Dalian Ocean University (121°56′ E, 38°87′ N) on 2 March 2022. Small sea cucumbers were cultured in fiberglass tanks (length × width × height: 115 cm × 75 cm × 60 cm) at a density of 400 g/m^2 [22,23] and with aeration. Water temperature was increased 1 °C each day from 8 °C to 25 °C (ambient temperature) by using a temperature-controlled system (Huixin Co., Dalian, China). Sea cucumbers were maintained at 25 °C for two weeks to be acclimatized to the experimental conditions. Sand-filtered seawater was replaced every three days with feces and uneaten food removed. Salinity, pH, and dissolved oxygen were tested weekly using a water quality detector (YSI Incorporated, Yellow Springs, OH, USA). The results were 31.36 ± 0.19, 8.01 ± 0.03 and 9.34 ± 0.21 mg/L, respectively. Small sea cucumbers were fed *ad libitum* with sea mud and commercial feed (Anyuan Industrial Co., Ltd., Yantai, China) at a ratio of 1:4 until the experiments started in April 2022.

2.2. Experiment I

This experiment was carried out to investigate whether artificial reefs improve the survival and fitness-related behaviors (crawling, feeding, and adhesion) of sea cucumbers exposed to the air at 25 °C. The presence or absence of artificial reefs was the experimental factor. An artificial reef (length × width × height: 50 mm × 60 mm × 55 mm, Figure 1A) consists of seven plastic cylinders (diameter × length: 15 mm × 50 mm, Figure 1A), in which only one sea cucumber is accessible. The plastic boxes (top length × bottom length × height: 20 cm × 16.5 cm × 14 cm) with and without artificial reefs were named group O and group C, respectively (*n* = 6). Twelve plastics boxes were placed in two temperature-controlled tanks (length × width × height: 115 cm × 75 cm × 60 cm) to maintain the water temperature at 25 °C. Two groups of sixteen sea cucumbers (~1 g of wet body weight) without ulcerated body walls were randomly selected and placed in each box with aeration (Figure 1B). Groups C and O each had six parallel groups. Sea cucumbers in each box were fed 5 g of a mixture of sea mud and commercial feed (Anyuan Industrial Co., Ltd. Yantai, China) at a 1:4 ratio every day. Sand-filtered seawater was changed daily in plastic boxes. Sea cucumbers were exposed to the air for 30 min after the seawater was removed from each box. Sea cucumbers were cultured in the above experimental conditions for 5 days.

We identified the sea cucumber with ulcerated skin as a diseased individual (Figure 1D). The numbers of dead and diseased sea cucumbers in each group were counted after the experiment. Sea cucumbers without ulcerated body walls (Figure 1D) were placed in plastics boxes (top length × bottom length × height: 20 cm × 16.5 cm × 14 cm) in group C. Consistently, individuals without ulcerated body walls in group O were divided into two groups, among which sea cucumbers in one group were inside the artificial reefs and sea cucumbers in the other group were outside the artificial reefs. Then, the two groups were placed in two different cubic plastic boxes (top length × bottom length × height: 20 cm × 16.5 cm × 14 cm). Finally, we obtained three groups of sea cucumbers without ulcerated body walls, which were the ones from group C, the ones inside the artificial reefs in group O (group O—In), and the ones outside the artificial reefs in group O (group O—Out). Three sea cucumbers were randomly selected from each of the three groups for the measurement of fitness-related behaviors (feeding, crawling, and adhesion). Feeding behavior and crawling behavior of the sea cucumbers were recorded using a digital camera (FDR-AXP55, Shanghai Suoguang Electronics Co., Ltd., Shanghai, China) fixed above the plastic cubic boxes (length × width × height: 60 mm × 47 mm × 45 mm) within 1 h. Adhesion behavior was measured in cubic plastic boxes (length × width × height: 180 mm × 140 mm × 45 mm) with 18 compartments. We repeated this experiment 6 times using different sea cucumbers for each group (*n* = 6).

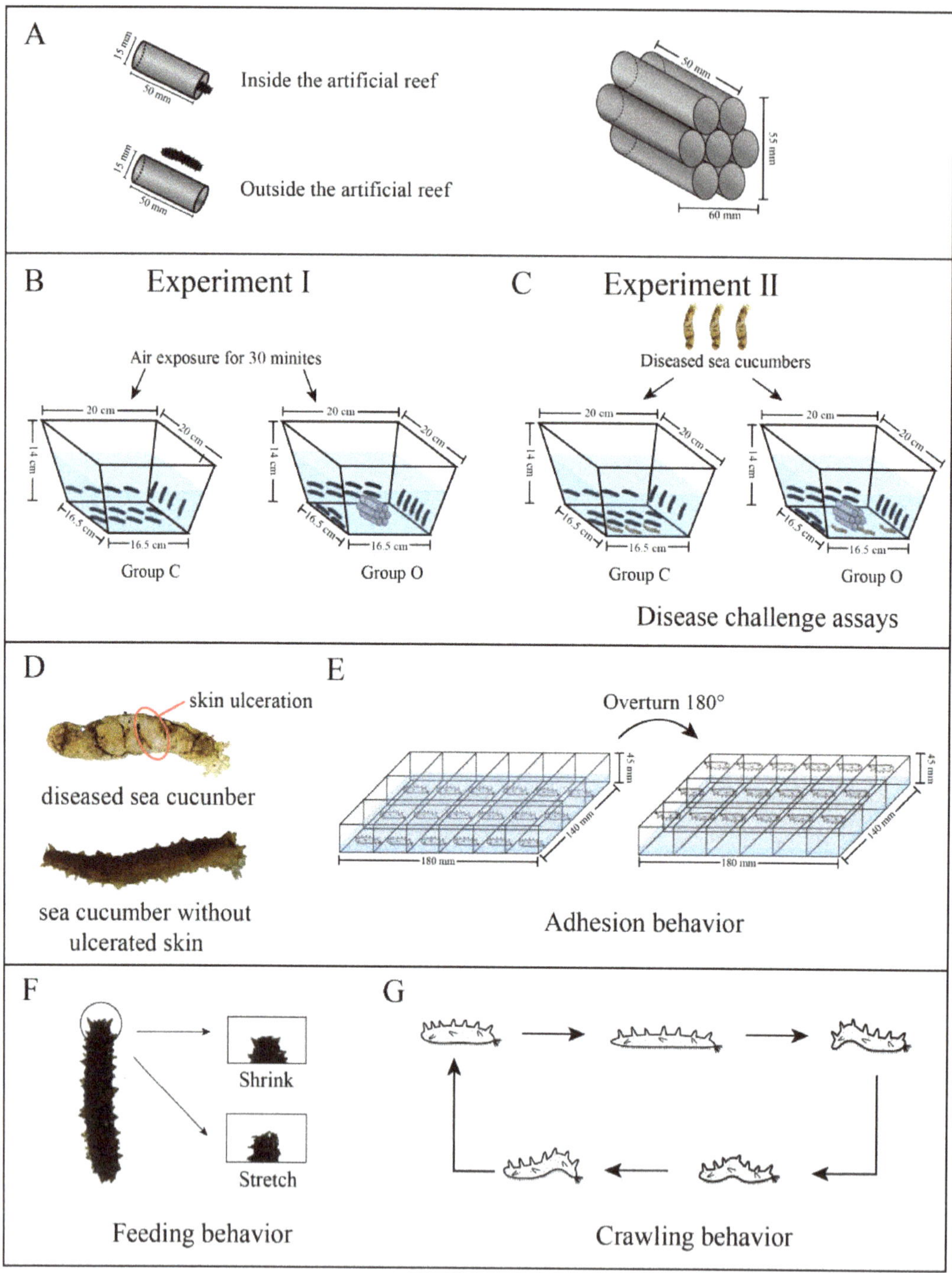

Figure 1. The diagrams show an artificial reef (**A**), the experiment designs for experiments I (**B**) and II (**C**), diseased and individuals without ulcerated skin (**D**), adhesion behavior (**E**), feeding behavior (**F**), and crawling behavior (**G**) of sea cucumbers.

2.3. Experiment II

We further investigated whether artificial reefs improve the survival and fitness-related behaviors (crawling, feeding, and adhesion) of sea cucumbers at 25 °C when disease outbreaks. The same artificial reefs as experiment I were consistently used in experiment II. Sixteen sea cucumbers without ulcerated body walls were placed in each box (Figure 1C). Diseased sea cucumbers were subsequently put in to create a disease outbreak. The plastic boxes (top length × bottom length × height: 20 cm × 16.5 cm × 14 cm) with and without artificial reefs were named as group O and group C, respectively ($n = 6$). Twelve plastic boxes (top length × bottom length × height: 20 cm × 16.5 cm × 14 cm) were placed in two temperature-controlled tanks (length × width × height: 115 cm × 75 cm × 60 cm). Each group was daily fed ad libitum with 5 g of a mixture of sea mud and commercial feed (Anyuan Industrial Co., Ltd., Yantai, China) at a 1:4 ratio. Sea cucumbers were cultured for three days without the seawater replacement in the same experimental conditions.

The numbers of dead and diseased sea cucumbers were counted after the experiment. Consistently, we obtained three groups of sea cucumbers without ulcerated body walls, which were the ones from group C, the ones inside the artificial reefs in group O (group O—In), and the ones outside the artificial reefs in group O (group O—Out). Crawling, feeding, and adhesion behaviors of sea cucumbers were recorded following the same procedures as described above for each group.

2.4. Mortality and Morbidity

Skin ulceration begins with one or more small white ulcerative spots, followed by the appearance of deep and enlarging ulcerative lesions, resulting in the exposure of the underlying muscles and spicules [24] (Figure 1D). Mortality and morbidity were calculated as follows:

$$M_t = \frac{D_e}{T} \times 100\%$$

$$M_b = \frac{D_i}{T} \times 100\%$$

where M_t is the mortality, M_b is the morbidity, D_e is the number of dead sea cucumbers, D_i is the number of diseased sea cucumbers, and T is the total number of sea cucumbers.

2.5. Crawling Behavior

Crawling behavior of sea cucumbers is composed of five stages (Figure 1G) [25]. A sea cucumber extends its body and then enters the contraction stage. The sea cucumber contracts from the back of the body to the middle and front in sequence. Finally, it returns to a static state [25]. Sea cucumbers complete each crawling cycle as described above, which is counted as one occurrence of crawling behavior. Sea cucumbers were randomly selected from each group and were placed into cubic plastic boxes (length × width × height: 60 mm × 47 mm × 45 mm). Two grams of mixture of sea mud and commercial feed at a ratio of 1:4 was put in the bottom of the box, and the number of occurrences of crawling behavior was counted within 1 h by using a digital camera (FDR-AXP55, Shanghai Suoguang Electronics Co., Ltd., Shanghai, China) ($n = 6$).

2.6. Feeding Behavior

Feeding behavior refers to the process of sea cucumbers ingesting food through its mouth tentacle activity [26]. Sea cucumbers stretch their tentacles out of their mouths, then extend their tentacles to grab the food, and finally capture the food into their mouths by contracting their tentacles [26]. One occurrence of tentacle activity was counted if sea cucumbers completed one tentacle activity cycle. Sea cucumbers were selected from each group and placed into cubic plastic boxes (length × width × height: 60 mm × 47 mm × 45 mm; Figure 1F). Two grams of mixture of sea mud and commercial feed at a ratio of 1:4 was put in the bottom of the box, and the duration of tentacle activity of all sea cucumbers within one hour was recorded by using a digital camera (FDR-AXP55, Shanghai Suoguang Electronics Co., Ltd., Shanghai, China) ($n = 6$).

2.7. Adhesion Behavior

Movement and ingestion of sea cucumbers greatly relies on the adhesion of mouth tentacles and tube feet. Adhesion behavior was measured according to Tian et al. [11]. Eighteen sea cucumbers from each group were randomly selected and placed in cubic devices (length × width × height: 180 mm × 140 mm × 45 mm; Figure 1E) with eighteen compartments. Each sea cucumber was placed in one compartment. Seawater was added to the devices until the seawater level reached a height of 2 cm. The devices were slowly turned to 180° after 10 min. We recorded the time that sea cucumbers fell from the top of the devices, since their adhesion abilities were disabled to support their body weight. The adhesion time was recorded as 600 s if sea cucumbers still adhered to the top of the box after 10 min. Wet body weight (g) of each sea cucumber was measured by using an electronic balance (G & G Co., San Diego, CA, USA) after the recording ($n = 6$). Data are accurate to one decimal place. Adhesion index (A_i) was formulated, as described by Tian et al. [11]:

$$A_i = \frac{T}{W}$$

where A_i is the adhesion index, T is the adhesion time (s), and W is the wet weight of an individual sea cucumber (g).

2.8. Statistical Analysis

All data were subjected to the analysis of variance distribution and homogeneity of variance using the Kolmogorov–Smirnov test and the Levene test, respectively. The mortality and morbidity of experiments I and II were non-normally distributed and/or heterogeneity of variance. Thus, the data were compared by using the Mann–Whitney U test. Crawling frequency, tentacle activity frequency, and the adhesion index were analyzed by using Kruskal–Wallis H in both experiments. All statistical analyses were performed using SPSS 22.0 statistical software. The level of significance was considered as $p < 0.05$.

3. Results

3.1. Experiment I

3.1.1. Mortality and Morbidity

Mortality was significantly higher in group C (43.06 ± 2.56%) than that in group O (5.56 ± 4.12%; Mann–Whitney U = 2.961, p = 0.003; Figure 2A). Consistently, group C showed significantly higher morbidity (48.61 ± 1.39%) than group O (6.94 ± 3.98%; Mann–Whitney U = 3.017, p = 0.003; Figure 2B).

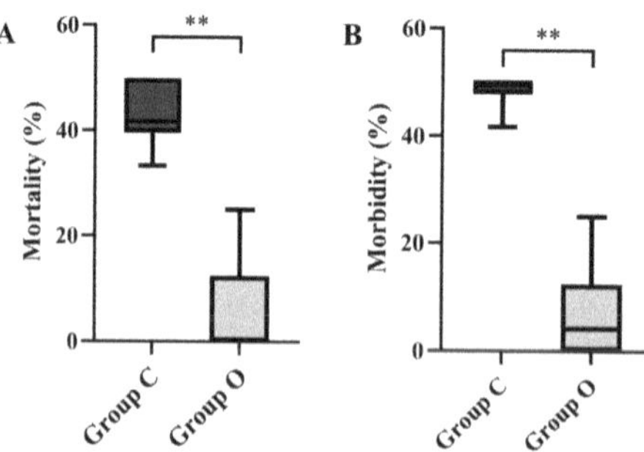

Figure 2. Mortality (**A**) and morbidity (**B**) of sea cucumbers between groups C and O. The asterisks ** mean $p < 0.01$ (mean ± SE, n = 6). Group C: sea cucumbers without artificial reefs; Group O: sea cucumbers with artificial reefs.

3.1.2. Crawling Frequency

There was no significant difference in crawling frequency between sea cucumbers of group C (4.44 ± 0.88 times) and individuals outside the artificial reef of group O (3.22 ± 0.50 times; Kruskal–Wallis H = 14.332, p = 1.000; Figure 3A). Crawling frequency was significantly higher in sea cucumbers inside the artificial reefs of group O (10.83 ± 1.52 times) than that in sea cucumbers outside the artificial reefs of group O (Kruskal–Wallis H = 14.332, p = 0.001) and group C (Kruskal–Wallis H = 14.332, p = 0.008; Figure 3A).

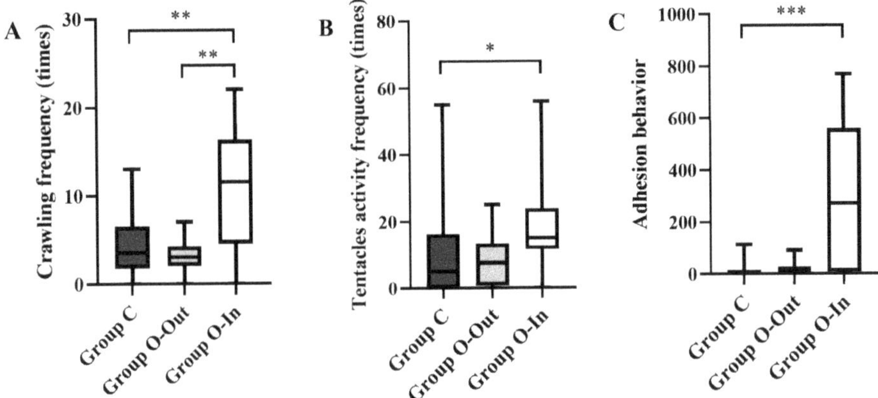

Figure 3. Crawling frequency (**A**), tentacles activity frequency (**B**) and adhesion behavior (**C**) of sea cucumbers. The asterisks *, **, *** mean p < 0.05, p < 0.01 and p < 0.001, respectively (mean ± SE, n = 6). Group C: sea cucumbers without artificial reefs in group C; Group O-Out: sea cucumbers outside the artificial reefs in group O; Group O-In: sea cucumbers inside the artificial reefs in group O.

3.1.3. Feeding Behavior

There was no significant difference in tentacle activity frequency between sea cucumbers of group C (10.44 ± 3.51 times) and individuals outside the artificial reefs of group O (8.50 ± 1.87 times; Kruskal–Wallis H = 7.705, p = 1.000; Figure 3B). Consistently, tentacle activity frequency was not significantly different between sea cucumbers outside (8.50 ± 1.87 times) and inside (18.72 ± 3.26 times) the artificial reefs (Kruskal–Wallis H = 7.705, p = 0.067; Figure 3B). Tentacle activity frequency was significantly higher in sea cucumbers inside the artificial reefs in group O than that in individuals of group C (Kruskal–Wallis H = 7.705, p = 0.036; Figure 3B).

3.1.4. Adhesion Behavior

The adhesion index was not significantly different between sea cucumbers outside the artificial reefs of group O (16.66 ± 5.50) and individuals of group C (9.75 ± 6.28; Kruskal–Wallis H = 0.561, p = 0.065; Figure 3C). Consistently, the adhesion index was not significantly different in sea cucumbers outside and inside the artificial reefs (283.62 ± 61.4) in group O (Kruskal–Wallis H = 0.561, p = 0.076; Figure 3C). However, sea cucumbers inside the artificial reefs of group O showed significantly higher adhesion index than individuals of group C (Kruskal–Wallis H = 0.561, p < 0.001; Figure 3C).

3.2. Experiment II

3.2.1. Mortality and Morbidity

Mortality was significantly higher in group C (30.56 ± 3.51%) than that in group O (4.17 ± 2.84%; Mann–Whitney U = 2.879 p = 0.004; Figure 4A). Morbidity was significantly higher in group C (43.06 ± 2.56%) than that in group O (6.94 ± 4.52%; Mann–Whitney U = 2.961, p = 0.003; Figure 4B).

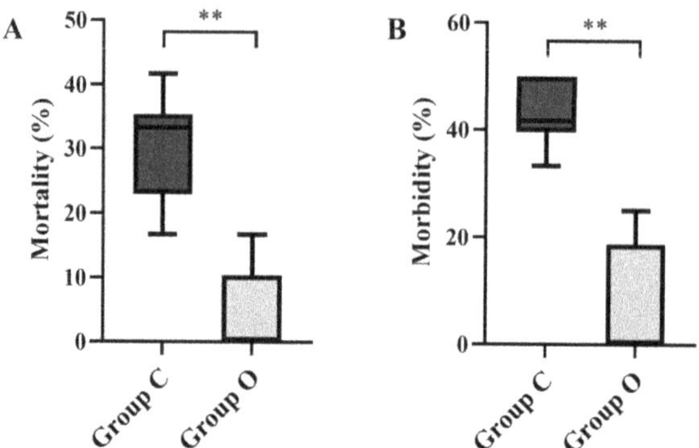

Figure 4. Mortality (**A**) and morbidity (**B**) of sea cucumbers between groups C and O. The asterisks ** mean $p < 0.01$ (mean $\pm$ SE, $n = 6$). Group C: sea cucumbers without artificial reefs; Group O: sea cucumbers with artificial reefs.

3.2.2. Crawling Frequency

There was no significant difference in crawling frequency between sea cucumbers of group C (1.67 ± 0.35 times) and individuals outside the artificial reefs in group O (2.56 ± 0.56 times; Kruskal–Wallis $H = 6.908$, $p = 0.762$; Figure 5A). Consistently, crawling frequencies of sea cucumbers outside (2.56 ± 0.56 times) and inside (3.67 ± 0.60 times) the artificial reefs in group O (Kruskal–Wallis $H = 6.908$, $p = 0.417$; Figure 5A) were not significantly different in the experiments. Significantly higher crawling frequency occurred in sea cucumbers inside the artificial reefs in group O than individuals in group C (Kruskal–Wallis $H = 6.908$, $p = 0.026$; Figure 5A).

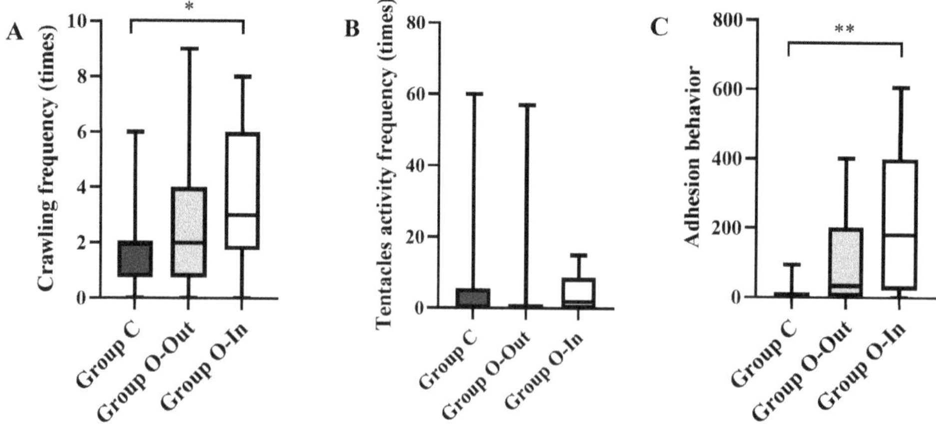

Figure 5. Crawling frequency (**A**), tentacle activity frequency (**B**), and adhesion behavior (**C**) of sea cucumbers. The asterisks *, ** mean $p < 0.05$ and $p < 0.01$, respectively (mean $\pm$ SE, $n = 6$). Group C: sea cucumbers without the artificial reefs in group C; Group O-Out: sea cucumbers outside the artificial reefs in group O; Group O-In: sea cucumbers inside the artificial reefs in group O.

3.2.3. Feeding Behavior

There was no significant difference in feeding behavior of sea cucumbers cultured inside artificial reefs of group O, outside artificial reefs of group O, and in group C ($p > 0.05$; Figure 5B).

3.2.4. Adhesion Behavior

The adhesion index was significantly higher in sea cucumbers inside the artificial reefs of group O (206.44 ± 47.12) than that in the individuals of group C (11.41 ± 5.37; Kruskal–Wallis $H = 13.007$, $p = 0.001$; Figure 5C). The adhesion index of sea cucumbers of group C (11.41 ± 5.37) was not significantly different from that of those outside the artificial reefs in group O (105.28 ± 28.26) (Kruskal–Wallis $H = 13.007$, $p = 0.053$; Figure 5C). Consistently, no significant difference was found in the adhesion indexes between sea cucumbers outside and inside the artificial reefs in group O (Kruskal–Wallis $H = 13.007$, $p = 0.732$; Figure 5C).

4. Discussion

The survival of sea cucumbers is constrained by high temperatures. A temperature of 25 °C is the threshold for severe heat stress in sea cucumbers, leading to a significantly increased risk of death and disease in sea cucumbers [6]. Air exposure is unavoidable in seed production, which not only causes mechanical damage [27] but also disrupts the immune system of sea cucumbers [9]. Furthermore, our previous study found that the negative effects of air exposure and high temperature on sea cucumbers were synergistic [11]. This present study found that artificial reefs significantly reduced mortality and morbidity of sea cucumbers exposed to the air at 25 °C. Specifically, in such conditions, the morbidity and mortality of sea cucumbers not cultured in artificial reefs were 7-fold and 8-fold the morbidity and mortality of those cultured in artificial reefs, respectively. This suggests that artificial reefs are an effective approach to improving the survival of sea cucumbers under the combined stress of high temperature and air exposure. Fitness-related behavior is a common indicator for evaluating the condition of sea cucumbers. The typical characteristics of sea cucumbers in poor conditions are anorexia, decreased adhesion capacity, and poor movements ability [17]. Tian et al. [11] found that artificial reefs improved fitness-related behaviors of sea cucumbers after they were exposed to high temperatures and the air. However, it is unknown how artificial reefs improve fitness-related behaviors. This present study found that sea cucumbers cultured inside the artificial reefs had significantly better feeding, crawling, and adhesion behaviors than those cultured without artificial reefs. Sea cucumbers exposed to the air and/or high temperatures reduce feeding or even stop feeding behavior [11,27,28]. Sea cucumbers cultured inside artificial reefs may acquire more energy by frequent feeding [28] and, consequently, may better cope with combined stressors. The adhesion of the tube feet of sea cucumbers affects the subsequent movement such as crawling behavior [29]. It is important to note that the fitness-related behaviors (feeding, crawling, and adhesion behaviors) of sea cucumbers cultured outside artificial reefs are not significantly different from those of sea cucumbers without artificial reefs. This suggests that the condition of the sea cucumbers inside the artificial reefs is better under the combined stressors. Therefore, artificial reef improves the survival and fitness-related behaviors of sea cucumbers after they are exposed to high temperatures and the air, thereby increasing the survival rate.

Skin ulcer syndrome is one of the common diseases with high infectivity and high mortality in seed production of sea cucumbers [17]. High temperatures reduce the immunity and disease resistance of sea cucumbers [30,31] and, thus, greatly increases the possibility of infection. The present study found that artificial reefs significantly reduced the mortality and morbidity of sea cucumbers during disease outbreaks at high temperatures. Specifically, during disease outbreaks at high temperatures, the morbidity and mortality of sea cucumbers not cultured in artificial reefs were 6-fold and 7-fold the morbidity and mortality of those cultured in artificial reefs. This suggests that artificial reefs improve the survival in disease environments and probably affect the spread of skin ulcer syndrome. In disease challenge assays, sea cucumbers cultured inside artificial reefs had significantly better adhesion and crawling behaviors, while there was no significant difference in the fitness-related behaviors between the sea cucumbers cultured outside and without artificial reefs. Decreased adhesion and movement ability are the symptoms of skin ulcer syndrome [17]. Hence, we speculated that sea cucumbers cultured outside and without artificial reefs were more affected by diseased sea cucumbers, although their skins were not ulcerated. Chemosensory cues are important for a variety of fundamental behavioral processes [32,33]. Lobsters use chemical cues from diseased individuals to determine whether the shelter is safe, and healthy lobsters rarely share the shelters with diseased lobsters [34]. Diseased sea cucumbers emit certain chemical cues for warning [33,35]. Sea cucumbers show escaping behavior after receiving alarm cues from diseased individuals [35,36]. Artificial reefs are possible places for escape due to their attraction to sea cucumbers [20,37]. Frequent physical contact with diseased individuals increases the probability of disease transmission [19]. Diseased sea cucumbers probably release chemical cues that promote the escape of individuals with strong movement to the crevices of artificial reefs. In the limited space, sea cucumbers that did not enter the reef inevitably contacted with diseased individuals, which increased the extent to which they were affected. In the present study, each crevice is accessible to only one sea cucumber. Sea cucumbers cultured with artificial reefs probably showed less contact with diseased and affected individuals after staying in porous artificial reefs, which reduced the extent of the affection. Seed production is an intensive aquaculture mode, and isolation of diseased sea cucumbers is expensive. Therefore, the artificial reef is a cost-effective approach to increasing the possibility of the survival of sea cucumbers by reducing the frequency of contact with diseased and affected individuals in seed production. Notably, our experiment is based on a laboratory experiment. Therefore, conducting a further field experiment is necessary.

In conclusion, we encourage aqua-farmers to use artificial reefs to decrease disease transmission and, thus, reduce the mortality and morbidity of small sea cucumbers in seed production at high temperatures. The present novel finding provides valuable insights into the improved management for the seed production of sea cucumbers in summer.

Author Contributions: Conceptualization, H.W., C.Z. and Y.S.; methodology, H.W., G.W., J.D. and C.Z.; software, H.W. and G.W.; validation, H.W., G.W., F.H. and R.T.; formal analysis, H.W., G.W. and F.H.; investigation, H.W., G.W. and R.T.; data curation, H.W. and G.W.; writing—original draft preparation, H.W. and G.W.; writing—review and editing, C.Z. and Y.S.; visualization, H.W.; supervision, C.Z., Y.S., J.D. and Y.C.; project administration, C.Z. and Y.S.; funding acquisition, C.Z., J.D., Y.S. and Y.C. All authors have read and agreed to the published version of the manuscript.

Funding: This research was funded by the National Natural Science Foundation of Liaoning Province (2022-MS-352), High-level talent support grant for innovation in Dalian (2020RD03), Modern Fisheries Technology Mission in Changhai, Liaoning (2022JH5/10400015) and Liaoning Province "Xingliao Talents Plan" project (XLYC2002107).

Institutional Review Board Statement: Not applicable.

Informed Consent Statement: Not applicable.

Data Availability Statement: The data presented in this study is available on request from the corresponding authors.

Acknowledgments: We appreciate Wei Tang for her editorial suggestions. We thank Mingfang Yang, Xiaomei Chi, Yushi Yu, and Xiang Li for their assistance in this experiment.

Conflicts of Interest: The authors declare no conflict of interest.

References

1. Liu, X.F. Study of Regional Differences in Nutrient Compositions and Bioactivities of Phospholipids in Sea Cucumber (*Apostichopus japonicus*). Ph.D. Thesis, Ocean University China, Qingdao, China, 2014. (In Chinese with an English Abstract)
2. Yang, H.; Hamel, J.F.; Mercier, A. *The Sea Cucumber Apostichopus japonicus: History, Biology and Aquaculture*; Academic Press: Cambridge, MA, USA, 2015.
3. Jiang, S.H.; Ren, Y.C.; Tang, B.P.; Li, C.F.; Jiang, C.B. Development status and countermeasures of *Apostichopus japonicus* culture industry in China. *J. Agric. Sci. Technol.* **2017**, *019*, 15–23.
4. Liu, X.Z. 2022 *China Fishery Statistical Yearbook*; China Agriculture Press: Beijing, China, 2022; p. 23. (In Chinese)

5. Chang, Y.; Yu, C.; Song, X. Pond Culture of Sea Cucumber, *Apostichopus japonicus* in Dalian. In *Advances in Sea Cucumber Aquaculture and Management*; Lovatelli, A., Conand, C., Purcell, S., Uthicke, S., Hamel, J.F., Mercier, A., Eds.; FAO: Rome, Italy, 2004; pp. 269–272.

6. Li, C.; Feng, H.; Xu, D. Effect of seasonal high temperature on the immune response in *Apostichopus japonicus* by transcriptome analysis. *Fish Shellfish Immunol.* **2019**, *92*, 765–771. [CrossRef] [PubMed]

7. Wang, C.; Song, S.; Li, D.; Liu, X.; Song, J. Research progress of effects and mechanism of environmental factors on physiology of sea cucumber *Apostichopus japonicus*. *Hebei Fish.* **2022**, *11*, 34–40. (In Chinese with an English Abstract)

8. Hou, S.Y.; Jin, Z.W.; Jiang, W.W.; Chi, L.; Xia, B.; Chen, J.H. Physiological and immunological responses of sea cucumber *Apostichopus japonicus* during desiccation and subsequent resubmersion. *PeerJ* **2019**, *7*, e7427. [CrossRef]

9. Cui, Y.; Hou, Z.; Ren, Y.; Men, X.; Zheng, B.; Liu, P.; Xia, B. Effects of aerial exposure on oxidative stress, antioxidant and non-specific immune responses of juvenile sea cucumber *Apostichopus japonicus* under low temperature. *Fish Shellfish Immunol.* **2020**, *101*, 58–65. [CrossRef]

10. Saeij, J.; Kemenade, V.V.; Muiswinkel, W.; Wiegertjes, G.F. Daily handling stress reduces resistance of carp to *Trypanoplasma borreli*: In vitro modulatory effects of cortisol on leukocyte function and apoptosis. *Dev. Comp. Immunol.* **2003**, *27*, 233–245. [CrossRef]

11. Tian, R.; Hu, F.; Wu, G.; Wang, H.; Ding, J.; Chang, Y.; Zhao, C. An effective approach to improving fitness-related behavior and digestive ability of small sea cucumbers *Apostichopus japonicus* at high temperature: New insights into seed production. *Aquaculture* **2023**, *562*, 738755. [CrossRef]

12. Wagner-Döbler, I.; Rheims, H.; Felske, A.; Pukall, R.; Tindall, B.J. Jannaschia helgolandensis, gen. nov., sp. nov., a novel abundant member of the marine Roseobacter clade from the North Sea. *Int. J. Syst. Evol. Microbiol.* **2003**, *53*, 731–738. [CrossRef]

13. Wang, L.; Wei, C.; Chang, Y.; Ding, J. Response of bacterial community in sea cucumber *Apostichopus japonicus* intestine, surrounding water and sediment subjected to high-temperature stress. *Aquaculture* **2021**, *535*, 736353. [CrossRef]

14. Eeckhaut, I.; Wayenberghe, K.V.; Nicolas, F.; Delroisse, J. Skin ulcerations in *Holothuria scabra* can be induced by various types of food. *SPC Beche-De-Mer. Inf. Bull.* **2019**, *39*, 31–35.

15. Delroisse, J.; Wayneberghe, K.V.; Flammang, P.; Gillan, D.; Gerbaux, P.; Opina, N.; Todinanahary, G.G.B.; Eeck-haut, I. Epidemiology of a Skin Ulceration Disease (SKUD) in the Sea Cucumber *Holothuria scabra* with a Re-view on the SKUDs in Holothuroidea (Echinodermata). *Sci. Rep.* **2020**, *10*, 22150. [CrossRef] [PubMed]

16. Wang, Y.G.; Rong, X.J.; Zhang, C.Y.; Sun, S.F. Main diseases of cultured *Apostichopus japonicus*: Prevention and treatment. *Mar. Sci.* **2005**, *9*, 1–7. (In Chinese with an English Abstract)

17. Deng, H.; He, C.; Zhou, Z.; Liu, C.; Tan, K.; Wang, N.; Jiang, B.; Gao, X.; Liu, W. Isolation and pathogenicity of pathogens from skin ulceration disease and viscera ejection syndrome of the sea cucumber *Apostichopus japonicus*. *Aquaculture* **2009**, *287*, 18–27. [CrossRef]

18. Hu, F.; Zhao, C.; Ding, P.; Li, Y.; Tian, R.; Qiao, Y.; Chang, Y. An effective facility decreases disease transmission and promotes resistance ability of small sea urchins *Strongylocentrotus intermedius*: A potential application in the longline culture. *Aquaculture* **2021**, *547*, 737542. [CrossRef]

19. Hu, F.; Zhao, Z.; Tian, R.; Ding, P.; Yin, D.; Li, X.; Qiao, Y.; Chang, Y.; Zhao, C. A cost-effective approach to decreasing the disease transmission of the sea urchin *Strongylocentrotus* intermedius: New information for seed production and longline culture. *Aquaculture* **2021**, *548*, 737569. [CrossRef]

20. Zhang, J.B. Experimental Studies on the Attractive Effects of Sea Cucumber Artificial Reefs on *Apostichopus japonicus*. Master's Thesis, Ocean University China, Qingdao, China, 2011. (In Chinese with an English Abstract)

21. Minami, K.; Masuda, R.; Takahashi, K.; Sawada, H.; Shirakawa, H.; Yamashita, Y. Seasonal and interannual variation in the density of visible *Apostichopus japonicus* (Japanese sea cucumber) in relation to sea water temperature. *Estuar. Coast. Shelf Sci.* **2019**, *229*, 106384. [CrossRef]

22. Pei, S.; Dong, S.; Wang, F.; Tian, X.; Gao, Q. Effects of density on variation in individual growth and differentiation in endocrine response of Japanese sea cucumber (*Apostichopus japonicus* Selenka). *Aquaculture* **2012**, *356–357*, 398–403. [CrossRef]

23. Xia, B.; Ren, Y.; Wang, Y.; Sun, Y.; Zhang, Z. Effects of feeding frequency and density on growth, energy budget and physiological performance of sea cucumber *Apostichopus japonicus* (Selenka). *Aquaculture* **2017**, *466*, 26–32. [CrossRef]

24. Lv, Z.; Guo, M.; Li, C.; Shao, Y.; Zhao, X.; Zhang, W. Divergent proteomics response of *Apostichopus japonicus* suffering from skin ulceration syndrome and pathogen infection. *Comp. Biochem.* **2019**, *30*, 196–205. [CrossRef]

25. Lin, C.G. Effects of Four Physical Environment Factors on the Movement and Feeding Behavior of Sea Cucumber *Apostichopus japonicus* (Selenka). Ph.D. Thesis, University of Chinese Academy of Sciences, Beijing, China, 2014. (In Chinese with an English Abstract)

26. Sun, J.M.; Zhang, L.B.; Pan, Y.; Lin, C.G.; Wang, F.; Yang, H.S. Feeding behavior and digestive physiology in sea cucumber *Apostichopus japonicus*. *Physiol. Behav.* **2015**, *139*, 336–343. [CrossRef]

27. Yang, M.F.; Li, X.; Hu, F.Y.; Ning, Y.C.; Tian, R.H.; Ding, P.; Zhao, C. Effects of handling stresses on fitness related behaviors of small sea cucumbers *Apostichopus japonicus*: New insights into seed production. *Aquaculture* **2022**, *546*, 737321. [CrossRef]

28. Chen, M.; Sun, S.; Xu, Q.; Gao, F.; Wang, H.; Wang, A. Influence of water temperature and flow velocity on locomotion behavior in tropical commercially important sea cucumber *Stichopus monotuberculatus*. *Front. Mar. Sci.* **2022**, *9*, 931430. [CrossRef]

29. Clement, J.C.; Schagerström, E.; Dupont, S.; Jutfelt, F.; Ramesh, K. Roll, right, repeat: Short-term repeatability in the self-righting behaviour of a cold-water sea cucumber. *J. Mar. Bio Assoc.* **2020**, *100*, 115–120. [CrossRef]

30. Yang, H.; Yuan, X.; Zhou, Y.; Mao, Y.; Zhang, T.; Liu, Y. Effects of body size and water temperature on food consumption and growth in the sea cucumber *Apostichopus japonicus* (selenka) with special reference to aestivation. *Aquac. Res.* **2005**, *36*, 1085–1092. [CrossRef]
31. Wang, F.Y.; Yang, H.S.; Gao, F.; Liu, G.B. Effects of acute temperature or salinity stress on the immune response in sea cucumber, *Apostichopus japonicus. Comp. Biochem. Physiol. Part A Mol. Integr. Physiol.* **2008**, *151*, 491–498. [CrossRef]
32. Ross, E.; Behringer, D. Changes in temperature, pH, and salinity affect the sheltering responses of Caribbean spiny lobsters to chemosensory cues. *Sci. Rep.* **2019**, *9*, 4375. [CrossRef] [PubMed]
33. Chi, X.; Hu, F.; Qin, C.; Huang, X.; Sun, J.; Cui, Z.; Ding, J.; Yang, M.; Chang, Y.; Zhao, C. Conspecific alarm cues are a potential effective barrier to regulate foraging behavior of the sea urchin *Mesocentrotus nudus. Mar. Environ. Res.* **2021**, *171*, 105476. [CrossRef]
34. Behringer, D.C.; Butler, M.J.; Shields, J.D. Avoidance of disease by social lobsters. *Nature* **2006**, *441*, 421. [CrossRef]
35. Sun, J.; Yu, Y.; Zhao, Z.; Tian, R.; Li, X.; Chang, Y.; Zhao, C. Macroalgae and interspecific alarm cues regulate behavioral interactions between sea urchins and sea cucumbers. *Sci. Rep.* **2022**, *12*, 3971. [CrossRef] [PubMed]
36. Vadas, R.L.; Elner, R.W. Responses to predation cues and food in two species of sympatric, tropical sea urchins. *Mar. Ecol.* **2003**, *24*, 101–121. [CrossRef]
37. Cui, Y.; Guan, C.; Wang, R.; Tan, J.; Huang, B.; Li, J. The study of attractive effects of artificial reef models on *Apostichopus japonicas. Prog. Mater. Sci.* **2010**, *31*, 109–113. (In Chinese with an English Abstract)

Journal of
*Marine Science
and Engineering*

Article

FOXO-like Gene Is Involved in the Regulation of 20E Pathway through mTOR in *Eriocheir sinensis*

Jiaming Li [1,†], Yuhan Ma [1,†], Zhichao Yang [1], Fengchi Wang [1], Jialin Li [1], Yusheng Jiang [1], Dazuo Yang [1,2], Qilin Yi [1,2,*] and Shu Huang [1,*]

1 College of Aquaculture and Life Science, Dalian Ocean University, Dalian 116026, China; 17866561266@163.com (J.L.)
2 Key Laboratory of Marine Bio-Resources Restoration and Habitat Reparation in Liaoning Province, Dalian Ocean University, Dalian 116023, China
* Correspondence: yiqilin@dlou.edu.cn (Q.Y.); huangshu@dlou.edu.cn (S.H.)
† These authors contributed equally to this work.

Abstract: The Forkhead Box O (FOXO) gene plays a key role in various biological processes, such as growth, metabolism, development, immunity and longevity. Molting is an essential process for crustacean growth, which is mainly regulated by 20-hydroxyecdysone (20E) and molt-inhibiting hormone (MIH). Although the role of FOXO in regulating the immune response of crustaceans is well documented, its involvement in controlling crustacean molting remains unclear. In this study, a FOXO-like gene (designed as *Es*FOXO-like) was identified in *Eriocheir sinensis*, and the regulation of the 20E pathway by *Es*FOXO-like was also investigated. The coding sequence of *Es*FOXO-like was 852 bp, which consisted of 283 amino acids including a conserved Forkhead (FH) domain. *Es*FOXO-like shared high similarity with FOXO genes from other crustaceans, and the mRNA expression levels of the *Es*FOXO-like gene were highest in the hepatopancreas and lowest in the hemocytes. However, transcription and protein expression of the *Es*FOXO-like gene were found to be up-regulated only during the pre-molt stage in the hepatopancreas, with lower expression levels observed at the post-molt stage. To explore the role of *Es*FOXO-like in the 20E pathway, *Es*FOXO-like was firstly inhibited by a specific FOXO inhibitor (AS1842856) and then through an *Es*FOXO-like dsRNA injection, respectively, and the results showed that the relative expression levels of *Es*FOXO-like were notably decreased in the hepatopancreas after both the inhibitor and dsRNA treatments. The 20E concentration, the mRNA expression levels of the 20E receptors including the ecdysone receptor (EcR) and the retinoid-X receptor (RXR) and *Es*mTOR transcription in the AS1842856 group or the *Es*FOXO-RNAi group were all significantly higher than that in the control group, while the mRNA expression level of *Es*MIH was significantly decreased after *Es*FOXO-like inhibition. To further investigate whether the *Es*FOXO-like acts through mTOR or not, Rapamycin was administered to inhibit mTOR activity in *Es*FOXO-like inhibited crabs. The results revealed a significant reduction in the concentration of 20E and the expression level of *Es*MIH in the AS1842856 + Rapamycin group compared to the AS1842856 + DMSO group, accompanied by an increase in *Es*EcR and *Es*RXR expression. These findings collectively suggest that *Es*FOXO-like regulates the 20E pathway through mTOR, which offered valuable insights into the understanding of the molting process in crustaceans.

Keywords: FOXO; 20-hydroxyecdysone (20E) pathway; mTOR; molting; *Eriocheir sinensis*

Citation: Li, J.; Ma, Y.; Yang, Z.; Wang, F.; Li, J.; Jiang, Y.; Yang, D.; Yi, Q.; Huang, S. FOXO-like Gene Is Involved in the Regulation of 20E Pathway through mTOR in *Eriocheir sinensis*. *J. Mar. Sci. Eng.* **2023**, *11*, 1225. https://doi.org/10.3390/jmse11061225

Academic Editor: Nguyen Hong Nguyen

Received: 26 April 2023
Revised: 9 June 2023
Accepted: 12 June 2023
Published: 14 June 2023

1. Introduction

The Forkhead Box O (FOXO) protein, which belongs to the Forkhead transcription factor family, was initially identified as a significant downstream molecular target of the insulin pathway [1,2]. FOXO is known to participate in numerous physiological responses that determine development, metabolism, cell cycle, apoptosis and longevity [3,4]. The

FOXO gene is evolutionarily conserved with a typical Forkhead (FH) DNA domain, which consists of three α-vehicles, three β-sands and two wing-like loops [5,6]. Until now, four FOXO homologous genes, including FOXO1, FOXO3α, FOXO4 and FOXO6, were identified in higher animals [7,8]. The distribution characteristics of the four FOXO genes are different as they are involved in the regulation of various biological processes [9]. However, only one FOXO gene was identified in invertebrates [10].

In recent studies of insects, the FOXO gene has been demonstrated to be involved in the regulation of development and growth [11]. For instance, FOXO overexpression in the adipose body could prolong the lifespan of *Drosophila melanogaster* [12]. Additionally, in *Bactrocera doralis*, the weight of the larvae body was significantly increased after FOXO inhibition [13]. Moreover, in *Blattella germanica*, *Bg*FOXO had an inhibitory effect on juvenile hormone (JH) biosynthesis in the case of nutrient deficiency [14]. The activation of FOXO through inhibition of insulin signaling led to reproductive diapause and, ultimately, resulted in the cessation of JH production in *Culex pipiens* [15]. In addition, it has been demonstrated that FOXO is involved in the metamorphosis of insects. Knocking down the expression of FOXO in *Helicoverpa armigera* larvae led to a molting failure [16]. Similarly, *Tribolium castaneum* larvae with silenced FOXO genes exhibited a significant delay in pupation [17]. Ecdysone, also known as 20-hydroxyecdysone (20E), has been found to stimulate FOXO transcription factor activity, resulting in increased expression of acid lipase-1 and subsequent promotion of fat decomposition in *Bombyx mori* [18]. In *H. armigera*, it has been observed that 20E can activate FOXO to facilitate protein hydrolysis during the molting cycle [16]. In larvae of *Tribolium castaneum*, RNA interference of the FOXO gene also led to delayed pupation, reduced levels of 20E and decreased expression of both the prothymotropic hormone (PTTH) and the spook (spo) gene, which are crucial for ecdysone biosynthesis [17]. Studies on the physiological role of FOXO in crustaceans are mainly focused on the immune response. It has been reported that the expression of FOXO was significantly decreased in the intestine of crabs after hepatopancreatic necrosis disease (HPND) stimulation [19]. In the Chinese mitten crab, FOXO has also been found to have a positive impact on the expression levels of genes coding antimicrobial peptides (AMPs) [19]. In *Marsupenaeus japonicas*, FOXO was discovered to upregulate the expression of AMPs via the IMD pathway as well as enhance the phagocytosis of hemocytes against pathogenic bacteria [1,3]. Despite the known involvement of FOXO in various physiological processes in crustaceans, it remained unclear whether the gene plays a role in the regulation of molting.

Molting is an important biological process closely related to the growth and development of crustaceans [20]. The molting cycle could be divided into three vital stages including pre-molt, post-molt and inter-molt [21,22]. As the primary component of ecdysteroids, 20-hydroxyecdysone (20E) plays a crucial role in mediating the changes that occur during the molting process [23]. The signals mediated by 20E are transduced via the binding of an isomeric dimer complex consisting of an Ecdysone receptor (EcR) and a Retinoid-X receptor (RXR) [24–26]. In addition, 20E and molt-inhibiting hormone (MIH) are mutually antagonistic and jointly regulate the molting process [27,28]. Recent studies demonstrated that the mammalian target of rapamycin (mTOR) is essential for the production of ecdysteroids in arthropods [29]. The mTOR gene has been shown to stimulate the synthesis of 20E and simultaneously downregulate the expression of molting inhibiting hormone (MIH) signaling genes [30]. Inhibition of mTOR expression by Rapamycin has been shown to impair ecdysteroid secretion in the prothoracic gland of insects [31,32]. Moreover, it is worth noting that FOXO could block rapamycin complex 1 (mTORC1) signal transduction in mammals, and the inactivation of FOXO alleviated mTORC1 inhibition [33]. However, whether FOXO regulates 20E synthesis and expression of molting-related genes through mTOR still remains unknown. In insects, it has been discovered that mTOR can enhance 20E production by regulating the size of the prothoracic and the molting glands [34–36]. Here, we speculated that FOXO might activate the ecdysone signaling pathway through mTOR activation, thereby regulating the occurrence of molting.

The Chinese mitten crab, *Eriocheir sinensis*, is one of the most important aquaculture crustaceans in China [37]. So far, the roles of 20E and its receptors, i.e., *EsEcR*, *EsRXR* as well as *EsMIH*, in the regulation of molting have been studied in this species [22,38–40]. It was found that *EsEcR* and *EsRXR* were highly expressed in the hepatopancreas at pre-molt and lowly expressed at post-molt, while *EsMIH* was highly expressed in eyestalk and showed the opposite expression pattern to that of *EsEcR* and *EsRXR* [40]. In this study, a FOXO-like molecule containing an FH domain was identified and characterized in *E. sinensis* (designated as *EsFOXO-like*), with the objectives to (1) examine its expression pattern at three molting stages, (2) investigate the impact of *EsFOXO-like* on the concentration of 20E and the expression levels of genes related to molting (*EsEcR*, *EsRXR* and *EsMIH*) and (3) explore the involvement of mTOR in the regulation of *EsFOXO-like* and its impact on the 20E pathway. Overall, these findings would be helpful for understanding the role of FOXO in the molting of crustaceans.

2. Materials and Methods

2.1. Crabs and Sample Preparation

Healthy Chinese mitten crabs (*E. sinensis*) weighing about 10 g were provided by a crab farm in Lianyungang, Jiangsu province. The crabs were cultivated in aerated water at 25–26 °C for at least one week to acclimate to the test conditions.

Nine crabs were selected for measuring the tissue distribution of *EsFOXO* mRNA expression levels. Hemocytes, heart, hepatopancreas, stomach, muscles, gills, and eyestalks were collected. Meanwhile, the hepatopancreas was collected from three crabs at each molting stage, including the post-molt stage, the inter-molt stage and the pre-molt stage [21,22]. Additionally, hemolymph (about 750 μL) was collected from the walking legs of the Chinese mitten crabs with a pre-cooling anticoagulant solution [41]. The hemolymph from three crabs was pooled into one sample and then centrifuged at $800 \times g$, 4 °C for 10 min to collect the hemocytes and serum. In total, three replicate samples were processed for the analyses of tissue distribution of the *EsFOXO-like* mRNA, and three replicate samples were processed for each molting stage (made from 9 individual crab samples for each molting stage). The tissues were stored at −80 °C in TRIzol reagent (Invitrogen, Carlsbad, CA, USA). Three crabs were pooled together as one parallel, and there were three parallels (including 9 individuals) for sampling at each molting stage and in various tissues [42,43].

2.2. Identification and Sequence Analysis of EsFOXO-like

A gene encoding FOXO (designed as *EsFOXO-like*) was identified by screening and downloading the genome of *E. sinensis* from NCBI (NCBI accession No. CL100111224_L02). Briefly, all the gene sequences annotated as FOXO were searched and obtained based on the *E. sinensis* genome annotation results. Then, the sequence alignment and domain analysis was performed, and, finally, the *EsFOXO-like* gene was screened. The primers *EsFOXO-F* and *EsFOXO-R* were designed to clone the ORF of *EsFOXO-like* according to the sequence of *EsFOXO-like* (Table 1). The Takara Ex Taq® DNA Polymerase (RR001Q, Takara, Otsu, Shiga, Japan) was used as the polymerase in the PCR reaction system. PCR amplification of hepatopancreas cDNA was performed as follows: one cycle at 95 °C for 5 min and 35 cycles at 95 °C for 30 s, 57 °C for 30 s, 72 °C for 1 min and 72 °C for 10 min. The PCR product was gel-purified and inserted into a pMD19-T simple vector (Takara) and verified by DNA sequencing. The recombinant plasmid (pMD19-T-*EsFOXO-like*) was transformed into competent cells of *Escherichia coli* Trans5α (TransGen Biotech, Beijing, China), as follows: The recombinant plasmid (5 μL) and *Escherichia coli* Trans5α (100 μL) were mixed and then placed on ice for 45 min, at 42 °C in a water bath for 90 s and on ice again for 2 min immediately. The transformants were incubated in Luria–Bertani (LB) medium, and three positive clones were selected and sequenced. The validated sequence of *EsFOXO-like* has been submitted with the GenBank accession number OR115551.

Table 1. Sequences of the primers used in the study.

Primer Name	Sequences (5′–3′)	Tm (°C)	Size (bp)	Efficiency (%)
Cloning primers				
*Es*FOXO-F	ATGACAAGTTTCTTCTCGCT	52.5	852	
*Es*FOXO-R	CTAGAGCAGGGGCAGGGG	62.3		
qRT-PCR primers				
*Es*FOXO-like-F	GGCTACGTGGAGAGCGAGGA	60.6	142	99%
*Es*FOXO-like-R	CCTGGGCGATCAGGTCTGC	60.0		
*Es*EcR-F	GAGAGAACAGAAAAAGGCACGA	57.3	105	102%
*Es*EcR-R	ATGGCTGACATTGGACTAATGG	58.8		
*Es*MIH-F	TGAAGACTGCGCCAACATCT	56.2	90	103%
*Es*MIH-R	CGTGAGGTCGTCCTTCTGTG	60.4		
*Es*RXR-F	AGGCTTCAGGTTCCACTCGC	58.3	131	98%
*Es*RXR-R	GTGTACGCTGCCCTGGAGGA	60.1		
*Es*mTOR-F	CTTGAGGAGTTCTACCCTGCGT	60.5	115	101%
*Es*mTOR-R	GGACCACCTGGGCAAGGTAT	56.9		
*Es*β-actin-F	GCATCCACGAGACCACTTACA	56.4	223	102%
*Es*β-actin-R	CTCCTGCTTGCTGATCCACATC	60.1		
RNAi primers				
EGFP-RNAi-F	TAATACGACTCACTATAGGGCGACGTAAACGGCCACAAGT			
EGFP-RNAi-R	TAATACGACTCACTATAGGGCTTGTACAGCTCGTCCATGC			
*Es*FOXO-RNAi-F	TAATACGACTCACTATAGGGATGACAAGTTTCTTCTCGCTGGTGA			
*Es*FOXO-RNAi-R	TAATACGACTCACTATAGGGATCAGGTCTGCATAGGACATGTTGC			

T_m: Annealing temperature; Size: Amplicon size.

The cDNA sequence of *Es*FOXO-like was blasted against the GenBank database (www.ncbi.nlm.nih.gov/blast) (accessed on 8 November 2022). The amino acid sequence and protein domain of *Es*FOXO-like were analyzed by the Expert Protein Analysis System (https://web.expasy.org/translate/) (accessed on 15 November 2022) and the online SMART tool (http://smart.embl-heidelberg.de/) (accessed on 20 November 2022), respectively. Multiple sequence alignment was performed with the Clustal X multiple alignment program. The phosphorylation sites were predicted by NetPhos-3.1 tool (https://services.healthtech.dtu.dk/services/NetPhos-3.1/) (accessed on 1 March 2023). The phylogenetic tree of the FOXO gene was constructed by using the maximum likelihood method and the MEGA 11.0 software. The reliability of the branching was tested using 1000 bootstrap samples.

2.3. FOXO Inhibitor and Rapamycin Treatment

In order to investigate the influence of *Es*FOXO-like on both the concentration of 20E and the expression levels of genes associated with molting, the expression level of *Es*FOXO-like was experimentally suppressed. For this purpose, the FOXO inhibitor AS1842856 (S8222, Selleck, UT, USA) was used, for which the inhibitory effect time was set to 24 h [44,45]. AS1842856 was dissolved in dimethyl sulfoxide (DMSO) (1% diluted in PBS) (Beyotime). Eighteen crabs at the inter-molt stage were divided into AS1842856 and DMSO groups. The crabs from each group were injected with 100 μL AS1842856 (1 μg/μL) and DMSO, respectively. The inhibitor and DMSO were injected into the hemolymph from the membrane of the third posterior walking leg on the right with slight modifications to the previously described methods [46]. Then, at 24 h post-injection of AS1842856 and DMSO, the hepatopancreas was collected and used for the detection of the mRNA expression of *Es*FOXO, *Es*EcR, *Es*RXR and *Es*mTOR. The eyestalks were also collected for detecting the mRNA expression of *Es*MIH. Finally, the upper liquid of hemolymph after centrifugation was obtained as serum, and the serum was isolated for the detection of ecdysone concentration.

Rapamycin (HY-10219, MCE, Dallas, TX, USA) was used to inhibit mTOR activity [47]. According to a previous study, it has been confirmed that mTOR could stimulate 20E synthesis [30]. Moreover, the mTOR inhibitor Rapamycin can cause impaired secretion of

20E [29]. Therefore, to explore whether *EsFOXO*-like regulates the 20E pathway through *EsmTOR* or not, rapamycin was administrated to *EsFOXO*-like inhibited crabs. Rapamycin was also dissolved in DMSO [48], and the effect time for Rapamycin was set to 12 h [49]. Eighteen inter-molt crabs were divided into the AS1842856 + Rapamycin group and the AS1842856 + DMSO group. Overall, 100 μL AS1842856 (1 μg/μL) was administered to *EsFOXO*-like inhibited crabs 24 h after the AS1842856 injection. Then, the crabs in the AS1842856 + Rapamycin group and the AS1842856 + DMSO group received 100 μL Rapamycin (2 μg/μL) and DMSO injection, respectively. At 12 h after injection, the mRNA expression levels of *EsEcR* and *EsRXR* in the hepatopancreas, *EsMIH* expression levels in the eyestalk and the 20E concentration in serum were collectively detected.

2.4. RNA Isolation, cDNA Synthesis and Quantitative Real-Time PCR Analysis

Total RNA was extracted from different tissues using the TRIzol reagent (Invitrogen, Carlsbad, CA, USA). The cDNA was synthesized with 1 μg of total RNA using the Prime Script™ RT reagent Kit with gDNA Eraser (Takara, Otsu, Shiga, Japan) according to the protocol of the manufacturer. The cDNA was synthesized by a primer mix containing Oligo dT and random hexamers, and the synthesis was conducted at 37 °C for 15 min, 85 °C for 5 s. The cDNA was stored at −80 °C for quantitative real-time PCR (qRT-PCR) analysis.

The qRT-PCR reactions were performed using the SYBR® Premix Ex Tap™ (Takara, Otsu, Shiga, Japan) with a PCR amplification procedure of 95 °C for 30 s, 40 cycles at 95 °C for 5 s and 60 °C for 30 s on the ABI PRISM 7500 Sequence Detection System. An incremental increase of 0.5 °C/5 s was conducted for melting curve analyses. The standard curve was performed using ten-fold dilutions of the cDNA templates from the hepatopancreas for each primer pair to determine the efficiency of each primer. In addition, non-template controls were tested to check for primer-dimers and contaminated samples. The gene-specific primers of *EsEcR* (GenBank accession No. KF732874.1) and *Es*β-actin (No. HM053699) were designed according to our previous studies [42,50] (Table 1). The gene-specific primers of *EsFOXO*-like, *EsmTOR* (No. MT920347.1), *EsMIH* (No. DQ341280.1) and *EsRXR* (No. MK604180.1) were designed using the qRT-PCR primer design website (https://www. genscript.com/tools/real-time-pcr-taqman-primer-design-tool) (accessed on 13 December 2022). Additionally, in the study, *Es*β-actin acted as an internal control [46,51–53]. The specificity of primers was evaluated by 1% agarose gel electrophoresis and melting curve analysis. The relative expression levels were normalized to the control samples using the comparative threshold cycle ($2^{-\triangle\triangle Ct}$) method [54].

2.5. Detection of 20-Hydroxyecdysone (20E) Concentration

The separated serum was used to detect ecdysone concentration, and it was measured using the crab ecdysone ELISA Kit according to the manufacturer's protocol [42]. The plates were first precoated with purified ecdysone antibodies. After that, 10 μL of serum (diluted 1:5) and standard samples were incubated in the thermostat bath at 37 °C for 30 min, the plate was washed five times and then the chromogenic agent and stop buffer were added. The absorbance (450 nm) was measured using a microtiter plate reader (BioTek, Winooski, VT, USA). Lastly, the 20E concentration of the samples was calculated by the standard curve that was constructed based on absorbance and standard concentration.

2.6. RNA Interference Assay

The double-stranded RNA (dsRNA) of *EsFOXO*-like and EGFP were synthesized according to the method described in a previous report [42]. T7 promoter-linked primers, including *EsFOXO*-RNAi-F and *EsFOXO*-RNAi-R and EGFP-RNAi-F and EGFP-RNAi-R (Table 1), were used to amplify the DNA fragment of *EsFOXO*-like and EGFP, respectively. The fragments were used as templates to synthesize dsRNA. The dsRNA was synthesized by using the in vitro Transcription T7 Kit (for siRNA synthesis) (6140, Takara) according to the instructions. Additionally, the RNA integrity was examined by electrophoresis, and the concentration was quantified using the NanoDrop 2000 spectrophotometer (Thermo Fisher

Scientific, Wilmington, DE, USA). The dsRNAs of *Es*FOXO-like and EGFP were dissolved in PBS at a final concentration of 1 μg μL^{-1}.

Twenty-seven crabs were randomly divided into three groups (PBS group, EGFP-RNAi group and *Es*FOXO-like-RNAi group) with nine individuals in each group to investigate the RNAi efficacy. The EGFP-RNAi group was employed as the control group according to previous studies [42]. The crabs in the PBS group, the EGFP-RNAi group and the *Es*FOXO-like-RNAi group received an injection of 100 μL PBS, EGFP dsRNA and *Es*FOXO-like dsRNA, respectively. The hepatopancreas, eyestalk and serum were collected from three crabs in each group at 24 h after the dsRNA or PBS injection. There were three replicates for each group, and the tissues from three crabs were taken as one replicate. To evaluate the RNAi efficacy, the mRNA and protein expression levels the in hepatopancreas were detected by qRT-PCR and western blotting, respectively. In addition, the mRNA expression levels of *Es*EcR, *Es*RXR and *Es*mTOR in the hepatopancreas and the *Es*MIH expression level in the eyestalk together with 20E concentration in serum were also analyzed after knocking down *Es*FOXO-like mRNA expression.

2.7. Western Blotting Analysis

Western blotting was used to detect the protein expression levels of *Es*FOXO-like with the commercial FOXO1 antibody (ab52857, Abcam, Cambridge, MA, USA). The proteins were extracted from the hepatopancreas using the Protein Extraction Kit (Biyotime, Beijing, China) according to the protocol. Samples were separated by 12% sodium dodecyl sulphate-polyacrylamide gel electrophoresis (SDS-PAGE) and transferred onto nitrocellulose membranes. The membranes were blocked with 3% Bovine Serum Albumin (BSA) (Sangon Biotech, Shanghai, China) in TBST (10 mM Tris-HCl, pH 7.5, 150 mM NaCl, 0.2% Tween-20) at 4 °C overnight. Then, the membranes were incubated with 1/3000 diluted FOXO1A antibody or beta-tubulin rabbit antibody (Beyotime Biotechnology, Beijing, China) in TBST with 3% BSA for 2 h at room temperature, and this was followed with washing in TBST three times. Next, the membranes were incubated with HRP-labeled goat anti-rabbit IgG (Beyotime Biotechnology, Beijing, China) at a ratio of 1:1000 (diluted in 3% BSA) for 1 h at room temperature. After three washes, the membranes were incubated in the western blot chemiluminescence HRP substrate (Beyotime Biotechnology, Beijing, China) in the dark for 2 min, and the protein bands were visualized using the chemiluminescent imaging system (Amersham Imager 600, USA). The relative protein expression levels of *Es*FOXO-like were analyzed by ImageJ software (http://rsb.info.nih.gov/ij/) (accessed on 17 March 2023). All bands were digitalized by using the ImageJ software.

2.8. Statistical Analysis

Data were firstly checked for normality of distribution and homogeneity of variances using the Shapiro–Wilk and Levene's tests, respectively. The Mann–Whitney U test was also used for the non-normally distributed data. The Student's *t* test and one-way ANOVA analysis were used for the normally distributed data. A Turkey multiple comparison test was also used after the one-way ANOVA analysis. Finally, the data were graphed by Origin 8.1. Significant differences were accepted at $p < 0.05$, and the data were shown as mean $\pm$ SD (N = 3) (* $p < 0.05$ and ** $p < 0.01$).

3. Results

3.1. Sequence and Phylogenetic Analysis of EsFOXO-like Gene

The ORF sequence of *Es*FOXO-like was found to be 852 bp long and encoded a polypeptide consisting of 283 amino acids (Figure 1a). Notably, the *Es*FOXO-like protein was found to contain a Forkhead (FH) domain between amino acids 140 and 208 (Figure 1b), which bears similar protein kinase A (PKA) phosphorylation modification sites. The *Es*FOXO-like protein has also been found to have a glycogen synthase kinase 3 (GSK3) phosphorylation modification site and a PKA phosphorylation modification site outside the FH domain (Figure 2). The deduced amino acid sequence of *Es*FOXO-like shared high similarity with those FOXOs from other invertebrates species (Figure 2), such as *Penaeus japonicus* (MW080526.1, 83.51%), *Penaeus monodon* (XP_037792098.1, 88.37%), *Portunus trituberculatus* (MPC 39137.1, 88.55%), *Penaeus vannamei* (XP_027228067.1, 75.00%), *Clupea harengus* (XP_031428678.1, 83.33%) and *Salvelinus alpinus* (XP_023824765.1, 85.45%) (Table 2).

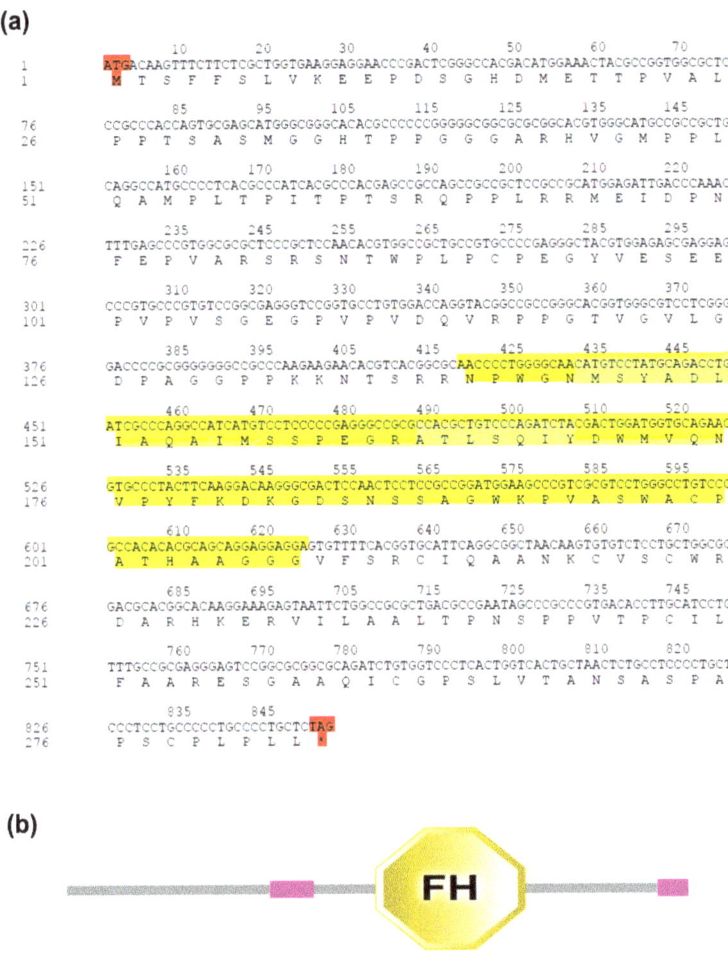

Figure 1. Sequence features and domain architecture of *Es*FOXO-like. (**a**) Nucleotide sequence and deduced amino acid sequence of *Es*FOXO-like. The number of amino acids and nucleotides are listed on the left of the sequence. The start codon (ATG) and stop codon (TAG) are marked in red. The FH domain is marked in yellow. (**b**) SMART analysis of *Es*FOXO-like. The *Es*FOXO-like protein domain contained two low complexity regions (pink boxes) and a conserved FH domain (yellow polygon).

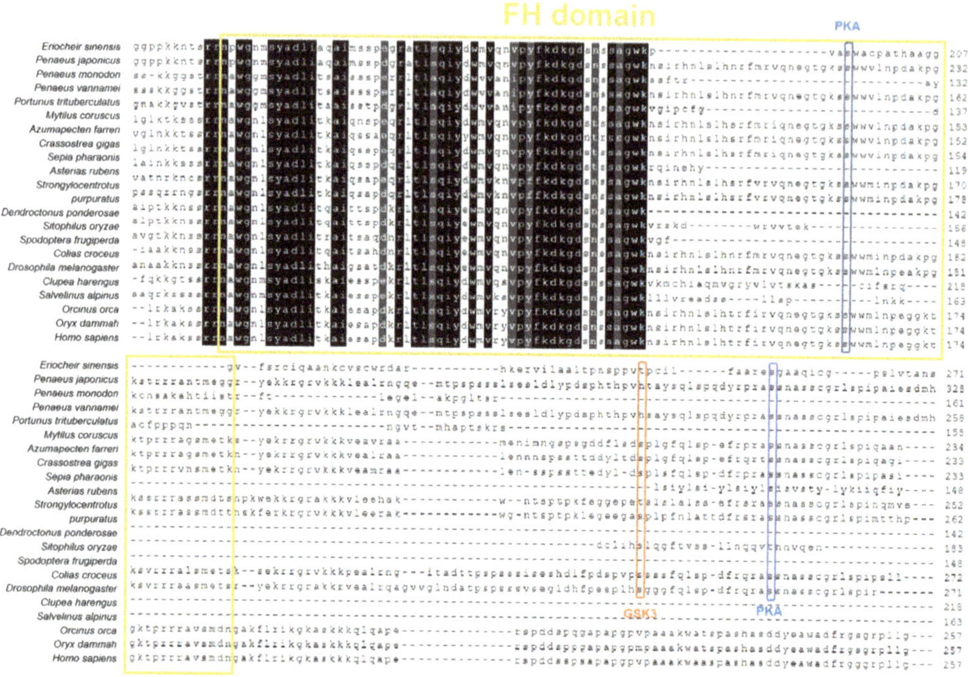

Figure 2. Multiple sequence alignment of the FOXOs from both vertebrates and invertebrates. The amino acids shaded in dark are identical residues between all analyzed sequences, and relatively conserved amino acids are shown in gray. The FH domain is marked in the yellow box. Sequence information of the FOXO proteins is listed in Table 2. PKA: protein kinase A; GSK3: glycogen synthase kinase 3.

Table 2. The FOXO genes used in the alignment of *Es*FOXO-like.

Species	Gene Name	Accession Number	Query Cover %	Identity %
Azumapecten farreri	FOXO-like protein	QFR 39803.1	43	53.66
Crassostrea gigas	forkhead box protein O	XM 011416057.3	43	60.16
Mytilus coruscus	FOXO3	CAC 5392548.1	43	57.03
Sepia pharaonis	FOXO3	CAE 1327826.1	39	60.53
Asterias rubens	forkhead box protein O3-like	XP 033635399.1	43	54.48
Strongylocentrotus purpuratus	forkhead transcription factor O	DQ 286746.2	43	49.30
Dendroctonus ponderosae	forkhead box protein O isoform X3	XP 048524044.1	52	51.23
Sitophilus oryzae	forkhead box protein O isoform X2	XP 030755757.1	50	59.17
Drosophila melanogaster	forkhead box, sub-group O isoform C	NP 996204.1	42	55.37
Spodoptera frugiperda	forkhead box protein O-like isoform X2	XP 035439085.1	44	54.33
Colias croceus	forkhead box protein O isoform X1	XP 045494104.1	54	47.67

Table 2. *Cont.*

Species	Gene Name	Accession Number	Query Cover %	Identity %
Penaeus vannamei	forkhead box protein O-like isoform X2	XP 027228067.1	40	75.00
Penaeus monodon	forkhead box protein O4-like	XP 037792098.1	40	88.37
Portunus trituberculatus	Forkhead box protein O	MPC 39137.1	42	88.55
Penaeus japonicus	Fork box protein O	MW080526.1	67	83.51
Clupea harengus	forkhead box protein O1-B-like	XP 031428678.1	30	83.33
Salvelinus alpinus	forkhead box protein O1	XP 023824765.1	45	85.45
Orcinus orca	forkhead box protein O6	XP 033277871.1	43	52.67
Oryx dammah	forkhead box protein O6	XP 040087454.1	43	52.67
Homo sapiens	forkhead box protein O6	NP 001278210.2	43	49.63

In the phylogenetic tree of FOXOs, all the selected FOXO members were divided into invertebrate and vertebrate branches. *Es*FOXO-like was firstly clustered with FOXO from *Penaeus japonicus*, and shared a closer evolutionary relationship with FOXO from *Portunus trituberculatus*, *Penaeus vannamei* and *Penaeus monodon*, enabling categorization into the invertebrate FOXO branch. The other FOXOs from *Clupea harengus*, *Salvelinus alpinus*, *Orcinus orca*, *Orys dammah* and *Homo sapiens* were clustered into the vertebrate clade (Figure 3).

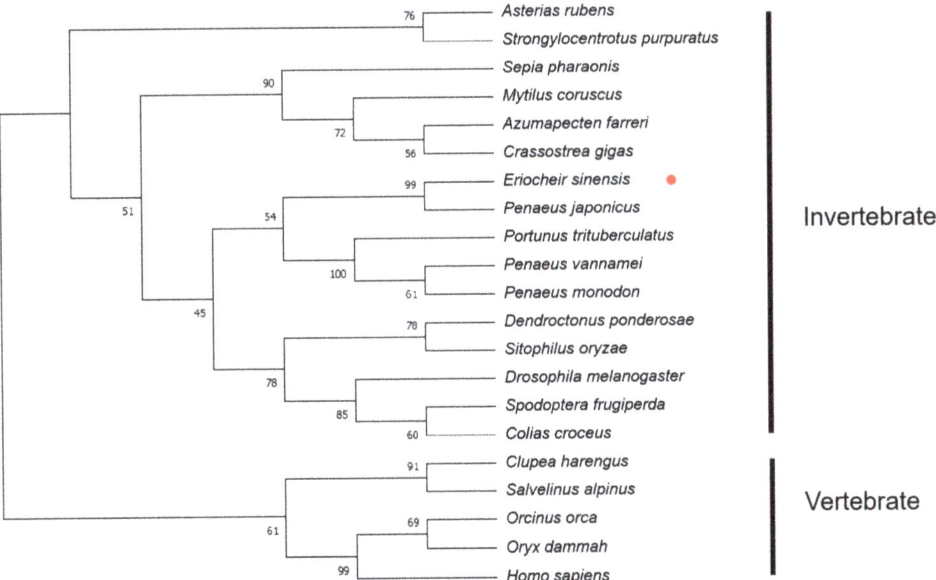

Figure 3. Phylogenetic relationship of FOXO sequences. The phylogenetic tree of FOXO gene was constructed based on the amino acid sequences of *Es*FOXO-like and another twenty FOXOs (Table 2) using maximum likelihood method implemented in the Mega 11.0 software. The bootstrap values of the branches were obtained by testing the tree 1000 times and are shown as percentage numbers.

3.2. The Distribution of EsFOXO-like Gene in Different Tissues

The *Es*FOXO-like gene mRNA transcripts were detected in different tissues, including the hepatopancreas, hemocytes, stomach, heart, gill, muscle and eyestalk. The highest levels of *Es*FOXO-like mRNA transcript were detected in the hepatopancreas (33.65-fold more than that in hemocytes, $p < 0.05$). The recorded mRNA relative expression levels of

*Es*FOXO-like in the heart, stomach, muscles, gills, and eyestalks were 4.06-fold, 1.92-fold, 2.86-fold, 3.12-fold and 6.58-fold of that in hemocytes ($p < 0.05$), respectively (Figure 4).

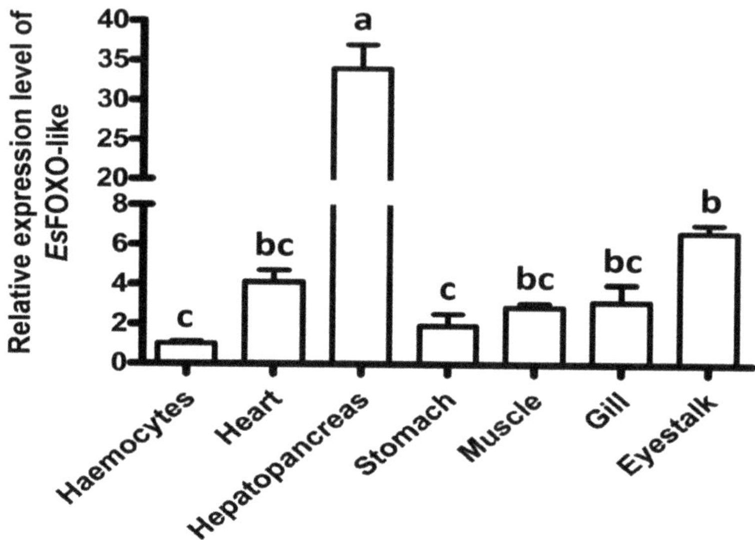

Figure 4. The mRNA expression level of *Es*FOXO-like in different tissues. The data were analyzed statistically using one-way ANOVA analysis. Each bar represents the mean ± standard deviation of three independent biological replicates (Supplementary Table S1). Different lowercase letters indicate significant differences ($p < 0.05$).

3.3. The EsFOXO-like mRNA and Protein Expression Characteristic in Hepatopancreas at Three Molting Stages

The mRNA and protein expression of *Es*FOXO-like in hepatopancreas at pre-molt, post-molt and inter-molt were measured. It was shown that *Es*FOXO-like was highly expressed at the pre-molt stage and lowly expressed at the post-molt stage (Figure 5a). The expression level at pre-molt was 22.01-fold ($p < 0.01$) and 3.58-fold ($p < 0.05$) of that at post-molt stage and inter-molt stage, respectively.

The results showed that there was a specific band of about 70 kDa (Figure 5b) detected with FOXO1 antibody in the hepatopancreas by western blotting (Figure 5b), indicating that the FOXO1 antibody was relatively specific. Additionally, the *Es*FOXO-like protein in the hepatopancreas was increased at the pre-molt stage and decreased at the post-molt stage. Specifically, the protein expression levels of *Es*FOXO-like at the pre-molt stage were 12.62-fold ($p < 0.01$) and 3.03-fold ($p < 0.01$) of that at the post-molt stage and the inter-molt stage, respectively (Figure 5c,d).

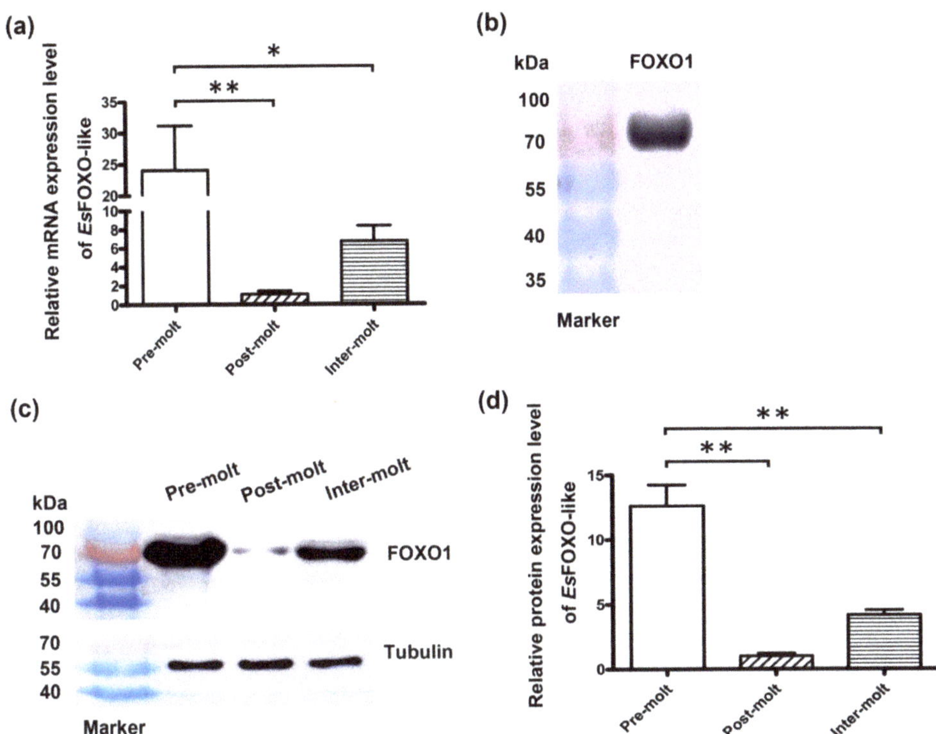

Figure 5. The mRNA and protein expression patterns of *Es*FOXO-like in hepatopancreas at different molt stages. (**a**) The relative mRNA expression level of *Es*FOXO-like in hepatopancreas at different molt stages by qRT-PCR. (**b**) The specific antibody detection of FOXO1 antibody in hepatopancreas. (**c**) The protein expression levels of *Es*FOXO-like in hepatopancreas at different molt stages using western blotting. (**d**) Statistical analysis of western blot performed in (**c**). Each bar represents the mean ± standard deviation of three independent biological replicates (Supplementary Table S1). The results were analyzed using one-way ANOVA. Asterisks indicate significant differences (* $p < 0.05$ and ** $p < 0.01$).

3.4. The EsFOXO-like mRNA and Protein Expression Levels after Inhibition of EsFOXO-like

The inhibitory effect of the FOXO inhibitor (AS1842856) on *Es*FOXO-like expression in hepatopancreas was evaluated by qRT-PCR and western blotting, respectively. After AS1842856 injection, the *Es*FOXO-like mRNA transcripts in the AS1842856 group were significantly decreased, representing 0.11-fold ($p < 0.05$) of that in the DMSO group (Figure 6a). The western blotting assay showed that the band against the *Es*FOXO-like antibody in AS1842856 injected crabs was thinner compared to that in the PBS group (Figure 6b). The protein expression level of *Es*FOXO-like in the hepatopancreas was also decreased (Figure 6c) in the AS1842856 group (0.14-fold of that in the PBS group, $p < 0.05$).

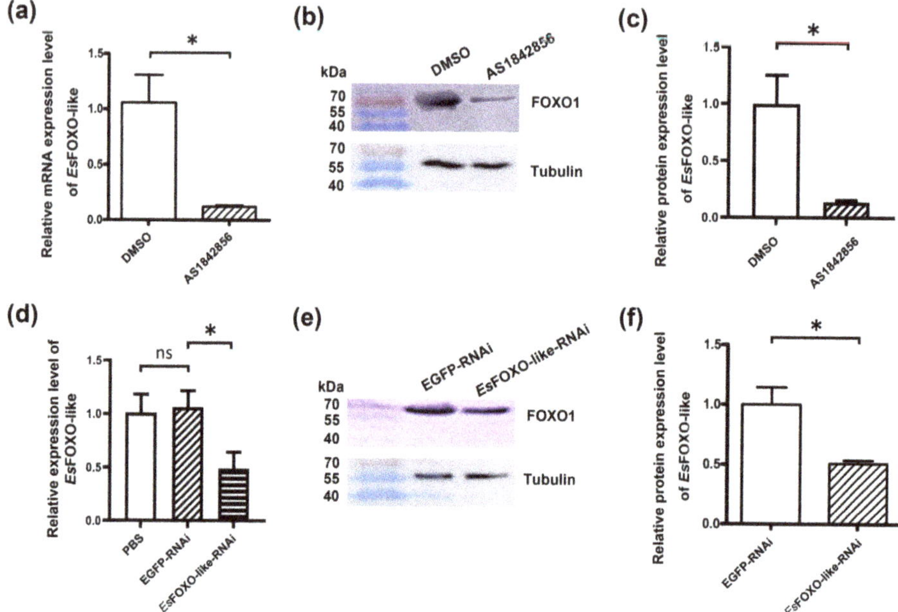

Figure 6. The mRNA and protein expression levels of *Es*FOXO-like gene in hepatopancreas after being injected with AS1842856 and *Es*FOXO-like dsRNA. (**a**) The *Es*FOXO-like mRNA expression level in hepatopancreas in AS1842856 injected crabs. (**b,c**) The protein expression levels of *Es*FOXO-like in hepatopancreas in AS1842856 and DMSO group. (**c**) Statistical analysis of blot performed in (**b**). (**d**) The mRNA expression level of *Es*FOXO-like in hepatopancreas after knockdown of *Es*FOXO-like by RNAi. (**e,f**) The protein expression levels of *Es*FOXO-like in hepatopancreas in *Es*FOXO-like-RNAi group and EFPG-RNAi group. (**f**) Statistical analysis of blot performed in (**e**). Each bar represents the mean ± standard deviation of three independent biological replicates (Supplementary Table S1). Asterisks indicate significant differences (* $p < 0.05$), "ns" = not significant. The results were analyzed using Student's *t* test.

Furthermore, the RNAi assay was employed to reduce the expression of *Es*FOXO-like mRNA, and the silencing efficiency of *Es*FOXO-like was detected by qRT-PCR and western blotting, respectively. The mRNA expression of *Es*FOXO-like in the hepatopancreas was significantly decreased 24 h after the injection of *Es*FOXO-like dsRNA, which was 0.46-fold compared with that of the EGFP-RNAi group ($p < 0.05$) (Figure 6d). No significant difference in the *Es*FOXO-like expression level was observed between the EGPF-RNAi group and the PBS group (Figure 6d). Similarly, the western blotting assay demonstrated that the intensity values of the *Es*FOXO-like band in *Es*FOXO-like-RNAi crabs decreased significantly (0.51-fold of that in the EGFP-RNAi group; $p < 0.05$) (Figure 6e,f).

3.5. The 20E Concentration and Molting-Related Genes mRNA Expression Levels after EsFOXO-like Inhibition

To investigate the regulation of *Es*FOXO-like gene in the 20E pathway, the 20E concentration and mRNA transcripts of *Es*EcR, *Es*RXR and *Es*MIH were detected after AS1842856 and *Es*FOXO-like dsRNA injection, respectively. The 20E concentration was significantly increased in the AS1842856 group, which was 1.65-fold ($p < 0.05$) of that in the DMSO group (Figure 7a). Similarly, the *Es*EcR and *Es*RXR mRNA expression levels in the hepatopancreas were both significantly increased in the AS1842856 group (5.03- and 5.18-fold compared to that in the DMSO group; $p < 0.05$) (Figure 7b,c). Conversely, the *Es*MIH mRNA expression level in eyestalk decreased significantly, which was 0.12-fold compared to that in the DMSO group ($p < 0.01$) (Figure 7d).

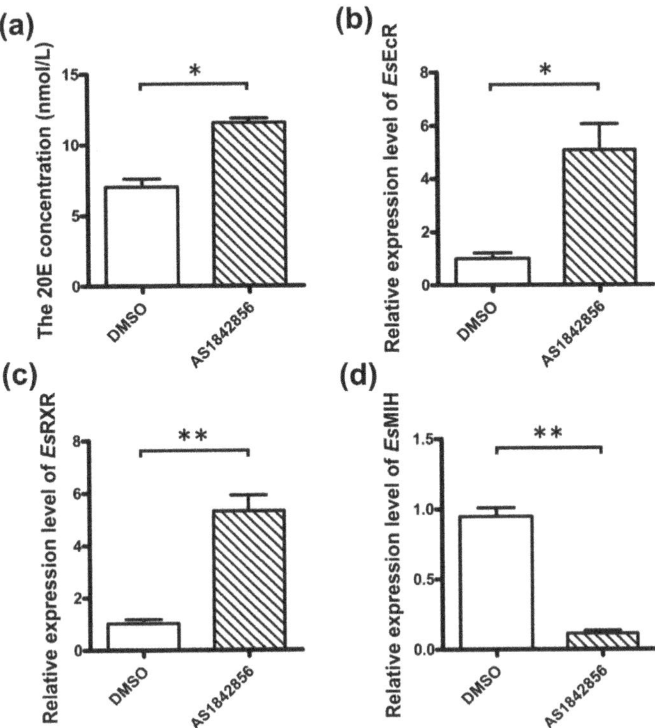

Figure 7. The 20E concentration and mRNA expression levels of *Es*EcR, *Es*RXR and *Es*MIH in the AS1842856 group. (**a**) The 20E concentration in serum in AS1842856 injected crabs. The data were analyzed using Mann–Whitney U test. (**b–d**) Relative expression levels of *Es*EcR and *Es*RXR in the hepatopancreas and *Es*MIH expression level in the eyestalk of crabs injected with AS1842856. The DMSO group was the control group. The data were analyzed by Student's *t* test. Each bar represents the mean $\pm$ standard deviation of three independent biological replicates (Supplementary Table S1). Asterisks indicate significant differences (* $p < 0.05$, ** $p < 0.01$).

After *Es*FOXO-like was knocked down by RNAi, the 20E concentration and the mRNA expression levels of *Es*EcR and *Es*RXR in the hepatopancreas were significantly up-regulated, which were 1.49-fold ($p < 0.01$), 5.59-fold ($p < 0.01$) and 3.88-fold ($p < 0.01$) of that in the EGFP-RNAi group, respectively (Figure 8a–c). On the contrary, the *Es*MIH mRNA expression in eyestalk was significantly lower than that (0.39-fold, $p < 0.01$) in the EGFP-RNAi group (Figure 8d).

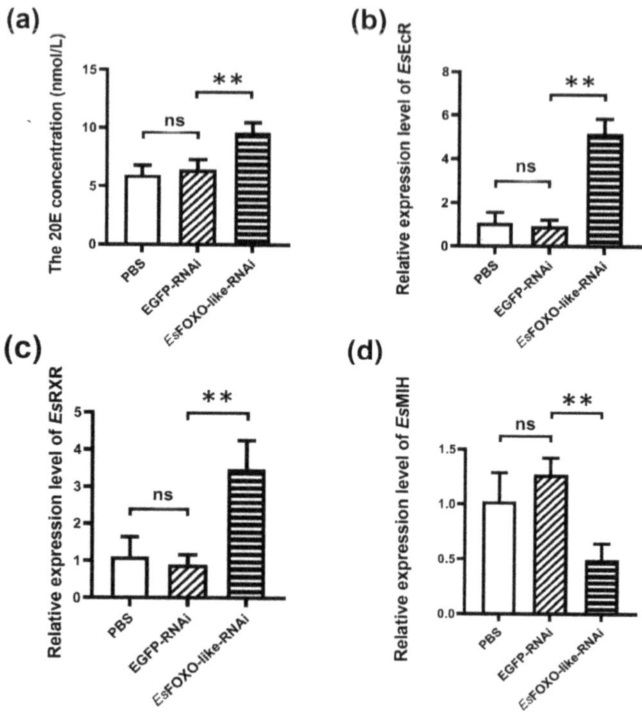

Figure 8. The 20E concentration and mRNA expression levels of *Es*EcR, *Es*RXR and *Es*MIH after *Es*FOXO-like was knocked down by RNAi. (**a**) The 20E concentration in serum in *Es*FOXO-like-RNAi group. (**b–d**) Relative expression levels of *Es*EcR and *Es*RXR in the hepatopancreas and *Es*MIH expression level in the eyestalk of *Es*FOXO-like-RNAi crabs group. The EFPG-RNAi group was the control group. Each bar represents the mean ± standard deviation of three independent biological replicates. Asterisks indicate significant differences (** $p < 0.01$), "ns" = not significant. The data were analyzed by one-way ANOVA.

3.6. The mRNA Expression Level of EsmTOR after EsFOXO-like Inhibition

In order to study whether or whether not *Es*FOXO-like regulates the 20E pathway through *Es*mTOR, the expression level of *Es*mTOR was first detected after *Es*FOXO-like inhibition by AS1842856 treatment or dsRNA-*Es*FOXO-like injection. The *Es*mTOR mRNA expression levels in the hepatopancreas were up-regulated after injection of both AS1842856 and dsRNA-*Es*FOXO-like. The mRNA transcript of *Es*mTOR in the AS1842856 group was 4.49-fold ($p < 0.01$) that of the DMSO group (Figure 9a). Similarly, the relative expression level of *Es*mTOR in the hepatopancreas was significantly ($p < 0.05$) increased in the *Es*FOXO-RNAi group, which was 2.91-fold compared to that in the EGFP-RNAi group (Figure 9b).

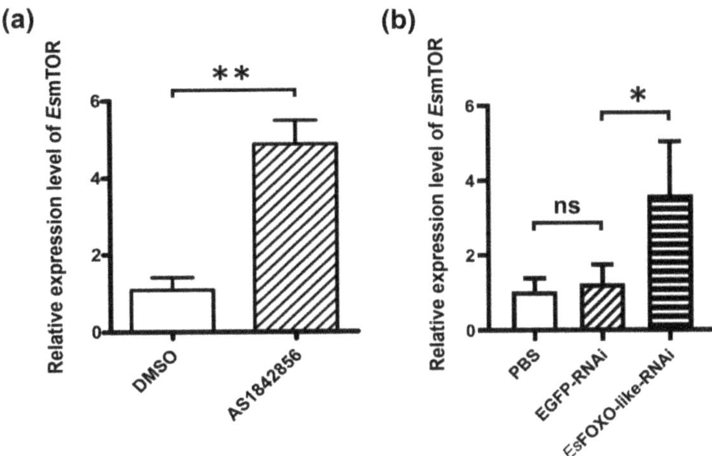

Figure 9. The expression level of *Esm*TOR gene in hepatopancreas after *Es*FOXO-like inhibition. (a) The *Esm*TOR expression level in AS1842856-injected crabs. The data were analyzed by Student's *t* test. (b) The *Esm*TOR expression level in *Es*FOXO-like-RNAi group. The data were analyzed by one-way ANOVA. Each bar represents the mean ± standard deviation of three independent biological replicates (Supplementary Table S1). Asterisks indicate significant differences (* $p < 0.05$ and ** $p < 0.01$), "ns" = not significant.

3.7. The EsmTOR mRNA Transcripts after AS1842856 and Rapamycin Injection

To further investigate whether *Es*FOXO-like regulates the 20E pathway through *Esm*TOR or not, we evaluated the expression level of *Esm*TOR in the hepatopancreas after inhibiting *Es*FOXO-like by AS1842856 injection, followed by administering Rapamycin. The mRNA expression level of *Esm*TOR in the hepatopancreas was decreased after injection of AS1842856 + Rapamycin, and the *Esm*TOR transcript in the AS1842856 + Rapamycin group was 0.21-fold of that in the AS1842856 + DMSO group ($p < 0.05$) (Figure 10).

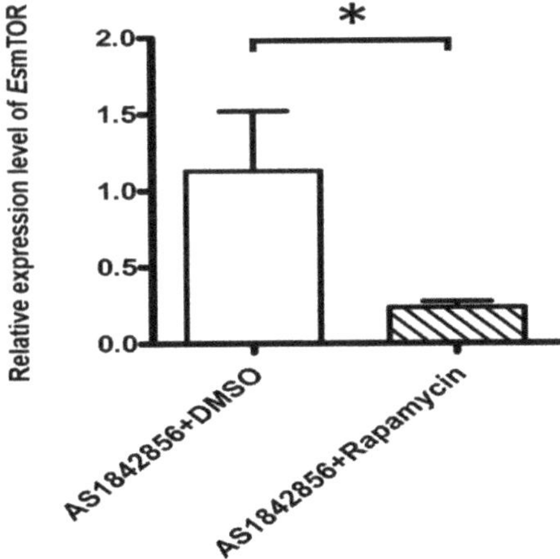

Figure 10. The expression level of *Esm*TOR gene in hepatopancreas after injection with AS1842856 and Rapamycin. AS1842856 + DMSO were injected and treated as the control group. Each bar represents the mean ± standard deviation of three independent biological replicates (Supplementary Table S1). Asterisks indicate significant differences (* $p < 0.05$). The data were analyzed by Student's *t* test.

3.8. The 20E Concentration and EsEcR, EsRXR and EsMIH mRNA Expression Levels in AS1842856 + Rapamycin Group after Inhibiting mTOR

To investigate the role of *EsmTOR* in the regulation of the 20E pathway through *EsFOXO*-like, the concentration of 20E and the mRNA expression levels of *EsEcR*, *EsRXR*, and *EsMIH* were measured in the crabs following AS1842856 and Rapamycin injections. The results showed that the 20E concentration in serum, as well as the *Es*MIH mRNA expression level in eyestalk, were significantly decreased in the AS1842856 + Rapamycin group, which was 0.81-fold ($p < 0.05$) and 0.23-fold ($p < 0.05$) of that in AS1842856 + DMSO group for the respective components analyzed (Figure 11a,d). On the contrary, the expression levels of *Es*EcR (5.78-fold, $p < 0.05$) and *Es*RXR (3.69-fold, $p < 0.05$) in the hepatopancreas in the AS1842856 + Rapamycin group were significantly higher than that in the AS1842856 + DMSO group (Figure 11b,c).

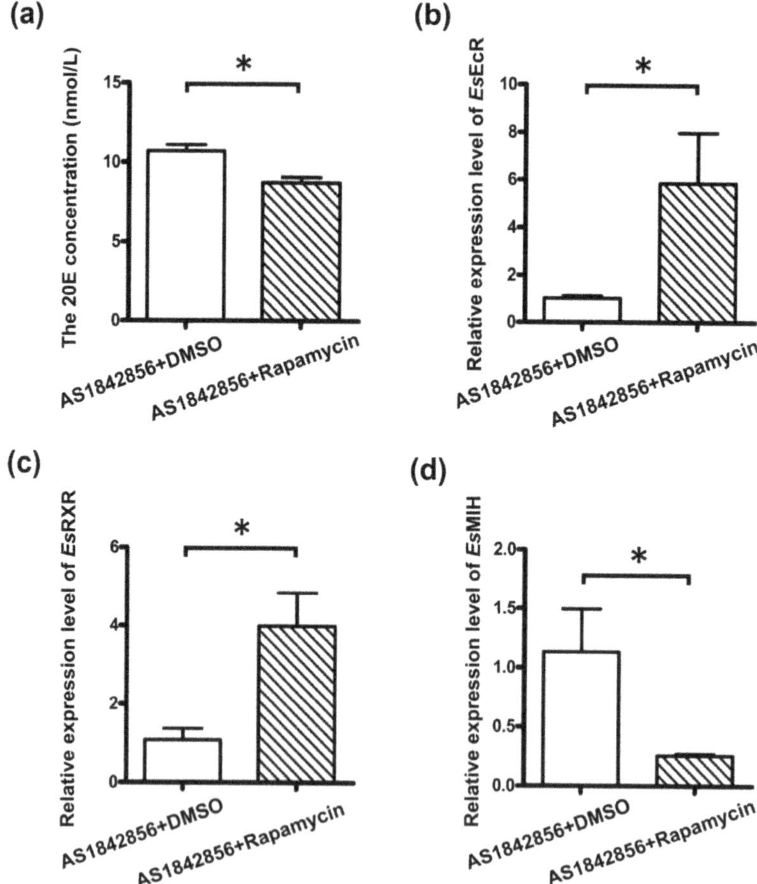

Figure 11. The 20E concentration and mRNA expression levels of *Es*EcR, *Es*RXR and *Es*MIH in AS1842856 + Rapamycin group. (**a**) The 20E concentration in serum in AS1842856- and Rapamycin-injected crabs. (**b**–**d**) The relative expression levels of *Es*EcR and *Es*RXR in hepatopancreas and *Es*MIH expression level in eyestalk in AS1842856 + Rapamycin group. The AS1842856 + DMSO group was treated as the control group. Each bar represents the mean ± standard deviation of three independent biological replicates (Supplementary Table S1). Asterisks indicate significant differences (* $p < 0.05$). The data were analyzed by Student's *t* test.

4. Discussion

FOXO plays a key regulatory role in many physiological, metabolic and immunoregulatory responses [55–57]. The FOXO gene is conserved from lower yeast to higher humans [58]. So far, there are four FOXO members found in mammals [2], and only one FOXO member has been identified in invertebrates [59]. In this study, a FOXO-like transcription factor (*Es*FOXO-like) was identified and characterized in *E. sinensis*. The ORF of *Es*FOXO-like was 852 bp and encoded a 283 amino acid polypeptide. The *Es*FOXO-like sequence perfectly contains a typical Forkhead (FH) domain [19]. The DBD domain, also named the FH domain, was found to be highly conserved in different species [60,61]. The FH domain has been confirmed to interact with p53 to stabilize it from degradation, which is required to induce apoptosis [62]. Phylogenetic analysis showed that *Es*FOXO-like exhibits high similarity to crustacean FOXO genes and clusters together with them, suggesting that EsFOXO-like is a member of the FOXO family in crustaceans.

In mammals, different FOXO genes had various tissue expression patterns. For instance, FOXO1 and FOXO3α are expressed in multiple tissues, while FOXO4 is predominantly expressed in the kidney and muscle, and FOXO6 exhibits high expression in the liver [9]. Our results showed that the *Es*FOXO-like transcripts were expressed in various tissues, such as hepatopancreas, hemocytes, heart, stomach, gills, muscles, and eyestalks, with the highest level of expression in the hepatopancreas, which is similar to the distribution of FOXO genes in crustaceans [3,19]. The hepatopancreas of crustaceans has been identified to serve as a vital immunologic and metabolic organ [63]. Furthermore, it is considered to be one of the crucial organs involved in the molting process of crustaceans [22,64,65]. Moreover, the function of four different FOXO genes is also varied in vertebrates. For instance, the global loss of FOXO1 could cause the death of embryonic cells [66]. Furthermore, the overall deletion of FOXO3 can affect lymphatic proliferation and extensive organ inflammation [67]. Additionally, the loss of FOXO4 can exacerbate colitis induced by inflammatory stimuli [68]. Lastly, FoxO6 is preferentially enriched in the hippocampus, and the deletion of the FOXO6 gene can lead to impaired memory consolidation that showed a reduced ability to form long-term contextual and object recognition memories in mice [69]. Recently, many studies have demonstrated that FOXO regulates molting and metamorphosis in insects [16]. Although FOXO has been identified in several crustacean species, most studies only focus on its regulatory role in the immune response against bacterial invasions [3,19]. Thus, the potential involvement of FOXO in growth still needs to be further investigated.

Molting is a typical biological trait that directly determines the behavior and physiological processes in arthropods [70–73]. Generally, the molting cycle of crustaceans has been divided into three main stages: inter-molt, pre-molt and ecdysis [21,22]. We found that *Es*FOXO-like mRNA expression was different at each molting stage, with the highest level at pre-molt and the lowest level at post-molt. In *H.armigera*, the FOXO protein in the fat body was up-regulated during the fifth-larvae molting stage and metamorphosis when compared to the feeding stage. This difference indicated that the expression of FOXO also increased during molting and metamorphosis [16]. In insects, 20E and juvenile hormone are mutually antagonistic and jointly regulate molting processes. Usually, the juvenile hormone (JH) concentration decreases sharply in the juvenile developmental stage, and the increase in 20E concentration causes pupation or adult morphogenesis [74]. It has been discovered that FOXO plays a role in regulating the degradation of juvenile hormone (JH) to control the growth of *B. mori* [11]. Additionally, in *Tribolium castaneum*, it has been observed that 20E up-regulates FOXO expression, promoting its nuclear migration, where FOXO subsequently regulates protease factors [75]. In *H. armigera*, 20E also up-regulates FOXO expression which, in turn, induces the expression of Carboxypeptidase A to regulate the final proteolysis step during the molting process [16]. In our study, it was observed that the mRNA transcripts and protein expression of *Es*FOXO-like reached their highest levels during the pre-molt stage, suggesting its potential involvement in the molting process.

Although FOXO has been demonstrated to play a key role in the regulation of metamorphosis and growth of insects, the involvement of FOXO in the regulation of molting in crustaceans is unclear. The role of *Es*FOXO-like in regulating crab molting was explored in our research by evaluating the 20E concentration and expression levels of molting-related genes after AS1842856 [44] and *Es*FOXO-like dsRNA injections. In this study, the *Es*FOXO-like mRNA expression was significantly decreased after injections of AS1842856 or *Es*FOXO-like dsRNA. In mammals, AS1842856, which served as a FOXO1 inhibitor, could inhibit the protein activity of FOXO1 but had no effect on the mRNA transcription of FOXO1 [44,45,76]. It has also been demonstrated that treatment with AS1842856 displayed similar effects to FOXO1 knockdown in pancreatic progenitors [76]. Until now, the potential role of AS1842856 on FOXO genes has not been reported in crustaceans. In this study, our results showed that the mRNA and protein expression levels of EsFOXO-like were both significantly decreased after AS1842856 injection. This observation might be related to the differences in metabolism between crustaceans and mammals. Similarly, in the study of 3T3-L1 preadipocytes cells, it has also been found that the level of total FOXO1 protein was reduced after AS1842856 treatment [77]. Additional investigation revealed a significant increase in 20E concentration after the inhibition of *Es*FOXO-like. In addition, our study showed that when *Es*FOXO-like was inhibited, the *Es*EcR and *Es*RXR expressions were significantly up-regulated, while the *Es*MIH expression was significantly down-regulated. In crustaceans, molting is promoted by 20E through its dependence on EcR and RXR, which leads to the activation of molting-related genes [78]. EcR and RXR proteins bind together to form a complex, which then interacts with the 20E hormone to create an active trimer. This trimer is responsible for regulating the growth and molting process in crustaceans [79]. On the other hand, the MIH hormone has been shown to have a negative effect on the synthesis of 20E [23]. Although the studies on the regulation of FOXO on EcR, RXR and MIH expressions have not been reported in crustaceans, it has been studied in insects. For example, it has been found that FOXO could directly interact with Ultraspiracle (Usp), the ecdysone receptor, and FoxO/Usp complexes suppress 20E biosynthesis in *Drosophila* [80]. Additionally, JH regulated the growth rate via inhibition of the 20E pathway, and it is also dependent on FOXO [81]. While it has been reported that FOXO inhibition does not affect the expression levels of EcRB1 and USP1, it could significantly block the molting of larvae in *H. armigera* [16]. These results collectively suggested that FOXO was closely related to the physiological process of molting in *E. sinensis* by inhibiting the 20E signal. However, whether FOXO affects the 20E signal through related genes or signaling pathways remains to be further explored.

Multiple studies have shown that mTOR, a key regulator of cellular metabolism, plays a role in controlling the synthesis of 20E [30–32]. Research has shown that the activation of Y-organ ecdysteroidogenesis is dependent on the activity of mTOR and that mTORC1 is primarily responsible for driving the increase in 20E concentration [48,82]. Therefore, in order to explore the mechanism of FOXO regulation in the 20E pathway in crabs, we first measured the *Es*mTOR mRNA expression level after FOXO inhibition. The results showed that *Es*mTOR transcription was significantly increased in the AS1842856-injected group and the *Es*FOXO-like-RNAi group. In mammals, the PI3K–AKT–FOXO–mTOR pathway has been verified to be a regulator of metabolism and somatic growth [83]. It has also been found that loss of FOXO function resulted in precocious differentiation in tissues with high mTOR activity on nutrient restriction in *Drosophila* [84]. Moreover, the cells that lost the function of FOXO have higher mTOR activity and are more sensitive to 20E response, possibly due to a similar inhibitory action of FOXO on the 20E receptor complex [84]. Moreover, Rapamycin, an effective and specific mTOR activity inhibitor, was injected into crabs to explore the function of mTOR in the FOXO regulation of the 20E pathway in this study [85]. Injections of AS1842856 and Rapamycin resulted in reduced expression of the *Es*mTOR transcript, and the AS1842856 + Rapamycin group exhibited a significantly lower concentration of 20E compared to the AS1842856 + DMSO group. Similarly, in *Manduca sexta*, Rapamycin resulted in delayed larval molting and reduced 20E production [86]. In

*B. mori*it, it has been found that rapamycin could also inhibit 20E synthesis and secretion [32]. Although accumulating evidence has suggested that mTOR could stimulate 20E synthesis and inhibit the expression levels of MIH signaling-related genes [29,30]. 20E is transduced via binding of an isomeric dimer complex consisting of EcR and RXR [25], but the impact of mTOR on EcR and RXR is unknown. Notably, mTOR signaling plays an essential role in the regulation of growth, body size and aging of arthropods. It has also been confirmed that the insulin receptor (InR) and phosphoinositide 3-kinase (PI3K) genes, involved in growth and development progress, showed progressively increasing mRNA levels in rapamycin-treated crabs [29]. Similarly, the InR and PI3K could promote EcR and USP (RXR in crustaceans) expression in insects [87]. Therefore, we speculate that the increase in *EsEcR* and *EsRXR* expression observed in the AS1842856- + Rapamycin-inhibited group in this study might be due to up-regulation of the insulin and PI3K pathway. Furthermore, 20E and its receptors have been confirmed to negatively affect MIH expression [27,28]. Thus, the decrease of *EsMIH* expression also observed in the AS1842856 + Rapamycin inhibited group might be due to the increased expression of *EsEcR* and *EsRXR*. It might be also associated with the sampling time after rapamycin injection. For instance, in *Gecarcinus lateralis*, rapamycin did not block or reverse the decrease in MIH signaling genes until the seventh day after rapamycin injection [29]. Overall, these results suggest that FOXO could regulate the 20E pathway through mTOR in mitten crabs.

5. Conclusions

In summary, a Forkhead Box O-like (named *EsFOXO*-like) was identified in *E. sinensis*, with the highest expression in the hepatopancreas at pre-molt stage and lowest expression at post-molt stage. It was also found that the 20E and *EsMIH* mRNA expressions were both significantly down-regulated after *EsFOXO*-like inhibition, while the expression of *EsEcR* and *EsRXR* showed the opposite expression pattern. Taken together, these results suggest that *EsFOXO*-like has a negative effect on 20E signaling by inhibiting mTOR. Molting is an essential process for the growth of crustaceans, and FOXO served as a key transcription factor. Its role in regulating molting had not yet been studied in crustaceans. Our current results reveal the inhibitory function of FOXO in the molting of crustaceans. Future research should be performed on the regulatory mechanisms involved in the promotion of molting and growth by inhibiting FOXO.

Supplementary Materials: The following supporting information can be downloaded at: https://www.mdpi.com/article/10.3390/jmse11061225/s1, Table S1: The original data for analysis in this manuscript.

Author Contributions: Conceptualization, J.L. (Jiaming Li) and S.H.; methodology, J.L. (Jiaming Li) and Y.M.; software, Z.Y. and F.W.; validation, J.L. (Jialin Li) and Y.M.; formal analysis, J.L. and F.W.; investigation, J.L. (Jialin Li) and Z.Y.; writing—original draft preparation, J.L. (Jiaming Li) and Y.M.; writing—review and editing, Y.J., D.Y., Q.Y. and S.H.; supervision, Q.Y. and S.H.; project administration, S.H.; funding acquisition, Q.Y. and S.H. All authors have read and agreed to the published version of the manuscript.

Funding: This work was supported by research grants from Educational Department of Liaoning Province (20220078), Natural Resources Department of Liaoning Province (20220001-18) and Educational Department of Liaoning Province (S202210158002X).

Institutional Review Board Statement: The animal study protocol conformed with the Animal Care Committee of Dalian Ocean University.

Informed Consent Statement: Not applicable.

Data Availability Statement: Data is contained within the article. All the data are available from the corresponding author upon request.

Acknowledgments: We are grateful to all the members for their help and support in this experiment.

Conflicts of Interest: The authors declare no conflict of interests.

References

1. Calissi, G.; Lam, E.W.; Link, W. Therapeutic strategies targeting FOXO transcription factors. *Nat. Rev. Drug Discov.* **2021**, *20*, 21–38. [CrossRef] [PubMed]
2. Jiramongkol, Y.; Lam, E.W. FOXO transcription factor family in cancer and metastasis. *Cancer Metastasis Rev.* **2020**, *39*, 681–709. [CrossRef]
3. Li, C.; Hong, P.P.; Yang, M.C.; Zhao, X.F.; Wang, J.X. FOXO regulates the expression of antimicrobial peptides and promotes phagocytosis of hemocytes in shrimp antibacterial immunity. *PLoS Pathog.* **2021**, *17*, e1009479. [CrossRef] [PubMed]
4. Arden, K.C. FOXO animal models reveal a variety of diverse roles for FOXO transcription factors. *Oncogene* **2008**, *27*, 2345–2350. [CrossRef] [PubMed]
5. Boura, E.; Silhan, J.; Herman, P.; Vecer, J.; Sulc, M.; Teisinger, J.; Obsilova, V.; Obsil, T. Both the N-terminal loop and wing W2 of the forkhead domain of transcription factor Foxo4 are important for DNA binding. *J. Biol. Chem.* **2007**, *282*, 8265–8275. [CrossRef]
6. Clark, K.L.; Halay, E.D.; Lai, E.; Burley, S.K. Co-crystal structure of the HNF-3/fork head DNA-recognition motif resembles histone H5. *Nature* **1993**, *364*, 412–420. [CrossRef]
7. Jacobs, F.M.; van der Heide, L.P.; Wijchers, P.J.; Burbach, J.P.; Hoekman, M.F.; Smidt, M.P. FoxO6, a novel member of the FoxO class of transcription factors with distinct shuttling dynamics. *J. Biol. Chem.* **2003**, *278*, 35959–35967. [CrossRef]
8. Arden, K.C.; Biggs, W.H. Regulation of the FoxO family of transcription factors by phosphatidylinositol-3 kinase-activated signaling. *Arch. Biochem. Biophys.* **2002**, *403*, 292–298. [CrossRef]
9. van der Vos, K.E.; Coffer, P.J. The extending network of FOXO transcriptional target genes. *Antioxid. Redox Signal.* **2011**, *14*, 579–592. [CrossRef]
10. Dong, X.; Zhai, Y.; Zhang, J.; Sun, Z.; Chen, J.; Chen, J.; Zhang, W. Fork head transcription factor is required for ovarian mature in the brown planthopper, *Nilaparvata lugens* (Stål). *BMC Mol. Biol.* **2011**, *12*, 53. [CrossRef]
11. Zeng, B.; Huang, Y.; Xu, J.; Shiotsuki, T.; Bai, H.; Palli, S.R.; Huang, Y.; Tan, A. The FOXO transcription factor controls insect growth and development by regulating juvenile hormone degradation in the silkworm, *Bombyx mori*. *J. Biol. Chem.* **2017**, *292*, 11659–11669. [CrossRef]
12. Giannakou, M.E.; Goss, M.; Jünger, M.A.; Hafen, E.; Leevers, S.J.; Partridge, L. Long-lived Drosophila with overexpressed dFOXO in adult fat body. *Science* **2004**, *305*, 361. [CrossRef]
13. Wu, Y.B.; Yang, W.J.; Xie, Y.F.; Xu, K.K.; Tian, Y.; Yuan, G.R.; Wang, J.J. Molecular characterization and functional analysis of *Bd*FoxO gene in the oriental fruit fly, *Bactrocera dorsalis* (Diptera: Tephritidae). *Gene* **2016**, *578*, 219–224. [CrossRef]
14. Süren-Castillo, S.; Abrisqueta, M.; Maestro, J.L. FoxO inhibits juvenile hormone biosynthesis and vitellogenin production in the German cockroach. *Insect Biochem. Mol. Biol.* **2012**, *42*, 491–498. [CrossRef]
15. Sim, C.; Denlinger, D.L. Insulin signaling and FOXO regulate the overwintering diapause of the mosquito *Culex pipiens*. *Proc. Natl. Acad. Sci. USA* **2008**, *105*, 6777–6781. [CrossRef]
16. Cai, M.J.; Zhao, W.L.; Jing, Y.P.; Song, Q.; Zhang, X.Q.; Wang, J.X.; Zhao, X.F. 20-Hydroxyecdysone activates Forkhead box O to promote proteolysis during *Helicoverpa armigera* molting. *Development* **2016**, *143*, 1005–1015. [CrossRef]
17. Lin, X.; Yu, N.; Smagghe, G. FoxO mediates the timing of pupation through regulating ecdysteroid biosynthesis in the red flour beetle, *Tribolium castaneum*. *Gen. Comp. Endocrinol.* **2018**, *258*, 149–156. [CrossRef]
18. Hossain, M.S.; Liu, Y.; Zhou, S.; Li, K.; Tian, L.; Li, S. 20-Hydroxyecdysone-induced transcriptional activity of FoxO upregulates brummer and acid lipase-1 and promotes lipolysis in *Bombyx* fat body. *Insect Biochem. Mol. Biol.* **2013**, *43*, 829–838. [CrossRef]
19. Zhao, Y.; Nie, X.; Han, Z.; Liu, P.; Xu, H.; Huang, X.; Ren, Q. The forkhead box O transcription factor regulates lipase and anti-microbial peptide expressions to promote lipid catabolism and improve innate immunity in the *Eriocheir sinensis* with hepatopancreatic necrosis disease. *Fish Shellfish Immunol.* **2022**, *124*, 107–117. [CrossRef]
20. Yu, Y.Q.; Ma, W.M.; Yang, W.J.; Yang, J.S. The complete mitogenome of the lined shore crab *Pachygrapsus crassipes* Randall 1840 (Crustacea: Decapoda: Grapsidae). *Mitochondrial DNA* **2014**, *25*, 263–264. [CrossRef]
21. Gao, Y.; Zhang, X.; Wei, J.; Sun, X.; Yuan, J.; Li, F.; Xiang, J. Whole Transcriptome Analysis Provides Insights into Molecular Mechanisms for Molting in *Litopenaeus vannamei*. *PLoS ONE* **2015**, *10*, e0144350. [CrossRef] [PubMed]
22. Huang, S.; Wang, J.; Yue, W.; Chen, J.; Gaughan, S.; Lu, W.; Lu, G.; Wang, C. Transcriptomic variation of hepatopancreas reveals the energy metabolism and biological processes associated with molting in Chinese mitten crab, *Eriocheir sinensis*. *Sci. Rep.* **2015**, *5*, 14015. [CrossRef] [PubMed]
23. Chang, E.S.; Mykles, D.L. Regulation of crustacean molting: A review and our perspectives. *Gen. Comp. Endocrinol.* **2011**, *172*, 323–330. [CrossRef] [PubMed]
24. Aranda, A.; Pascual, A. Nuclear hormone receptors and gene expression. *Physiol. Rev.* **2001**, *81*, 1269–1304. [CrossRef]
25. Nakagawa, Y.; Henrich, V.C. Arthropod nuclear receptors and their role in molting. *FEBS J.* **2009**, *276*, 6128–6157. [CrossRef]
26. Techa, S.; Chung, J.S. Ecdysone and retinoid-X receptors of the blue crab, *Callinectes sapidus*: Cloning and their expression patterns in eyestalks and Y-organs during the molt cycle. *Gene* **2013**, *527*, 139–153. [CrossRef]
27. Techa, S.; Chung, J.S. Ecdysteroids regulate the levels of Molt-Inhibiting Hormone (MIH) expression in the blue crab, *Callinectes sapidus*. *PLoS ONE* **2015**, *10*, e0117278. [CrossRef]
28. Mykles, D.L.; Chang, E.S. Hormonal control of the crustacean molting gland: Insights from transcriptomics and proteomics. *Gen. Comp. Endocrinol.* **2020**, *294*, 113493. [CrossRef]

29. Shyamal, S.; Das, S.; Guruacharya, A.; Mykles, D.L.; Durica, D.S. Transcriptomic analysis of crustacean molting gland (Y-organ) regulation via the mTOR signaling pathway. *Sci. Rep.* **2018**, *8*, 7307. [CrossRef]

30. Mykles, D.L. Signaling Pathways That Regulate the Crustacean Molting Gland. *Front. Endocrinol.* **2021**, *12*, 674711. [CrossRef]

31. Gu, S.H.; Young, S.C.; Lin, J.L.; Lin, P.L. Involvement of PI3K/Akt signaling in PTTH-stimulated ecdysteroidogenesis by prothoracic glands of the silkworm, *Bombyx mori. Insect Biochem. Mol. Biol.* **2011**, *41*, 197–202. [CrossRef]

32. Gu, S.H.; Yeh, W.L.; Young, S.C.; Lin, P.L.; Li, S. TOR signaling is involved in PTTH-stimulated ecdysteroidogenesis by prothoracic glands in the silkworm, *Bombyx mori. Insect Biochem. Mol. Biol.* **2012**, *42*, 296–303. [CrossRef]

33. Lin, A.; Yao, J.; Zhuang, L.; Wang, D.; Han, J.; Lam, E.W.; Gan, B. The FoxO-BNIP3 axis exerts a unique regulation of mTORC1 and cell survival under energy stress. *Oncogene* **2014**, *33*, 3183–3194. [CrossRef]

34. Teleman, A.A. Molecular mechanisms of metabolic regulation by insulin in *Drosophila. Biochem. J.* **2009**, *425*, 13–26. [CrossRef]

35. Mirth, C.K.; Shingleton, A.W. Integrating body and organ size in *Drosophila*: Recent advances and outstanding problems. *Front. Endocrinol.* **2012**, *3*, 49. [CrossRef]

36. Rewitz, K.F.; Yamanaka, N.; O'Connor, M.B. Developmental checkpoints and feedback circuits time insect maturation. *Curr. Top. Dev. Biol.* **2013**, *103*, 1–33.

37. He, J.; Wu, X.; Li, J.Y.; Huang, Q.; Huang, Z.H.; Cheng, Y.X. Comparison of the culture performance and profitability of wild-caught and captive pond-reared Chinese mitten crab (*Eriocheir sinensis*) juveniles reared in grow-out ponds: Implications for seed selection and genetic selection programs. *Aquaculture* **2014**, *434*, 48–56. [CrossRef]

38. Zhang, Y.; Sun, Y.; Liu, Y.; Geng, X.; Wang, X.; Wang, Y.; Sun, J.; Yang, W. Molt-inhibiting hormone from Chinese mitten crab (*Eriocheir sinensis*): Cloning, tissue expression and effects of recombinant peptide on ecdysteroid secretion of YOs. *Gen. Comp. Endocrinol.* **2011**, *173*, 467–474. [CrossRef]

39. Chen, H.; Gu, X.; Zeng, Q.; Mao, Z.; Liang, X.; Martyniuk, C.J. Carbamazepine disrupts molting hormone signaling and inhibits molting and growth of *Eriocheir sinensis* at environmentally relevant concentrations. *Aquat. Toxicol.* **2019**, *208*, 138–145. [CrossRef]

40. Li, C.; Huang, L.; Zhang, Y.; Guo, X.; Cao, N.; Yao, C.; Duan, L.; Li, X.; Pang, S. Effects of triazole plant growth regulators on molting mechanism in Chinese mitten crab (*Eriocheir sinensis*). *Fish Shellfish Immunol.* **2022**, *131*, 646–653. [CrossRef]

41. Söderhäll, K.; Smith, V.J. Separation of the haemocyte populations of *Carcinus maenas* and other marine decapods, and prophenoloxidase distribution. *Dev. Comp. Immunol.* **1983**, *7*, 229–239. [CrossRef] [PubMed]

42. Huang, S.; Yi, Q.; Lian, X.; Xu, S.; Yang, C.; Sun, J.; Wang, L.; Song, L. The involvement of ecdysone and ecdysone receptor in regulating the expression of antimicrobial peptides in Chinese mitten crab, *Eriocheir sinensis. Dev. Comp. Immunol.* **2020**, *111*, 103757. [CrossRef] [PubMed]

43. Yang, W.; Liu, C.; Xu, Q.; Qu, C.; Sun, J.; Huang, S.; Kong, N.; Lv, X.; Liu, Z.; Wang, L.; et al. Beclin-1 is involved in the regulation of antimicrobial peptides expression in Chinese mitten crab *Eriocheir sinensis. Fish Shellfish Immunol.* **2019**, *89*, 207–216. [CrossRef] [PubMed]

44. Xiao, Q.; Liu, H.; Wang, H.S.; Cao, M.T.; Meng, X.J.; Xiang, Y.L.; Zhang, Y.Q.; Shu, F.; Zhang, Q.G.; Shan, H.; et al. Histone deacetylase inhibitors promote epithelial-mesenchymal transition in Hepatocellular Carcinoma via AMPK-FOXO1-ULK1 signaling axis-mediated autophagy. *Theranostics* **2020**, *10*, 10245–10261. [CrossRef] [PubMed]

45. Shi, Y.; Fan, S.; Wang, D.; Huyan, T.; Chen, J.; Chen, J.; Su, J.; Li, X.; Wang, Z.; Xie, S.; et al. FOXO1 inhibition potentiates endothelial angiogenic functions in diabetes via suppression of ROCK1/Drp1-mediated mitochondrial fission. *Biochim. Biophys. Acta Mol. Basis Dis.* **2018**, *1864*, 2481–2494. [CrossRef] [PubMed]

46. Zhou, K.; Qin, Y.; Song, Y.; Zhao, K.; Pan, W.; Nan, X.; Wang, Y.; Wang, Q.; Li, W. A novel Ig domain-containing C-type lectin triggers the intestine-hemocyte axis to regulate antibacterial immunity in crab. *J. Immunol.* **2022**, *208*, 2343–2362. [CrossRef]

47. Bulut-Karslioglu, A.; Biechele, S.; Jin, H.; Macrae, T.A.; Hejna, M.; Gertsenstein, M.; Song, J.S.; Ramalho-Santos, M. Inhibition of mTOR induces a paused pluripotent state. *Nature* **2016**, *540*, 119–123. [CrossRef]

48. Abuhagr, A.M.; MacLea, K.S.; Mudron, M.R.; Chang, S.A.; Chang, E.S.; Mykles, D.L. Roles of mechanistic target of rapamycin and transforming growth factor-β signaling in the molting gland (Y-organ) of the blackback land crab, *Gecarcinus lateralis. Comp. Biochem. Physiol. A Mol. Integr. Physiol.* **2016**, *198*, 15–21. [CrossRef]

49. Zhao, C.; Peng, C.; Wang, P.; Yan, L.; Fan, S.; Qiu, L. Identification of a Shrimp E3 Ubiquitin Ligase TRIM50-like involved in restricting White Spot Syndrome virus proliferation by its mediated autophagy and ubiquitination. *Front. Immunol.* **2021**, *12*, 682562. [CrossRef]

50. Chen, X.; Wang, J.; Yue, W.; Huang, S.; Chen, J.; Chen, Y.; Wang, C. Structure and function of the alternatively spliced isoforms of the ecdysone receptor gene in the Chinese mitten crab, *Eriocheir sinensis. Sci. Rep.* **2017**, *7*, 12993. [CrossRef]

51. Sun, Q.; Lin, S.; Zhang, M.; Gong, Y.; Ma, H.; Tran, N.T.; Zhang, Y.; Li, S. SpRab11a-regulated exosomes inhibit bacterial infection through the activation of antilipopolysaccharide factors in crustaceans. *J. Immunol.* **2022**, *209*, 710–722. [CrossRef]

52. Wang, F.; Yang, Z.; Li, J.; Ma, Y.; Tu, Y.; Zeng, X.; Wang, Q.; Jiang, Y.; Huang, S.; Yi, Q. The involvement of hypoxia inducible factor-1α on the proportion of three types of haemocytes in Chinese mitten crab under hypoxia stress. *Dev. Comp. Immunol.* **2023**, *140*, 104598. [CrossRef]

53. Wang, M.; Zhou, J.; Su, S.; Tang, Y.; Xu, G.; Li, J.; Yu, F.; Li, H.; Song, C.; Liang, M.; et al. Comparative transcriptome analysis on the regulatory mechanism of thoracic ganglia in *Eriocheir sinensis* at post-molt and inter-molt Stages. *Life* **2022**, *12*, 1181. [CrossRef]

54. Schmittgen, T.D.; Livak, K.J. Analyzing real-time PCR data by the comparative C(T) method. *Nat. Protoc.* **2008**, *3*, 1101–1108. [CrossRef]

55. van der Vos, K.E.; Coffer, P.J. FOXO-binding partners: It takes two to tango. *Oncogene* **2008**, *27*, 2289–2299. [CrossRef]
56. Wang, S.; Xia, P.; Huang, G.; Zhu, P.; Liu, J.; Ye, B.; Du, Y.; Fan, Z. FoxO1-mediated autophagy is required for NK cell development and innate immunity. *Nat. Commun.* **2016**, *7*, 11023. [CrossRef]
57. Iyer, S.; Ambrogini, E.; Bartell, S.M.; Han, L.; Roberson, P.K.; de Cabo, R.; Jilka, R.L.; Weinstein, R.S.; O'Brien, C.A.; Manolagas, S.C.; et al. FOXOs attenuate bone formation by suppressing Wnt signaling. *J. Clin. Investig.* **2013**, *123*, 3409–3419. [CrossRef]
58. Barbieri, M.; Bonafè, M.; Franceschi, C.; Paolisso, G. Insulin/IGF-I-signaling pathway: An evolutionarily conserved mechanism of longevity from yeast to humans. *Am. J. Physiol. Endocrinol. Metab.* **2003**, *285*, E1064–E1071. [CrossRef]
59. Park, D.; Hahm, J.H.; Park, S.; Ha, G.; Chang, G.E.; Jeong, H.; Kim, H.; Kim, S.; Cheong, E.; Paik, Y.K. A conserved neuronal DAF-16/FoxO plays an important role in conveying pheromone signals to elicit repulsion behavior in *Caenorhabditis elegans*. *Sci. Rep.* **2017**, *7*, 7260. [CrossRef]
60. Obsil, T.; Obsilova, V. Structure/function relationships underlying regulation of FOXO transcription factors. *Oncogene* **2008**, *27*, 2263–2275. [CrossRef]
61. Mazet, F.; Yu, J.K.; Liberles, D.A.; Holland, L.Z.; Shimeld, S.M. Phylogenetic relationships of the Fox (Forkhead) gene family in the Bilateria. *Gene* **2003**, *316*, 79–89. [CrossRef] [PubMed]
62. Wang, F.; Marshall, C.B.; Yamamoto, K.; Li, G.Y.; Plevin, M.J.; You, H.; Mak, T.W.; Ikura, M. Biochemical and structural characterization of an intramolecular interaction in FOXO3a and its binding with p53. *J. Mol. Biol.* **2008**, *384*, 590–603. [CrossRef] [PubMed]
63. Röszer, T. The invertebrate midintestinal gland ("hepatopancreas") is an evolutionary forerunner in the integration of immunity and metabolism. *Cell Tissue Res.* **2014**, *358*, 685–695. [CrossRef] [PubMed]
64. Xu, Z.; Liu, A.; Li, S.; Wang, G.; Ye, H. Hepatopancreas immune response during molt cycle in the mud crab, *Scylla paramamosain*. *Sci. Rep.* **2020**, *10*, 13102. [CrossRef] [PubMed]
65. Liu, M.; Ni, H.; Zhang, X.; Sun, Q.; Wu, X.; He, J. Comparative transcriptomics reveals the immune dynamics during the molting cycle of swimming crab *Portunus trituberculatus*. *Front. Immunol.* **2022**, *13*, 1037739. [CrossRef]
66. Hosaka, T.; Biggs, W.H.; Tieu, D.; Boyer, A.D.; Varki, N.M.; Cavenee, W.K.; Arden, K.C. Disruption of forkhead transcription factor (FOXO) family members in mice reveals their functional diversification. *Proc. Natl. Acad. Sci. USA* **2004**, *101*, 2975–2980. [CrossRef]
67. Lin, L.; Hron, J.D.; Peng, S.L. Regulation of NF-kappaB, Th activation, and autoinflammation by the forkhead transcription factor Foxo3a. *Immunity* **2004**, *21*, 203–213. [CrossRef]
68. Zhou, W.; Cao, Q.; Peng, Y.; Zhang, Q.J.; Castrillon, D.H.; DePinho, R.A.; Liu, Z.P. FoxO4 inhibits NF-kappaB and protects mice against colonic injury and inflammation. *Gastroenterology* **2009**, *137*, 1403–1414. [CrossRef]
69. Salih, D.A.M.; Rashid, A.J.; Colas, D.; de la Torre-Ubieta, L.; Zhu, R.P.; Morgan, A.A.; Santo, E.E.; Ucar, D.; Devarajan, K.; Cole, C.J.; et al. FoxO6 regulates memory consolidation and synaptic function. *Genes Dev.* **2012**, *26*, 2780–2801. [CrossRef]
70. Panganiban, G.; Sebring, A.; Nagy, L.; Carroll, S. The development of crustacean limbs and the evolution of arthropods. *Science* **1995**, *270*, 1363–1366. [CrossRef]
71. Hopkins, P.M.; Chung, A.C.-K.; Durica, D.S. Limb Regeneration in the Fiddler Crab, Uca pugilator: Hormonal and Growth Factor Control. *Am. Zool.* **2001**, *41*, 389–398. [CrossRef]
72. Morris, S.; Postel, U.; Mrinalini; Turner, L.M.; Palmer, J.; Webster, S.G. The adaptive significance of crustacean hyperglycaemic hormone (CHH) in daily and seasonal migratory activities of the Christmas Island red crab *Gecarcoidea natalis*. *J. Exp. Biol.* **2010**, *213 Pt 17*, 3062–3073. [CrossRef]
73. Jung, H.; Lyons, R.E.; Hurwood, D.A.; Mather, P.B. Genes and growth performance in crustacean species: A review of relevant genomic studies in crustaceans and other taxa. *Rev. Aquacult.* **2013**, *5*, 77–110. [CrossRef]
74. Li, K.; Jia, Q.Q.; Li, S. Juvenile hormone signaling—A mini review. *Insect. Sci.* **2019**, *26*, 600–606. [CrossRef]
75. Ji, C.; Zhang, N.; Jiang, H.; Meng, X.; Ge, H.; Yang, X.; Xu, X.; Qian, K.; Park, Y.; Zheng, Y.; et al. 20-hydroxyecdysone regulates expression of methioninesulfoxide reductases through transcription factor FOXO in the red flour beetle, *Tribolium castaneum*. *Insect Biochem. Mol. Biol.* **2021**, *131*, 103546. [CrossRef]
76. Yu, F.; Wei, R.; Yang, J.; Liu, J.; Yang, K.; Wang, H.; Mu, Y.; Hong, T. FoxO1 inhibition promotes differentiation of human embryonic stem cells into insulin producing cells. *Exp. Cell Res.* **2018**, *362*, 227–234. [CrossRef]
77. Zou, P.; Liu, L.; Zheng, L.; Liu, L.; Stoneman, R.E.; Cho, A.; Emery, A.; Gilbert, E.R.; Cheng, Z. Targeting FoxO1 with AS1842856 suppresses adipogenesis. *Cell Cycle* **2014**, *13*, 3759–3767. [CrossRef]
78. Durica, D.S.; Arthur, C.-K.C.; Hopkins, P.M. Characterization of EcR and RXR gene homologs and receptor expression during the molt cycle in the crab, *Uca pugilator*. *Am. Zool.* **1999**, *39*, 758–773. [CrossRef]
79. Riddiford, L.M.; Hiruma, K.; Zhou, X.; Nelson, C.A. Insights into the molecular basis of the hormonal control of molting and metamorphosis from *Manduca sexta* and *Drosophila melanogaster*. *Insect Biochem. Mol. Biol.* **2003**, *33*, 1327–1338. [CrossRef]
80. Koyama, T.; Rodrigues, M.A.; Athanasiadis, A.; Shingleton, A.W.; Mirth, C.K. Nutritional control of body size through FoxO-Ultraspiracle mediated ecdysone biosynthesis. *eLife* **2014**, *3*, e03091. [CrossRef]
81. Mirth, C.K.; Tang, H.Y.; Makohon-Moore, S.C.; Salhadar, S.; Gokhale, R.H.; Warner, R.D.; Koyama, T.; Riddiford, L.M.; Shingleton, A.W. Juvenile hormone regulates body size and perturbs insulin signaling in *Drosophila*. *Proc. Natl. Acad. Sci. USA* **2014**, *111*, 7018–7023. [CrossRef] [PubMed]

82. Abuhagr, A.M.; Maclea, K.S.; Chang, E.S.; Mykles, D.L. Mechanistic target of rapamycin (mTOR) signaling genes in decapod crustaceans: Cloning and tissue expression of mTOR, Akt, Rheb, and p70 S6 kinase in the green crab, *Carcinus maenas*, and blackback land crab, *Gecarcinus lateralis*. *Comp. Biochem. Physiol. A Mol. Integr. Physiol.* **2014**, *168*, 25–39. [CrossRef] [PubMed]
83. Goldbraikh, D.; Neufeld, D.; Eid-Mutlak, Y.; Lasry, I.; Gilda, J.E.; Parnis, A.; Cohen, S. USP1 deubiquitinates Akt to inhibit PI3K-Akt-FoxO signaling in muscle during prolonged starvation. *EMBO Rep.* **2020**, *21*, e48791. [CrossRef] [PubMed]
84. Nowak, K.; Gupta, A.; Stocker, H. FoxO restricts growth and differentiation of cells with elevated TORC1 activity under nutrient restriction. *PLoS Genet.* **2018**, *14*, e1007347. [CrossRef]
85. Benjamin, D.; Colombi, M.; Moroni, C.; Hall, M.N. Rapamycin passes the torch: A new generation of mTOR inhibitors. *Nat. Rev. Drug Discov.* **2011**, *10*, 868–880. [CrossRef]
86. Kemirembe, K.; Liebmann, K.; Bootes, A.; Smith, W.A.; Suzuki, Y. Amino acids and TOR signaling promote prothoracic gland growth and the initiation of larval molts in the tobacco hornworm *Manduca sexta*. *PLoS ONE* **2012**, *7*, e44429. [CrossRef]
87. Deng, P.; Xu, Q.Y.; Fu, K.Y.; Guo, W.C.; Li, G.Q. RNA interference against the putative insulin receptor substrate gene chico affects metamorphosis in *Leptinotarsa decemlineata*. *Insect. Biochem. Mol. Biol.* **2018**, *103*, 1–11. [CrossRef]

Journal of
*Marine Science
and Engineering*

Article

Comparative Analysis of Genetic Structure and Diversity in Five Populations of Yellowtail Kingfish (*Seriola aureovittata*)

Aijun Cui [1,2], Yongjiang Xu [1,2,*], Kiyoshi Kikuchi [3], Yan Jiang [1,2], Bin Wang [1,2], Takashi Koyama [3] and Xuezhou Liu [1,2]

[1] Yellow Sea Fisheries Research Institute, Chinese Academy of Fishery Sciences, Qingdao 266071, China; cuiaj@ysfri.ac.cn (A.C.); jiangyan@ysfri.ac.cn (Y.J.); wangbin@ysfri.ac.cn (B.W.); liuxz@ysfri.ac.cn (X.L.)
[2] Joint Laboratory for Deep Blue Fishery Engineering, Qingdao 266071, China
[3] Fisheries Laboratory, University of Tokyo, Maisaka 431-0214, Shizuoka, Japan; akikuchi@mail.ecc.u-tokyo.ac.jp (K.K.); takashi.k1201@gmail.com (T.K.)
* Correspondence: xuyj@ysfri.ac.cn; Tel.: +86-0532-8582-3368

Abstract: To clarify the population genetic structure, intrapopulation diversity, and interpopulation differentiation of yellowtail kingfish (*Seriola aureovittata*), we sampled 143 individuals from five collections of yellowtail kingfish: farmed (n = 30) and wild (n = 33) collections in China, a wild collection in Japan (n = 20), and farmed (n = 31) and wild (n = 29) collections in Australia. Using 2b-RAD simplified genome sequencing, we obtained an average of 287,594 unique tags per population, with an average sequencing depth of 27.13×. Our final genotype dataset included 48,710 SNPs (Single Nucleotide Polymorphisms). The five collections were all in Hardy–Weinberg equilibrium, and the interpopulation differentiation varied among the sample collections. The genetic differentiation coefficients (*Fst*) between the Chinese and Japanese yellowtail kingfish collections were low and the gene flow (*Nm*) values were high. These results suggest continuous gene flow occurs frequently between the collections, indicating that they belong to the same population. In contrast, genetic differentiation was high between the Australian collections and the Chinese and Japanese populations, suggesting different evolutionary origins and belonging to different populations. The farmed and wild Australian collections fell into distinct clades in a neighbor-joining phylogeny tree, suggesting farmed fish have begun to differentiate from the wild collection. A similar level of genetic diversity between the wild collections in China and Japan suggests that they originated from the same spawning ground. This, therefore, reminds us that in future aquaculture processes attention is needed regarding implementing targeted breeding strategies. In addition, our data will contribute to Chinese yellowtail kingfish genetic breeding and the sustainable use of Chinese yellowtail kingfish germplasms.

Keywords: *Seriola aureovittata*; population genetics; single nucleotide polymorphisms; 2b-RAD sequencing

Citation: Cui, A.; Xu, Y.; Kikuchi, K.; Jiang, Y.; Wang, B.; Koyama, T.; Liu, X. Comparative Analysis of Genetic Structure and Diversity in Five Populations of Yellowtail Kingfish (*Seriola aureovittata*). *J. Mar. Sci. Eng.* **2023**, *11*, 1583. https://doi.org/10.3390/jmse11081583

Academic Editor: Nguyen Hong Nguyen

Received: 19 June 2023
Revised: 3 August 2023
Accepted: 7 August 2023
Published: 12 August 2023

1. Introduction

The yellowtail kingfish, *Seriola aureovittata* Temminck and Schlegel (1845) (Perciformes: Carangidae), is a long-distance migratory oceanic fish that inhabits temperate and subtropical waters worldwide [1]. The yellowtail kingfish is large, grows rapidly, and is highly favored by the international consumption market due to its excellent taste, nutritional quality, and economic value [2]. The yellowtail kingfish is the most valuable sashimi product after bluefin tuna in Japan [3]. Several countries have attempted to culture the yellowtail kingfish in recent years, including China, Japan, Australia, New Zealand, Chile, the Netherlands, and South Africa [1], and the current annual global aquaculture production of the yellowtail kingfish is 7000–8000 tons. In China, populations of wild yellowtail kingfish are mainly found in the Yellow Sea, Bohai Sea, and East China Sea [4,5]. In addition, China has cultivated yellowtail kingfish since 2012, and a breakthrough in artificial breeding in 2017 promoted the rapid development of the Chinese yellowtail kingfish aquaculture industry [6]. At present, the Chinese annual production of yellowtail kingfish is around 300–400 tons. As the aquaculture industry develops and the market demand for this species increases, with increasingly frequent exchanges of embryos and larvae, as well as international import and exports of products, careful attention to and control of the associated genetic risks to local populations of yellowtail kingfish will be required [7]. Moreover, while aquaculture is seen to reduce fishing pressure on natural populations, cultured fish can also have a significant impact on the health and stability of wild populations [8]. Therefore, it is necessary to investigate population genetic structure, intrapopulation genetic diversity, and interpopulation differentiation to protect the health of local fish populations and ecosystems.

In previous studies, a significant genetic divergence was determined between yellowtail kingfish populations from Japan and Australia/New Zealand, whereas no significant differentiation was found between Australian and New Zealand populations using microsatellite DNA and mitochondrial DNA (mtDNA) control region markers [9]. Martinez-Takeshita et al. (2015) used mitochondrial and nuclear genetic markers and morphometric analysis to reveal the significant genetic and morphological divergence among yellowtail kingfish from the Northeast Pacific, Northwest Pacific, and Southern Hemisphere. The yellowtail kingfish was shown to be a complex of three cryptic species. The name proposed for the Northwest Pacific species is *Seriola aureovittata* Temminck and Schlegel (1845), for the Northeast Pacific species is *Seriola dorsalis* Gill (1863), and for the Southern Hemisphere species, *Seriola lalandi* Valenciennes (1833) [10]. To demonstrate this, Ai et al. (2021) collected wild samples from the Bohai Sea (China) and Southern Ocean (Australia) and evaluated genetic diversity based on 17,690 nuclear loci. Their analyses showed that yellowtail kingfish from China and Australia formed two distinct clusters, and that there was no genetic introgression from the Australian population into the Chinese population [11]. To date, there have been no comparative studies of the genetic structure in wild/farmed yellowtail kingfish populations in the northwestern Pacific (i.e., China and Japan) and southern (e.g., Australia) populations, or in relation to wild/farmed yellowtail kingfish populations, using SNP molecular markers at the genome level. The available research on the genetic characteristics of wild yellowtail kingfish in China, have analyzed karyotype banding patterns and mitochondrial DNA sequence variation [12–14]. Cui et al. (2020) developed 100 high-quality SNP markers using the 2b-RAD sequencing of 33 wild yellowtail kingfish individuals, which were collected in a database for resource conservation research [15].

Studies on the genetic structure of yellowtail kingfish populations from Chinese and other international waters have become important for genetic breeding and large-scale farming. Moreover, the genetic differentiation between the wild and farmed populations of yellowtail kingfish has sparked interest in the genetic improvement of this species for sustainable aquaculture. Compared with traditional molecular markers, such as mitochondrial DNA and microsatellite, 2b-RAD technology can obtain numerous genome-wide SNP markers in a species without an available reference genome. 2b-RAD-constructed genomes are repeatable, with uniform coverage and sequencing depth, good label independence, high typing accuracy, and other advantages [16]. Therefore, 2b-RAD has been widely used in various genetic studies of aquatic animals, including high-density genetic map constructions, QTL (quantitative trait locus) positioning, genome-wide association analysis, population evolution characterization, assisted genome assembly, and whole-genome selective breeding [17–20]. Therefore, we aimed to characterize wild and farmed Chinese, wild Japanese, and wild and farmed Australian populations of yellowtail kingfish with respect to genetic structure, interpopulation differentiation, and intrapopulation diversity by using SNP molecular markers through 2b-RAD genomic sequencing. Our data will provide theoretical and technical support for the selective breeding of yellowtail kingfish to ensure the sustainable culture of this species in China.

2. Materials and Methods

2.1. Fish Sampling

We assessed five collections of yellowtail kingfish from China (wild and farmed), Japan (wild), and Australia (wild and farmed) (Figure 1): wild yellowtail kingfish (n = 33) from the Yellow Sea, off the coast of Dalian, China (39°57′ N, 123°04′ E); F1 generation yellowtail kingfish (n = 30) were provided by the Dalian Fugu Food Co., Ltd. (Zhuanghe, Dalian, China); wild yellowtail kingfish (n = 20) were collected off the Gotō Islands, Japan (32°45′ N, 128°27′ E); wild yellowtail kingfish (n = 29) were collected from the Tasman sea off of New South Wales, Australia (33°44′ S, 151°45′ E); and F2 generation, farmed yellowtail kingfish (n = 31) were provided by the South Australian Department of Primary Industries and Regions. Pectoral fin clips were collected from the wild and farmed fish from China and Australia; the fin tissues were immersed in 95% ethanol and stored in a refrigerator at 4 °C before genomic DNA was extracted. Genomic DNA from the fin clips of wild fish from Japan was provided by a collaborator at the University of Tokyo.

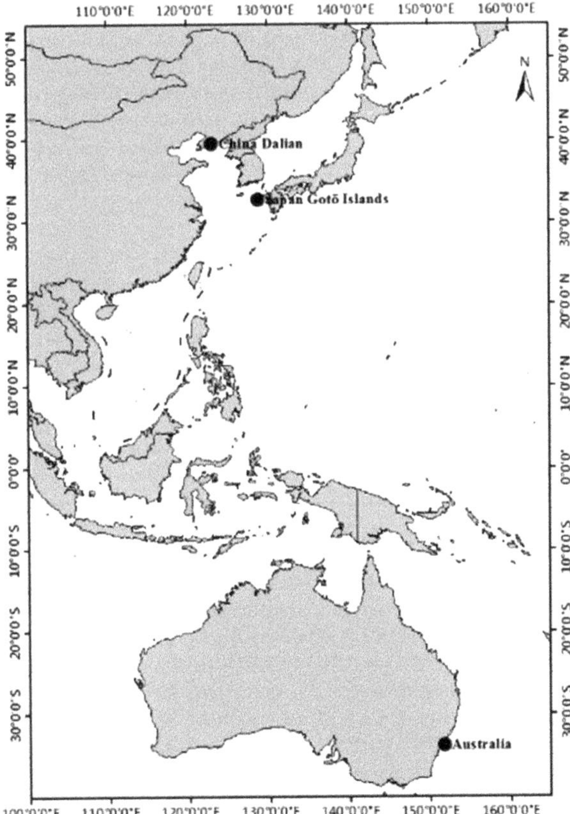

Figure 1. Geographical distribution of the yellowtail kingfish collections sampled in this study.

2.2. DNA Extraction, SNP Library Construction, and SNP Marker Typing

DNA from fin clips was extracted using animal-tissue genomic DNA Extraction Kits (DP121221; Qiagen, Hilden, Germany). DNA concentration was measured using a NanoDrop 2000 (Thermo Fisher, Waltham, MA, USA), and DNA integrity was checked using 1% agarose gel electrophoresis. Extracted genomic DNA was stored at −20 °C until use. The 2b-RAD libraries were constructed by Qingdao OE Biotechnology Co., Ltd. (Qingdao, China), following the methods of Wang et al. [16]. In brief, 100–200 ng of each DNA sample was digested with a Type IIB restriction enzyme (Bsa XI), and the digested products were ligated to restriction-site-specific adaptor sequences using T4 DNA Ligase (New England Biolabs Inc., Ipswich, MA, USA). Sample-specific barcodes were incorporated by PCR. PCR products were purified using the MinElute PCR Purification Kit and then pooled for paired-end sequencing using the Illumina Hiseq Xten sequencing platform (Illumina, CA, USA). Raw reads were cleaned by removing adaptor sequences, reads containing more than 8% unknown (N) bases, and low-quality reads (those where more than 15% of all bases had quality values lower than Q30). We used PEAR v0.9.6 (Heidelberg, Germany) [21] to stitch the pairs of clean reads together, and then reads corresponding to each sample were extracted based on sample location at the time of library building. Reads containing enzyme recognition sites were extracted. The selected samples were CW13, CW23, CW26, CW29, CW32, CW3, AF13, AF23, AF3, AW13, AW23, AW3, CF13, CF23, CF3. The ustacks module in the Stacks software package v1.34 (Eugene OR, USA) [22] was used to cluster the sequences and identify the reference sequences. The extracted reads containing enzyme recognition sites were aligned to the reference sequences using the SOAP software (Shenzhen, China) [23] with the following requirements: unique alignment, optimal alignment, and with a maximum of two mismatches allowed ($-r\ 0\ -M\ 4\ -v\ 2$). We performed 2b-RAD genotyping using the maximum likelihood (ML) method in RADtyping software [24]. To ensure the accuracy of subsequent analyses, we excluded the following SNP sites from all samples: those for which less than 80% of the associated individuals could be typed; those with MAF (minor allele frequency) values less than 0.01; those with one or four alleles; and those with more than one SNP within the tag.

2.3. Population Genetic Diversity

We calculated the probability that each collection was in Hardy–Weinberg equilibrium (*HW-P*) using the chi-square test in the R package Genepop v1.0.5 (Montpellier, France) [25] with probabilities >0.05; for collections in Hardy–Weinberg equilibrium, expected heterozygosity (*He*) and observed heterozygosity (*Ho*) were then calculated. We calculated nucleotide diversity (*Pi*) for each of the five collections using VCFtools v0.1.14 (Cambridge, UK) [26]. We also used Genepop v1.0.5 (Montpellier, France) [25] to calculate the inbreeding coefficient within each collection (*Fis*), the inbreeding coefficient across all collections (*Fit*), and the pairwise genetic differentiation coefficient between collections (*Fst*). We calculated Reynolds's genetic distance (*DR*) and gene flow (*Nm*) between pairs of collections as $-\ln (1 - Fst)$ and $(1 - Fst)/4 \times Fst$, respectively.

2.4. Population Genetic Structure

We used PLINK 1.9 (Massachusetts, USA) [27] to assess SNPs across the entire constructed reference sequence of the yellowtail kingfish genome and to identify SNPs with no close linkages (using the main parameter "indep-pairwise 50 10 0.6", where 0.6 is the r^2 threshold). We used ADMIXTURE v1.3.0 (Houston, USA) [28], with 10 different seeds for 10 repeated analyses, to determine the most likely number of subpopulations (K) among the 143 individuals, with tested values of K from 1 to 10. Using PLINK v2.0 [29], we performed a PCA (principal component analysis) of the obtained SNP markers. Finally, we constructed a phylogenetic tree based on neighbor-joining (NJ) analyses using bottom-up clustering for 143 individual yellowtail kingfish, using TreeBest v1.9.2 (Hinxton, UK) [30] with 1000 bootstrap replicates.

3. Results

3.1. Yellowtail Kingfish Population Genetics

The average number of unique tags per sample was 287,594, and the average sequencing depth was 27.13×. Across all samples, the percentage of unique tags was 94.61–96.25%. After quality control filtering, 48,710 sites remained.

In the genetic diversity indicators, *He* ranges from 0 to 1, where 0 represents no polymorphism, and 1 represents an infinite number of alleles with the same frequency.

A higher *Pi* value indicates greater genetic diversity between sequences, meaning there is a larger degree of variation among the samples or populations. Conversely, a smaller *Pi* value indicates higher genetic similarity between sequences, implying a lower level of variation among the samples or populations. *Fis* represents the inbreeding coefficient within populations. *Fit* represents the total population inbreeding coefficient, and their values range from −1 to 1. A significantly positive value of *Fis* indicates a high level of inbreeding within the population, while a significantly negative value suggests the presence of outbreeding. *Fst* ranges from 0 to 1, and a larger value indicates a more pronounced genetic differentiation between subpopulations. Across all five yellowtail kingfish collections, the average probability of the Hardy–Weinberg equilibrium was 0.4112 ($p > 0.05$; Table 1), which indicated that each of the collections were in Hardy–Weinberg equilibrium. The average *Ho* was 0.0824, average *He* was 0.2013, average *Pi* was 0.2020, average *Fis* was 0.0627, average *Fit* was 0.3298, and average *Fst* was 0.2898.

Table 1. Genetic diversity of five yellowtail kingfish collections.

Collection	*Ho*	*He*	*Pi*	*HW-P*
All	0.0824	0.2013	0.2020	0.4112
JW	0.0833	0.0814	0.0837	0.9239
CW	0.0718	0.0771	0.0785	0.8872
CF	0.0828	0.0814	0.0829	0.9093
AW	0.0902	0.0955	0.0974	0.8591
AF	0.0940	0.0927	0.0943	0.8996

Notes: A Japanese wild collection (JW), Chinese farmed collection (CF), Chinese wild collection (CW), Australian farmed collection (AF), and Australian wild collection (AW). *Ho*: observed heterozygosity; *He*: expected heterozygosity; *Pi*: nucleotide diversity; HW-P: *p*-value for the Hardy–Weinberg equilibrium test.

In general, the *Ho*, *He*, and *Pi* values for the farmed Chinese yellowtail kingfish collection (0.0828, 0.0814, and 0.0829, respectively) were greater than those for the wild Chinese yellowtail kingfish collection (0.0718, 0.0771 and 0.0785, respectively, $p < 0.05$). In contrast, the *He* and *Pi* values for the farmed Australian yellowtail kingfish collection (0.0927 and 0.0943, respectively) were lower than those for the wild Australian yellowtail kingfish population (0.0955 and 0.0974, respectively), although the value of Ho was greater for the farmed collection (0.0940) for the wild (0.0902; Table 1).

Among the Chinese and Japanese collections, *Fst* was 0.00097–0.01888 and *DR* was 0.0010–0.0191; similarly, *Fst* and *DR* were 0.02569 and 0.0260, respectively, between the two Australian collections (Table 2), suggesting no genetic differentiation (*Fst* < 0.05). However, the *Fst* values between the Chinese or Japanese collections and the Australian collections were 0.7256–0.7447 (*DR*: 1.2932–1.3653; Table 2), suggesting substantial genetic differentiation (*Fst* > 0.25).

Table 2. Pairwise measures of genetic differentiation among five collections of yellowtail kingfish.

Collection	JW	CW	CF	AW	AF
JW	-	0.0010	0.0168	1.2932	1.3205
CW	0.00097	-	0.0191	1.3409	1.3653
CF	0.01662	0.01888	-	1.3164	1.3417
AW	0.7256	0.7384	0.7319	-	0.0260
AF	0.7330	0.7447	0.7386	0.02569	-

Notes: A Japanese wild collection (JW), Chinese farmed collection (CF), Chinese wild collection (CW), Australian farmed collection (AF), and Australian wild collection (AW). The genetic differentiation coefficient (*Fst*) is shown below the diagonal, and the genetic distance (*DR*) is shown above the diagonal.

The *Nm* values among the Chinese and Japanese collections were high (12.9287–256.4499), as was the *Nm* value between the Australian collections (9.9002; Table 3). This suggested frequent gene exchange between the wild Chinese and farmed Chinese and wild Japanese collections, as well as between the wild Australian and farmed Australian collections.

Indeed, the highest *Nm* was calculated between the Chinese and Japanese wild collections, and the gene flow may be frequent and/or recent. In contrast, the *Nm* values between the Chinese or Japanese and Australian collections were low (0.0864–0.0953; Table 3), indicating little to no gene exchange between the Asian and Australian yellowtail kingfish collections.

Table 3. Values of the gene flow parameter (*Nm*) among the five collections of yellowtail kingfish.

Collection	JW	CW	CF	AW	AF
JW	-				
CW	256.4499	-			
CF	14.7921	12.9287	-		
AW	0.0953	0.0894	0.0923	-	
AF	0.0911	0.0864	0.0892	9.9002	-

Notes: A Japanese wild collection (JW), Chinese farmed collection (CF), Chinese wild collection (CW), Australian farmed collection (AF), and Australian wild collection (AW).

3.2. Yellowtail Kingfish Population Structure

The ADMIXTURE cross-validation error levels indicated that the most likely number of subpopulations (parameter *K*) was 2 (Figure S1). When *K* = 2, the Australian wild and farmed collections are grouped together in one cluster, while the Chinese wild and farmed collections, as well as the Japanese wild collections, form another cluster. When *K* = 3 or *K* = 4 (Figure 2B,C), individual outliers from the cultured population cluster separately within the Australian and Chinese farmed collections, but did not show distinct separations. Based on the current analysis, the origins of the Chinese and Japanese populations are similar, while the Australian collections origins are different (Figure 2A). Similarly, our PCA showed that the Chinese and Japanese collections clustered together, distinct from the Australian cluster (Figure 3A); PC1 and PC2 explained 41.16% and 3.73% of the variance across populations, respectively. On PC1, there were individual outliers within the wild Australian collection. PC3 revealed a weak clustering pattern in the Australian cultured collection, and PC3 explained only 1.26% of the variance across populations (Figure 3B). Similarly, PC4 indicated a weak clustering pattern in the Chinese cultured collection, and PC4 explained only 1.16% of the variance across populations (Figure 3C).

The phylogenetic tree shows that Australian collections and Chinese and Japanese collections are divided into two distinct clades. The farmed and wild Australian collections formed two distinct clades. Whereas, the wild Chinese and Japanese collection formed a branch, with the wild Japanese collection branching off from the farmed Chinese collection, though no clear clade divergence was observed among these three collections (Figure 4).

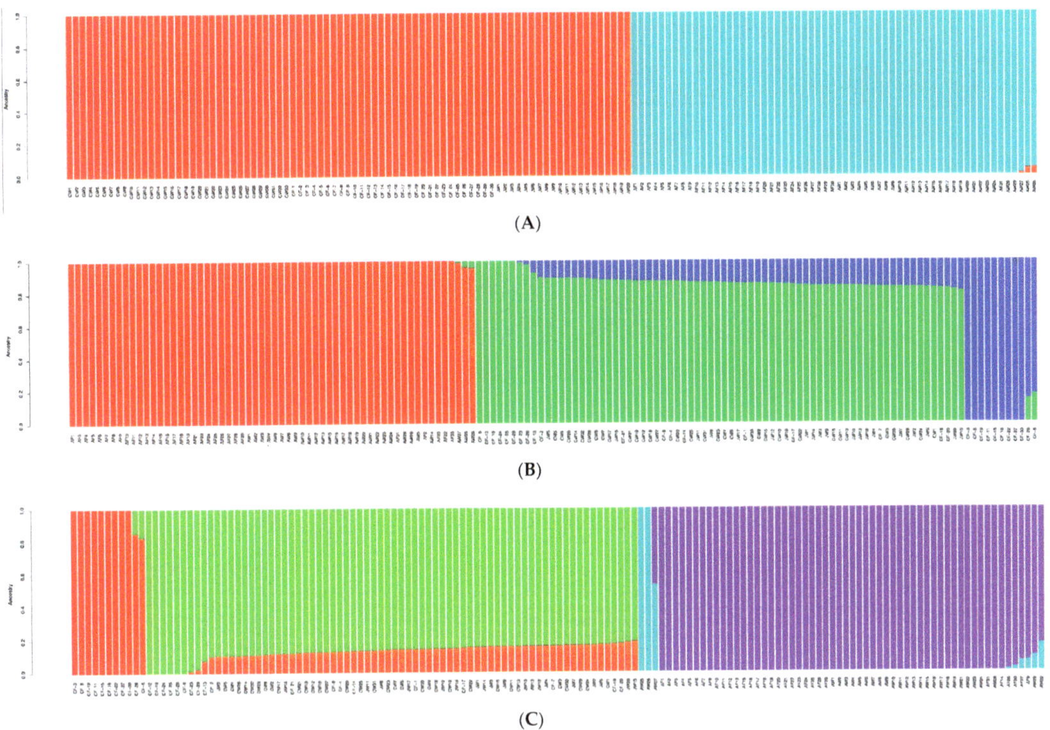

Figure 2. (**A**). The genetic structure of the five yellowtail kingfish collections for *K* = 2. (**B**). The genetic structure of the five yellowtail kingfish collections for *K* = 3. (**C**). The genetic structure of the five yellowtail kingfish collections for *K* = 4. Notes: Each bar represents one individual. Different colors suggest different origins and show the proportion of each genotype belonging to each genetic cluster.

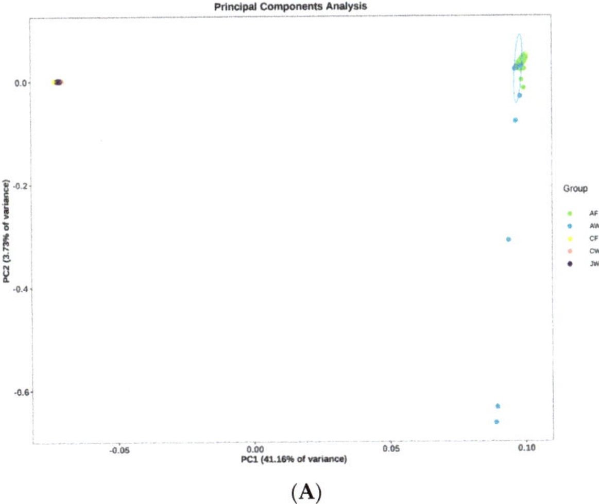

(**A**)

Figure 3. *Cont.*

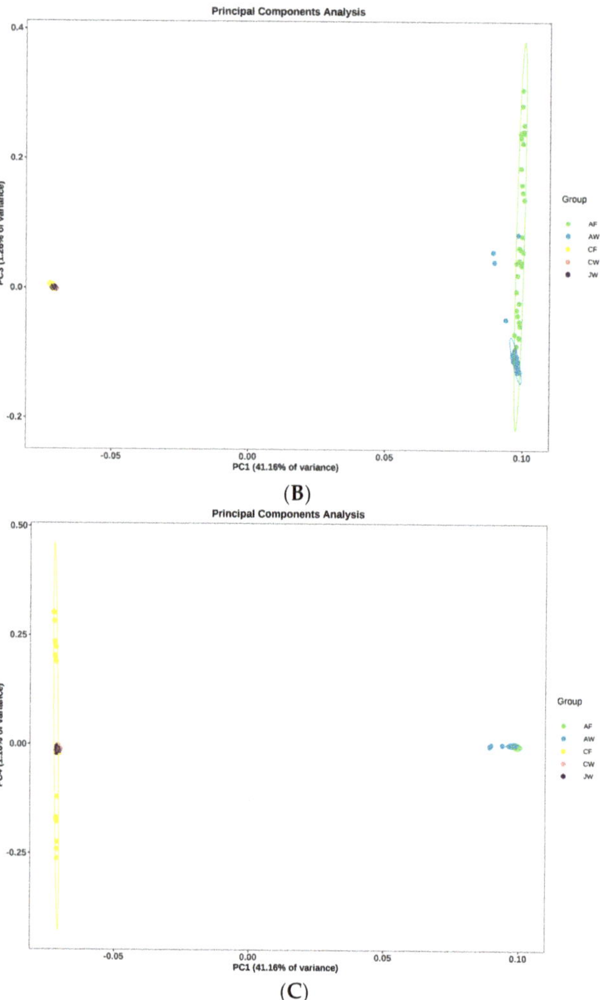

Figure 3. (**A–C**). Principal components analysis (PCA) showing the genetic structure of five yellowtail kingfish collections. Notes: A Japanese wild collection (JW), Chinese farmed collection (CF), Chinese wild collection (CW), Australian farmed collection (AF), and Australian wild collection (AW). Each symbol represents an individual fish.

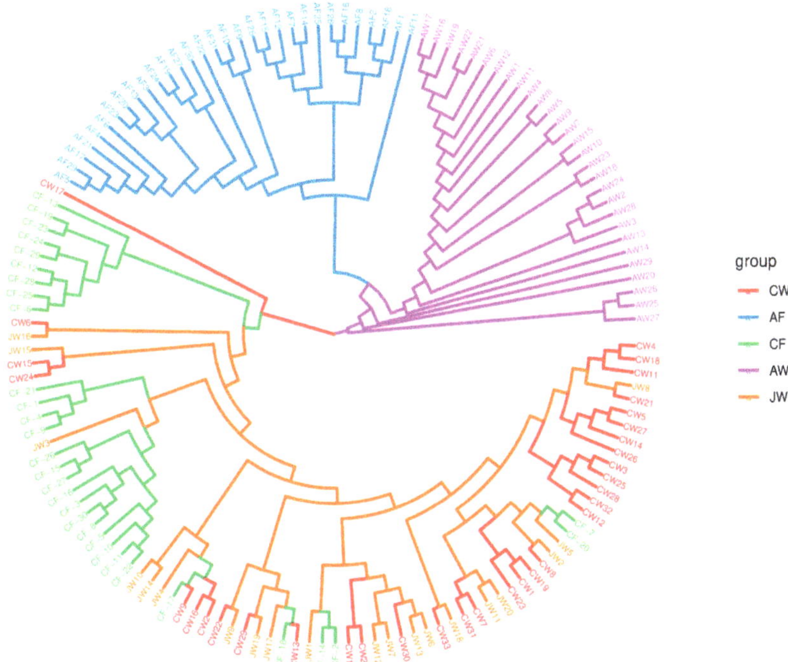

Figure 4. Neighbor-joining phylogenetic tree of the five yellowtail kingfish collections. Notes: The reliability of the phylogenetic tree was assessed using the bootstrap method with 1000 replicates. Branch lengths reflect the genetic distance. A Japanese wild collection (JW), Chinese farmed collection (CF), Chinese wild collection (CW), Australian farmed collection (AF), and Australian wild collection (AW).

4. Discussion

A large number of molecular markers were obtained by the 2b-RAD genome sequencing of yellowtail kingfish samples from five collections. A comparative analysis of the genetic characters of the wild and farmed Chinese collections, the wild Japanese collection, and the wild and farmed Australian collections was performed using these markers. The results of the population genetic structure analysis, PCA, and a neighbor-joining phylogenetic tree all supported the same conclusions. Genetic differentiation of the wild and farmed Chinese populations, wild Japanese populations, and the wild and farmed Australian populations was high. The calculated gene flow values for the wild and farmed Australian populations compared with the wild and farmed Chinese populations and wild Japanese population were much less than 1, which indicates almost no gene exchange. These results are consistent with the findings of Ai et al. [11], who showed, by comparing nuclear loci and morphological characters in samples from the Bohai Sea, that the Chinese and Australian yellowtail kingfish formed two completely distinct clusters and that there was no genetic introgression from the Australian yellowtail kingfish into the Chinese yellowtail kingfish population based on nuclear gene markers and morphological characteristics.

Although yellowtail kingfish are an oceanic species that can migrate more than 2000 km, their migration routes are confined to continental shelf waters within the same hemisphere, and adult fish mostly migrate at regional scales; thus, the likelihood of migration across the open ocean is low [31,32]. In addition, the equator acts as a natural geographic barrier to yellowtail kingfish migration in the Pacific Ocean because the equatorial water temperatures (18–24 °C in continental shelf areas) exceed those tolerated by yellowtail kingfish [33]. Therefore, the gene flow between Northern Hemisphere and

Southern Hemisphere yellowtail kingfish is blocked [10]. The equator has also been shown to act as a natural barrier to gene flow in other wide-ranging or migratory teleosts. For example, mtDNA analyses indicated a lack of genetic communication between Pacific jack mackerel (*Trachurus symmetricus*) populations from the North and South Pacific Ocean [34], and genetic differentiation between the North Pacific and South Pacific populations of yellowfin tuna (*Thunnas albacares*) in the two oceans was attributed to equatorial isolation [35]. Similarly, our data showed high genetic differentiation between the Chinese, Japanese and Australian populations, indicating that these populations are geographically (and therefore reproductively) isolated. Sepúlveda et al. (2017) pointed out that the temporal movements of yellowtail kingfish in the southeastern Pacific revealed a particular life strategy (i.e., reproductive or habitat segregation) for this species [36]. This was consistent with previous studies of genetic differentiation among Australian, Japanese, and/or Chinese yellowtail kingfish populations analyzed by mtDNA or nuclear genes [9,31]. Using mitochondrial and nuclear genetic markers and morphometric analysis, Martinez-Takeshita et al. (2015) revealed significant genetic and morphological divergence between yellowtail kingfish (*Seriola lalandi*) specimens collected from the Northeast Pacific, Northwest Pacific, and Southern Hemisphere [10]. Even if it were possible for yellowtail kingfish to migrate across the equator, gene exchange remains unlikely. Yellowtail kingfish in the Northern Pacific spawn between March and October (e.g., the Chinese wild populations spawn from April to June) [37,38], but in the Southern Hemisphere they typically spawn between October and January [39].

Although previous studies speculated that the Chinese and Japanese yellowtail kingfish populations are indistinct based on their geographic distributions [1], genetic and quantitative morphological evidence for this was lacking. Previously, we found that the mitochondrial genome structure of Chinese yellowtail kingfish was highly similar to that of Japanese yellowtail kingfish, except that the mitochondrial genome of the Chinese population was slightly longer, and that variation in the COX1 and NAD5 genes was observed between Chinese and Japanese populations [14,40]. This awakened our interest in exploring the population genetic structure of Chinese and Japanese yellowtail kingfish. Here, our population genetic structure analysis showed no genetic differentiation between the wild and farmed Chinese populations and the wild Japanese population, suggesting that they are derived from the same population. The phylogenetic analysis showed that the first branch of the Chinese wild population was closely related to the wild Japanese population, while the wild Japanese population was linked to the farmed Chinese population. Interestingly, gene flow between the Chinese and Japanese wild populations (*Nm* 256.4499) was much higher than that between the Chinese wild and farmed populations. The *Nm* value between the farmed Chinese and wild Japanese populations was slightly higher than that between the wild Chinese and wild Japanese populations. The calculated *Fst* and *DR* values also suggested that the wild populations in China and Japan are most closely related to each other. This may be because China and Japan are located in Current and Kuroshio Current, of which are relatively high. However, the Northern Pacific Ocean currents are colder and thus more suitable for the migration of yellowtail kingfish adults, juveniles, and larvae. Indeed, two of the main factors affecting the migratory routes of yellowtail kingfish are ocean currents and temperature [41–43]. Our preliminary surveys indicated that the Dasha fishing ground in the Yellow Sea (~32°00′–34°00′ N, ~122°30′–125°00′ E) is the spawning grounds of the wild Chinese yellowtail kingfish population. Furthermore, fishermen in various coastal Chinese cities report catching yellowtail kingfish adults, juveniles, and larvae at staggered periods throughout the year: April–June at Qingdao, May–August at Dalian, July–August at Weihai, and August–October at Zhoushan. From this we inferred that yellowtail kingfish migrate south to north along the Chinese coast, following food resources and the Kuroshio Current, and then swim either northwestward into the Yellow Sea with the Yellow Sea Warm Current or northwestward around the Korean peninsula with the Tsushima Current. However, further genetic data from yellowtail kingfish populations in the East China Sea and the Yellow Sea are required to verify this speculation.

Therefore, in our next study, we plan to establish 8–10 large-scale sampling stations in various locations along the hypothesized yellowtail kingfish migration route (e.g., the East China Sea, Yellow Sea, Bohai Sea, Japan Sea, and Korean Peninsula coastal waters) to capture individuals at various spatial and temporal points during their migration. These data will also help to further clarify the phylogenetic relationship between the Chinese and Japanese populations.

When a fish population is influenced by factors such as mutation, selection pressure, migration, random genetic drift or non-random mating, the genotype frequencies within the population can be affected. All populations conform to the Hardy-Weinberg equilibrium, but the Hardy–Weinberg equilibrium (*HW-P*) value for the total sample was lower than the *HW-P* values for each cluster. This is consistent with the result of *He* (0.2013) being higher than *Ho* (0.0824). All clusters were divided into two populations, with gene flow occurring within each. Gene flow between the Chinese and Japanese populations displayed an especially high level at *Nm* 256.449. The Northwest Pacific yellowtail kingfish has a common spawning ground and follows similar migration routes. We calculated the *F*-statistics, *Fit* (0.3298) is greater than *Fst* (0.2898), which is greater than *Fis* (0.0627). There was no inbreeding between individuals and there was genetic differentiation between each population. Fernández et al. (2015) studied the genetic structure of the Chilean population of *Seriola lalandi* to detect moderate inbreeding (*Fis* = 0.12). Previous studies [9,44] consistently yielded higher *Ho* and *He* values for yellowtail kingfish populations in comparison to the present findings. Significantly higher values than obtained in this study might be an outcome of different sequencing methods. The size of the sample may also be one of the factors that influence the results. It has been suggested that sample sizes of more than 50 are best for population analysis.

In the Chinese populations, the genetic diversity values were only slightly higher for the farmed versus wild fish. This could reflect the influence of aquaculture domestication on Chinese yellowtail kingfish, even though the culture of this species in China is still in an initial stage and the culture effort has not yet undertaken direct breeding. As a result, there is no significant genetic differentiation attributed to kinship within the population. In general, cultured fish populations frequently originate from a limited number of individuals, leading to a constrained genetic background. This species tends to show a small degree of schooling behavior [36]. In contrast, aquaculture efforts in Australia have focused on the impact of escaped farmed fish on the wild population [8], although no significant genetic differentiation has yet been detected between wild and farmed populations there. Although the phylogenetic results show distinct distributions between the wild and farmed fish in Australia. In the PCA, individual outliers could be observed, but their impact seems minimal. Structure to *K* = 3 or 4, there are no clear clusters within the two populations. Genetic diversity is a valuable piece of information for fish farmers to use to improve and manage their fish stocks [36]. Based on the results of our low genetic diversity, the underlying reasons for this outcome could be complex and may also be influenced by the experimental methods used. Nevertheless, this serves as a reminder for us to pay attention to improving the genetic diversity of farmed populations and preserving the integrity of the habitats and wild populations. This highlights supplementing new individuals to farmed populations to increase gene flow between individuals, and should also remind us of the benefit of implementing targeted breeding strategies. Additionally, comprehensive genetic management plans should be developed and implemented, including monitoring and maintaining genotype frequencies, genetic diversity, and genetic health, along with appropriate intervention measures.

5. Conclusions

We isolated and identified SNP markers in yellowtail kingfish populations from China (wild and farmed), Japan (wild), and Australia (wild and farmed). The genetic structure of each population was analyzed using these SNP markers. Our analyses show that: (1) no significant genetic differentiation was detected between the wild and farmed

Chinese and wild Japanese populations, and both of those populations showed significant genetic differentiation from the wild and farmed Australian populations; (2) no obvious genetic differentiation was observed between the wild and farmed Chinese populations, whereas the wild and farmed Australian populations are likely becoming genetically differentiated because of longer farming generations or genetic bottleneck. Moreover, the genetic relationship between the Chinese and Japanese wild populations was closer than that between the wild and farmed Chinese populations. Our data will provide theoretical and technical support for genetic breeding, ensuring the sustainable production of wild and farmed yellowtail kingfish in China.

Supplementary Materials: The following supporting information can be downloaded at: https://www.mdpi.com/article/10.3390/jmse11081583/s1, Figure S1: Boxplot showing cross-validation error levels for various numbers of subpopulations ($K = 1$–10).

Author Contributions: Conceptualization, Y.X. and A.C.; methodology, Y.X. and A.C.; validation, Y.X., K.K. and Y.J.; formal analysis, A.C. and Y.X.; investigation, A.C. and Y.X.; data curation, K.K., Y.J., B.W., T.K. and X.L.; writing—original draft preparation, A.C.; writing—review and editing, A.C. and Y.X.; supervision, Y.X.; funding acquisition, Y.X. All authors have read and agreed to the published version of the manuscript.

Funding: This work was supported by grants from National Key R&D Program of China (2022YFD2401102), Marine Science and Technology Fund of Shandong Province for Qingdao Marine Science and Technology Center (2022QNLM030001), Taishan Industrial Experts Program, Central Public-interest Scientific Institution Basal Research Fund, CAFS (2020TD47), and Earmarked Fund for China Agriculture Research System (CARS-47).

Institutional Review Board Statement: The animal study protocol was approved by the Institutional Animal Care and Use Committee of Yellow Sea Fisheries Research Institute, Chinese Academy of Fishery Sciences (Approval code: YSFRI-2022025, and date: 3 July 2019).

Informed Consent Statement: Not applicable.

Data Availability Statement: The data in this study are available from the authors upon reasonable request.

Conflicts of Interest: The authors declare no conflict of interest.

References

1. Sicuro, B.; Luzzana, U. The state of *Seriola* spp. other than yellowtail (*S. quinqueradiata*) farming in the world. *Rev. Fish Sci. Aquac.* **2016**, *24*, 314–325. [CrossRef]
2. Liu, X.; Xu, Y.; Li, R.; Lu, Y.; Shi, B.; Ning, J.; Wang, B. Analysis and evaluation of nutritional composition of the muscle of yellowtail kingfish (*Seriola aureovittata*). *Prog. Fish. Sci.* **2017**, *1*, 128–135.
3. Whatmore, P.; Nguyen, N.H.; Miller, A.; Lamont, R.; Powell, D.; D'Antignana, T.; Bubner, E.; Elizur, A.; Knibb, W. Genetic parameters for economically important traits in yellowtail kingfish *seriola lalandi*. *Aquaculture* **2013**, *400–401*, 77–84. [CrossRef]
4. Zhang, C. *Investigation Report on Fishes in the Yellow Sea and Bohai Sea*; Science Press: Beijing, China, 1995; pp. 116–119.
5. Liu, J.; Chen, Y.; Ma, L. *Fish Atlas of the Yellow Sea and Bohai Sea*; Science Press: Beijing, China, 2015; pp. 171–173.
6. Liu, X. Great breakthrough has been made in breeding technology of yellowtail kingfish. *Ocean. Fish.* **2017**, *8*, 19.
7. Premachandra, H.K.A.; La Cruz, L.D.; Takeuchi, Y.; Miller, A.; Fielder, S.; O'Connor, W. Genomic DNA variation confirmed *seriola lalandi* comprises three different populations in the pacific, but with recent divergence. *Sci. Rep.* **2017**, *7*, 9386. [CrossRef]
8. Purcell, C.M.; Chabot, C.L.; Craig, M.T.; Martinez-Takeshita, N.; Allen, L.G. Developing a genetic baseline for the yellowtail amberjack species complex, *Seriola lalandi* sensu lato, to assess and preserve variation in wild populations of these globally important aquaculture species. *Conserv. Genet.* **2015**, *16*, 1475–1488. [CrossRef]
9. Nugroho, E.; Ferrell, D.J.; Smith, P.; Taniguchi, N. Genetic divergence of kingfish from Japan, Australia and New Zealand inferred by microsatellite DNA and mitochondrial DNA control region markers. *Fish Sci.* **2001**, *67*, 843–850. [CrossRef]
10. Martinez-Takeshita, N.; Purcell, C.M.; Chabot, C.L.; Craig, M.T.; Paterson, C.N.; Hyde, J.R.; Allen, L.G. A tale of three tails: Cryptic speciation in a globally distributed marine fish of the genus *seriola*. *Copeia* **2015**, *103*, 357–368. [CrossRef]
11. Ai, Q.; Song, L.; Tan, H.; Huang, X.; Bao, B.; Li, C. Genetic and morphological differences between yellowtail kingfish (*Seriola lalandi*) from the Bohai Sea, China and the Southern Ocean, Australia. *Aquac. Fish.* **2021**, *6*, 260–266. [CrossRef]
12. Shi, B.; Liu, Y.; Liu, X.; Xu, Y.; Li, R.; Song, X.; Zhou, L. Study on the karyotype of yellowtail kingfish (*Seriola aureovittata*). *Prog. Fish. Sci.* **2017**, *1*, 136–141.

13. Liu, Y.; Liu, X.; Shi, B.; Xu, Y.; Li, R.; Lv, Y.; Song, X.; Wang, B.; Jiang, Y. Analysis of the banding patterns of *Seriola aureovittata*. *J. Fish. China* **2018**, *9*, 1338–1347.

14. Shi, B.; Liu, X.; Liu, Y.; Zhang, Y.; Gao, Q.; Xu, Y.; Wang, B.; Jiang, Y.; Song, X. Complete sequence and gene organization of the mitochondrial genome of *Seriola aureovittata*. *J. Fish. Sci. China* **2019**, *26*, 405–415. [CrossRef]

15. Cui, A.; Wang, B.; Jiang, Y.; Liu, X.; Xu, Y. Development of SNP markers for yellowtail kingfish (*Seriola lalandi*) by 2b-RAD simplified genome sequencing. *Conserv. Genet. Resour.* **2020**, *12*, 403–407. [CrossRef]

16. Wang, S.; Meyer, E.; Mckay, J.K.; Matz, M.V. 2b-rad: A simple and flexible method for genome-wide genotyping. *Nat. Methods* **2012**, *9*, 808–810. [CrossRef]

17. Pecoraro, C.; Babbucci, M.; Villamor, A.; Franch, R.; Papetti, C.; Leroy, B.; Cariani, A. Methodological assessment of 2b-RAD genotyping technique for population structure inferences in yellowfin tuna (*Thunnus albacares*). *Mar. Genom.* **2016**, *25*, 43–48. [CrossRef]

18. Dou, J.; Li, X.; Fu, Q.; Jiao, W.; Li, Y.; Li, T. Evaluation of the 2b-rad method for genomic selection in scallop breeding. *Sci. Rep.* **2016**, *6*, 19244. [CrossRef]

19. Yi, S.; Li, Y.; Shi, L.; Zhang, L.; Li, Q.; Chen, J. Characterization of population genetic structure of red swamp crayfish, *Procambarus clarkii*, in China. *Sci. Rep.* **2018**, *8*, 5586. [CrossRef]

20. Manuzzi, A.; Zane, L.; Muñoz-Merida, A.; Griffiths, A.M.; Veríssimo, A. Population genomics and phylogeography of a benthic coastal shark (*Scyliorhinus canicula*) using 2b-RAD single nucleotide polymorphisms. *Biol. J. Linn. Soc.* **2018**, *126*, 289–303. [CrossRef]

21. Zhang, J.; Kobert, K.; Flouri, T.; Stamatakis, A. PEAR: A fast and accurate Illumina Paired-End reAd mergeR. *Bioinformatics* **2013**, *30*, 614–620. [CrossRef]

22. Catchen, J.; Hohenlohe, P.A.; Bassham, S.; Amores, A.; Cresko, W.A. Stacks: An analysis tool set for population genomics. *Mol. Ecol.* **2013**, *22*, 3124–3140. [CrossRef] [PubMed]

23. Li, R.; Li, Y.; Kristiansen, K. SOAP: Short oligonucleotide alignment program. *Bioinformatic* **2008**, *24*, 713–714. [CrossRef] [PubMed]

24. Fu, X.; Dou, J.; Mao, J.; Su, H.; Jiao, W.; Zhang, L.; Hu, X.; Huang, X.; Wang, S.; Bao, Z. RADtyping: An integrated package for accurate de novo codominant and dominant RAD genotyping in mapping populations. *PLoS ONE* **2013**, *8*, e79960. [CrossRef] [PubMed]

25. Rousset, F. GENEPOP' 007: A complete re-implementation of the GENEPOP software for Windows and Linux. *Mol. Ecol. Resour.* **2008**, *8*, 103–106. [CrossRef]

26. Danecek, P.; Auton, A.; Abecasis, G. The variant call format and VCFtools. *Bioinformatics* **2011**, *27*, 2156–2158. [CrossRef] [PubMed]

27. Purcell, S.; Neale, B.; Toddbrown, K. PLINK: A tool set for whole-genome association and population-based linkage analyses. *Am. J. Hum. Genet.* **2007**, *81*, 559–575. [CrossRef]

28. Chakraborty, R.; Weiss, K.M. Admixture as a tool for finding linked genes and detecting that difference from allelic association between loci. *Proc. Natl. Acad. Sci. USA* **1988**, *85*, 9119–9123. [CrossRef] [PubMed]

29. Chang, C.C.; Chow, C.C.; Cam, T.L.; Shashaank, V.; Purcell, S.M.; Lee, J.J. Second-generation plink: Rising to the challenge of larger and richer datasets. *Giga Sci.* **2015**, *4*, 7. [CrossRef]

30. Vilella, A.; Severin, J.; Ureta-Vidal, A.; Heng, L.; Durbin, R.; Birney, E. Ensembl Compara Gene Trees: Complete, duplication-aware phylogenetic trees in vertebrates. *Genome Res.* **2009**, *19*, 327–335. [CrossRef]

31. Gillanders, B.M.; Ferrell, D.J.; Andrew, N.L. Estimates of movement and life-history parameters of yellowtail kingfish (*seriola lalandi*): How useful are data from a cooperative tagging programme? *Mar. Freshw. Res.* **2001**, *52*, 179–192. [CrossRef]

32. Miller, P.A.; Fitch, A.J.; Gardner, M.; Hutson, K.S.; Mair, G. Genetic population structure of yellowtail kingfish (*Seriola lalandi*) in temperate Australasian waters inferred from microsatellite markers and mitochondrial DNA. *Aquaculture* **2011**, *319*, 328–336. [CrossRef]

33. Wolvaardt, C. *Initiation of Company Coverage—IPO, Western Kingfish Ltd.*; State One Stockbroking Ltd.: Sydney, Australia, 2007; pp. 1–9.

34. Poulin, E.; Cardenas, L.; Hernandez, C.E.; Kornfield, I.; Ojeda, F.P. Resolution of the taxonomic status of Chilean and Californian jack mackerels using mitochondrial DNA sequence. *J. Fish Biol.* **2004**, *65*, 1160–1164. [CrossRef]

35. Díaz-Jaimes, P.; Uribe-Alcocer, M.; Rocha-Olivares, A.; García-de-León, F.J.; Nortmoon, P.; Durand, J.D. Global phylogeography of the dolphinfish (coryphaena hippurus): The influence of large effective population size and recent dispersal on the divergence of a marine pelagic cosmopolitan species. *Mol. Phylogenet Evol.* **2010**, *57*, 1209–1218. [CrossRef]

36. Sepúlveda, F.A. and González, M.T. Spatio-temporal patterns of genetic variations in populations of yellowtail kingfish *Seriola lalandi* from the south-eastern Pacific Ocean and potential implications for its fishery management. *J. Fish Biol.* **2017**, *90*, 249–264. [CrossRef]

37. Sala, E.; Aburtooropeza, O.; Paredes, G.; Thompson, G. Spawning aggregations and reproductive behavior of reef fishes in the gulf of California. *B Mar. Sci.* **2003**, *72*, 103–121.

38. Erisman, B.; Mascarenas, I.; Paredes, G.; Mitcheson, Y.S.D.; Aburto-Oropeza, O.; Hastings, P. Seasonal, annual, and long-term trends in commercial fisheries for aggregating reef fishes in the gulf of California, Mexico. *Fish Res.* **2010**, *106*, 279–288. [CrossRef]

39. Poortenaar, C.; Hooker, S.; Sharp, N. Assessment of yellowtail kingfish (*Seriola lalandi*) reproductive physiology, as a basis for aquaculture development. *Aquaculture* **2001**, *201*, 271–286. [CrossRef]

40. Iguchi, J.; Takashima, Y.; Namikoshi, A.; Yamashita, M. Species identification method for marine products of *seriola* and related species. *Fish Sci.* **2012**, *78*, 197–206. [CrossRef]
41. Moran, D.; Smith, C.K.; Gara, B.; Poortenaar, C.W. Reproductive behaviour and early development in yellowtail kingfish (*Seriola lalandi* Valenciennes 1833). *Aquaculture* **2007**, *262*, 95–104. [CrossRef]
42. Moran, D.; Gara, B.; Wells, R.M.G. Energetics and metabolism of yellowtail kingfish (*Seriola lalandi* Valenciennes 1833) during embryogenesis. *Aquaculture* **2007**, *265*, 359–369. [CrossRef]
43. National Oceanic and Atmospheric Administration. *NOAA Optimum Interpolation Sea Surface Temperature Analysis*; NOAA/OAR/ESRL PSD (United States Department of Commerce): Boulder, CO, USA, 2011.
44. Fernández, G.; Cichero, D.; Patel, A.; Martínez, V. Genetic structure of Chilean populations of *Seriola lalandi* for the diversification of the national aquaculture in the north of Chile. *Lat. Am. J. Aquat. Res.* **2015**, *43*, 374–379. [CrossRef]

MDPI

St. Alban-Anlage 66

4052 Basel

Switzerland

www.mdpi.com

Journal of Marine Science and Engineering Editorial Office

E-mail: jmse@mdpi.com

www.mdpi.com/journal/jmse